Springer Series in Materials Science

Volume 326

The Springer Series in Materials Science covers the complete spectrum of materials research and technology, including fundamental principles, physical properties, materials theory and design. Recognizing the increasing importance of materials science in future device technologies, the book titles in this series reflect the state-of-the-art in understanding and controlling the structure and properties of all important classes of materials.

Adam Morawiec

Indexing of Crystal Diffraction Patterns

From Crystallography Basics to Methods of Automatic Indexing

Springer

Adam Morawiec
Polish Academy of Sciences
Institute of Metallurgy and Materials
Science
Kraków, Poland

ISSN 0933-033X ISSN 2196-2812 (electronic)
Springer Series in Materials Science
ISBN 978-3-031-11076-4 ISBN 978-3-031-11077-1 (eBook)
https://doi.org/10.1007/978-3-031-11077-1

This Springer imprint is published by the registered company Springer Nature Switzerland AG
The registered company address is: Gewerbestrasse 11, 6330 Cham, Switzerland

Preface

Most solids consist of crystallites. Properties of crystalline and polycrystalline materials are, to a large extent, determined by crystal structures and crystal orientations. Description of the structures and their orientations is based on the simple geometric notion of a lattice. This book is about investigation into the geometry of crystal lattices.

The determination of crystal lattices (their parameters and orientations) is carried out mainly by diffraction experiments. Linking diffraction data to a crystal lattice is commonly referred to as indexing. More precisely, indexing is an assignment of indices of stacks of planes in a crystal to corresponding peaks in a diffraction pattern. There are two distinct types of indexing. The ab initio indexing is an early step in structure determination when little is known about the architecture of the crystal; it amounts to the determination of unknown lattice parameters. Besides that, there is indexing of diffraction patterns originating from crystals of known structures, which arises, in particular, when crystal orientation is to be determined. The key application of such indexing is in the so-called orientation mappings: the reconstruction of microstructures of polycrystalline materials based on spatially resolved determination of crystallite orientations. Closely related to indexing and orientation determination are the refinement of lattice parameters and the evaluation of elastic lattice strains.

"Manual" analysis of diffraction patterns is time-consuming and prone to errors. Therefore, pattern indexing and orientation determination are usually carried out by software supplied with experimental instruments. Frequently, these elements of experimental systems are treated as "black boxes" and regarded in terms of their output without understanding their internal mechanisms. The "black-box" approach is convenient, but relying on it is not always a proper scientific practice, and it is reasonable to invest in learning how particular functions of a system are implemented. Moreover, the complexity of the diffraction data calls for further automation of the lattice parameter and orientation determination methods. Software for automatic indexing and orientation determination is constantly being developed in numerous laboratories and in various configurations, and still, there is an ever-growing demand for faster and more reliable computer programs of this kind.

The literature on indexing of diffraction patterns and determination/refinement of lattice parameters and orientations is huge, diverse, and widely scattered in time and across various fields of interest. This book is intended to be an introduction to automatic indexing and crystal orientation determination bringing the information on the fundamentals of these subjects together in a single volume. Yet, only selected aspects of indexing are considered in detail. In some cases (indexing of single-crystal diffraction patterns, indexing of Laue patterns, indexing for orientation determination), descriptions are quite thorough and include some algorithms. Other chapters contain reviews of existing methods (indexing of powder diffraction, multigrain indexing) or just definitions of indexing schemes (quasicrystals). Parts of the book (e.g., sections on distributions or Fourier transformation) are not expositions of theories but rather informal sketches intended to be useful for understanding other sections. The book only touches on some aspects of refinement of lattice parameters.

Pattern indexing, orientation determination, and strain determination are, to some extent, geometric problems. Numerous issues in these areas can be resolved within the geometric description of diffraction, and the book relies mainly on this approach. On the other hand, for a more complete analysis of diffraction patterns, one needs to compute the intensities of reflections, which requires some physical aspects of diffraction. Therefore, two chapters on crystal diffraction are included with the goal of making the book more self-contained.

Although most of the considered topics are in a sense classical, the book sometimes goes beyond standard descriptions of subjects. Besides scientific breakthroughs, understanding also comes through formulating ideas in various ways. The most natural and concise descriptions are usually preferred, but it may take some time for concepts to reach optimal forms and to be absorbed by communities. A number of topics are described below significantly differently than in textbooks. There are also few departures from the notation and conventions used by crystallographers. The most notable is the use of the symbols $\boldsymbol{a}_{\mathrm{i}}$ and $\boldsymbol{a}^{\mathrm{i}}$ $(i = 1, 2, 3)$ for basis vectors of direct and reciprocal lattices instead of the conventional $\mathbf{a}$, $\mathbf{b}$, $\mathbf{c}$ and $\mathbf{a}^*$, $\mathbf{b}^*$, $\mathbf{c}^*$. With the standard symbols, many expressions would be longer and more cumbersome.

Indexing arises in numerous applications, and a crystallographer, chemist, mineralogist, biologist, physicist, and mathematician may have different views on methods to present the subject. In this book, indexing is presented from a viewpoint of a materials scientist which is both practical and formal. The book is addressed to readers already having some basic knowledge on indexing, orientation determination, and refinement of lattice parameters who would like to see these subjects from a perspective which, at least in some aspects, may be new to them. It is intended equally for a materials scientist curious about the "nuts and bolts" of indexing or orientation mapping systems, and for a programmer involved in the development of software for the analysis of diffraction patterns. The former may learn about theoretical foundations of an experimental setup in use, whereas the programmer may find it helpful in constructing algorithms and software of practical use in the analysis of crystal diffraction data. The text will be relatively easy to read for researchers involved in crystallographic computing. It may be challenging, but still accessible, for doctoral students who would like to understand the functioning of indexing engines.

I would like to express my gratitude to Marc De Graef for correcting and commenting on Chaps. 1 through 5 and to Emmanuel Bouzy and Henning Poulsen for their remarks on Chaps. 9 and 11, respectively. I am also much indebted to coworkers who have allowed me to use their experimental diffraction patterns as acknowledged in the figure captions. Work on the book was supported by the Institute of Metallurgy and Materials Science of the Polish Academy of Sciences.

Giebułtów, Poland

Adam Morawiec

Contents

Preliminaries

Points and Vectors in Space

Locations, orientations, and some properties of physical objects are conveniently described using the three-dimensional Euclidean vector space. We consider objects (lattices, crystals, crystallites) in a point space. To each pair, say, (p, q) of points p and q, corresponds to a vector. The correspondence satisfies the rules that the pair (p, p) corresponds to the zero vector $\mathbf{0}$, and the difference of vectors corresponding to (p, q) and (p, r) corresponds to the pair (r, q). The vector corresponding to (p, q) is unique. If there were two such vectors $\boldsymbol{v}$ and $\boldsymbol{w}$, then their difference as vectors corresponding to (p, q) and (p, q) would correspond to (q, q); hence the difference is $\mathbf{0}$ and $\boldsymbol{w} = \boldsymbol{v}$.

With a point p_0 selected as an origin in the point space, each vector $\boldsymbol{v}$ of the Euclidean vector space determines a unique point p such that v corresponds to the pair (p_0, p). In other words, points of the space are identified by the terminal points of vectors from the origin $\mathbf{0}$.

Points and vectors will be used interchangeably. The point space is equipped with the (Euclidean) distance function. The distance of a point indicated by vector $\boldsymbol{v}$ to the origin $\mathbf{0}$ is equal to the magnitude (length) $|\boldsymbol{v}|$ of $\boldsymbol{v}$. The distance between points indicated by vectors $\boldsymbol{v}$ and $\boldsymbol{w}$ is given by $|\boldsymbol{v} - \boldsymbol{w}|$. The scalar product of vectors $\boldsymbol{v}$ and $\boldsymbol{w}$ is given by the product of their lengths and the cosine of the angle θ between the vectors.

$$\boldsymbol{v} \cdot \boldsymbol{w} = |\boldsymbol{v}|\,|\boldsymbol{w}|\cos\theta. \tag{1}$$

The magnitude of $\boldsymbol{v}$ can be expressed as $|\boldsymbol{v}| = \sqrt{\boldsymbol{v} \cdot \boldsymbol{v}}$. For a *unit* vector $\boldsymbol{v}$, there occurs $|\boldsymbol{v}| = 1$. With a non-zero vector $\boldsymbol{w}$, the dash over $\boldsymbol{w}$ will indicate the unit vector $\hat{\boldsymbol{w}} = \boldsymbol{w}/|\boldsymbol{w}|$. The cross product $\boldsymbol{v} \times \boldsymbol{w}$ of vectors $\boldsymbol{v}$ and $\boldsymbol{w}$ is defined as a vector $\boldsymbol{u}$ that is perpendicular to both $\boldsymbol{v}$ and $\boldsymbol{w}$ has the length

$$|\boldsymbol{u}| = |\boldsymbol{v}|\,|\boldsymbol{w}|\sin\theta, \tag{2}$$

and its direction is given by the right-hand rule.

Index Notation

It is assumed that the reader is familiar with basic algebraic notions like a group and a vector space. In almost all cases, vectors will have basic geometric interpretation. Classical index-based notation for vectors is used in most of the chapters. It allows for easy almost "mechanical" manipulation of formulas. The notation applies the summation convention: if an index appears twice in a single term, then summation over all allowed values of that index is assumed, e.g., with $i = 1, 2, 3$, one has $v^i \boldsymbol{e}_i = v^1 \boldsymbol{e}_1 + v^2 \boldsymbol{e}_2 + v^3 \boldsymbol{e}_3$. There is no summation if the indices are taken in parentheses, e.g., $v^{(i)} w_{(i)}$. With the Kronecker symbol.

$$\delta^{ij} = \delta_{ij} = \delta^i_j = \delta_j^{\ i} = \begin{cases} 1 \text{ if } i = j \\ 0 \text{ otherwise,} \end{cases}$$

there occurs $\delta^i_j v^j = v^i$. The permutation symbol is defined as

$$\varepsilon^{ijk} = \varepsilon_{ijk} = \begin{cases} 1 & \text{if } (i, j, k) \text{ is an even permutation of } (1, 2, 3) \\ -1 & \text{if } (i, j, k) \text{ is an odd permutation of } (1, 2, 3) \\ 0 & \text{otherwise.} \end{cases}$$

One has

$$\varepsilon^{ijk}\varepsilon_{lmn} = \det \begin{bmatrix} \delta^i_l & \delta^i_m & \delta^i_n \\ \delta^j_l & \delta^j_m & \delta^j_n \\ \delta^k_l & \delta^k_m & \delta^k_n \end{bmatrix}, \quad \varepsilon^{ijm}\varepsilon_{klm} = \delta^i_k\delta^j_l - \delta^i_l\delta^j_k \quad \text{and} \quad \varepsilon^{ikl}\varepsilon_{jkl} = 2\delta^i_j.$$

Moreover, for a 3×3 matrix X, there occurs

$$\varepsilon^{ijk} X_{il} X_{jm} X_{kn} = \det(X)\varepsilon_{lmn}. \tag{3}$$

In the orthonormal basis $\boldsymbol{e}_i = \boldsymbol{e}^i$ (i.e., $\boldsymbol{e}_i \cdot \boldsymbol{e}_j = \delta_{ij}$), the scalar product $\boldsymbol{v} \cdot \boldsymbol{w}$ of vectors $\boldsymbol{v} = v^i \boldsymbol{e}_i$ and $\boldsymbol{w} = w^i \boldsymbol{e}_i$ is

$$\boldsymbol{v} \cdot \boldsymbol{w} = \delta_{ij} v^i w^j,$$

and the components of the cross product $\boldsymbol{u} = u_i \boldsymbol{e}^i = \boldsymbol{v} \times \boldsymbol{w}$ are $u_i = \varepsilon_{ijk} v^j w^k$.

Symbols are used consistently throughout individual chapters. Only selected symbols are consistent throughout the whole book.

List of Selected Symbols

$\boldsymbol{e}_i = \boldsymbol{e}^i$	i-th vector of an orthonormal (Cartesian) basis
$\boldsymbol{a}_i$	i-th vector of a basis in direct space
$\boldsymbol{a}^i$	i-th vector of a basis in reciprocal space; $\boldsymbol{a}_i \cdot \boldsymbol{a}^j = \delta_i^{\ j}$
$\mathcal{T}_i^{\ j}$	j-th Cartesian component of $\boldsymbol{a}_i = \mathcal{T}_i^{\ j}\boldsymbol{e}_j$; $\mathcal{T}_i^{\ j} = \boldsymbol{a}_i \cdot \boldsymbol{e}^j$
$\mathcal{T}^i_{\ j}$	j-th Cartesian component of $\boldsymbol{a}^i = \mathcal{T}^i_{\ j}\boldsymbol{e}^j$; $\mathcal{T}^i_{\ j} = \boldsymbol{a}^i \cdot \boldsymbol{e}_j$
I	identity matrix
I_n	$n \times n$ identity matrix
Λ	lattice
Λ^*	lattice reciprocal to Λ
$\wedge$	accent marking unit vectors; $\hat{\boldsymbol{v}} = \boldsymbol{v}/\lvert\boldsymbol{v}\rvert$ for $\boldsymbol{v} \neq \boldsymbol{0}$
$\vee$	accent marking lattice vectors; e.g., $\check{\boldsymbol{h}} = h_i\boldsymbol{a}^i$, where h_i are integers
$\lceil x \rceil$	ceiling of real x; the smallest integer not smaller than x
$\lfloor x \rfloor$	floor of real x; the greatest integer not greater than x
$\lfloor x \rceil$	integer nearest to real x
$\lvert\boldsymbol{x}\rvert$	magnitude of vector $\boldsymbol{x}$; $\lvert\boldsymbol{x}\rvert = \sqrt{\boldsymbol{x} \cdot \boldsymbol{x}}$
$\lVert M \rVert$	Frobenius norm of matrix M; $\lVert M \rVert = \sqrt{\mathrm{tr}\, M^T M}$
M^{-T}	inverse of transposed matrix M; $M^{-T} = (M^T)^{-1}$
∇	gradient
$\mathcal{F}[f]$	Fourier transform of function f
$\mathcal{F}_x[f](\xi)$	Fourier transform of f with respect to its argument x at ξ
$\mathcal{F}^{-1}$	inverse Fourier transform
δ	Dirac delta "function"
Ш	Dirac comb, Shah "function"
$Ш_\Lambda$	Shah "function" with support at nodes of lattice Λ
Ш	sum of δ "functions"
$1_{\lvert A}$	characteristic function of set A; the function equals 1 in interior of A, 0 in its exterior, and 1/2 on its boundary
Π	rectangular function; $\Pi(x)$ equals 1 if $\lvert x\rvert < 1/2$, 1/2 if $\lvert x\rvert = 1/2$, and 0 otherwise
λ	radiation wavelength
θ	Bragg angle
$\boldsymbol{k}_0$	wave vector of the incident beam
$\boldsymbol{k}$ or $\boldsymbol{k}_{\check{h}}$	wave vector of a diffracted beam
$\rho(\boldsymbol{x})$	charge density at point $\boldsymbol{x}$
$V(\boldsymbol{x})$	electrostatic potential at point $\boldsymbol{x}$
f_n	scattering factor of n-th atom
F	structure factor

NIST Values of Physical Constants

speed of light in vacuum	299792458 m/s (c)
electron mass	$9.1093837015 \times 10^{-31}$ kg (m_0)
elementary charge	$1.602176634 \times 10^{-19}$ C (e)
Planck constant	$6.62607015 \times 10^{-34}$ J s; $4.135667696 \times 10^{-15}$ eV s (h)
electric constant	$8.8541878128 \times 10^{-12}$ C/(V m) (ε_0)

Chapter 1
Elements of Geometric Crystallography

This chapter is mainly about elementary geometric crystallography. For most readers, it will be a just a reminder of simple facts known from geometry and crystallography courses. A detailed account on geometric crystallography can be found in the International Tables for Crystallography [1]. (Further on, the Tables are referred to as [*ITC*].)

Briefly, an ideal crystal is a triply periodic arrangement of a motif of atoms in space. Although atoms vibrate about equilibrium locations, with time averaging, one can assume that they are located at equilibrium points. The idealized crystal (with unlimited extensions) can be seen as a static structure of atoms or ions allocated in space in such a way that the distances between them are not smaller than a certain finite limit, and the arrangement is invariant under a group of translations. It is convenient to discuss the combinations against the background of linear oblique coordinate systems. Therefore, we begin with a brief introduction to such systems.

1.1 Linear Oblique Coordinate Systems

There are a number of different conventions of formal description of crystallographic coordinate systems. The most common approach is to use *lattice parameters* described in Sect. 1.4 below. This approach, however, is not convenient for general considerations, as some expressions (e.g., for angles between crystal planes or directions) become complicated when written for low symmetry cases. Here we use a formalism which may look scary but in reality is relatively simple and easy for implementation, because formulas are expressed directly in vector components or point coordinates. A careful reader will see the links between co- and contra-variance of vectors, co- and contra-variance of Fourier transformation and the duality between coordinate and momentum (wave vector) spaces.

A. Morawiec, *Indexing of Crystal Diffraction Patterns*, Springer Series in Materials Science 326, https://doi.org/10.1007/978-3-031-11077-1_1

Let $\boldsymbol{a}_i$ $(i = 1, 2, 3)$ denote three linearly independent vectors of the three-dimensional Euclidean vector space. These vectors constitute a basis of a coordinate system. The basis may be non-orthogonal, and the coordinate system may be oblique (non-Cartesian). An arbitrary vector $\boldsymbol{v}$ can be expressed as a unique linear combination of basis vectors $\boldsymbol{v} = v^i \boldsymbol{a}_i$. In particular, vectors $\boldsymbol{a}'_i$ of another basis can be expressed as linear combinations of $\boldsymbol{a}_i$

$$\boldsymbol{a}'_i = T_i{}^j \boldsymbol{a}_j \,. \tag{1.1}$$

T represents the matrix of transformation between $\boldsymbol{a}_i$ and $\boldsymbol{a}'_i$.

With $\boldsymbol{v} = v^j \boldsymbol{a}_j = v'^i \boldsymbol{a}'_i = v'^i T_i{}^j \boldsymbol{a}_j$, based on the linear independence of the vectors $\boldsymbol{a}_i$, the (contravariant) components v^i of $\boldsymbol{v}$ satisfy $v^j = v'^i T_i{}^j$, or they are transformed according to

$$v'^i = v^j (T^{-1})_j{}^i \,, \tag{1.2}$$

where (T^{-1}) denotes the inverse of T, i.e., $T_k{}^j (T^{-1})_j{}^i = \delta_k{}^i = (T^{-1})_k{}^j \, T_j{}^i$. Sometimes, instead of $(T^{-1})_j{}^i$, it is convenient to use the symbol $T^i{}_j = (T^{-1})_j{}^i$ as some relationships become more elegant; e.g., (1.2) takes the form $v'^i = T^i{}_j v^j$. With this shorter notation, one needs to be careful about the positions of indices. Therefore, it will be used sporadically (e.g., in Sect. 1.1.2). Moreover, T will denote the matrix with the entries $T_i{}^j$. Consequently, the one with the entries $T^i{}_j$ corresponds to $\left(T^{-1}\right)^T = T^{-T}$.

The scalar product of vectors $\boldsymbol{v}$ and $\boldsymbol{w}$ can be expressed as $\boldsymbol{v} \cdot \boldsymbol{w} = \boldsymbol{a}_i \cdot \boldsymbol{a}_j \, v^i w^j = g_{ij} v^i w^j$, where

$$g_{ij} = \boldsymbol{a}_i \cdot \boldsymbol{a}_j \tag{1.3}$$

are (covariant) components of the *metric tensor*. The components g_{ij} constitute a symmetric, positive definite (Gram) matrix. For brevity, it will be denoted by g.

The orthonormal basis $\boldsymbol{e}_i$ has particularly simple metric: it satisfies $g_{ij} = \boldsymbol{e}_i \cdot \boldsymbol{e}_j = \delta_{ij}$. For denoting the transformation linking a given non-orthonormal basis to an orthonormal basis, instead of T, the special symbol $\mathcal{T}$ will be used; i.e., $\mathcal{T}$ is the transformation between the bases $\boldsymbol{e}_i$ and $\boldsymbol{a}_i$, one has $\boldsymbol{a}_i = \mathcal{T}_i{}^j \boldsymbol{e}_j$, and $\mathcal{T}_i{}^j$ is the j-th component of $\boldsymbol{a}_i$ in the orthonormal basis.

The transformation rule for the entries g_{ij} of the metric g is

$$g'_{ij} = \boldsymbol{a}'_i \cdot \boldsymbol{a}'_j = T_i{}^k T_j{}^l \boldsymbol{a}_k \cdot \boldsymbol{a}_l = T_i{}^k T_j{}^l \, g_{kl} \,. \tag{1.4}$$

This implicates the transformation rule for $\det(g)$. Using (1.3), one obtains $\det(g') = \varepsilon^{ijk} g'_{i1} g'_{j2} g'_{k3} = (\varepsilon^{ijk} T_i{}^m T_j{}^p T_k{}^s)(g_{mn} g_{pq} g_{st})(T_1{}^n T_2{}^q T_3{}^t) = \det(T)(\varepsilon^{mps} g_{mn} g_{pq} g_{st})(T_1{}^n T_2{}^q T_3{}^t)$ and, consequently,

$$\det(g') = \det(T)^2 \det(g) \,. \tag{1.5}$$

In particular, with $\mathcal{T}$ linking $\boldsymbol{a}_i$ to an orthonormal basis $\boldsymbol{e}_i$ ($\boldsymbol{a}_i = \mathcal{T}_i{}^j \boldsymbol{e}_j$), one has from (1.4) and (1.5)

$$g_{ij} = \mathcal{T}_i{}^k \mathcal{T}_j{}^k \text{ and } \det(g) = \det(\mathcal{T})^2 . \tag{1.6}$$

The determinant $\det(g)$ is related to the volume V of a parallelepiped (comprised of all points $x^i \boldsymbol{a}_i$ such that $0 \leq x^i < 1$) spanned by the basis vectors $\boldsymbol{a}_i$. In the Cartesian coordinate system based on the vectors $\boldsymbol{e}_i$, the volume of a parallelepiped spanned on $\boldsymbol{a}_1, \boldsymbol{a}_2, \boldsymbol{a}_3$ equals $V = |\boldsymbol{a}_1 \cdot (\boldsymbol{a}_2 \times \boldsymbol{a}_3)| = |\det(\mathcal{T})|$. Hence, the second of (1.6) leads to

$$\det(g) = V^2 , \tag{1.7}$$

i.e., the determinant of the matrix g is equal to the squared volume of the parallelepiped spanned by the basis vectors.

The components of the matrix inverse to g are denoted by the same letter but with upper indices g^{ij} ($g^{ij} g_{jk} = \delta^i{}_k$). By taking inverses in (1.4), one obtains the transformation rule for the (contravariant) components g^{ij}

$$g'^{ij} = (T^{-1})_k{}^i (T^{-1})_l{}^j g^{kl} = T^i{}_k T^j{}_l \, g^{kl} . \tag{1.8}$$

The vectors $\boldsymbol{a}^i$ ($i = 1, 2, 3$) defined by

$$\boldsymbol{a}^i = g^{ij} \boldsymbol{a}_j \tag{1.9}$$

are linearly independent and serve as a (dual) basis. It is easy to see that they satisfy the relationships

$$g_{ij} \boldsymbol{a}^j = \boldsymbol{a}_i , \quad \boldsymbol{a}^i \cdot \boldsymbol{a}_k = \delta^i{}_k \text{ and } \boldsymbol{a}^i \cdot \boldsymbol{a}^j = g^{ij} . \tag{1.10}$$

If the basis $\boldsymbol{a}_i$ is transformed by T according to (1.1), then the transformation rule for the vectors $\boldsymbol{a}^i$ is

$$\boldsymbol{a}'^i = g'^{ij} \boldsymbol{a}'_j = (T^{-1})_k{}^i (T^{-1})_l{}^j g^{kl} T_j{}^s \boldsymbol{a}_s = (T^{-1})_k{}^i g^{kl} \boldsymbol{a}_l = (T^{-1})_k{}^i \boldsymbol{a}^k = T^i{}_k \boldsymbol{a}^k . \tag{1.11}$$

The (covariant) vector components in the basis $\boldsymbol{a}^i$ are distinguished from the contravariant components by lower indices, e.g., $\boldsymbol{v} = v_i \boldsymbol{a}^i$. With $\boldsymbol{v} = v_j \boldsymbol{a}^j = v'_i \boldsymbol{a}'^i = v'_i (T^{-1})_j{}^i \boldsymbol{a}^j$, the covariant components v_i are transformed via

$$v'_i = T_i{}^j v_j . \tag{1.12}$$

Moreover, one has

$$\boldsymbol{v} \cdot \boldsymbol{a}_i = v_j \boldsymbol{a}^j \cdot \boldsymbol{a}_i = v_i \text{ and } \boldsymbol{v} \cdot \boldsymbol{a}^i = v^j \boldsymbol{a}_j \cdot \boldsymbol{a}^i = v^i .$$

Table 1.1 A two-dimensional example of a vector ($\boldsymbol{v}$) given in two mutually dual bases $\boldsymbol{a}_i$ and $\boldsymbol{a}^i$.

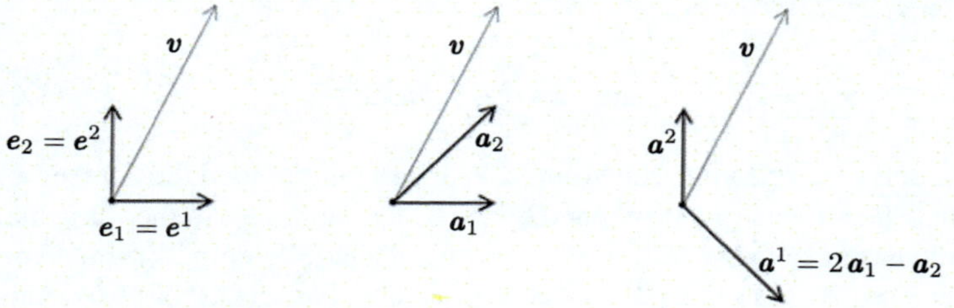

$\mathbf{e}_i \cdot \mathbf{e}_j = \delta_{ij}$	$g = [g_{ij}] = \begin{bmatrix} 1 & 1 \\ 1 & 2 \end{bmatrix}$	$g^{-1} = [g^{ij}] = \begin{bmatrix} 2 & -1 \\ -1 & 1 \end{bmatrix}$
$\boldsymbol{v} = 1\,\mathbf{e}_1 + 2\,\mathbf{e}_2$	$\boldsymbol{v} = -1\,\boldsymbol{a}_1 + 2\,\boldsymbol{a}_2$	$\boldsymbol{v} = 1\,\boldsymbol{a}^1 + 3\,\boldsymbol{a}^2$
	$v^1 = -1\ ,\quad v^2 = 2$	$v_1 = 1\ ,\quad v_2 = 3$

Hence, $v^i = \boldsymbol{v} \cdot \boldsymbol{a}^i = v_j \boldsymbol{a}^j \cdot \boldsymbol{a}^i = g^{ij} v_j$ and, similarly, $v_i = g_{ij} v^j$. Thus, the metric tensor raises and lowers vector indices. Using covariant components, the scalar product of vectors $\boldsymbol{v}$ and $\boldsymbol{w}$ can be written as $\boldsymbol{v} \cdot \boldsymbol{w} = g^{ij} v_i w_j = v^i w_i = v_i w^i$. In particular, with the notation $\boldsymbol{v} \cdot \boldsymbol{v} = \boldsymbol{v}^2 = |\boldsymbol{v}|^2$, the length of a vector equals $|\boldsymbol{v}| = \sqrt{\boldsymbol{v} \cdot \boldsymbol{v}} = \sqrt{v^i v_i}$. For a simple example of a vector given in bases $\boldsymbol{a}_i$ and $\boldsymbol{a}^i$ see Table 1.1.

It is easy to verify that the components of the cross product $\boldsymbol{v} \times \boldsymbol{w} = \boldsymbol{u}$ can be written as

$$u_i = \varepsilon_{ijk} v^j w^k \det(\mathcal{T}) \quad \text{or} \quad u^i = \varepsilon^{ijk} v_j w_k / \det(\mathcal{T}) \tag{1.13}$$

with $\det(\mathcal{T})$ linked to $\det(g)$ and V by (1.6) and (1.7). In particular, since the k-th contravariant component of $\boldsymbol{a}_i$ $(= \delta_i{}^k \boldsymbol{a}_k)$ is $\delta_i{}^k$, the cross product of the basis vectors $\boldsymbol{a}_i$ and $\boldsymbol{a}_j$ is

$$\boldsymbol{a}_i \times \boldsymbol{a}_j = \varepsilon_{ijk} \boldsymbol{a}^k \det(\mathcal{T})\ . \tag{1.14}$$

Hence, to get the vectors $\boldsymbol{a}^i$ from $\boldsymbol{a}_i$, one can use

$$\boldsymbol{a}^k = \varepsilon^{ijk} \boldsymbol{a}_i \times \boldsymbol{a}_j / (2 \det(\mathcal{T})) \tag{1.15}$$

instead of (1.9). Analogous expressions involving $\boldsymbol{a}^i \times \boldsymbol{a}^j$ and $\boldsymbol{a}_k$ can be easily obtained from the second of (1.13).

Let A be the matrix of a linear transformation of vectors within the same basis $\boldsymbol{a}_i$; for $\boldsymbol{v} = v^i \boldsymbol{a}_i$ and $\boldsymbol{w} = w^i \boldsymbol{a}_i$, one has

$$w^i = v^j A_j{}^i\ .$$

In the basis $\boldsymbol{a}'_i = T_i{}^j \boldsymbol{a}_j$, the relationship between the contravariant components of the vectors $\boldsymbol{w} = w'^i \boldsymbol{a}'_i$ and $\boldsymbol{v} = v'^i \boldsymbol{a}'_i$ is

$$w'^i = w^j \,(T^{-1})_j{}^i = v^k \, A_k{}^j \,(T^{-1})_j{}^i = v'^l \, T_l{}^k \,(T^{-1})_j{}^i \, A_k{}^j \; .$$

Thus, the matrix A' of the transformation in the basis $\boldsymbol{a}'_i$ is related to A via

$$A' = TAT^{-1} \; , \tag{1.16}$$

which is an abbreviation for $A'^{\,j}_i = T_i{}^k A_k{}^l \,(T^{-1})_l{}^j = T_i{}^k T^j_{\;l} \, A_k{}^l$.

If the transformation A is an isometry, then $|\boldsymbol{u} - \boldsymbol{w}|^2 = |A\boldsymbol{u} - A\boldsymbol{w}|^2 = |A(\boldsymbol{u} - \boldsymbol{w})|^2$ or, with the substitution $\boldsymbol{u} - \boldsymbol{w} \to \boldsymbol{v}$, one has $(A\boldsymbol{v}) \cdot (A\boldsymbol{v}) = \boldsymbol{v} \cdot \boldsymbol{v}$. Using components, this relationship takes the form $g_{ij} v^k A_k{}^i v^l A_l{}^j = g_{kl} v^k v^l$, and it is true for all $\boldsymbol{v}$. Hence, the entries $A_i{}^j$ in the basis with the metric g must satisfy

$$A_i{}^k A_j{}^l \, g_{kl} = g_{ij} \; , \tag{1.17}$$

and the determinant of A must be equal to ± 1. In the orthonormal basis (with $g_{ij} = \delta_{ij}$), (1.40) is the condition for A to be an orthogonal matrix.

This formalism is applicable to the transformation by a proper rotation. The matrix representing a rotation about $\widehat{\boldsymbol{n}} = n^i \boldsymbol{e}_i$ by the angle ω in a Cartesian frame $\boldsymbol{e}_i$ is known to be given by

$$R_k^{\,l} = \delta_k^{\,l} \cos\omega + n_k n^l (1 - \cos\omega) - \delta^{ls} \varepsilon_{ksm} n^m \;\sin\omega \; . \tag{1.18}$$

In a non-Cartesian frame with the metric g', by (1.16), the representation is $R'_i{}^j = \mathcal{T}_i{}^k R_k^{\,l} (\mathcal{T}^{-1})_l{}^j$. With $R_k^{\,l}$ substituted by (1.18), using the rules collected in Table 1.2 and (1.3), the matrix R' can be expressed as

$$R'_i{}^j = \delta_i{}^j \cos\omega + n'_i \, n'^j \,(1 - \cos\omega) - g'^{jk} \,\varepsilon_{ikl} \, n'^l \;\det(\mathcal{T}) \;\sin\omega \; , \tag{1.19}$$

where n'_i and n'^i are, respectively, co- and contravariant components of $\boldsymbol{n}$ in the oblique frame, and $\det(\mathcal{T})^2 = \det(g')$. Clearly, a proper rotation is an isometry, i.e., the elements $R'_i{}^j$ satisfy (1.17), i.e., $R'_i{}^k R'_j{}^l \, g'_{kl} = g'_{ij}$.

For those less familiar with the subject matter of this section, it is worth noting that it is a part of the tensor formalism used standardly in physics. Tensors have components transforming according to rules analogous to (1.2, 1.4, 1.8, 1.12). For a summary of formulas, see Table 1.2. The metric tensor allows for passing between co- and contra-variant tensor components. With $\boldsymbol{a}_i = g_{ij} \boldsymbol{a}^j$, g can be seen as a matrix linking two particular bases of the same vector space, but it is convenient to consider vectors expressed in these bases as elements of separate spaces; in the crystallographic context these are the *direct space* based on $\boldsymbol{a}_i$ and the *reciprocal space* based on vectors $\boldsymbol{a}^i$.

Table 1.2 Juxtaposition of formulas for contravariant and covariant vector components. The bases and the metrics are related by $\boldsymbol{a}^i \cdot \boldsymbol{a}_j = \delta^i{}_j$ and $g^{ik} g_{kj} = \delta^i{}_j$, respectively.

$\boldsymbol{a}'_i = T_i{}^j \boldsymbol{a}_j$	$\boldsymbol{a}'^i = (T^{-1})_j{}^i \boldsymbol{a}^j = T^i{}_j \boldsymbol{a}^j$
$\boldsymbol{v} = v^j \boldsymbol{a}_j = v'^i \boldsymbol{a}'_i$	$\boldsymbol{v} = v_i \boldsymbol{a}^i = v'_i \boldsymbol{a}'^i$
$v'^i = v^j (T^{-1})_j{}^i = T^i{}_j v^j$	$v'_i = T_i{}^j v_j$
$\boldsymbol{v} \cdot \boldsymbol{a}^i = v^i$	$\boldsymbol{v} \cdot \boldsymbol{a}_i = v_i$
$g_{ij} = \boldsymbol{a}_i \cdot \boldsymbol{a}_j$	$g^{ij} = \boldsymbol{a}^i \cdot \boldsymbol{a}^j$
$g'_{ij} = T_i{}^k T_j{}^l g_{kl}$	$g'^{ij} = (T^{-1})_k{}^i (T^{-1})_l{}^j g^{kl} = T^i{}_k T^j{}_l g^{kl}$
$\boldsymbol{v} \cdot \boldsymbol{w} = g_{ij} v^i w^j$	$\boldsymbol{v} \cdot \boldsymbol{w} = g^{ij} v_i w_j$
$\boldsymbol{a}_i = g_{ij} \boldsymbol{a}^j$	$\boldsymbol{a}^i = g^{ij} \boldsymbol{a}_j$
$v^i = g^{ij} v_j$	$v_i = g_{ij} v^j$

1.1.1 Component-free Tensor Notation

A more abstract component-free tensor notation is sometimes used. Below is a brief explanation limited to second order tensors and parts of the formalism described above.

⊡ A second order tensor is a linear transformation of the vector space onto itself. Linearity means that the tensor T acting on the linear combination $p\boldsymbol{v} + q\boldsymbol{w}$ of vectors $\boldsymbol{v}$ and $\boldsymbol{w}$ (p and q are real numbers) satisfies

$$T(p\boldsymbol{v} + q\boldsymbol{w}) = pT(\boldsymbol{v}) + qT(\boldsymbol{w}) .$$

For brevity, one usually writes $T\boldsymbol{v}$ instead of $T(\boldsymbol{v})$. By definition, the identity tensor I satisfies $I\boldsymbol{v} = \boldsymbol{v}$ for all $\boldsymbol{v}$, and the zero tensor transforms every vector to $\boldsymbol{0}$. Second order tensors T_1 and T_2 are multiplied using the rule

$$(T_1 T_2)\boldsymbol{v} = T_1(T_2\boldsymbol{v}) .$$

Addition of tensors and multiplication by a number are based on

$$((pT_1) + (qT_2))\boldsymbol{v} = p(T_1\boldsymbol{v}) + q(T_2\boldsymbol{v}) .$$

Tensor T^T satisfying $(T^T\boldsymbol{v}) \cdot \boldsymbol{w} = \boldsymbol{v} \cdot (T\boldsymbol{w})$ for arbitrary $\boldsymbol{v}$ and $\boldsymbol{w}$ is referred to as the transpose of T. If $T^T = +T$ ($T^T = -T$), the tensor T is called symmetric (anti-symmetric). The tensor inverse to T is defined as T^{-1} such that $T^{-1}(T\boldsymbol{v}) = T(T^{-1}\boldsymbol{v}) = \boldsymbol{v}$ for every $\boldsymbol{v}$. This condition is briefly written as $T^{-1}T = TT^{-1} = I$.

The tensor product $\boldsymbol{u} \otimes \boldsymbol{v}$ of two vectors $\boldsymbol{u}$ and $\boldsymbol{v}$ is a second order tensor acting on a given vector $\boldsymbol{w}$ via $(\boldsymbol{u} \otimes \boldsymbol{v})\boldsymbol{w} = \boldsymbol{u}(\boldsymbol{v} \cdot \boldsymbol{w})$. It is easy to verify that $(p\boldsymbol{u} + q\boldsymbol{v}) \otimes \boldsymbol{w} = p\boldsymbol{u} \otimes \boldsymbol{w} + q\boldsymbol{v} \otimes \boldsymbol{w}$, $(\boldsymbol{u} \otimes \boldsymbol{v})^T = \boldsymbol{v} \otimes \boldsymbol{u}$ and $(\boldsymbol{u} \otimes \boldsymbol{v})(\boldsymbol{w} \otimes \boldsymbol{s}) = (\boldsymbol{v} \cdot \boldsymbol{w})(\boldsymbol{u} \otimes \boldsymbol{s})$.

With $\boldsymbol{a}_i$ being vectors of a direct basis and $\boldsymbol{a}^i$ constituting the dual basis determined by the conditions $\boldsymbol{a}_i \cdot \boldsymbol{a}^j = \delta_i^j$, the tensor $g = \boldsymbol{a}_i \otimes \boldsymbol{a}^i$ transforms the basis $\boldsymbol{a}_i$ onto itself, and due to linearity, it transforms an arbitrary vector onto itself. Hence, $g = \boldsymbol{a}_i \otimes \boldsymbol{a}^i = I$, and similarly $\boldsymbol{a}^i \otimes \boldsymbol{a}_i = I$.[1]

[1] It is worth noting that the tensor $G = \delta^{ij} \boldsymbol{a}_i \otimes \boldsymbol{a}_j = \boldsymbol{a}_i \otimes \boldsymbol{a}_i$ transforms $\boldsymbol{a}^i$ on $\boldsymbol{a}_i$, i.e., $G\boldsymbol{a}^i = \boldsymbol{a}_i$. In other words, this tensor links the dual basis to the direct basis. Clearly, G is symmetric. Similarly, the symmetric tensor $\boldsymbol{a}^i \otimes \boldsymbol{a}^i$ transforms the direct basis $\boldsymbol{a}_i$ on its dual $\boldsymbol{a}^i$, and the tensor $\boldsymbol{a}^i \otimes \boldsymbol{a}^i$ is inverse to G, i.e., $\boldsymbol{a}^i \otimes \boldsymbol{a}^i = G^{-1}$.

A tensor, say, T, can be expressed as

$$T = T_{ij}\boldsymbol{a}^i \otimes \boldsymbol{a}^j$$

where the components T_{ij} are given by

$$T_{ij} = \boldsymbol{a}_i \cdot (T\boldsymbol{a}_j) .$$

This can be easily seen as for an arbitrary vector $\boldsymbol{v} = v^k\boldsymbol{a}_k$ the following string of equalities occurs: $(T_{ij}\boldsymbol{a}^i \otimes \boldsymbol{a}^j)\boldsymbol{v} = v^k T_{ij}(\boldsymbol{a}^i \otimes \boldsymbol{a}^j)\boldsymbol{a}_k = v^k T_{ij}\boldsymbol{a}^i \delta_k^j = v^j T_{ij}\boldsymbol{a}^i = v^j \boldsymbol{a}^i(\boldsymbol{a}_i \cdot (T\boldsymbol{a}_j)) = \boldsymbol{a}^i(\boldsymbol{a}_i \cdot (T\boldsymbol{v})) = (\boldsymbol{a}^i \otimes \boldsymbol{a}_i)(T\boldsymbol{v}) = IT\boldsymbol{v} = T\boldsymbol{v}$. Similarly, one can write $T = T^{ij}\boldsymbol{a}_i \otimes \boldsymbol{a}_j$, where $T^{ij} = \boldsymbol{a}^i \cdot (T\boldsymbol{a}^j)$. The components g_{ij} of $g = \boldsymbol{a}_i \otimes \boldsymbol{a}^i$ in the basis $\boldsymbol{a}_i$ are

$$g_{ij} = \boldsymbol{a}_i \cdot (g\boldsymbol{a}_j) = \boldsymbol{a}_i \cdot ((\boldsymbol{a}_k \otimes \boldsymbol{a}^k)\boldsymbol{a}_j) = \boldsymbol{a}_i \cdot (\boldsymbol{a}_k \delta_j^{\,k}) = \boldsymbol{a}_i \cdot \boldsymbol{a}_j .$$

Similarly, the components g^{ij} of g in the basis $\boldsymbol{a}^i$ are $g^{ij} = \boldsymbol{a}^i \cdot (g\boldsymbol{a}^j) = \boldsymbol{a}^i \cdot \boldsymbol{a}^j$. Based on the above, one has $g = g^{ij}\boldsymbol{a}_i \otimes \boldsymbol{a}_j = g_{ij}\boldsymbol{a}^i \otimes \boldsymbol{a}^j$.

Let $\boldsymbol{a}_i$ and $\boldsymbol{a}'_i$ be vectors of two bases. With the basis $\boldsymbol{a}^i$ dual to $\boldsymbol{a}_i$, the tensor T defined as

$$T = \boldsymbol{a}'_i \otimes \boldsymbol{a}^i \tag{1.20}$$

transforms the basis $\boldsymbol{a}_i$ to the basis $\boldsymbol{a}'_i$

$$T\boldsymbol{a}_i = (\boldsymbol{a}'_k \otimes \boldsymbol{a}^k)\boldsymbol{a}_i = \boldsymbol{a}'_k(\boldsymbol{a}^k \cdot \boldsymbol{a}_i) = \boldsymbol{a}'_k \delta^k_{\ i} = \boldsymbol{a}'_i .$$

The vector $\boldsymbol{a}'_i$ can be expressed as a combination of basis vectors $\boldsymbol{a}_j$ with coefficients $T_i^{\ j}$; one has

$$\boldsymbol{a}'_i = T\boldsymbol{a}_i = T_i^{\ j}\boldsymbol{a}_j . \tag{1.21}$$

Hence, T can be written in the form referring only to the basis $\boldsymbol{a}_i$ and its dual basis $\boldsymbol{a}^i$

$$T = T_i^{\ j}\boldsymbol{a}_j \otimes \boldsymbol{a}^i$$

with coefficients

$$T_i^{\ j} = (T\boldsymbol{a}_i) \cdot \boldsymbol{a}^j = \boldsymbol{a}'_i \cdot \boldsymbol{a}^j . \tag{1.22}$$

For the tensor $\mathcal{T}$ transforming an orthonormal basis $\boldsymbol{e}_i = \boldsymbol{e}^i$ to $\boldsymbol{a}_i$, (1.20), (1.21) and (1.22) take the particular forms $\mathcal{T} = \boldsymbol{a}_i \otimes \boldsymbol{e}^i$, $\boldsymbol{a}_i = \mathcal{T}\boldsymbol{e}_i = \mathcal{T}_i^{\ j}\boldsymbol{e}_j$ and $\mathcal{T}_i^{\ j} = (\mathcal{T}\boldsymbol{e}_i) \cdot \boldsymbol{e}^j = \boldsymbol{a}_i \cdot \boldsymbol{e}^j$, respectively.

For arbitrary vectors $\boldsymbol{u}$, $\boldsymbol{v}$ and $\boldsymbol{w}$ in three-dimensional space, one has $T\boldsymbol{u} \cdot (T\boldsymbol{v} \times T\boldsymbol{w}) = \mathcal{I}_3\, \boldsymbol{u} \cdot (\boldsymbol{v} \times \boldsymbol{w})$ with the scalar $\mathcal{I}_3$ depending only on T; $\det T = \mathcal{I}_3$ is referred to as determinant of tensor T. Similarly, one has $T\boldsymbol{u} \cdot (\boldsymbol{v} \times \boldsymbol{w}) + \boldsymbol{u} \cdot (T\boldsymbol{v} \times \boldsymbol{w}) + \boldsymbol{u} \cdot (\boldsymbol{v} \times T\boldsymbol{w}) = \mathcal{I}_1\, \boldsymbol{u} \cdot (\boldsymbol{v} \times \boldsymbol{w})$ with the scalar $\mathcal{I}_1$ depending only on T; $\mathrm{tr}T = \mathcal{I}_1$ is referred to as the trace of T.

1.1.2 Frames—Overcomplete Sets of Vectors

A more general representation of a vector is needed for an in-depth explanation of 'symmetric' Bravais–Miller and Weber indices in Sect. 1.5.2 below or quasilattices in Chap. 13. It relies on a frame—a generalization of a basis of a vector space [2]. Frames can be useful for accounting for symmetries [3]. As the number of vectors in a frame can be larger than three (vectors in a basis), using frames instead of bases can improve stability of some computations.

Besides three linearly independent basis vectors $\boldsymbol{a}_i$, one may consider a set (frame) of $N \geq 3$ vectors $\boldsymbol{a}'_\mu$. By abuse of notation, the prime in $\boldsymbol{a}'_\mu$ will be omitted, and the vectors $\boldsymbol{a}'_\mu$ will be distinguished from vectors $\boldsymbol{a}_i$ only by their Greek index, i.e., $\boldsymbol{a}'_\mu = \boldsymbol{a}_\mu$ and in general $\boldsymbol{a}_\mu \neq \boldsymbol{a}_i$ even if $\mu = i$. (Analogously, in the case of quantities used below, $v^\mu \neq v^i$ even if $\mu = i$, and $g_{\mu\nu} \neq g_{ij}$ even if $\mu = i$ and $\nu = j$.) Clearly, a vector of the frame can be expressed as

$$\boldsymbol{a}_\mu = T_\mu{}^j \boldsymbol{a}_j \,, \quad \mu = 1, 2, ..., N \,, \quad j = 1, 2, 3 \,.$$

It is assumed here that no two vectors $\boldsymbol{a}_\mu$ are collinear, and the rank of the matrix $T_\mu{}^j = \boldsymbol{a}_\mu \cdot \boldsymbol{a}^j$ equals 3. Thus, the vectors $\boldsymbol{a}_\mu$ span the three-dimensional vector space, and an arbitrary vector can be expressed as a linear combination of $\boldsymbol{a}_\mu$. Let the matrix with entries $T^\mu{}_j$ be the transposed pseudoinverse[2] of $T_\mu{}^j$. With the full rank of $T_\mu{}^j$, one has $T_\mu{}^i T^\mu{}_j = \delta^i{}_j$, but the products with summation over Latin indices

$$T_\mu{}^j \, T^\nu{}_j = g_\mu{}^\nu \quad \text{and} \quad T^\mu{}_j \, T_\nu{}^j = g^\mu{}_\nu$$

are generally different from the Kronecker delta. These matrices satisfy the symmetry condition $g_\mu{}^\nu = g^\nu{}_\mu$, they are idempotent ($g_\mu{}^\kappa g_\kappa{}^\nu = g_\mu{}^\nu$ and $g^\mu{}_\kappa g^\kappa{}_\nu = g^\mu{}_\nu$), and $g_\mu{}^\mu = 3$. Using $T^\mu{}_j$, one defines the (canonically) dual frame

$$\boldsymbol{a}^\mu = T^\mu{}_j \boldsymbol{a}^j$$

and the analogue of the metric tensor

$$g_{\mu\nu} = \boldsymbol{a}_\mu \cdot \boldsymbol{a}_\nu \quad \text{and} \quad g^{\mu\nu} = \boldsymbol{a}^\mu \cdot \boldsymbol{a}^\nu \,.$$

Based on the above definitions, it can be shown that $g^{\mu\nu}$ is the pseudoinverse of $g_{\mu\nu}$. Moreover, one has

$$\begin{aligned} g^\mu{}_\nu \boldsymbol{a}^\nu &= \boldsymbol{a}^\mu \,, \quad & g_\mu{}^\nu \boldsymbol{a}_\nu &= \boldsymbol{a}_\mu \,, \\ g_{\mu\nu} \boldsymbol{a}^\nu &= \boldsymbol{a}_\mu \,, \quad & g^{\mu\nu} \boldsymbol{a}_\nu &= \boldsymbol{a}^\mu \,, \\ \boldsymbol{a}_\mu \cdot \boldsymbol{a}^\nu &= g_\mu{}^\nu \quad \text{and} \quad & g_{\mu\rho}\, g^{\rho\nu} &= g_\mu{}^\nu \,. \end{aligned}$$

If $N > 3$, the set of vectors $\boldsymbol{a}_\mu$ is not a basis, and the decomposition of a given vector into a linear combination of $\boldsymbol{a}_\mu$ is ambiguous. It is made unique by additional constraints which follow from a particular designation of vector components. With $v^j \boldsymbol{a}_j = \boldsymbol{v} = v_j \boldsymbol{a}^j$, and vector components v^μ and v_μ defined as

$$v^\mu = T^\mu{}_j \, v^j \quad \text{and} \quad v_\mu = T_\mu{}^j v_j \,,$$

[2] Readers unfamiliar with generalized matrix inverses are referred to [4].

one has a set of relationships analogous to those given for linearly independent basis vectors

$$v^\mu \boldsymbol{a}_\mu = \boldsymbol{v} = v_\mu \boldsymbol{a}^\mu \ , \tag{1.23}$$

$$v^\mu = \boldsymbol{v} \cdot \boldsymbol{a}^\mu \ , \quad v_\mu = \boldsymbol{v} \cdot \boldsymbol{a}_\mu \ , \tag{1.24}$$

$$g_{\mu\nu} v^\nu = v_\mu \ , \quad g^{\mu\nu} v_\nu = v^\mu \ , \tag{1.25}$$

and the additional conditions

$$g^\mu{}_\nu v^\nu = v^\mu \ , \quad g_\mu{}^\nu v_\nu = v_\mu \ . \tag{1.26}$$

The properties (1.26) are general in the sense that they are applicable to other tensor quantities, e.g., $g_{\mu\kappa} g^\kappa{}_\nu = g_{\mu\nu}$, $g^{\mu\kappa} g_\kappa{}^\nu = g^{\mu\nu}$, $g_\mu{}^\nu T_\nu{}^j = T_\mu{}^j$ and $g^\mu{}_\nu T^\nu{}_j = T^\mu{}_j$. The relationships (1.23–1.26) are illustrated in Fig. 1.1.

Rotation matrices in frames

With the Cartesian basis $\boldsymbol{e}_i$, the general T is replaced by $\mathcal{T}$, i.e., in particular, $\mathcal{T}_\mu{}^i = \boldsymbol{a}_\mu \cdot \boldsymbol{e}^i$. Proceeding as in the case of (1.18) and (1.19), one has

$$R_\mu{}^\nu = \mathcal{T}_\mu{}^i \mathcal{T}^\nu{}_j R_i{}^j \ . \tag{1.27}$$

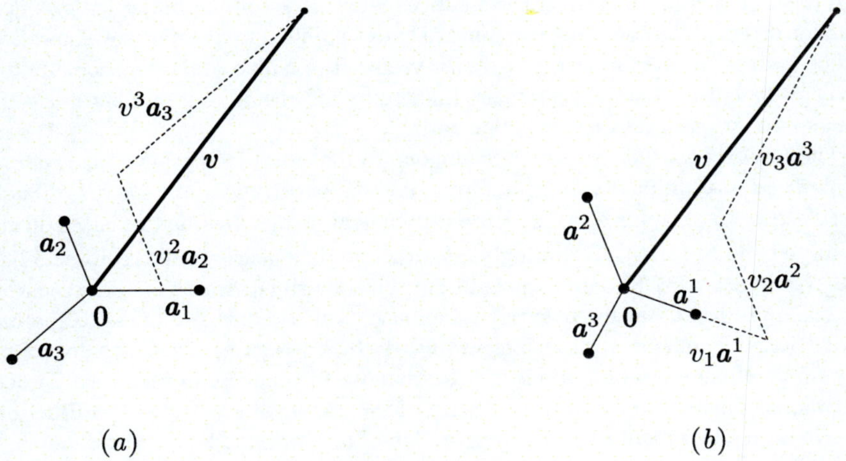

Fig. 1.1 Schematic two-dimensional illustration of decompositions $v^\mu \boldsymbol{a}_\mu$ and $v_\mu \boldsymbol{a}^\mu$ ($\mu = 1, 2, 3$) of vector $\boldsymbol{v}$. With $|\boldsymbol{a}_1| = 1$, the set of vectors $\boldsymbol{a}^\mu$ in (*b*) is dual to $\boldsymbol{a}_\mu$ shown in (*a*). The coefficients v^μ and v_μ satisfy the relationships (1.26).

Hence, based on (1.18), the rotation by θ about $n^\mu = \mathcal{T}^\mu_{\ j} n^j$ is represented by

$$R_\mu^{\ \nu} = g_\mu^{\ \nu} \cos\theta + n_\mu n^\nu (1 - \cos\theta) - g^{\nu\rho} \epsilon_{\mu\rho\kappa} n^\kappa \sin\theta \,, \tag{1.28}$$

where $\epsilon_{\kappa\mu\nu} = \mathcal{T}_\kappa^{\ i} \mathcal{T}_\mu^{\ j} \mathcal{T}_\nu^{\ k} \varepsilon_{ijk}$. The quantity $\epsilon_{\kappa\mu\nu}$ is antisymmetric in each pair of indices. Moreover, $\epsilon_{\kappa\mu\nu}$ is equal to the signed volume of the parallelepiped spanned on the vectors $\boldsymbol{a}_\kappa$, $\boldsymbol{a}_\mu$, $\boldsymbol{a}_\nu$. In particular, if vectors $\boldsymbol{a}_\kappa$, $\boldsymbol{a}_\mu$, $\boldsymbol{a}_\nu$ are linearly dependent, and one has $\epsilon_{\kappa\mu\nu} = 0$.

Analogously to the 3×3 matrices $R_i^{\ j}$, the $N \times N$ matrices $R_\mu^{\ \nu}$ of (1.28) constitute a faithful representation of proper rotations. If $N > 3$, this is a singular-matrix representation. The null rotation is represented by $R_\mu^{\ \nu} = g_\mu^{\ \nu}$. The matrices satisfy the analogue of the orthogonality condition (1.17); one has $R_\mu^{\ \kappa} R_\nu^{\ \lambda} g_{\kappa\lambda} = g_{\mu\nu}$.

1.2 Lattices

A n-dimensional lattice is the set of terminal points of all vectors

$$\boldsymbol{x}_0 + m^i \boldsymbol{a}_i \,, \quad \text{with } i = 1, ..., n \text{ and } \textit{integer} \text{ coefficients } m^i \,,$$

where $\boldsymbol{a}_i$ are linearly independent vectors in Euclidean space.[3] Here, mainly three-dimensional ($n = 3$) lattices will be considered, but also the cases of $n = 1$ and $n = 2$ will arise. It is convenient to discuss lattices with the assumption that $\boldsymbol{x}_0 = \boldsymbol{0}$, i.e., with the origin of the coordinate system at a lattice point. Lattice points are also called lattice nodes. The vector $m^i \boldsymbol{a}_i$ is a lattice vector. The lattice vectors $\boldsymbol{a}_i$ constitute a basis of the lattice. Lattice vectors are marked by the check$\check{}$ (e.g., $\check{\boldsymbol{v}}$), but this does not apply to vectors selected as lattice basis.

The notion of *lattice metric* is analogous to that used in the previous section: $g_{ij} = \boldsymbol{a}_i \cdot \boldsymbol{a}_j$. Additionally, in crystallography, the lattice based on vectors $\boldsymbol{a}^i$ such that $\boldsymbol{a}^i \cdot \boldsymbol{a}_j = \delta^i_{\ j}$ is referred to as *reciprocal* (or dual) to the *direct* lattice based on $\boldsymbol{a}_i$; cf. the second of (1.10). Accordingly, $g^{ij} = \boldsymbol{a}^i \cdot \boldsymbol{a}^j$ is a metric of the reciprocal lattice. Basis vectors of the reciprocal lattice are customarily defined by (1.15). Clearly, the scalar product of the direct lattice vector $m^i \boldsymbol{a}_i$ with a vector $n_i \boldsymbol{a}^i$ of the reciprocal lattice (i.e., with integer n_i) is an integer $(m^i \boldsymbol{a}_i) \cdot (n_i \boldsymbol{a}^i) = m^i n_i$. Thanks to this property (in combination with Fourier transformation), the concept of reciprocal lattice is essential for interpretation of diffraction patterns, as an ideal crystal diffraction pattern is an experiment-dependent projection of the (weighted) reciprocal lattice.[4]

---

[3] The notion of lattice (or quasilattice of Sect. 13.1.5) corresponds to $\mathbb{Z}$-module, an algebraic structure analogous to classic vector space with the field of real numbers replaced by the ring of integers.

[4] The concept of the reciprocal (polar) lattice appeared first in works of Auguste Bravais. In the present form, reciprocal systems of three vectors became widely recognized thanks to the seminal *Vector analysis* by J. Willard Gibbs and Edwin B. Wilson [5]. The reciprocal lattice was linked to

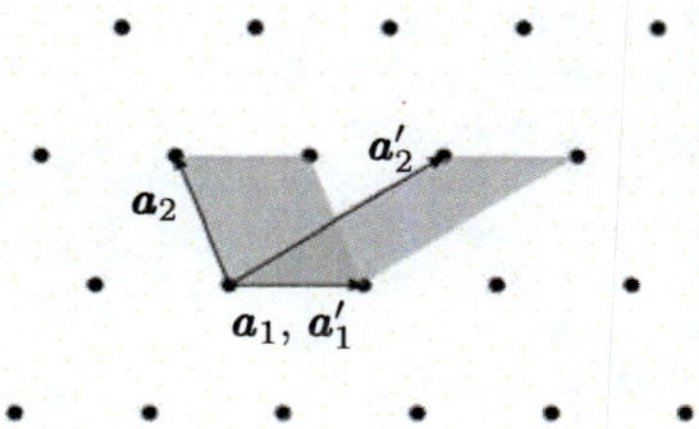

Fig. 1.2 Two different bases $\boldsymbol{a}_i$ and $\boldsymbol{a}'_i$ $(i = 1, 2)$ of a two-dimensional lattice. Primitive cells spanned by the bases are marked in gray.

The choice of lattice basis is not unique; e.g., if the basis vectors $\boldsymbol{a}_1$ and $\boldsymbol{a}_2$ are replaced by $\boldsymbol{a}'_1 = \boldsymbol{a}_1$ and $\boldsymbol{a}'_2 = \boldsymbol{a}_2 + m\boldsymbol{a}_1$, where m is an integer, the new vectors also constitute a basis (Fig. 1.2). For a three-dimensional lattice, there are an infinite number of different bases. Three independent lattice vectors $\boldsymbol{a}'_i$ of the lattice based on $\boldsymbol{a}_i$ also constitute its basis if and only if they span a parallelepiped of the same volume as the one spanned by $\boldsymbol{a}_i$. This last statement can be easily proved.

□ Since $\boldsymbol{a}'_i$ are lattice vectors, one has $\boldsymbol{a}'_i = T_i{}^j \boldsymbol{a}_j$ with T containing integer entries.
⇒ If the vectors $\boldsymbol{a}'_i$ constitute a basis, then $(T^{-1})_j{}^i \boldsymbol{a}'_i = \boldsymbol{a}_j$, and (T^{-1}) contains integer entries. Thus, both $\det(T)$ and $\det(T^{-1}) = 1/\det(T)$ are integers. Hence, T is unimodular ($|\det(T)| = 1$), and due to (1.5) and (1.7), the volumes of parallelepipeds are equal.
⇐ Assuming equal volumes of the parallelepipeds, T must be unimodular. Hence, its inverse (T^{-1}) contains integer entries. It follows then that an arbitrary vector (say, $m^i \boldsymbol{a}_i$ with integer m^i) of the lattice can be expressed as a linear combination of $\boldsymbol{a}'_i$: $m^i \boldsymbol{a}_i = m^i (T^{-1})_i{}^j \boldsymbol{a}'_j$, with integer coefficients $m^i (T^{-1})_i{}^j$, i.e., the vectors $\boldsymbol{a}'_j$ constitute a basis.

In crystallography, parallelepipeds spanned by the lattice bases are referred to as primitive cells. Any two primitive cells of the same lattice have equal volumes; the squared volume of a primitive cell is equal to the determinant of the lattice metric tensor; see, (1.7). Coordinates of points with respect to a lattice basis are sometimes called *lattice coordinates*. Lattice coordinates of points of the primitive cell spanned by this basis are within the range between 0 and 1. Clearly, lattice coordinates of lattice nodes are integers.

1.2.1 Lagrange-Gauss Reduction

With the possibility of selecting a basis of a given lattice in an infinite number of ways, the question arises about a method to substitute an arbitrary initial basis by an equivalent 'canonical' basis, such that the result is unique (apart from orientation), i.e., independent of the initial basis. Clearly, such a method would simplify comparisons and classifications of lattices. Another point is that some bases are more convenient

---

diffraction by Paul P. Ewald [6, 7]. The general reciprocal lattice in application to crystallography was first presented by Max von Laue at the second Solvay Conference in 1913 [8]. The conference proceedings were published in 1921, but the same material is contained in [9] published in 1914.

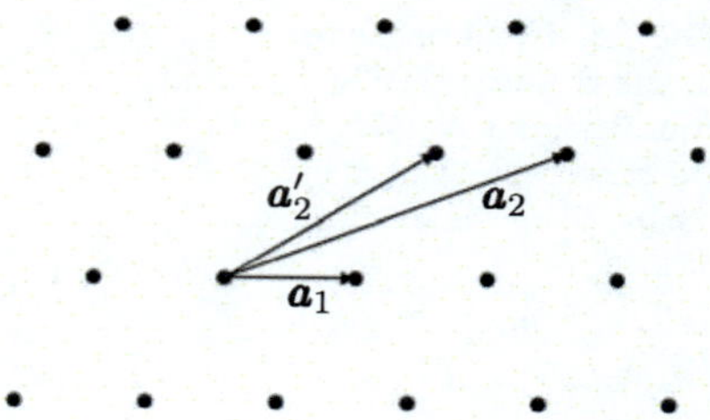

Fig. 1.3 Illustration of the Lagrange-Gauss reduction. With $|\boldsymbol{a}_1| \leq |\boldsymbol{a}_2|$ and $2\boldsymbol{a}_1 \cdot \boldsymbol{a}_2 > \boldsymbol{a}_1^2$, the vector $\boldsymbol{a}_2' = \boldsymbol{a}_2 - \mathrm{sign}(\boldsymbol{a}_2 \cdot \boldsymbol{a}_1)\,\boldsymbol{a}_1$ is shorter than $\boldsymbol{a}_2$, and the pair $(\boldsymbol{a}_1, \boldsymbol{a}_2')$ is a basis of the lattice based on $(\boldsymbol{a}_1, \boldsymbol{a}_2)$.

than others. Particularly convenient are bases made of short nearly orthogonal vectors. The process of determining such bases is known as basis reduction. One can formulate the (shortest basis) problem: given a lattice basis, find a basis such that the length of its longest vector is the shortest possible.

In two-dimensional space, one can select the shortest independent lattice vectors $\boldsymbol{a}_1$ and $\boldsymbol{a}_2$ such that $|\boldsymbol{a}_1| \leq |\boldsymbol{a}_2|$ and $0 \leq 2\boldsymbol{a}_1 \cdot \boldsymbol{a}_2 \leq \boldsymbol{a}_1^2$, and the corresponding metric tensor is unique. Reduction of an arbitrary basis in two-dimensional space can be easily performed using the following property: given two lattice vectors $\check{\boldsymbol{v}}$ and $\check{\boldsymbol{w}}$ and $\check{\boldsymbol{w}}' = \check{\boldsymbol{w}} - \mathrm{sign}(\check{\boldsymbol{v}} \cdot \check{\boldsymbol{w}})\,\check{\boldsymbol{v}}$, if the vectors $\check{\boldsymbol{v}}$ and $\check{\boldsymbol{w}}$ satisfy $2|\check{\boldsymbol{v}} \cdot \check{\boldsymbol{w}}| > \check{\boldsymbol{v}}^2$ then

$$\check{\boldsymbol{w}}^2 - \check{\boldsymbol{w}}'^2 = 2\mathrm{sign}(\check{\boldsymbol{v}} \cdot \check{\boldsymbol{w}})\,\check{\boldsymbol{w}} \cdot \check{\boldsymbol{v}} - \check{\boldsymbol{v}}^2 = 2|\check{\boldsymbol{v}} \cdot \check{\boldsymbol{w}}| - \check{\boldsymbol{v}}^2 > 0\,,$$

i.e., the magnitude of $\check{\boldsymbol{w}}'$ is smaller than that of $\check{\boldsymbol{w}}$. Thus, the consecutive replacement of the longer basis vector $\boldsymbol{a}_2$ by $\boldsymbol{a}_2 - \mathrm{sign}(\boldsymbol{a}_2 \cdot \boldsymbol{a}_1)\,\boldsymbol{a}_1$ ultimately gives the shortest basis vectors (Fig. 1.3). Since the minimum of $|\boldsymbol{a}_2 - x\boldsymbol{a}_1|^2$ is reached at $x = (\boldsymbol{a}_1 \cdot \boldsymbol{a}_2)/\boldsymbol{a}_1^2$, one can speed up the process by replacing the longer basis vector $\boldsymbol{a}_2$ by $\boldsymbol{a}_2 - \lfloor x \rceil \boldsymbol{a}_1$, where $\lfloor x \rceil$ is the integer nearest to x. The procedure is known as Lagrange-Gauss reduction.

⊡ In spaces of higher dimension the problem of lattice reduction is more complicated. Actually, many lattice-related problems are hard to solve (and hence lattice-based cryptography and cryptanalysis). Although most of the theory concerning lattices is far beyond the needs of a crystallographer, it is worth to be aware of some results. Generally, in spaces of arbitrary dimension n, one might consider the following approach: With $\check{\boldsymbol{v}}_1$ being the shortest non-zero vector of the lattice, $\check{\boldsymbol{v}}_i$ is a shortest lattice vector linearly independent of the vectors $\check{\boldsymbol{v}}_k$, where $k < i \leq n$. Such vectors with 'successive minimal' $|\check{\boldsymbol{v}}_i|$ will definitively constitute a basis only if the space dimension n does not exceed four.[5] This complication is circumvented by modifying of the above assertion: A lattice basis $\boldsymbol{a}_1, \boldsymbol{a}_2, \ldots, \boldsymbol{a}_n$ is (Minkowski) reduced if for each $i \leq n$ the vector $\boldsymbol{a}_i$ is the shortest vector such that the set $\boldsymbol{a}_1, \boldsymbol{a}_2, \ldots, \boldsymbol{a}_i$ can be extended to a basis. Computation of Minkowski reduced bases turns out to be costly. Therefore, the weaker but faster Lenstra-Lenstra-Lovász (LLL) algorithm and its variants have become important tools of lattice reduction. The LLL algorithm is a generalization

[5] The simplest example the breakdown of the above scheme in spaces of higher dimensions is as follows: Let $\boldsymbol{e}_i$ $(i = 1, \ldots, 5)$ denote unit vectors along Cartesian axes. With the five-dimensional lattice based on $\boldsymbol{a}_1 = \boldsymbol{e}_1$, $\boldsymbol{a}_2 = \boldsymbol{e}_2$, $\boldsymbol{a}_3 = \boldsymbol{e}_3$, $\boldsymbol{a}_4 = \boldsymbol{e}_4$ and $\boldsymbol{a}_5 = \sum_{i=1}^{5} \boldsymbol{e}_i/2$, the lattice vector $\boldsymbol{e}_5$ is shorter than $\boldsymbol{a}_5$, but the set of independent vectors $\boldsymbol{e}_1, \boldsymbol{e}_2, \boldsymbol{e}_3, \boldsymbol{e}_4, \boldsymbol{e}_5$ is not a basis of the lattice.

of the Lagrange-Gauss reduction supported by the Gram-Schmidt orthogonalization[6] to estimate the quality of a basis. Most studies of lattices are occupied with the fully-dimensional case (with the number of basis vectors equal to the dimension of the space), but from our viewpoint, it is interesting to consider the following (Pohst) problem: Given some lattice vectors which exceed in number the space dimension (and, consequently, are linearly dependent), find a lattice basis made of short vectors. This problem is solved by a modified version of LLL lattice reduction algorithm (sometimes referred to as MLLL) by Pohst [10, 11]. All this is mentioned here to shed light on indexing of diffraction patterns. As will be shown below, the task of indexing ideal diffraction data can be formulated as a Pohst problem: having $m > 3$ (reciprocal) lattice vectors $\check{\boldsymbol{h}}_1, \check{\boldsymbol{h}}_2, \ldots, \check{\boldsymbol{h}}_m$ (from an ideal diffraction experiment), get a reduced basis $(\boldsymbol{a}^1, \boldsymbol{a}^2, \boldsymbol{a}^3)$ of the lattice.

1.2.2 Buerger- and Niggli-Reduced Bases

For an arbitrary basis of a three-dimensional lattice, the basis vectors $\boldsymbol{a}_i$ can be re-labeled so that

$$\boldsymbol{a}_1^2 \leq \boldsymbol{a}_2^2 \leq \boldsymbol{a}_3^2\,, \quad \text{if}\ \boldsymbol{a}_1^2 = \boldsymbol{a}_2^2\ \text{then}\ |\boldsymbol{a}_2 \cdot \boldsymbol{a}_3| \leq |\boldsymbol{a}_3 \cdot \boldsymbol{a}_1|\,, \quad \text{if}\ \boldsymbol{a}_2^2 = \boldsymbol{a}_3^2\ \text{then}\ |\boldsymbol{a}_3 \cdot \boldsymbol{a}_1| \leq |\boldsymbol{a}_1 \cdot \boldsymbol{a}_2|\,. \tag{1.29}$$

By changing the signs of the re-labeled vectors, one can enforce that all $\boldsymbol{a}_1 \cdot \boldsymbol{a}_2$, $\boldsymbol{a}_2 \cdot \boldsymbol{a}_3, \boldsymbol{a}_3 \cdot \boldsymbol{a}_1$ are positive, or all of them are non-positive; i.e.,

$$\text{either}\ \boldsymbol{a}_1 \cdot \boldsymbol{a}_2,\ \boldsymbol{a}_2 \cdot \boldsymbol{a}_3,\ \boldsymbol{a}_3 \cdot \boldsymbol{a}_1 > 0 \quad \text{or} \quad \boldsymbol{a}_1 \cdot \boldsymbol{a}_2,\ \boldsymbol{a}_2 \cdot \boldsymbol{a}_3,\ \boldsymbol{a}_3 \cdot \boldsymbol{a}_1 \leq 0\,. \tag{1.30}$$

Hence, one has a classification of lattice cells into two types

$$\text{cell type} = \begin{cases} I & \text{if}\ (\boldsymbol{a}_1 \cdot \boldsymbol{a}_2)(\boldsymbol{a}_2 \cdot \boldsymbol{a}_3)(\boldsymbol{a}_3 \cdot \boldsymbol{a}_1) > 0 \\ II & \text{otherwise.} \end{cases} \tag{1.31}$$

The above requirements and those given below involve only the scalar products of the basis vectors. Therefore, they can be (and frequently are) expressed using the components of the metric tensor $g_{ij} = \boldsymbol{a}_i \cdot \boldsymbol{a}_j$.

If $\boldsymbol{a}_1$ is the shortest lattice vector, $\boldsymbol{a}_2$ is the shortest lattice vector not colinear with $\boldsymbol{a}_1$, and $\boldsymbol{a}_3$ is the shortest lattice vector not coplanar with $\boldsymbol{a}_1$ and $\boldsymbol{a}_2$, then these vectors constitute a lattice basis; e.g., [12]. In crystallography, a basis made of the shortest linearly independent vectors satisfying the condition (1.30) is referred to as *Buerger-reduced.*[7] A parallelepiped spanned by a Buerger-reduced basis is called a Buerger (-reduced) cell. Clearly, Buerger cells are primitive.

It is easy to verify that a two-dimensional lattice has a basis such that

$$2|\boldsymbol{a}_1 \cdot \boldsymbol{a}_2| \leq \boldsymbol{a}_1^2 \leq \boldsymbol{a}_2^2\,. \tag{1.32}$$

---

[6] The standard method for orthogonalizing vectors: With linearly independent vectors $\boldsymbol{v}_i$ ($i = 1, \ldots, n$) and $\boldsymbol{w}_1 = \boldsymbol{v}_1$, calculate vectors $\boldsymbol{w}_i = \boldsymbol{v}_i - \sum_{i=1}^{j-1} \mu_{i(j)} \boldsymbol{w}_{(j)}$, where $\mu_{ij} = \boldsymbol{v}_i \cdot \boldsymbol{w}_j / \boldsymbol{w}_j^2$. The vectors $\boldsymbol{w}_i$ are orthogonal.

[7] Martin Julian Buerger (1903–1986).

(These relationships are satisfied by $\boldsymbol{a}_1$ being the shortest lattice vector, and $\boldsymbol{a}_2$ being the shortest vector not collinear with $\boldsymbol{a}_1$. One clearly has $\boldsymbol{a}_1^2 \leq \boldsymbol{a}_2^2$. If $2|\boldsymbol{a}_1 \cdot \boldsymbol{a}_2|$ were larger than $\boldsymbol{a}_1^2$, then $(\boldsymbol{a}_2 - \text{sign}(\boldsymbol{a}_1 \cdot \boldsymbol{a}_2)\,\boldsymbol{a}_1)^2 - \boldsymbol{a}_2^2 = \boldsymbol{a}_1^2 - 2|\boldsymbol{a}_1 \cdot \boldsymbol{a}_2|$ would be smaller than zero, i.e., the vector $\boldsymbol{a}_2 - \text{sign}(\boldsymbol{a}_1 \cdot \boldsymbol{a}_2)\,\boldsymbol{a}_1$ not collinear with $\boldsymbol{a}_1$ would be shorter than $\boldsymbol{a}_2$.) Now, having an arbitrary lattice basis $\boldsymbol{a}_1$ and $\boldsymbol{a}_2$, one can obtain a basis satisfying (1.32) using the Lagrange-Gauss reduction: the vectors are swapped if $\boldsymbol{a}_1^2 > \boldsymbol{a}_2^2$, and if $2\,|\boldsymbol{a}_1 \cdot \boldsymbol{a}_2| > \boldsymbol{a}_1^2$, $\boldsymbol{a}_2$ is replaced by the shorter vector $\boldsymbol{a}_2 - \text{sign}(\boldsymbol{a}_1 \cdot \boldsymbol{a}_2)\,\boldsymbol{a}_1$; these steps are repeated until (1.32) are satisfied.

The Lagrange-Gauss reduction is the main step in the determination of Buerger-reduced bases of three-dimensional lattices. If two vectors $\boldsymbol{a}_i$ and $\boldsymbol{a}_j$ of a three-dimensional basis satisfy $2\,|\boldsymbol{a}_i \cdot \boldsymbol{a}_j| > \boldsymbol{a}_i^2$, then the vector $\boldsymbol{a}_j$ is replaced by the shorter vector $\boldsymbol{a}_j - \text{sign}(\boldsymbol{a}_{(i)} \cdot \boldsymbol{a}_j)\,\boldsymbol{a}_{(i)}$, and the new set of vectors constitutes a basis. However, such reductions of vector magnitudes in two-dimensional lattices determined by the pairs $(\boldsymbol{a}_1, \boldsymbol{a}_2)$, $(\boldsymbol{a}_2, \boldsymbol{a}_3)$ and $(\boldsymbol{a}_3, \boldsymbol{a}_1)$ do not guarantee that the final three-dimensional basis will be Buerger-reduced. For the reduction of the three-dimensional basis, it is necessary that besides (1.29), (1.30) and

$$2\,|\boldsymbol{a}_2 \cdot \boldsymbol{a}_3| \leq \boldsymbol{a}_2^2\,, \quad 2\,|\boldsymbol{a}_3 \cdot \boldsymbol{a}_1| \leq \boldsymbol{a}_1^2\,, \quad 2\,|\boldsymbol{a}_1 \cdot \boldsymbol{a}_2| \leq \boldsymbol{a}_1^2\,, \tag{1.33}$$

also the condition

$$\left(\sum\nolimits_{i=1}^{3} \boldsymbol{a}_i\right)^2 - \boldsymbol{a}_3^2 = (\boldsymbol{a}_1 + \boldsymbol{a}_2) \cdot (\boldsymbol{a}_1 + \boldsymbol{a}_2 + 2\boldsymbol{a}_3) \geq 0 \tag{1.34}$$

is satisfied; $\sum_i \boldsymbol{a}_i$ is the diagonal of the parallelepiped spanned on the vectors $\boldsymbol{a}_i$, and it must be at least as long as the longest basis vector $\boldsymbol{a}_3$.

In general, a Buerger-reduced basis of a lattice is not unique. (An example is given below.) For one lattice there may exists up to five non-congruent Buerger cells [13]. For a unique specification of a basis, and thus, for a unique description of the lattice, additional 'special' conditions are needed. A set of conditions of this kind is contained in the following algorithm given after [13, 14] (see also [15]):

1. Re-label the vectors $\boldsymbol{a}_1, \boldsymbol{a}_2, \boldsymbol{a}_3$, and change their signs so the conditions (1.29) and (1.30) are satisfied.
2. If $2|\boldsymbol{a}_2 \cdot \boldsymbol{a}_3| > \boldsymbol{a}_2^2$
 $\{$or $(2\boldsymbol{a}_2 \cdot \boldsymbol{a}_3 = \boldsymbol{a}_2^2$ and $2\boldsymbol{a}_3 \cdot \boldsymbol{a}_1 < \boldsymbol{a}_1 \cdot \boldsymbol{a}_2)$ or $(2\boldsymbol{a}_2 \cdot \boldsymbol{a}_3 = -\boldsymbol{a}_2^2$ and $\boldsymbol{a}_1 \cdot \boldsymbol{a}_2 < 0)\}$,
 then replace $\boldsymbol{a}_3$ by $\boldsymbol{a}_3 - \text{sign}(\boldsymbol{a}_2 \cdot \boldsymbol{a}_3)\boldsymbol{a}_2$ and go to point 1.
3. If $2|\boldsymbol{a}_3 \cdot \boldsymbol{a}_1| > \boldsymbol{a}_1^2$
 $\{$or $(2\boldsymbol{a}_3 \cdot \boldsymbol{a}_1 = \boldsymbol{a}_1^2$ and $2\boldsymbol{a}_2 \cdot \boldsymbol{a}_3 < \boldsymbol{a}_1 \cdot \boldsymbol{a}_2)$ or $(2\boldsymbol{a}_3 \cdot \boldsymbol{a}_1 = -\boldsymbol{a}_1^2$ and $\boldsymbol{a}_1 \cdot \boldsymbol{a}_2 < 0)\}$,
 then replace $\boldsymbol{a}_3$ by $\boldsymbol{a}_3 - \text{sign}(\boldsymbol{a}_3 \cdot \boldsymbol{a}_1)\boldsymbol{a}_1$ and go to point 1.
4. If $2|\boldsymbol{a}_1 \cdot \boldsymbol{a}_2| > \boldsymbol{a}_1^2$
 $\{$or $(2\boldsymbol{a}_1 \cdot \boldsymbol{a}_2 = \boldsymbol{a}_1^2$ and $2\boldsymbol{a}_2 \cdot \boldsymbol{a}_3 < \boldsymbol{a}_3 \cdot \boldsymbol{a}_1)$ or $(2\boldsymbol{a}_1 \cdot \boldsymbol{a}_2 = -\boldsymbol{a}_1^2$ and $\boldsymbol{a}_3 \cdot \boldsymbol{a}_1 < 0)\}$,
 then replace $\boldsymbol{a}_2$ by $\boldsymbol{a}_2 - \text{sign}(\boldsymbol{a}_1 \cdot \boldsymbol{a}_2)\boldsymbol{a}_1$ and go to point 1.
5. If $(\boldsymbol{a}_1 + \boldsymbol{a}_2) \cdot (\boldsymbol{a}_1 + \boldsymbol{a}_2 + 2\boldsymbol{a}_3) < 0$
 $\{$or $((\boldsymbol{a}_1 + \boldsymbol{a}_2) \cdot (\boldsymbol{a}_1 + \boldsymbol{a}_2 + 2\boldsymbol{a}_3) = 0$ and $\boldsymbol{a}_1 \cdot (\boldsymbol{a}_1 + \boldsymbol{a}_2 + 2\boldsymbol{a}_3) > 0)\}$,
 then replace $\boldsymbol{a}_3$ by $\boldsymbol{a}_3 - \text{sign}(\boldsymbol{a}_3 \cdot (\boldsymbol{a}_1 + \boldsymbol{a}_2))(\boldsymbol{a}_1 + \boldsymbol{a}_2)$ and go to point 1.

This algorithm leads to a so called *Niggli-reduced*[8] basis. The ‘special’ conditions are given in parentheses; if they are omitted, the resulting basis is only Buerger-reduced. The Niggli-reduced basis is Buerger-reduced. A parallelepiped spanned on Niggli-reduced basis is called the Niggli (-reduced) cell. For a given lattice, the metric obtained from the Niggli-reduced basis is *unique*. (The Niggli-reduced basis is not unique. E.g., the bases $\boldsymbol{a}_1 = (0, 0, 1)$ $\boldsymbol{a}_2 = (2, 0, 0)$ $\boldsymbol{a}_3 = (-1, -2, 0)$, and $\boldsymbol{a}'_1 = \boldsymbol{a}_1$ $\boldsymbol{a}'_2 = -\boldsymbol{a}_2$ $\boldsymbol{a}'_3 = -\boldsymbol{a}_3$ of the same lattice are both Niggli-reduced. The basis ambiguity is a result of lattice symmetry; apart from its handedness and orientation, the Niggli-reduced basis is unique.)

□ Let $\gamma_1 = \angle(\boldsymbol{a}_2, \boldsymbol{a}_3), \gamma_2 = \angle(\boldsymbol{a}_3, \boldsymbol{a}_1)$ and $\gamma_3 = \angle(\boldsymbol{a}_1, \boldsymbol{a}_2)$. According to [16], of all Buerger bases of a lattice, the one maximizing the deviation $\Delta_G = \sum_i |\pi/2 - \gamma_i|$ (or $\sum_i |\cos\gamma_i|$ or $\prod_i |\cos\gamma_i|$) is Niggli-reduced and vice versa, if the basis is Niggli-reduced it has the largest deviation among Buerger bases of the lattice.

With the Niggli-reduced basis $\boldsymbol{a}_i$, the entries of the corresponding metric $g_{ij} = \boldsymbol{a}_i \cdot \boldsymbol{a}_j$ satisfy a uniquely determined set of equalities and/or inequalities. Based on these relationships and reduced-cell type, lattices are classified into 44 classes known as Niggli lattice ‘characters’. The relationships are conveniently expressed using the ‘reduced parameters’ [17]

$$
\begin{aligned}
u &= \boldsymbol{a}_1^2/\boldsymbol{a}_3^2\,, \quad & x &= 2\boldsymbol{a}_2 \cdot \boldsymbol{a}_3/\boldsymbol{a}_3^2\,, \quad & s &= u + v + x + y + z\,, \\
v &= \boldsymbol{a}_2^2/\boldsymbol{a}_3^2\,, \quad & y &= 2\boldsymbol{a}_3 \cdot \boldsymbol{a}_1/\boldsymbol{a}_3^2\,, \quad & w &= 2u + 2y + z\,, \\
& & z &= 2\boldsymbol{a}_1 \cdot \boldsymbol{a}_2/\boldsymbol{a}_3^2\,,
\end{aligned}
$$

where $\boldsymbol{a}_i$ are vectors of the Niggli-reduced basis. The relationships defining the Niggli classes are listed in Table 1.3. The first conditions on u and v in columns 3 and 4 are to eliminate non-uniqueness when basis vectors have equal magnitudes, and the remaining conditions are to eliminate non-uniqueness when there is more than one cell based on the shortest vectors.

It is worth noting that the reciprocal of a Buerger-reduced basis may not be Buerger-reduced. The reciprocal of a Niggli-reduced basis may not be Niggli-reduced or even not Buerger-reduced.

□ Example: For convenience, components of the vector $k^i \boldsymbol{e}_i$ will be written as (k^1, k^2, k^3). The lattice basis $\boldsymbol{a}_i$ given in the orthogonal basis by

$$
\boldsymbol{a}_1 = (0, 1, 1)\,,\ \boldsymbol{a}_2 = (1, 0, -1),\ \boldsymbol{a}_3 = (-1, 0, -1)
$$

$$
\left(\text{i.e., } \mathcal{T} = \left[\mathcal{T}_i{}^j\right] = \begin{bmatrix} 0 & 1 & 1 \\ 1 & 0 & -1 \\ -1 & 0 & -1 \end{bmatrix} \text{ and } g = \left[g_{ij}\right] = \begin{bmatrix} 2 & -1 & -1 \\ -1 & 2 & 0 \\ -1 & 0 & 2 \end{bmatrix}\right)
$$

is Buerger-reduced, whereas its reciprocal

---

[8] Paul Niggli (1888–1953).

Table 1.3 The relationships defining Niggli classes of lattices after [17]. The classes are numbered by 'characters' 1 to 44. 'Type' denotes the type of the Niggli-reduced cell. For a given class, each entry (other than the dot) of the table is equal to the reduced parameter at the top of the column. A lattice belongs to the first (from top) class for which the type is right and the conditions are satisfied.

Character	Type	u	v	x	y	z	s	w
1	*I*	1	1	1	1	1	.	.
2	*I*	1	1	.	x	x	.	.
3	*II*	1	1	0	0	0	.	.
5	*II*	1	1	.	x	x	0	.
4	*II*	1	1	.	x	x	.	.
6	*II*	1	1	.	x	.	0	.
7	*II*	1	1	.	.	y	0	.
8	*II*	1	1	.	.	.	0	.
9	*I*	.	u	u	u	u	.	.
10	*I*	.	u	.	x	.	.	.
11	*II*	.	u	0	0	0	.	.
12	*II*	.	u	0	0	$-u$	.	.
13	*II*	.	u	0	0	.	.	.
15	*II*	.	u	$-u$	$-u$	0	.	.
16	*II*	.	u	.	x	.	0	.
14	*II*	.	u	.	x	.	.	.
17	*II*	.	u	.	.	.	0	.
18	*I*	.	1	$u/2$	u	u	.	.
19	*I*	.	1	.	u	u	.	.
20	*I*	.	1	.	.	y	.	.
21	*II*	.	1	0	0	0	.	.
22	*II*	.	1	-1	0	0	.	.
23	*II*	.	1	.	0	0	.	.
24	*II*	.	1	.	.	y	0	0
25	*II*	.	1	.	.	y	.	.
26	*I*	.	.	$u/2$	u	u	.	.
27	*I*	.	.	.	u	u	.	.
28	*I*	.	.	.	u	$2x$	.	.
29	*I*	.	.	.	$2x$	u	.	.
30	*I*	.	.	v	.	$2y$	.	.
31	*I*	.	.	.	.	.	.	.
32	*II*	.	.	0	0	0	.	.
40	*II*	.	.	$-v$	0	0	.	.
35	*II*	.	.	.	0	0	.	.
36	*II*	.	.	0	$-u$	0	.	.
33	*II*	.	.	0	.	0	.	.
38	*II*	.	.	0	0	$-u$	.	.
34	*II*	.	.	0	0	.	.	.
42	*II*	.	.	$-v$	$-u$	0	.	.
41	*II*	.	.	$-v$	.	0	.	.
37	*II*	.	.	.	$-u$	0	.	.
39	*II*	.	.	.	0	$-u$	.	.
43	*II*	.	.	.	.	.	0	0
44	*II*	.	.	.	.	.	.	.

$$a^1 = (0,1,0)\,,\ \ a^2 = (1,1,-1)/2\,,\ \ a^3 = (-1,1,-1)/2\quad \left(\text{with the metric } \frac{1}{4}\begin{bmatrix} 4 & 2 & 2 \\ 2 & 3 & 1 \\ 2 & 1 & 3 \end{bmatrix}\right) \tag{1.35}$$

is not. The Buerger reduction of the reciprocal (1.35) gives

$$a'^1 = (-1,-1,1)/2\,,\ \ a'^2 = (-1,1,-1)/2\,,\ \ a'^3 = (1,1,1)/2\,.$$

The bases $\boldsymbol{a}^i$ and $\boldsymbol{a}'^i$ are related via $\boldsymbol{a}^1 = \boldsymbol{a}'^2 + \boldsymbol{a}'^3$, $\boldsymbol{a}^2 = -\boldsymbol{a}'^1$ and $\boldsymbol{a}^3 = \boldsymbol{a}'^2$. The basis reciprocal to $\boldsymbol{a}'^i$ is given by

$$a'_1 = (-1,0,1)\,,\ \ a'_2 = (-1,1,0)\,,\ \ a'_3 = (0,1,1)\,;$$

it is Buerger-reduced, and it is related to $\boldsymbol{a}_i$ via $\boldsymbol{a}_1 = \boldsymbol{a}'_3$, $\boldsymbol{a}_2 = -\boldsymbol{a}'_1$ and $\boldsymbol{a}_3 = \boldsymbol{a}'_2 - \boldsymbol{a}'_3$. The two bases $\boldsymbol{a}_i$ and $\boldsymbol{a}'_i$ of the same lattice are both Buerger-reduced but the corresponding metrics are different; $\boldsymbol{a}_2$ is perpendicular to $\boldsymbol{a}_3$, whereas the angles between $\boldsymbol{a}'_i$ vectors are all $\pi/3$. Their deviations Δ_G are $\pi/3$ and $\pi/2$, respectively. The basis $\boldsymbol{a}'_i$ $(i = 1, 2, 3)$ is Niggli-reduced with the cell of type I and 'reduced parameters' $u = v = x = y = z = 1$, i.e., the Niggli character of the lattice is 1.

1.2.3 Delaunay Reduction

If $\boldsymbol{a}_1, \boldsymbol{a}_2$ and $\boldsymbol{a}_3$ constitute a lattice basis, the vector $-(\boldsymbol{a}_1 + \boldsymbol{a}_2 + \boldsymbol{a}_3)$ forms a basis of the lattice with any two of the vectors $\boldsymbol{a}_i$ $(i = 1, 2, 3)$. A lattice basis supplemented by the negative sum of its vectors is used in Delaunay[9] reduction [18]. The reduction relies on a star of vectors $\boldsymbol{a}_\mu$ $(\mu = 1, 2, 3, 4)$ such that

$$\sum_{\mu=1}^{4} \boldsymbol{a}_\mu = \boldsymbol{0}\,, \tag{1.36}$$

and any three of the vectors $\boldsymbol{a}_\mu$ form a lattice basis. Here, the star (Delaunay's "Sellingisches Vierseit"[10]) will be called a Delaunay frame. Like Buerger and Niggli reductions, the Delaunay reduction is performed based on values of the scalar products $\boldsymbol{a}_\mu \cdot \boldsymbol{a}_\nu = g_{\mu\nu}$. Briefly, the Delaunay reduction of a Delaunay frame gives a Delaunay frame such that the trace $g_{\mu\mu}$ of its metric tensor is minimal. Based on (1.36), it is easy to see that $\sum_\mu g_{\mu\nu} = 0$, and each diagonal entry of the metric tensor can be expressed by off-diagonal entries $g_{\mu\nu}$ $(\mu \neq \nu)$ known as Selling parameters. One has

$$g_{\mu\mu} = g_{11} + g_{22} + g_{33} + g_{44} = -2(g_{12} + g_{23} + g_{31} + g_{14} + g_{24} + g_{34})\,,$$

---

[9] Boris Nikolaevich Delaunay (or Delone, 1890–1980).

[10] The subject of lattice reduction is closely related to reduction of quadratic forms, i.e., forms of the type $g_{ij}x^i x^j$. By definition, two quadratic forms are equivalent if one can be transformed into the other by a unimodular linear transformation with integer coefficients. The point is to provide a set of inequalities on the coefficients such that 1) every class of equivalent forms has a representative satisfying the inequalities, and 2) this representative is unique. Delaunay's work [18] was based on a scheme for reduction of ternary quadratic forms of E. Selling.

and only the Selling parameters are used in the reduction process. The key step of the Delaunay reduction is as follows: given two vectors $\boldsymbol{a}_\mu$ and $\boldsymbol{a}_\nu$ ($\mu \neq \nu$) such that $g_{\mu\nu} > 0$, change the sign of $\boldsymbol{a}_\mu$, keep $\boldsymbol{a}_\nu$ unchanged, and add $\boldsymbol{a}_\mu$ to the remaining two vectors of the frame. Such transformation of a Delaunay frame results in a different Delaunay frame, and it reduces $g_{\mu\mu}$ by $2g_{\mu\nu}$. One can check this by considering a concrete case. E.g., if the step is based on $g_{12} > 0$, then $\mu = 1$ and $\nu = 2$, and the transformation has the form

$$\boldsymbol{a}'_1 = -\boldsymbol{a}_1 , \quad \boldsymbol{a}'_2 = \boldsymbol{a}_2 , \quad \boldsymbol{a}'_3 = \boldsymbol{a}_3 + \boldsymbol{a}_1 , \quad \boldsymbol{a}'_4 = \boldsymbol{a}_4 + \boldsymbol{a}_1 .$$

Each triplet of distinct $\boldsymbol{a}'_\mu$ vectors is a basis. The corresponding Selling parameters $g'_{\mu\nu} = \boldsymbol{a}'_\mu \cdot \boldsymbol{a}'_\nu$ are

$$\begin{aligned}
&g'_{12} = -g_{12} , && g'_{14} = -g_{14} - g_{11} = g_{12} + g_{31} ,\\
&g'_{23} = g_{12} + g_{23} , && g'_{24} = g_{24} + g_{12} ,\\
&g'_{31} = -g_{31} - g_{11} = g_{12} + g_{14} , && g'_{34} = g_{34} + g_{31} + g_{14} + g_{11} = g_{34} - g_{12} ,
\end{aligned}$$

and the trace of the metric tensor changes from $g_{\mu\mu}$ to

$$\begin{aligned}
g'_{\mu\mu} &= -2(g'_{12} + g'_{23} + g'_{31} + g'_{14} + g'_{24} + g'_{34})\\
&= -2(2g_{12} + g_{23} + g_{31} + g_{14} + g_{24} + g_{34}) = g_{\mu\mu} - 2g_{12} ,
\end{aligned}$$

i.e., it is diminished by $g_{\mu\mu} - g'_{\mu\mu} = 2g_{12}$. There are no rules for selecting the two vectors $\boldsymbol{a}_\mu$ and $\boldsymbol{a}_\nu$ which should be used at a given step. A simple (but not necessarily most effective) approach is to use the pair with the largest scalar product $g_{\mu\nu}$.

□ *Double reduction*
A faster reduction step was proposed in [19]. Given three vectors indexed by κ, μ and ν ($\kappa \neq \mu \neq \nu \neq \kappa$) such that $g_{\mu\kappa} + g_{\kappa\nu} > 0$, change the sign of $\boldsymbol{a}_\kappa$ and add $2\boldsymbol{a}_\kappa$ to the fourth vector of the frame. Such transformation of a Delaunay frame results in a different Delaunay frame. It reduces $g_{\mu\mu}$ by $4(g_{\mu\kappa} + g_{\kappa\nu})$. E.g., if one uses $g_{12} + g_{23} > 0$, then $\mu = 1$, $\kappa = 2$, $\nu = 3$, and the frame $\boldsymbol{a}_\mu$ is replaced by

$$\boldsymbol{a}'_1 = \boldsymbol{a}_1 , \quad \boldsymbol{a}'_2 = -\boldsymbol{a}_2 , \quad \boldsymbol{a}'_3 = \boldsymbol{a}_3 , \quad \boldsymbol{a}'_4 = \boldsymbol{a}_4 + 2\boldsymbol{a}_2 .$$

With this transformation, the trace of the metric is diminished by $g_{\mu\mu} - g'_{\mu\mu} = 4(g_{12} + g_{23})$.

The Delaunay reduction of a frame is performed by recursive application of the Delaunay reduction step. Since the value of $g_{\mu\mu}$ decreases in each step, there is no risk of going into an infinite loop. If a reduction step is carried out based on a given $g_{\mu\nu} > 0$ ($\mu \neq \nu$), the sign of $g_{\mu\nu}$ is changed to negative. The reduction is stopped when

$$g_{\mu\nu} \leq 0 \qquad \text{for all } \mu \text{ and } \nu \text{ such that } \mu \neq \nu . \tag{1.37}$$

For a reduced frame, all products $\boldsymbol{a}_\mu \cdot \boldsymbol{a}_\nu$ ($\mu \neq \nu$) are non-positive, i.e., the angles between vectors of the Delaunay reduced frame are non-acute.

In general, the resulting frame is not unique. It is unique (up to the order of the vectors and the sign of all vectors) if none of the Selling parameters of the reduced frame is zero. The Delaunay reduction step based on the parameter $g_{\mu\nu} = 0$ $(\mu \neq \nu)$ changes the frame but it does not decrease $g_{\mu\mu}$, and the set of values taken by the off-diagonal entries $g_{\mu\nu}$ remains the same.

⊡ *Delaunay reduction of bases of two-dimensional lattices*
The principles of Delaunay reduction are applicable to bases of two-dimensional lattices [19]. If $\boldsymbol{a}_1$ and $\boldsymbol{a}_2$ are basis vectors, $-(\boldsymbol{a}_1 + \boldsymbol{a}_2)$ forms a basis of the lattice with any of these two vectors. As in the three-dimensional case, a lattice basis is supplemented by the negative sum of its vectors, i.e., one has the Delaunay frame $\boldsymbol{a}_\mu$ $(\mu = 1, 2, 3)$ such that $\sum_{\mu=1}^{3} \boldsymbol{a}_\mu = \boldsymbol{0}$, and any two of the vectors $\boldsymbol{a}_\mu$ form a lattice basis. The reduction of a Delaunay frame gives a Delaunay frame such that the trace $g_{\mu\mu}$ of its metric tensor is minimal. Based on $\sum_\mu \boldsymbol{a}_\mu = \boldsymbol{0}$, one has $\sum_\mu g_{\mu\nu} = 0$ and $g_{\mu\mu} = g_{11} + g_{22} + g_{33} = -2(g_{12} + g_{23} + g_{31})$. The reduction step is as follows: given two vectors $\boldsymbol{a}_\mu$ and $\boldsymbol{a}_\nu$ $(\mu \neq \nu)$ such that $g_{\mu\nu} > 0$, change the sign of $\boldsymbol{a}_\mu$, keep $\boldsymbol{a}_\nu$ unchanged, and add $2\boldsymbol{a}_\mu$ to the third vector of the frame. Such transformation of a Delaunay frame results in a different Delaunay frame, and it reduces $g_{\mu\mu}$ by $4g_{\mu\nu}$. E.g., if the step is based on $g_{12} > 0$, then $\mu = 1$ and $\nu = 2$, and the transformation has the form $\boldsymbol{a}'_1 = -\boldsymbol{a}_1$, $\boldsymbol{a}'_2 = \boldsymbol{a}_2$ and $\boldsymbol{a}'_3 = \boldsymbol{a}_3 + 2\boldsymbol{a}_1$. Each doublet of distinct $\boldsymbol{a}'_\mu$ vectors is a basis. The corresponding parameters $g'_{\mu\nu} = \boldsymbol{a}'_\mu \cdot \boldsymbol{a}'_\nu$ $(\mu \neq \nu)$ are $g'_{12} = -g_{12}$, $g'_{23} = 2g_{12} + g_{23}$, $g'_{31} = -g_{31} - 2g_{11} = 2g_{12} + g_{31}$, and the trace of the metric tensor changes from $g_{\mu\mu}$ to $g'_{\mu\mu} = g_{\mu\mu} - 4g_{12}$. Only one of $g_{\mu\nu}$ $(\mu \neq \nu)$ can be larger than zero, and is used in the reduction step. The reduction is performed by recursive application of the reduction step. The reduction is stopped when all $g_{\mu\nu}$ $(\mu \neq \nu)$ are non-positive, i.e., the angles between vectors of the reduced frame are non-acute.

Fedorov Parallelohedra

In the three-dimensional case, the three shortest vectors of the Delaunay-reduced frame may serve as a lattice basis. In general, such a basis is not Buerger-reduced. In particular, it cannot be Buerger-reduced when the shortest linearly independent vectors of the lattice violate conditions (1.37). A Buerger reduced basis can be obtained by taking the three shortest linearly independent vectors from among $\boldsymbol{a}_1, \boldsymbol{a}_2, \boldsymbol{a}_3, \boldsymbol{a}_4$, $\boldsymbol{a}_1 + \boldsymbol{a}_4$, $\boldsymbol{a}_2 + \boldsymbol{a}_4$ and $\boldsymbol{a}_3 + \boldsymbol{a}_4$ [20].

The above set of vectors determines a Voronoï cell of lattice points. The Voronoï cell of a lattice point consists of all points of the space which are not farther from that lattice point than from any other lattice point (Fig. 1.4). The Voronoï cells fill the space and their interiors do not intersect. For a three-dimensional lattice, the Voronoï cells are mutually congruent convex centrosymmetric polyhedra with centrosymmetric faces.

Let $\boldsymbol{a}_\mu$ be vectors of a Delaunay reduced frame. It turns out that the Voronoï cell of $\boldsymbol{0}$ is bounded by (all or some of) the planes orthogonally bisecting the vectors $\pm\boldsymbol{a}_1$, $\pm\boldsymbol{a}_2$, $\pm\boldsymbol{a}_3$, $\pm\boldsymbol{a}_4$, $\pm(\boldsymbol{a}_1 + \boldsymbol{a}_4)$, $\pm(\boldsymbol{a}_2 + \boldsymbol{a}_4)$, $\pm(\boldsymbol{a}_3 + \boldsymbol{a}_4)$. A vector from this family will be denoted by $\boldsymbol{a}_K$ $(K = 1, 2, \ldots, 14)$.

For a point $\boldsymbol{x}$ to be closer to $\boldsymbol{0}$ than to $\check{\boldsymbol{v}}$, $\boldsymbol{x}$ must satisfy the condition $\boldsymbol{x} \cdot \check{\boldsymbol{v}} < \check{\boldsymbol{v}}^2/2$. Thus, according to the above assertion, a point $\boldsymbol{x}$ in the interior of the Voronoï cell of $\boldsymbol{0}$ satisfies the inequalities $\boldsymbol{x} \cdot \boldsymbol{a}_K < \boldsymbol{a}_K^2/2$, and there are no other lattice points that

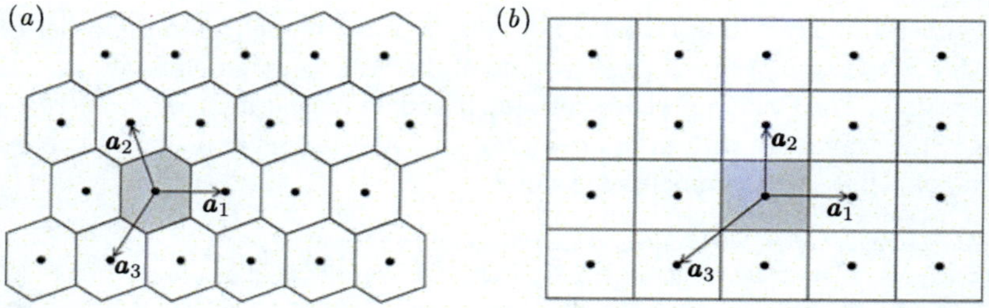

Fig. 1.4 Illustration of Voronoï tessellations of two-dimensional lattices. Voronoï cells of the lattice node at **0** are marked in gray. Unlike in *(a)*, the vector $\boldsymbol{a}_3$ in *(b)* does not induce a side of the cell.

would affect the shape of the cell. In other words, if $\boldsymbol{x}\cdot\boldsymbol{a}_K < \boldsymbol{a}_K^2/2$, then $\boldsymbol{x}\cdot\check{\boldsymbol{v}} < \check{\boldsymbol{v}}^2/2$ for arbitrary lattice vector $\check{\boldsymbol{v}}$.

⊡ To prove this claim, one needs to note that the vectors $\boldsymbol{a}_\mu$ of a Delaunay reduced frame can be rearranged in such a way that an arbitrary lattice vector $\check{\boldsymbol{v}}$ can be expressed as $\check{\boldsymbol{v}} = m^i\boldsymbol{a}_i$, where the integers m^i $(i = 1, 2, 3)$ satisfy the conditions $0 \le m^1 \le m^2 \le m^3$ and $|\boldsymbol{a}_2| \le |\boldsymbol{a}_3|$.

First, one needs to arrange $\boldsymbol{a}_\mu$ in such a way that m^i are non-negative. If all three m^i are negative, replace $\boldsymbol{a}_\mu$ by $-\boldsymbol{a}_\mu$. If one coefficient m^i is negative, eliminate $\boldsymbol{a}_i$ using $\sum\boldsymbol{a}_\mu = \boldsymbol{0}$; e.g., in the case of negative m^1, on gets $m^i\boldsymbol{a}_i = m^1(-\boldsymbol{a}_4-\boldsymbol{a}_2-\boldsymbol{a}_3)+m^2\boldsymbol{a}_2+m^3\boldsymbol{a}_3 = (-m^1)\boldsymbol{a}_4+(m^2-m^1)\boldsymbol{a}_2+(m^3-m^1)\boldsymbol{a}_3$, and all coefficients of the vectors are non-negative. If two coefficients m^i are negative, use both these steps.

Now, with non-negative m^i, one needs to reorder the vectors $\boldsymbol{a}_\mu$ so the conditions $m^1 \le m^2 \le m^3$ and $|\boldsymbol{a}_2| \le |\boldsymbol{a}_3|$ are satisfied. For this purpose, order the vectors $\boldsymbol{a}_\mu$ according to their magnitudes $|\boldsymbol{a}_1| \le |\boldsymbol{a}_2| \le |\boldsymbol{a}_3| \le |\boldsymbol{a}_4|$. If $m^2 < m^1 \le m^3$, simply exchange $\boldsymbol{a}_1$ with $\boldsymbol{a}_2$. If $m^1 \le m^3 < m^2$, eliminate $\boldsymbol{a}_2$ so $m^i\boldsymbol{a}_i = m^1\boldsymbol{a}_1+m^2(-\boldsymbol{a}_1-\boldsymbol{a}_3-\boldsymbol{a}_4)+m^3\boldsymbol{a}_3 = (m^2-m^3)(-\boldsymbol{a}_3)+(m^2-m^1)(-\boldsymbol{a}_1)+m^2(-\boldsymbol{a}_4)$; with m^2-m^3, m^2-m^1 and m^2 substituted by m^1, m^2 and m^3, respectively, and $\boldsymbol{a}_1, \boldsymbol{a}_2, \boldsymbol{a}_3$ and $\boldsymbol{a}_4$ substituted by $-\boldsymbol{a}_2$, $-\boldsymbol{a}_4$, $-\boldsymbol{a}_1$ and $-\boldsymbol{a}_3$, the conditions $m^1 \le m^2 \le m^3$ and $|\boldsymbol{a}_2| \le |\boldsymbol{a}_3|$ are satisfied. These steps need to be combined when $m^2 \le m^3 \le m^1$, $m^3 \le m^1 \le m^2$ or $m^3 \le m^2 \le m^1$; again, after reaching $m^1 \le m^2 \le m^3$, the final vectors satisfy the condition $|\boldsymbol{a}_2| \le |\boldsymbol{a}_3|$.

The lattice vector $\check{\boldsymbol{v}}$ can be expressed as

$$\check{\boldsymbol{v}} = m^i\boldsymbol{a}_i = m^1(\boldsymbol{a}_1+\boldsymbol{a}_2+\boldsymbol{a}_3)+(m^2-m^1)(\boldsymbol{a}_2+\boldsymbol{a}_3)+(m^3-m^2)\boldsymbol{a}_3 = k^i\boldsymbol{a}'_i\,,$$

where k^i $(i = 1, 2, 3)$ are non-negative integers, $\boldsymbol{a}'_1 = \boldsymbol{a}_1+\boldsymbol{a}_2+\boldsymbol{a}_3$, $\boldsymbol{a}'_2 = \boldsymbol{a}_2+\boldsymbol{a}_3$ and $\boldsymbol{a}'_3 = \boldsymbol{a}_3$, The vectors $\boldsymbol{a}'_i$ belong to the family of $\boldsymbol{a}_K$ vectors $(K = 1, 2, \ldots, 14)$, i.e., $\boldsymbol{x}$ satisfies the inequality $\boldsymbol{x}\cdot\boldsymbol{a}'_i < \boldsymbol{a}'^2_i/2$, and therefore

$$\boldsymbol{x}\cdot\check{\boldsymbol{v}} = \boldsymbol{x}\cdot(k^i\boldsymbol{a}'_i) = k^i(\boldsymbol{x}\cdot\boldsymbol{a}'_i) < \sum_i k^i\boldsymbol{a}'^2_i/2 \le \sum_i k^{i2}\boldsymbol{a}'^2_i/2\,.$$

Since $\boldsymbol{a}'_1\cdot\boldsymbol{a}'_2 = -\boldsymbol{a}_4\cdot(\boldsymbol{a}_2+\boldsymbol{a}_3) = -(g_{24}+g_{34}) \ge 0$, $\boldsymbol{a}'_2\cdot\boldsymbol{a}'_3 = (\boldsymbol{a}_2+\boldsymbol{a}_3)\cdot\boldsymbol{a}_3 \ge (\boldsymbol{a}_2+\boldsymbol{a}_3)^2/2 > 0$, and $\boldsymbol{a}'_3\cdot\boldsymbol{a}'_1 = -g_{31} \ge 0$, one has $\sum_i k^{i2}\boldsymbol{a}'^2_i \le (k^i\boldsymbol{a}'_i)^2 = \check{\boldsymbol{v}}^2$ and the postulated inequality $\boldsymbol{x}\cdot\check{\boldsymbol{v}} < \check{\boldsymbol{v}}^2/2$.

There are five types of Voronoï cells with distinct topologies of edges; two cells are of distinct types if one cannot be deformed into the other without adding or removing

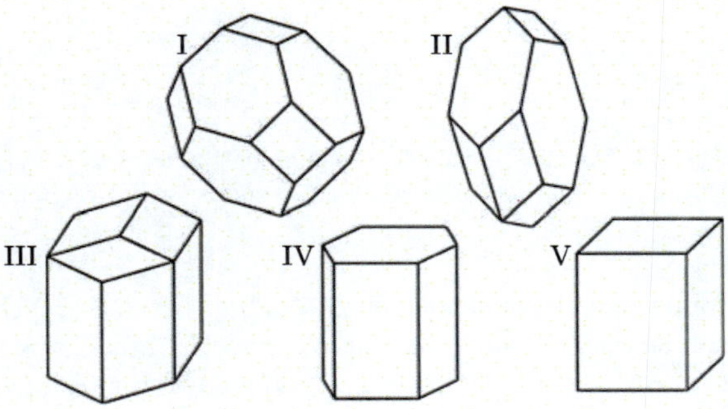

Fig. 1.5 Fedorov parallelohedra. **I** Truncated octahedron. **II** Elongated dodecahedron. **III** Rhombic dodecahedron. **IV** Right hexagonal prism. **V** Rectangular parallelopiped.

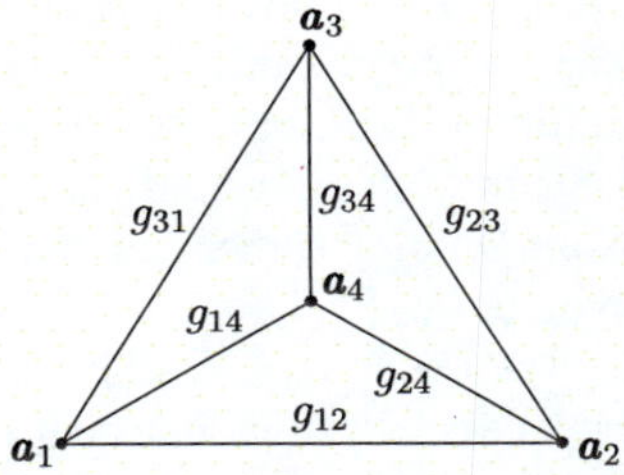

Fig. 1.6 Delaunay tetrahedron. Its vertices represent vectors $\boldsymbol{a}_\mu$ ($\mu = 1, 2, 3, 4$) of the Delaunay frame. The edge between the vertices representing $\boldsymbol{a}_\mu$ and $\boldsymbol{a}_\nu$ corresponds to the Selling parameter $g_{\mu\nu}$.

any edges. These distinct cells are known as Fedorov[11] parallelohedra (Fig. 1.5). The cells have the shapes of [18]:
– truncated octahedron—a tetradecahedron with eight hexagonal and six rhombic faces (I),
– 'elongated' dodecahedron with four hexagonal and eight rhombic faces (II),
– rhombic dodecahedron with twelve rhombic faces (III),
– right hexagonal prism (IV) and
– rectangular parallelopiped (V).
Lattices are classified based on the types of the Voronoï cells of their points. This classification is not compatible with that of Niggli: lattices of the same Niggli character may have different types of Voronoï cells [21].

Delaunay's Sorts of Lattices

Symmetry is discussed below, but getting ahead of the narrative, it is convenient to mention here Delaunay's symmetry-based classification of lattices. The symmetry of the Voronoï cell[12] is reflected in the presence of zeros among Selling parameters and equalities between them. The symmetries are conveniently visualized using Delaunay tetrahedron shown in Fig. 1.6. Its vertices correspond to vectors of the Delaunay

---

[11] Evgraf Stepanovich Fedorov (1853–1919).

[12] The symmetry of the Voronoï cell is the same as the point symmetry of the lattice; see, Sect. 1.3.3.

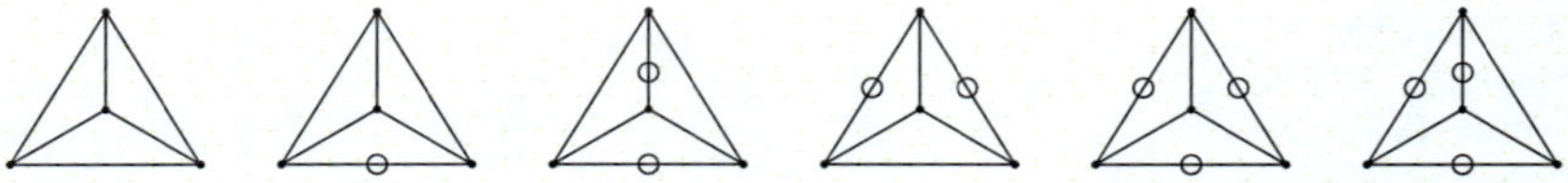

Fig. 1.7 Six non-equivalent configurations of zeros of Selling parameters. Circle on an edge of Delaunay tetrahedron indicates that the corresponding Selling parameter is zero.

frame and edges represent the Selling parameters. Up to three Selling parameters can be equal to zero, and there cannot be three zeros for the same index; cases with four zeros or $g_{\kappa\rho} = g_{\mu\rho} = g_{\nu\rho} = 0$ with distinct κ, μ, ν and ρ violate the condition of linear independence of every three vectors of the $\boldsymbol{a}_\mu$ frame. It is easy to see that there are six non-equivalent configurations of zeros shown in Fig. 1.7. Particular symmetries are identified by considering possible configurations of equalities between Selling parameters. Based on the types of the Voronoï cells and their symmetries, Delaunay classified lattices into 24 "Symmetriesorten". They are listed in Table 1.4. Configurations other than those in Table 1.4 are equivalent or do not imply additional symmetries.

The Delaunay classification is not compatible with that of Niggli [21]. For completeness, worth mentioning is the Gruber's lattice classification into 127 "genera" [22]; it subdivides both Niggli and Delaunay classes.

1.2.4 Sublattices and Superlattices

If a lattice Λ contains every point of a lattice Λ', then Λ' is said to be a sublattice of Λ; moreover, Λ' is called a proper sublattice of Λ if $\Lambda' \neq \Lambda$. With Λ' being a (proper) sublattice of Λ, Λ is referred to as a (proper) superlattice of Λ'. Clearly, the volume of the parallelepiped spanned on a basis of the sublattice is an integral multiple of the volume of the parallelepiped spanned on a basis of the lattice. A cell of the sublattice is called a supercell. Similarly, the volume of the parallelepiped spanned on a basis of the lattice is an integral multiple of the volume of the parallelepiped spanned on a basis of its superlattice. A cell of the superlattice is called a subcell. The (integer) ratio of the volumes is referred to as the *index*.

A lattice has a finite number of superlattices (sublattices) of a given index. For illustration, Fig. 1.8 shows index-two superlattices of a two-dimensional lattice. The enumeration and construction of superlattices in spaces of higher dimension is more complicated. Since each vector of the lattice belongs to the superlattice, the basis $\boldsymbol{a}_i$ of a superlattice is linked to the lattice basis $\boldsymbol{a}'_i$ by $\boldsymbol{a}'_i = T_i^{\ j} \boldsymbol{a}_j$, where T is an integer matrix with determinant equal to the index of the superlattice. The problem of constructing superlattices can be formulated in terms of these matrices. Two different matrices will give different bases of the same lattice if one of them is equal to the product of a unimodular integer matrix with the other matrix. Each

Table 1.4 Delaunay classification of lattices into 24 "Symmetriesorten". Roman numbers specify types of Voronoï cells. Edges with the same number of dashes correspond to equal Selling parameters. After [18] and [21].

1	I	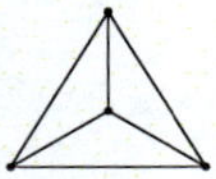
2	I	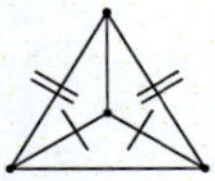
3	I	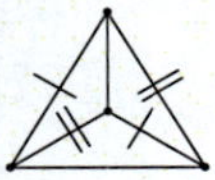
4	I	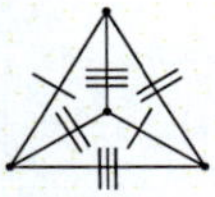
5	I	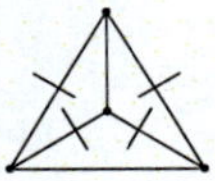
6	I	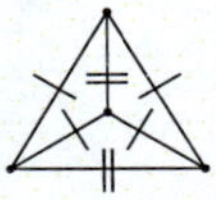
7	I	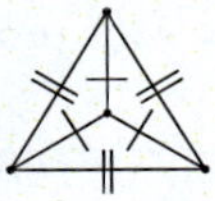
8	I	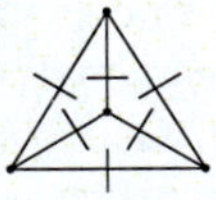
9	II	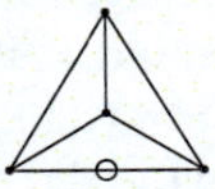
10	II	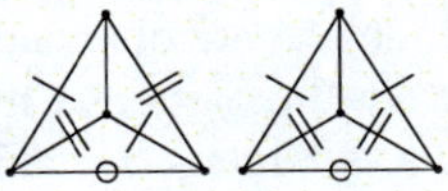
11	II	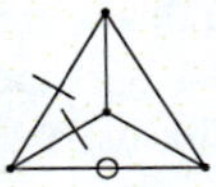
12	II	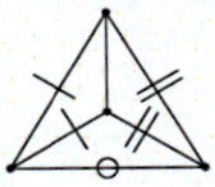
13	II	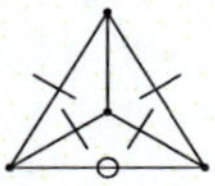
14	III	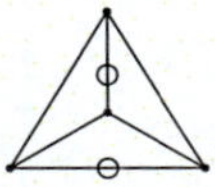
15	III	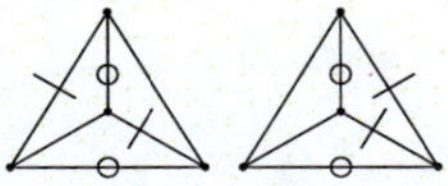
16	III	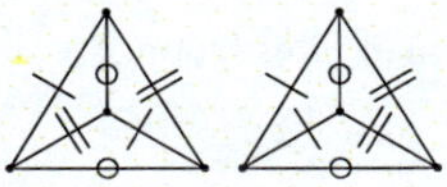
17	III	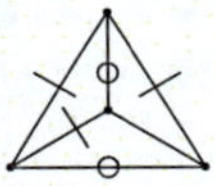
18	III	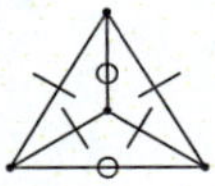
19	IV	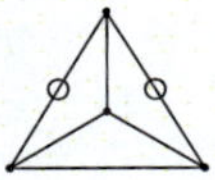
20	IV	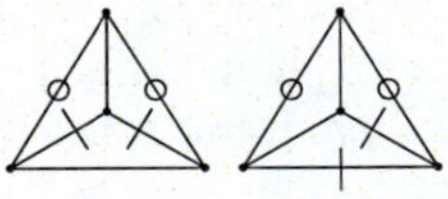
21	IV	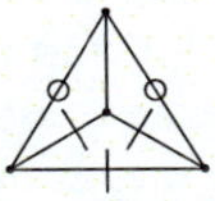
22	V	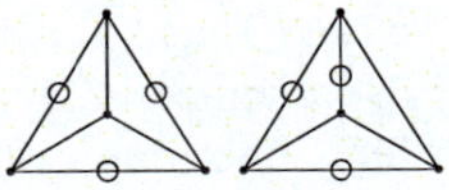
23	V	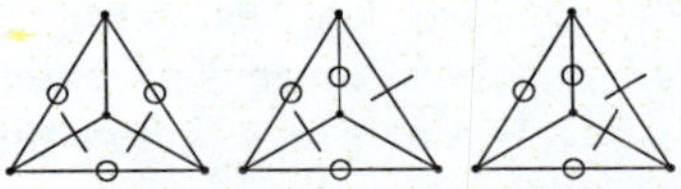
24	V	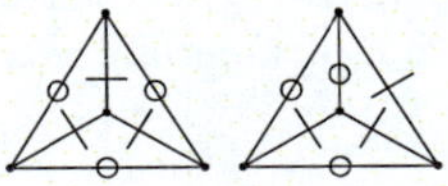

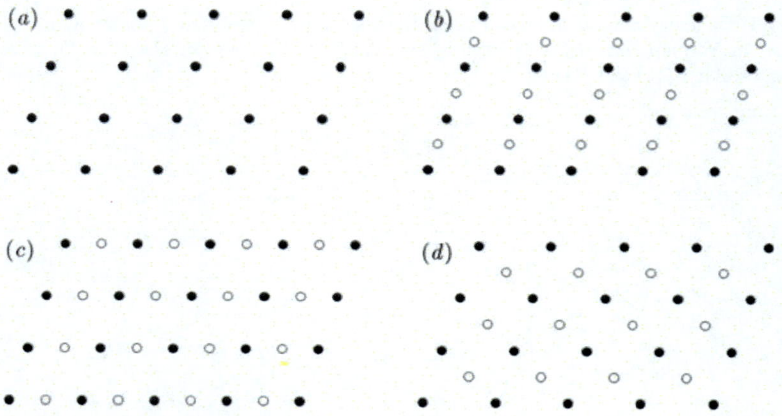

Fig. 1.8 Schematic illustration of (index 2) superlattices (*b–d*) of the two-dimensional lattice shown in (*a*).

Table 1.5 The number of low-index two-dimensional superlattices of index i. It is equal to the sum of the divisors of i. This is the Sloane's sequence A000203 [26].

i	1	2	3	4	5	6	7	8	9	10	11	12	13	14	15	16
	1	3	4	7	6	12	8	15	13	18	12	28	14	24	24	31

nonsingular integer matrix has a canonical form with respect to such products – the Hermite normal form.

A square integer matrix A is in Hermite normal form if it is upper triangular and its entries satisfy the conditions $A_{(i)(i)} > A_{ij} \geq 0$ for $j > i$. For an integer non-singular square matrix M, there exists a unimodular integer matrix U such that $MU = A$ is in Hermite normal form. A is a unique Hermite normal form of M.

Different superlattices of a given index correspond to different matrices in Hermite normal forms [23–25]. To construct all superlattices of a given index, it is enough to list all matrices in Hermite normal forms with the determinant equal to the index. The determinant of the matrix in Hermite normal form is equal to the products of its diagonal entries. Thus, all possible diagonals of interest are obtained as triplets of factors of the assumed index. For a given diagonal, the particular matrices are obtained by filling in the upper off-diagonal vacancies with integers smaller than the corresponding diagonal entries.[13] The total numbers of low-index superlattices of two- and three-dimensional lattices are listed in Tables 1.5 and 1.6, respectively.

Finally, one needs to note that an analogous scheme can be used to construct sublattices. The section is closed with the simplest non-trivial example illustrating the construction of superlattices.

[13] Some of these superlattices may turn out to be symmetrically equivalent.

Table 1.6 The number of low-index three-dimensional superlattices of index i. This is the Sloane's sequence A001001 [26].

i	1	2	3	4	5	6	7	8	9	10	11	12	13	14	15	16
	1	7	13	35	31	91	57	155	130	217	133	455	183	399	403	651

⊡ Construction of index-2 superlattices of a three-dimensional lattice. Since the determinant of the matrices in Hermite normal form is 2, only three matrix diagonals are possible (1, 1, 2), (1, 2, 1) and (2, 1, 1). Hence one can derive the following admissible forms of the T matrix:

$$\begin{bmatrix}1&0&0\\0&1&0\\0&0&2\end{bmatrix},$$
$$\begin{bmatrix}1&0&0\\0&2&0\\0&0&1\end{bmatrix},\ \begin{bmatrix}1&0&0\\0&2&1\\0&0&1\end{bmatrix},$$
$$\begin{bmatrix}2&0&0\\0&1&0\\0&0&1\end{bmatrix},\ \begin{bmatrix}2&0&1\\0&1&0\\0&0&1\end{bmatrix},\ \begin{bmatrix}2&1&0\\0&1&0\\0&0&1\end{bmatrix},\ \begin{bmatrix}2&1&1\\0&1&0\\0&0&1\end{bmatrix}.$$

Thus, there are seven superlattices of index 2.

1.2.5 *Centerings and Non-Primitive Lattice Cells*

In crystallography, to simplify the representation of symmetry operations, basis vectors are conveniently chosen along symmetry directions. (See Sect. 1.4.) These vectors may constitute a basis of a sublattice Λ' of the actual crystal lattice Λ. A proper superlattice Λ of the lattice Λ' is referred to as a centering of Λ'. Let the lattices Λ^*, Λ'^* be reciprocal to Λ, Λ' respectively. If Λ is a centering of Λ', then Λ'^* is a centering of Λ^*, some integer combinations of the basis vectors of Λ do not belong to Λ', and some integer combinations of the basis vectors of Λ'^* do not belong to Λ^*.

For brevity, the basis of a lattice centering will be referred to as *non-primitive basis*[14] of the lattice. Parallelepipeds spanned by such symmetry-dictated (not necessarily primitive) bases are referred to as *unit cells*. A unit cell is either equal to or a multiple of the primitive cell.

[14] The 'non-primitive basis' is not a basis of the lattice.

Let vectors $\boldsymbol{a}_i$ constitute a (primitive) basis of a lattice. The non-primitive basis $\boldsymbol{a}'_i$ defines a sublattice. The relationship between the bases is determined by the integer matrix T via $\boldsymbol{a}'_i = T_i{}^j \boldsymbol{a}_j$ (see (1.1)), and components of the reciprocal lattice vector $\check{\boldsymbol{h}} = h_i \boldsymbol{a}^i = h'_i \boldsymbol{a}'^i$ are transformed according to $h'_i = T_i{}^j h_j$ (1.12). Such T matrices (and their inverses) for three basic cases of non-primitive bases are listed below (together with comments relevant in the description of diffraction). See also Fig. 1.9.

1. Base-centered lattice (C)

$$T = \begin{bmatrix} 1 & -1 & 0 \\ 1 & 1 & 0 \\ 0 & 0 & 1 \end{bmatrix}, \quad \det(T) = 2\,, \quad T^{-1} = \frac{1}{2}\begin{bmatrix} 1 & 1 & 0 \\ -1 & 1 & 0 \\ 0 & 0 & 2 \end{bmatrix}$$

With $(h\,k\,l) = (h_1\,h_2\,h_3)$, $(h'\,k'\,l') = (h'_1\,h'_2\,h'_3)$ and the T matrix given above, one has $(h'\,k'\,l') = (h-k\;\; h+k\;\; l)$. Hence, $h'+k' = 2h$, and the sum $h'+k'$ is even for arbitrary $(h\,k\,l)$. Thus, if the non-primitive basis is used for the direct base-centered lattice, the integer combinations with odd $h'+k'$ do not belong to its reciprocal lattice.

2. Face-centered lattice (F)

$$T = \begin{bmatrix} -1 & 1 & 1 \\ 1 & -1 & 1 \\ 1 & 1 & -1 \end{bmatrix}, \quad \det(T) = 4\,, \quad T^{-1} = \frac{1}{2}\begin{bmatrix} 0 & 1 & 1 \\ 1 & 0 & 1 \\ 1 & 1 & 0 \end{bmatrix}$$

For the above T, the transformation rule (1.12) gives $(h'\,k'\,l') = (-h+k+l\;\; h-l+k\;\; h+k-l)$. Hence, for arbitrary $(h\,k\,l)$, the indices h' k' and l' are either all odd or all even. Thus, if the non-primitive basis is used for the direct face-centered lattice, the integer combinations with mixed (i.e., odd and even) indices h' k' and l' do not belong to its reciprocal lattice.

3. Body-centered lattice (I)

$$T = \begin{bmatrix} 0 & 1 & 1 \\ 1 & 0 & 1 \\ 1 & 1 & 0 \end{bmatrix}, \quad \det(T) = 2\,, \quad T^{-1} = \frac{1}{2}\begin{bmatrix} -1 & 1 & 1 \\ 1 & -1 & 1 \\ 1 & 1 & -1 \end{bmatrix}$$

From (1.12), $(h'\,k'\,l') = (k+l\;\; l+h\;\; h+k)$. Hence, $h'+k'+l' = 2(h+k+l)$, and for arbitrary $(h\,k\,l)$, the sum $h'+k'+l'$ is even. If the non-primitive basis is used for the direct body-centered lattice, the integer combinations with odd $h'+k'+l'$ do not belong to its reciprocal lattice.

Besides the primitive case (denoted by P), six types of centering are in use: A, B defined analogously to C (with the distinguished face 3 replaced by 1 and 2, respectively), I, F and R; the latter is described in Sect. 1.3.6 below.

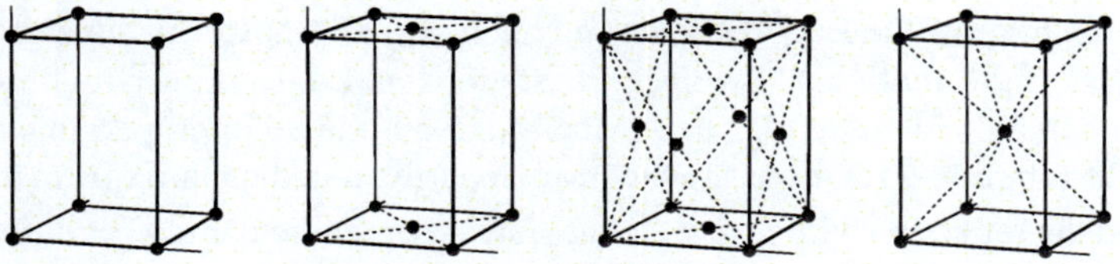

Fig. 1.9 Unit cells of a primitive (*P*), base-centered (*C*), face-centered (*F*) and body-centered (*I*) lattices.

1.3 Crystal Symmetry Groups

1.3.1 Euclidean Group

An object is symmetric if it demonstrates invariance under transformations. All crystals show evidence of nontrivial symmetry. The issue of crystal symmetry is one of the main topics in the field of crystallography.

By definition, a crystal symmetry operation transforms the crystal onto itself. The crystal symmetry operations preserve distances, i.e., they are isometric transformations of Euclidean space. Isometries in the three-dimensional Euclidean point space constitute a *Euclidean group*. An element of the Euclidean group consists of a rotation, say, R, followed by a translation, say, $\boldsymbol{t}$. This element of the Euclidean group is denoted by the (Seitz) symbol $(R, \boldsymbol{t})$. It transforms the point $\boldsymbol{v}$ to the point $R\boldsymbol{v} + \boldsymbol{t}$, i.e.,

$$(R, \boldsymbol{t})\, \boldsymbol{v} = R\boldsymbol{v} + \boldsymbol{t} \ .$$

The representation of an element of the Euclidean group by $(R, \boldsymbol{t})$ is unique. The composition of $(R, \boldsymbol{t})$ and $(R', \boldsymbol{t}')$ is given by

$$(R', \boldsymbol{t}')(R, \boldsymbol{t}) = (R'R,\ R'\boldsymbol{t} + \boldsymbol{t}') \ . \tag{1.38}$$

With the components of the identity rotation I given by $I_i{}^j = \delta_i{}^j$, the neutral element of the group is $(I, \boldsymbol{0})$, i.e. $I\boldsymbol{v} + \boldsymbol{0} = \boldsymbol{v}$. Moreover, the inverse of $(R, \boldsymbol{t})$ is

$$(R, \boldsymbol{t})^{-1} = (R^{-1}, -R^{-1}\boldsymbol{t}) \ , \tag{1.39}$$

where R^{-1} denotes rotation inverse to R. Since the transformation $(R, \boldsymbol{t})$ is an isometry, one has $|\boldsymbol{u} - \boldsymbol{w}|^2 = |(R, \boldsymbol{t})\boldsymbol{u} - (R, \boldsymbol{t})\boldsymbol{w}|^2 = |R\boldsymbol{u} - R\boldsymbol{w}|^2 = |R(\boldsymbol{u} - \boldsymbol{w})|^2$, and analogously to (1.17), the entries $R_i{}^j$ in the basis attached to the crystal satisfy

$$R_i{}^k R_j{}^l \, g_{kl} = g_{ij} \ , \tag{1.40}$$

where is g the metric tensor in that basis. The determinant of R is equal to ± 1.

The subgroup of the Euclidean group containing all transformations preserving handedness (rigid motions) involves translations and only proper rotations, i.e., such that $\det(R) = 1$. If $\det(R) = -1$, the rotation is called improper. In particular, the simplest improper rotations are the inversion (given in the matrix form by $R_i{}^j = -\delta_i{}^j$) and the reflection with respect to the mirror perpendicular to $\widehat{\boldsymbol{n}} = n_i \boldsymbol{a}^i = n^i \boldsymbol{a}_i$, which is represented by

$$R_i{}^j = \delta_i{}^j - 2n_i n^j \ . \tag{1.41}$$

Clearly, also a rotoreflection—a composition of a proper rotation and reflection with respect to the plane perpendicular to the rotation axis – is an improper rotation.

It is easy to see (cf. (1.38)) that both translations (i.e., transformations of the type $(I, \boldsymbol{t})$) and rotations (i.e., transformations of the type $(R, \boldsymbol{0})$) are subgroups of the Euclidean group. Our focus now will be on the group of rotations. In particular, we will recall classic results concerning enumeration of its finite subgroups known as point groups.

1.3.2 Finite Point Groups

Finite point groups containing only proper rotations are referred to as point groups of the first kind. Those containing also improper rotations are point groups of the second kind. The first step towards enumerating finite point groups is to enumerate point groups of the first kind.

The simplest groups of the first kind are cyclic groups, which consist of proper rotations about a given axis. Rotations by the angles $2k\pi/n$, where $n \geq 2$ and $k = 0, 1, ..., n-1$ constitute a cyclic group denoted by C_n.[15] The order (number of elements) of C_n is n. The axis of rotation is referred to as an n-fold rotation axis. In relation to the above, the symbol C_1 is used to denote the trivial group consisting only of the identity transformation.

It is more complicated to classify groups of rotations with two or more distinct axes. The question is, what are the allowed angles between the intersecting axes and what are the corresponding rotation angles? It can be relatively easily answered using the so-called Rodrigues-Hamilton theorem (see, e.g., [27], p. 5).

⊡ For a given spherical triangle (on the sphere centered at the fixed point) with vertices P_1, P_2, P_3 and vertex angles A_1, A_2, A_3, one considers three successive rotations about the axes determined by the points P_1, P_2, P_3, by the angles $2A_1$, $2A_2$, $2A_3$, respectively, with the sense of the rotations opposite to that indicated by the order of the vertices; according to the theorem, these rotations restore the object to its original position. Now, let a point group contain rotations about distinct m_1-fold and m_2-fold axes. The rotations about these axes by the angles $2\pi/m_1$ and $2\pi/m_2$ are taken to be the first two rotations of Rodrigues-Hamilton theorem ($2A_1$ and $2A_2$). Their product, which is inverse to the third rotation of Rodrigues-Hamilton theorem, must also belong to the point group. It is a rotation about a certain m_3-fold axis. The spherical cosine formula (see Fig. 1.10) applied to the triangle of Rodrigues-Hamilton theorem gives

---

[15] Here and below, the Schoenflies notation is used.

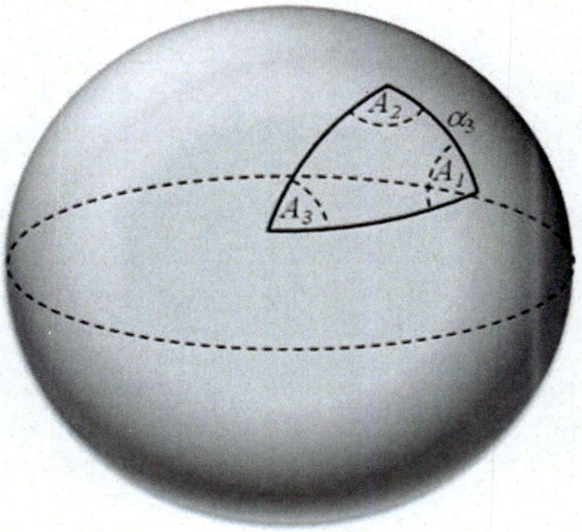

Fig. 1.10 Illustration of the (second) spherical law of cosines: given a spherical triangle with vertex angles A_1, A_2 and A_3, then $\cos(A_3) = -\cos(A_1)\cos(A_2) + \sin(A_1)\sin(A_2)\cos(\alpha_3)$, where α_3 is the side angle opposite to A_3.

$$\cos(k\pi/m_3) = -\cos(\pi/m_1)\cos(\pi/m_2) + \sin(\pi/m_1)\sin(\pi/m_2)\cos(\alpha_3)\ , \tag{1.42}$$

where α_3 is the angle between the m_1-fold and m_2-fold axes and $k = 1, ..., m_3 - 1$, and analogous relations can be written for the angle between the other axes.

There is a simple solution to (1.42) with $m_1 = 2 = m_2$ and $\alpha_3 = k\pi/m_3$. It results in the so-called dihedral groups D_n ($n = m_3$) with an n-fold (principal) axis perpendicular to n two-fold axes. Moreover, it is clear that some combinations of m_1, m_2 and m_3 follow from symmetries of regular polyhedra. They correspond to, e.g., $m_1 = 2, m_2 = 3, m_3 = 3, 4, 5$, and one obtains three groups of the first kind known as the tetrahedral group T, the octahedral group O, and the icosahedral group I.

It turns out that there are no combinations which would lead to other finite point groups. Thus, summarizing, the list of finite groups of proper rotations consists of cyclic groups C_n ($n = 1, 2, ...$), dihedral groups D_n ($n = 2, 3, ...$), and the tetrahedral T, octahedral O and icosahedral I groups.

The next step is to consider point groups of the second kind. They are generated from the groups of the first kind C_n, D_n, T, O and I: a specific improper rotation is added to the list of elements of the group of the first kind, and products of all operations are determined. The number of distinct cases obtained by this procedure turns out to be small. One obtains the following list:

- C_n appended with the reflection with respect to the plane containing the rotation axis of C_n. The resulting group is denoted by C_{nv}.
- C_n appended with the reflection with respect to the plane perpendicular to the axis. This gives the group C_{nh}. The symbol C_s is used to denote the geometrically identical groups C_{1h} and C_{1v}.
- C_n appended with the rotoreflection with the rotation by π/n about the axis of C_n and reflection with respect to the plane perpendicular to the axis. One obtains the group denoted by S_{2n}. The symbol C_i is usually used for S_2, and C_{3i} is sometimes used for S_6.

- D_n appended with the reflection with respect to the plane perpendicular to the principal axis. One gets the group D_{nh}.
- D_n appended with the reflection with respect to the plane containing the principal axis and bisecting the angle between neighboring twofold axes. This gives the group denoted by D_{nd}.
- T appended with the reflection with respect to the plane containing the twofold and threefold axes. The group is denoted by T_d.
- T, O and I appended with inversion. The groups are known as T_h, O_h and I_h, respectively.

For an extensive derivation of finite point groups in 3D space see, e.g., Appendices in [28].

Basics of Lattice Symmetry

The following simple properties of lattice symmetry are worth noting here. First, all centers of inversion of a lattice based on vectors $\boldsymbol{a}_i$ are at the points of the superlattice based on the vectors $\boldsymbol{a}_i/2$. This can be easily verified. If a center of inversion is at the point $\boldsymbol{x}$, then the inversion maps a lattice point $\check{\boldsymbol{v}}$ on the lattice point $\check{\boldsymbol{v}}' = \boldsymbol{x} + (\boldsymbol{x} - \check{\boldsymbol{v}})$; see, Fig. 1.11$a$. Hence, one has $\boldsymbol{x} = (\check{\boldsymbol{v}} + \check{\boldsymbol{v}}')/2$, i.e., $\boldsymbol{x}$ is a midpoint of a lattice vector $\check{\boldsymbol{h}} = \check{\boldsymbol{v}} + \check{\boldsymbol{v}}'$. There are eight centers of inversion per primitive cell; these are the cell vertices, the midpoints of cell edges, the centers of cell faces and the center of the cell.

Second, if a lattice has an n-fold symmetry axis not passing through $\mathbf{0}$, it has also a parallel n-fold symmetry axis passing through $\mathbf{0}$. The proof is simple: If $n > 1$ and $\mathbf{0}$ is not on the axis, the rotation of the initial lattice about the symmetry axis maps all lattice points on lattice points. In particular, $\mathbf{0}$ is mapped on a certain $\check{\boldsymbol{v}}$. Parallel translation of the rotated lattice by $-\check{\boldsymbol{v}}$ maps it on the initial lattice. The final lattice is rotated with respect to the initial lattice, $\mathbf{0}$ is at the same position in both lattices, and due to parallel translation, the new rotation axis is parallel to the initial one.

Third, with $n > 1$, an n-fold symmetry axis of a lattice is parallel to a lattice direction, i.e., to a line through multiple lattice nodes. (Cf. Sect. 1.5.1.) To see this, it suffices to take a lattice vector at an acute non-zero angle to the axis (Fig. 1.11b). The sum of this vector and all its images in the symmetry about the axis is a lattice vector, and it is located on the symmetry axis at a point different from the origin.

Fourth, with $n > 1$, an n-fold symmetry axis of a lattice is perpendicular to a (lattice) plane through three non-collinear lattice nodes. (Cf. Sect. 1.5.1.) With two lattice vectors $\check{\boldsymbol{v}}_1$ and $\check{\boldsymbol{v}}_2$ not coplanar with the axis, the vectors $S\check{\boldsymbol{v}}_i - \check{\boldsymbol{v}}_i$ $(i = 1, 2)$ are non-collinear lattice vectors perpendicular to the axis, and they span a lattice plane perpendicular to the axis.

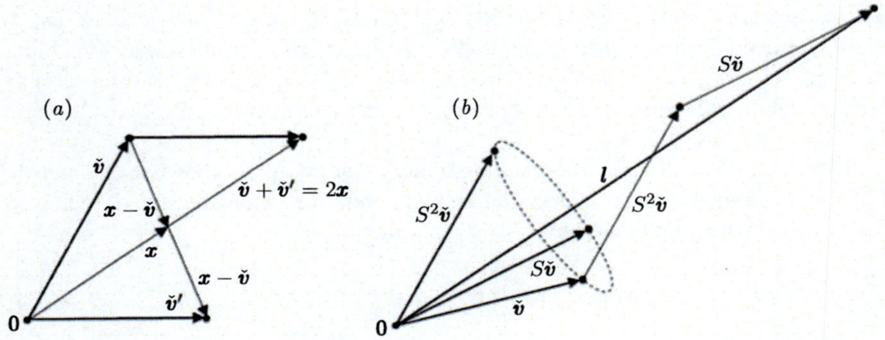

Fig. 1.11 (*a*) With the lattice point $\check{\boldsymbol{v}}'$ being the image of the lattice point $\check{\boldsymbol{v}}$ with respect to the inversion center at $\boldsymbol{x}$, the point $\boldsymbol{x}$ is the midpoint of the lattice vector $\check{\boldsymbol{v}} + \check{\boldsymbol{v}}'$. (*b*) The vector $\boldsymbol{l}$ is along a three-fold axis of a lattice symmetry operation S. The operation transforms the lattice point indicated by the vector $\check{\boldsymbol{v}}$ to lattice point $S\check{\boldsymbol{v}}$, and similarly, $S\check{\boldsymbol{v}}$ to $SS\check{\boldsymbol{v}} = S^2\check{\boldsymbol{v}}$. The sum of $\check{\boldsymbol{v}}$, $S\check{\boldsymbol{v}}$ and $S^2\check{\boldsymbol{v}}$ is a lattice vector and it is collinear with $\boldsymbol{l}$, i.e., $\boldsymbol{l}$ is along a lattice direction.

1.3.3 Crystallographic Point Groups

A point group is referred to as crystallographic if it transforms a three-dimensional lattice onto itself. Only some of the point groups listed in Sect. 1.3.2 are crystallographic. Since a symmetry operation transforms a lattice vector into a lattice vector, the entries of R in the basis of the lattice must be integers. Hence, the rotational symmetry operations correspond to integer matrices R satisfying (1.40), i.e., $R_i^{\ k} R_j^{\ l} g_{kl} = g_{ij}$, where g is the lattice metric.

⊡ For instance, if the metric is given by $g_{ij} \propto \delta_{ij}$ (cubic lattice), R must be an orthogonal matrix with integer entries. Thus, the only possible entries are ± 1 and 0, and there can be only one non-zero entry (i.e., $+1$ or -1) per column and per row. Such a matrix can be expressed as

$$R_i^{\ j} = \varepsilon_{iab}\varepsilon^{j12}\sigma^3 + \varepsilon_{ibc}\varepsilon^{j23}\sigma^1 + \varepsilon_{ica}\varepsilon^{j31}\sigma^2 \,, \tag{1.43}$$

where $\sigma^i = \pm 1$ and (a, b, c) is a permutation of $(1, 2, 3)$. With arbitrary parameters σ^i and (a, b, c), there are 48 matrices constituting a representation of the group O_h. Since $\det(R) = \sigma^1\sigma^2\sigma^3$, the matrices represent the group O if the condition $\sigma^1\sigma^2\sigma^3 = +1$ is satisfied. A representation of the group T_h (of the order of 24) is obtained if (a, b, c) is an even permutation of $(1, 2, 3)$; if additionally, $\sigma^1\sigma^2\sigma^3 = 1$, one gets a representation of the group T. Finally, the group T_d is represented by matrices satisfying $\sigma^1\sigma^2\sigma^3 = \varepsilon_{abc}$. The orders of T and T_d are 12 and 24, respectively.

With integer entries of R, the trace of the matrix $\det(R)R$ is also an integer. Since the trace does not depend on the basis, it has the same value as that in the orthonormal basis which is equal to $1 + 2\cos\omega$, where ω is the rotation angle. Hence, $\cos\omega = \pm 1, \pm 1/2,$ or 0, and ω can only be zero, a multiple of $\pi/3$ or a multiple of $\pi/2$. In other words, only 1, 2, 3, 4 and 6-fold rotation axes are allowed. Of all point groups listed in the previous section, thirty two satisfy this (crystallographic) restriction.

These are eleven point groups of the first kind ($C_1, C_2, C_3, C_4, C_6, D_2, D_3, D_4, D_6, T$ and O), eleven (Laue) groups generated from the groups of the first kind by adding inversion (C_i, C_{2h}, S_6, C_{4h}, C_{6h}, D_{2h}, D_{3d}, D_{4h}, D_{6h}, T_h and O_h) and ten groups which contain reflections or rotoreflections but no inversion (C_s, S_4, C_{3h}, C_{2v}, C_{3v}, C_{4v}, C_{6v}, D_{2d}, D_{3h} and T_d).

The above classification of crystallographic point groups is known as *geometric*. For a given point group, its representation by the matrices R depends on the selected basis. If two bases are related by $\boldsymbol{a}'_i = T_i{}^j \boldsymbol{a}_j$, then

$$R'_i{}^j = T_i{}^l R_l{}^k (T^{-1})_k{}^j = T_i{}^l T^j{}_k R_l{}^k \,, \tag{1.44}$$

where R and R' are the matrices representing an operation of the point group in the bases $\boldsymbol{a}_i$ and $\boldsymbol{a}'_i$, respectively; (1.44) must be true for all elements of the *geometrically* equivalent point groups. More abstractly, two matrix representations $G_{geom}(\boldsymbol{a}_i)$ and $G_{geom}(\boldsymbol{a}'_i)$ are geometrically equivalent if

$$G_{geom}(\boldsymbol{a}_i) = T\, G_{geom}(\boldsymbol{a}'_i)\, T^{-1} \,, \quad \text{where } T \text{ is an invertible matrix.} \tag{1.45}$$

In simple terms, by a suitable choice of bases, two geometrically equivalent point groups can be represented by the same set of matrices.

Besides the geometric classification, there are other divisions of crystallographic point groups. First, some of the point groups, which are distinct in the geometric sense, are isomorphic in the algebraic sense, i.e., there exists a group-structure preserving one-to-one mapping of one group onto the other. Groups of order two (C_2, C_i and C_s) may serve as a simple example. Second, when referred to the (primitive) basis of a lattice, elements of a point group leaving the lattice invariant are represented by integer matrices. There exists a partition of finite groups of integer matrices finer than the geometric classification. Two groups of integer matrices are *arithmetically* equivalent if there exists a unimodular[16] *integer* matrix T such that the relationship (1.44) occurs for all elements of the groups. (For more details, see Sect. 1.3.6 below.)

It is clear that arithmetically equivalent groups are equivalent in the geometric sense, and geometrically equivalent point groups are isomorphic. It turns out that in the three-dimensional space, there are 73 'arithmetic' classes of crystallographic integer matrix groups and 18 types of non-isomorphic point groups.

1.3.4 Space Groups

In general, symmetry operations of an arbitrary object constitute a group. The group of crystal symmetry operations is called crystal space group. It is a discrete subgroup of the Euclidean group.

[16] I.e., with absolute value of determinant equal to 1.

With periodic arrangement of atoms, some translations transform the crystal onto itself, and thus they are crystal symmetry operations. Such translations constitute a subgroup of the crystal space group. Formally, a space group is a subgroup of the Euclidean group such that its pure translations constitute a three-dimensional lattice. Translation vectors determine a lattice—it is a lattice of the particular crystal. All general information on lattices contained in Sect. 1.2 is applicable to *crystal* lattices.

Besides pure translations and pure rotations, a space group may contain combined operations $(R, \boldsymbol{t})$, in which the rotation R differs from I and the translation $\boldsymbol{t}$ is not a lattice translation (i.e., it does not transform the lattice onto itself). Presence of such operations involves composite 'symmetry elements' like screw axes of glide planes. The former is an axis of a proper rotation followed by a translation parallel to that axis. The latter is a plane of a reflection followed by a translation parallel to that plane.

Let $(R, \boldsymbol{t})$ be a symmetry operation of the crystal space group with a nontrivial rotational part $R \neq I$. For an arbitrary vector $\check{\boldsymbol{t}}'$ of the crystal lattice, $(I, \check{\boldsymbol{t}}')$ is an element of the space group, and (1.38) and (1.39) give

$$(R, \boldsymbol{t})(I, \check{\boldsymbol{t}}')\,(R, \boldsymbol{t})^{-1} = (I, R\check{\boldsymbol{t}}') = \text{pure translation}\ ,$$

i.e., $R\check{\boldsymbol{t}}'$ is a vector of the crystal lattice. Generally, R transforms the lattice vector into another vector of the lattice. Thus, the rotational parts of the symmetry operations transform the crystal lattice onto itself, and they constitute a crystallographic point group. The group is referred to as the *crystal* point group, but it must be stressed that a rotation R being an element of a crystal point group may not necessarily be a symmetry operation of the crystal, i.e., $(R, \mathbf{0})$ may not be an element of the crystal space group. One may ask here what is the point in considering such rotations. They are important because (macroscopic) observations involving large numbers of primitive cells reveal only the point group of the crystal.[17] In particular, in diffraction experiments (which reveal crystal reciprocal lattice weighted by structure factors, cf. Chap. 2) only the point symmetry is maintained. The description of an orientation of the crystal is also influenced only by the rotations of the crystal point group. Moreover, the translational parts of space group operations play no role in analyzes of lattice parameters and (external) lattice strains.

Space groups will not be considered any further, but before closing the subject, it is worth noting how these groups are enumerated. The process begins with construction of *symmorphic* space groups: With a given origin of the coordinate frame, the vector $\boldsymbol{t}$ in an element $(R, \boldsymbol{t})$ of a space group may belong to the group of translations or not; a space group is symmorphic if there exists an origin with respect to which all translation parts $\boldsymbol{t}$ of space group elements $(R, \boldsymbol{t})$ belong to the group of translations. With a symmorphic space group, there exists an origin, such that elements $(R, \mathbf{0})$ constitute its subgroup. It is possible to write all elements $(R, \boldsymbol{t})$ of a symmorphic space

[17] The above statement is related to Neumann's principle (Franz Ernst Neumann, 1798–1895), which states that "the symmetry of any physical property of a crystal must include all symmetry operations of the point group of the crystal".

group as the products $(I, \boldsymbol{t})(R, \mathbf{0})$. Enumeration of symmorphic groups is done based on classifications of crystallographic point groups and three-dimensional lattices. It turns out that there are 73 symmorphic space groups. As for the non-symmorphic space groups, they all contain some glide-plane or screw-axis operations. They are constructed using a property following directly from (1.38); since the product of space group elements $(R, \boldsymbol{\tau})(R', \boldsymbol{\tau}') = (RR', R\boldsymbol{\tau}' + \boldsymbol{\tau})$ is an element of the space group, the term $R\boldsymbol{\tau}' + \boldsymbol{\tau}$ is equal to some translation. Solving this equation with respect to pairs $(R, \boldsymbol{\tau})$ based on classifications of crystallographic point groups and three-dimensional lattices leads to the non-symmorphic space groups. The total number of space groups in three dimensions turns out to be 219 (or 230 if eleven groups which differ only by reversal of handedness of coordinate frames are added).

1.3.5 Crystal Systems

The (geometric) holohedry of a lattice is a group of all point symmetry operations of the lattice, or in other words, a group of all orthogonal transformations that leave the lattice invariant. Thus, the holohedry is the maximal point group transforming a lattice onto itself. As in the case of point groups (cf. (1.45)), what matters first are the classes of geometric holohedries: Let $H_{geom}(\boldsymbol{a}_i)$ denote a holohedry of the lattice with a basis $\boldsymbol{a}_i$; for lattices with geometrically equivalent geometric holohedries, one has $H_{geom}(\boldsymbol{a}_i') = T\, H_{geom}(\boldsymbol{a}_i)\, T^{-1}$, where T is an invertible matrix.

The inversion is a symmetry operation of every lattice (with points of the lattice and midpoints between points of the lattice being centers of symmetry). Thus, the potential list of holohedries is limited to the Laue groups (C_i, C_{2h}, S_6, C_{4h}, C_{6h}, D_{2h}, D_{3d}, D_{4h}, D_{6h}, T_h and O_h). However, a lattice with the tetrahedral symmetry T_h, has also the higher symmetry of O_h, and analogous statements apply to the pairs C_{6h} and D_{6h}, C_{4h} and D_{4h}, S_6 and D_{3d}. Thus, in three-dimensional space, there are seven distinct holohedries: C_i, C_{2h}, D_{2h}, D_{3d}, D_{4h}, D_{6h} and O_h. The orders of these groups are 2, 4, 8, 12, 16, 24 and 48, respectively.

Holohedries are used to classify point groups into *crystal systems*. If a point group is not a holohedry, it is assigned to a holohedry by requiring that it is its subgroup but not a subgroup of a holohedry of smaller order. The classification of crystallographic point groups into crystal systems (triclinic, monoclinic, orthorhombic, trigonal, tetragonal, hexagonal and cubic) is given in Table 1.7.

Chirality, Centrosymmetry and Related Terms

Additional, partly terminological remarks are in place as they are related to point group classification and Table 1.7. Point groups that are not holohedries are called merohedries, and a non-holohedral crystal is called merohedral.

Chirality is the property of a crystal structure (or, generally, an object) of being superposable by inversion but not by a proper rotation. A crystal structure is either

Table 1.7 Crystal systems in three dimensions and geometric classes of point groups. Holohedries are in the line marked by H, hemihedries (groups with only half the number of all lattice symmetry operations) by h. Laue groups are marked by L, and groups of the first kind are marked by e. Classes of the trigonal crystal system are split between rhombohedral and hexagonal lattice systems.

	Triclinic	Monoclinic	Orthorhombic	Trigonal	Tetragonal	Hexagonal	Cubic
H, L	C_i	C_{2h}	D_{2h}	D_{3d}	D_{4h}	D_{6h}	O_h
h, e	C_1	C_2	D_2	D_3	D_4	D_6	O
h			C_{2v}	C_{3v}	C_{4v}	C_{6v}	
h		C_s			D_{2d}	D_{3h}	T_d
h, L				S_6	C_{4h}	C_{6h}	T_h
e				C_3	C_4	C_6	T
					S_4	C_{3h}	

chiral or achiral. The point group of a chiral structure is of the first kind.[18] The point group of an achiral structure, besides proper rotations, contains also improper rotations. Chiral crystals have right-handed and left-handed forms. These two forms related by inversion are called enantiomorphs.

A crystal structure is centrosymmetric if its point group contains the inversion. As was already noted, there are eleven Laue point groups containing the inversion. A centrosymmetric structure is achiral. A non-centrosymmetric crystal structure is either chiral or achiral.

A crystal direction is polar if its one sense is not equivalent to the other. Polar directions are possible in the case of non-centrosymmetric point groups. The term 'polar' sometimes means non-centrosymmetric achiral. Alternatively, ten polar point groups are those having more than one point invariant under all symmetry operations; these are C_n and C_{nv} including $C_s = C_{1v}$ (point groups of pyroelectric crystals). Five point groups of non-centrosymmetric achiral crystals are not polar in the latter sense; these are C_{3h}, S_4, D_{2d}, D_{3h} and T_d.

1.3.6 Bravais Types

Is was mentioned in Sect. 1.3.3 that the finer arithmetic classification subdivides the geometric classes of point groups into arithmetic classes. Since the well known classification of lattices into Bravais types is based on the arithmetic classification, the subject needs to be discussed in more detail. For the rotation S being a symmetry operation of a lattice, there exists a unimodular matrix R with integer entries such that $S\boldsymbol{a}_i = R_i{}^j \boldsymbol{a}_j$. ($R$ is unimodular because S is an isometry, and $R_i{}^j$ are integers because $S\boldsymbol{a}_i$ are lattice vectors.) The integer matrices R represent particular point symmetry operations in the basis $\boldsymbol{a}_i$. They constitute an integer matrix representation of the point

[18] As for space groups of chiral crystals, there are 65 (Sohncke) space groups containing only symmetry operations of the first kind.

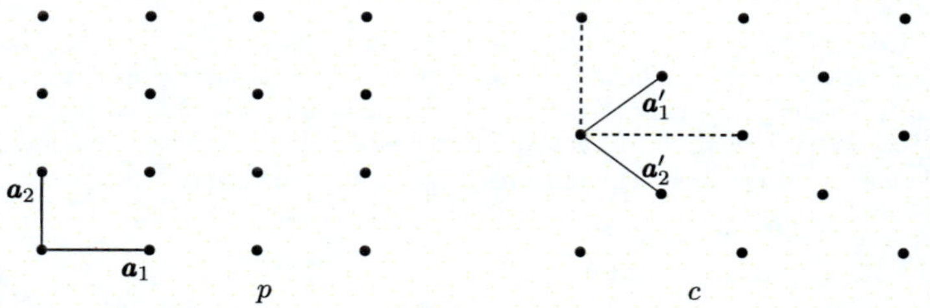

Fig. 1.12 Two-dimensional lattices of the rectangular system (i.e., of the same holohedry): primitive p and centered c. In the second case, the conventional cell is spanned by $a'_1 + a'_2$ and $a'_1 - a'_2$ (dashed lines), which are mutually perpendicular and generate a sublattice of the same type as p.

group. Briefly, two crystallographic point groups are arithmetically equivalent if there exists a *unimodular integer* matrix T relating their integer matrix representations. Representations of point groups of two lattices with the same symmetry (the same geometric holohedries) do not need to be arithmetically equivalent.

⊡ It is easy to see this by considering a two-dimensional example. The two-dimensional rectangular system involves two reflections and the rotation by π. These are the symmetries of the primitive lattice based on mutually perpendicular vectors $\boldsymbol{a}_1, \boldsymbol{a}_2$ of different magnitudes; see Fig. 1.12*a*. In this basis, the symmetry operations are represented by

$$G_{arithm}(\boldsymbol{a}_i): \begin{bmatrix} 1 & 0 \\ 0 & 1 \end{bmatrix}, \begin{bmatrix} -1 & 0 \\ 0 & -1 \end{bmatrix}, \begin{bmatrix} 1 & 0 \\ 0 & -1 \end{bmatrix}, \begin{bmatrix} -1 & 0 \\ 0 & 1 \end{bmatrix}. \tag{1.46}$$

Now, the second lattice shown in Fig. 1.12*b* is based on vectors $\boldsymbol{a}'_1, \boldsymbol{a}'_2$ of equal magnitude with an angle different from $\pi/2$ and $\pi/3$ or $2\pi/3$ (to avoid square and hexagonal lattices of higher symmetry). The second lattice exhibits the same symmetries as the first one. In the basis $\boldsymbol{a}'_i$, the symmetry operations are represented by

$$G_{arithm}(\boldsymbol{a}'_i): \begin{bmatrix} 1 & 0 \\ 0 & 1 \end{bmatrix}, \begin{bmatrix} -1 & 0 \\ 0 & -1 \end{bmatrix}, \begin{bmatrix} 0 & 1 \\ 1 & 0 \end{bmatrix}, \begin{bmatrix} 0 & -1 \\ -1 & 0 \end{bmatrix}. \tag{1.47}$$

Matrices of $G_{arithm}(\boldsymbol{a}_i)$ can be transformed onto those of $G_{arithm}(\boldsymbol{a}'_i)$ via $G_{arithm}(\boldsymbol{a}'_i) = T\, G_{arithm}(\boldsymbol{a}_i) T^{-1}$ by T of the form

$$\begin{bmatrix} a & b \\ -a & b \end{bmatrix} \quad \text{or} \quad \begin{bmatrix} a & -b \\ a & b \end{bmatrix}, \quad \text{with} \quad ab \neq 0\,.$$

Since $\det(T) = 2ab$, the matrix T cannot be integer unimodular. This means that the integer matrix representations (1.46) and (1.47) are not arithmetically equivalent.

The arithmetic equivalence between integer matrix groups of lattice symmetries leads to an algebraic classification finer than the geometric classification of point groups. Two integer matrix groups of lattice symmetries belong to the same *arithmetic* crystal class if the lattices have bases in which the groups are represented by the same matrices. Groups $G_{arithm}(\boldsymbol{a}_i)$ and $G_{arithm}(\boldsymbol{a}'_i)$ of the same arithmetic class are linked by $G_{arithm}(\boldsymbol{a}'_i) = T\, G_{arithm}(\boldsymbol{a}_i)\, T^{-1}$, where T is a unimodal integer matrix. There are 73 three-dimensional arithmetic classes of point groups, and there is one-to-one correspondence between these classes and symmorphic space groups. Clearly, arithmetically equivalent point groups are also geometrically equivalent, and hence, geometric classes consist of complete arithmetic classes.

Similarly to geometric holohedries, one defines arithmetic holohedries. The arithmetic holohedry of a lattice is the maximal point group leaving the lattice invariant, but the classes of arithmetic holohedries are linked by *unimodal integer* matrices. All *integer* matrices transforming the lattice to a congruent lattice constitute the aforementioned arithmetic holohedry H_{arithm} of the lattice.

Lattices with *arithmetically* equivalent holohedries

$$H_{arithm}(\boldsymbol{a}_i) = T\, H_{arithm}(\boldsymbol{a}'_i)\, T^{-1}\,, \quad \text{with integer unimodular } T$$

are said to belong to the same *Bravais*[19] *lattice type* (or *class*). There are 14 classes of arithmetic holohedries and, consequently, 14 three-dimensional Bravais lattices. (See Fig. 1.16 in Sect. 1.4.) In simple terms, lattices of the same Bravais type have the same point symmetry and the same type of centering. The Bravais classification agrees with that of Niggli in the sense that lattices of a Niggli class are contained in a Bravais class. Differently than the Niggli classification, the Bravais classification relies on symmetry. The Bravais classification is not compatible with that based on the types of the Voronoï cells: lattices of the same Bravais type have different types of Voronoï cells. For more details, see [*ITC-A*].

Two lattices belong to the same lattice system if their arithmetic holohedries are *geometrically* equivalent. There are seven lattice systems: triclinic, monoclinic, orthorhombic, rhombohedral, tetragonal, hexagonal and cubic; they are denoted by a, m, o, r, t, h and c, respectively. There is only one Bravais class per each: triclinic, rhombohedral and hexagonal lattice systems. As for other systems, there are two classes in the monoclinic (P and S) and the tetragonal (P and I) systems, three classes in the cubic system (P, I, F), and four in the orthorhombic system (P, S, I, F); as in Sect. 1.2.5, the symbols P, S, I, and F denote primitive lattice, and side-face, body and face centered lattices, respectively. The complete list of symbols for the Bravais lattice types is: aP, mP, mS, oP, oS, oI, oF, $rP = hR$, tP, tI, hP, cP, cI and cF.

Clearly, a crystal is classified to a crystal system based on its point group, and it is classified to a lattice system based on its lattice. When applied to crystals (or their space groups), the above classifications exhibit some incompatibilities. Bravais classification does not agree with the geometric classification, and the classification into lattice systems does not agree with the classification into crystal systems; see, Table 1.8. This is explained in detail in the example below.

---

[19] Auguste Bravais (1811–1863).

Table 1.8 Distribution of arithmetic classes of point groups with respect to geometric classes of point groups and Bravais classes of lattices. Stars indicate positions of arithmetic holohedries. Horizontal lines separate crystal systems and vertical lines separate lattice systems.

	aP	mP	mS	oP	oS	oI	oF	rP	tP	tI	hP	cP	cI	cF
C_i	1*													
C_1	1													
C_{2h}		1*	1*											
C_2		1	1											
C_s		1	1											
D_{2h}				1*	1*	1*	1*							
D_2				1	1	1	1							
C_{2v}				1	2	1	1							
D_{3d}								1*			2			
D_3								1			2			
C_{3v}								1			2			
S_6								1			1			
C_3								1			1			
D_{4h}									1*	1*				
D_4									1	1				
C_{4v}									1	1				
D_{2d}									2	2				
C_{4h}									1	1				
C_4									1	1				
S_4									1	1				
D_{6h}											1*			
D_6											1			
C_{6v}											1			
D_{3h}											2			
C_{6h}											1			
C_6											1			
C_{3h}											1			
O_h												1*	1*	1*
O												1	1	1
T_d												1	1	1
T_h												1	1	1
T												1	1	1

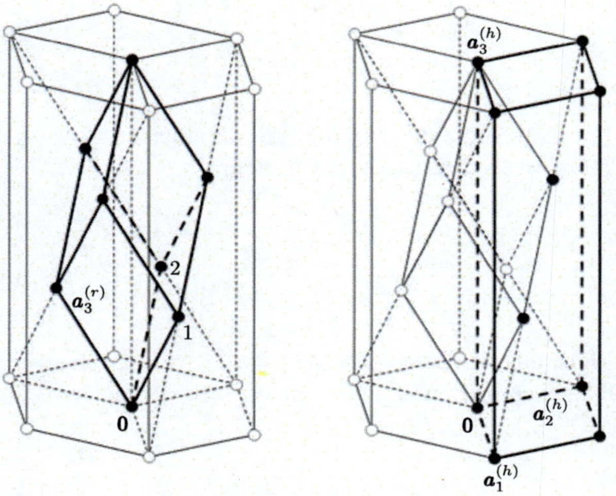

Fig. 1.13 Rhombohedral cell in R (rhombohedrally) centered hexagonal (obverse) setting. The symbols 1 and 2 in the left figure indicate terminal points of $\boldsymbol{a}_1^{(r)}$ and $\boldsymbol{a}_2^{(r)}$, respectively. Hexagonal prisms are drawn for clarity.

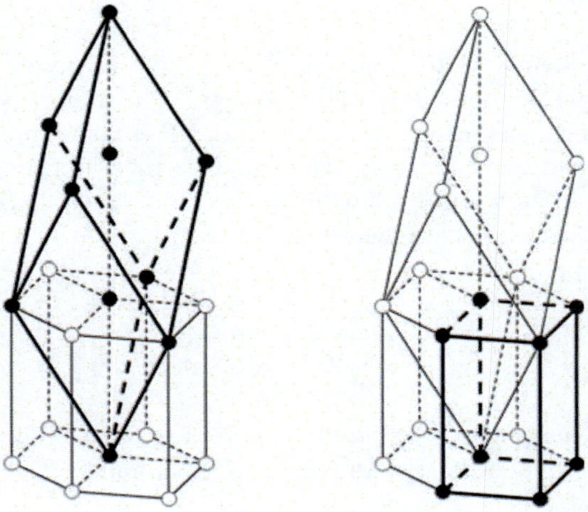

Fig. 1.14 Hexagonal unit cell in centered rhombohedral setting.

Example

A rhombohedral lattice can be seen as a centering of a hexagonal lattice (Fig. 1.13), but its symmetry is lower than that of the hexagonal lattice. Similarily, a hexagonal lattice can be seen as a centering of a rhombohedral lattice (Fig. 1.14), but its symmetry is higher than that of the rhombohedral lattice.

Based on Fig. 1.13, the hexagonal lattice basis $\boldsymbol{a}_i^{(h)}$ (which satisfies the conditions $\boldsymbol{a}_1^{(h)} \cdot \boldsymbol{a}_3^{(h)} = 0 = \boldsymbol{a}_2^{(h)} \cdot \boldsymbol{a}_3^{(h)}$ and $\boldsymbol{a}_1^{(h)} \cdot \boldsymbol{a}_1^{(h)} = -2\boldsymbol{a}_1^{(h)} \cdot \boldsymbol{a}_2^{(h)} = \boldsymbol{a}_2^{(h)} \cdot \boldsymbol{a}_2^{(h)}$) and the 'rhombohedral' basis $\boldsymbol{a}_i^{(r)}$ (satisfying $\boldsymbol{a}_i^{(r)} \cdot \boldsymbol{a}_j^{(r)} = a\delta_{ij} + b$ with the same non-zero a, b for all pairs i, j) are related via $\boldsymbol{a}_i^{(h)} = T_i{}^j \boldsymbol{a}_i^{(r)}$, where

$$T = \begin{bmatrix} 1 & -1 & 0 \\ 0 & 1 & -1 \\ 1 & 1 & 1 \end{bmatrix}, \quad \det(T) = 3\,, \quad T^{-1} = \frac{1}{3}\begin{bmatrix} 2 & 1 & 1 \\ -1 & 1 & 1 \\ -1 & -2 & 1 \end{bmatrix}$$

Rotations about the threefold axis permute the basis vectors $\boldsymbol{a}_i^{(r)}$. Therefore, with the rhombohedral basis, the point group C_3 is generated by the matrix

$$M_r = \begin{bmatrix} 0 & 1 & 0 \\ 0 & 0 & 1 \\ 1 & 0 & 0 \end{bmatrix}.$$

Thus, one has the group, say, $G(\boldsymbol{a}_i^{(r)})$, of integer matrices: $I(= M_r^3),\ M_r,\ M_r^2$. In the hexagonal setting, the same symmetry operations are generated by

$$M_h = TM_rT^{-1} = \begin{bmatrix} 0 & 1 & 0 \\ -1 & -1 & 0 \\ 0 & 0 & 1 \end{bmatrix},$$

where T is given above. Thus, one has the group $G(\boldsymbol{a}_i^{(h)}) = TG(\boldsymbol{a}_i^{(r)})T^{-1}$ of integer matrices: $I(= M_h^3),\ M_h,\ M_h^2$. The groups $G(\boldsymbol{a}_i^{(r)})$ and $G(\boldsymbol{a}_i^{(h)})$ belong to the same geometric class (C_3), but as T is not unimodular, they are not arithmetically equivalent, i.e., they belong to different arithmetic classes.

The arithmetic holohedry $H(\boldsymbol{a}_i^{(r)})$ of the group $G(\boldsymbol{a}_i^{(r)})$ is a group (of the geometric class D_{3d}) generated by the matrices

$$\begin{bmatrix} 0 & -1 & 0 \\ 0 & 0 & -1 \\ -1 & 0 & 0 \end{bmatrix}, \quad \begin{bmatrix} 0 & -1 & 0 \\ -1 & 0 & 0 \\ 0 & 0 & -1 \end{bmatrix},$$

which represent sixfold rotoreflection and twofold proper rotation, respectively. The arithmetic holohedry $H(\boldsymbol{a}_i^{(h)})$ of the group $G(\boldsymbol{a}_i^{(h)})$ is a group (of the geometric class D_{6h}) generated by

$$\begin{bmatrix} 1 & 1 & 0 \\ -1 & 0 & 0 \\ 0 & 0 & 1 \end{bmatrix}, \quad \begin{bmatrix} 0 & 1 & 0 \\ 1 & 0 & 0 \\ 0 & 0 & -1 \end{bmatrix}, \quad \begin{bmatrix} 1 & 0 & 0 \\ 0 & 1 & 0 \\ 0 & 0 & -1 \end{bmatrix},$$

representing sixfold and twofold proper rotations and reflection with respect to the plane perpendicular to the principal axis, respectively. As the groups $H(\boldsymbol{a}_i^{(r)})$ and $H(\boldsymbol{a}_i^{(h)})$ are not geometrically equivalent, the Bravais classes and the lattice systems in these two cases are different, despite the fact that the groups $G(\boldsymbol{a}_i^{(r)})$ and $G(\boldsymbol{a}_i^{(h)})$ belong to the same geometric class C_3 and, consequently, to the same trigonal crystal system.

The above example shows that the Bravais classification splits the geometric class C_3. Similar splits occur for the remaining groups of the trigonal crystal class, i.e., Bravais classification does not agree with the geometric classification. Moreover, the division into lattice systems splits the trigonal crystal class, i.e., the classification into lattice systems does not agree with the classification into crystal systems.[20] The trigonal crystal system comprises *all* crystals (or space groups, if the classification is applied to them) of the rhombohedral lattice system and *some* crystals (or space groups) of the hexagonal lattice system. All other crystals (or space groups) of the hexagonal lattice system are in the hexagonal crystal system.

To simplify things, the notion of a crystal family was introduced. A crystal family is the smallest set comprising complete crystal systems and complete lattice systems. The hexagonal crystal family comprises both rhombohedral and hexagonal lattice systems and both trigonal and hexagonal crystal systems. The other families are (triclinic, monoclinic, orthorhombic, tetragonal and cubic) contain corresponding crystal systems as well as lattice systems. The division into crystal families applies to all: crystals, lattices, space groups, point groups, geometric crystal classes and arithmetic crystal classes.

1.3.7 Symmetry of the Reciprocal Lattice

With a lattice ascribed to a given class, the question arises about allocation of the reciprocal lattice. It turns out that an operation of a point group of a direct lattice also leaves its reciprocal unchanged.

⊡ Vectors $\boldsymbol{a}^k$ of the basis reciprocal to $\boldsymbol{a}_k$ satisfy (1.14), i.e., $\boldsymbol{a}_i \times \boldsymbol{a}_j = \det(\mathcal{T})\varepsilon_{ijk}\boldsymbol{a}^k$, where $|\det(\mathcal{T})|$ is the volume of the cell spanned by the vectors $\boldsymbol{a}_i$. Let S be a proper orthogonal transformation leaving the direct lattice invariant. Since $S\boldsymbol{a}_i = R_i{}^m\boldsymbol{a}_m$ with integer unimodular R, and $(S\boldsymbol{a}_i) \times (S\boldsymbol{a}_j) = \det(\mathcal{T})\varepsilon_{ijk}(S\boldsymbol{a}^k)$, one has

$$(S\boldsymbol{a}_i) \times (S\boldsymbol{a}_j) = R_i{}^m R_j{}^n \boldsymbol{a}_m \times \boldsymbol{a}_n = R_i{}^m R_j{}^n \det(\mathcal{T})\varepsilon_{mnl}\, \boldsymbol{a}^l$$

and

$$S\boldsymbol{a}^k = (R_i{}^m R_j{}^n \varepsilon^{ijk} \varepsilon_{mnl}/2)\, \boldsymbol{a}^l\ .$$

For integer matrix R, the coefficients $\varepsilon^{ijk} R_i{}^m R_j{}^n \varepsilon_{mnl}/2$ are integers, and thus S transforms the vectors of the reciprocal basis to vectors of the reciprocal lattice. Because of the linearity of S, it transforms an arbitrary vector of the reciprocal lattice to a vector of this lattice. Since the inversion is a symmetry operation of every lattice, and every point symmetry operation of the lattice is either a proper rotation or can be expressed as a composition of inversion and proper rotations, the conclusion reached for S can be generalized: every symmetry operation of the point group of the direct lattice transforms the reciprocal lattice onto itself.

---

20 To give a concrete example, α–quartz (SiO_2) belongs to trigonal crystal system and hexagonal lattice system. On the other hand, calcite ($CaCO_3$) belongs to trigonal crystal system and rhombohedral lattice system; see, e.g., [29].

Table 1.9 Bravais type of a lattice of a given Niggli character.

1	cF	10	mS	19	oI	28	mS	37	mS
2	hR	11	tP	20	mS	29	mS	38	oS
3	cP	12	hP	21	tP	30	mS	39	mS
4	hR	13	oS	22	hP	31	aP	40	oS
5	cI	14	mS	23	oS	32	oP	41	mS
6	tI	15	tI	24	hR	33	mP	42	oI
7	tI	16	oF	25	mS	34	mP	43	mS
8	oI	17	mS	26	oF	35	mP	44	aP
9	hR	18	tI	27	mS	36	oS		

Consequently, the direct lattice and its reciprocal have the same holohedry. As geometry of crystal diffraction patterns reflects the reciprocal lattice (see Sect. 2.2), the symmetry of the crystal direct lattice is observable in the geometry of the patterns.

1.3.8 Bravais Type from Niggli Character or Delaunay Sort

Both Niggli characters and Delaunay sorts of lattices are subdivisions of the symmetry-based Bravais classification. Niggli and Bravais classifications are compatible in the sense that the former is a subdivision of the latter. The subdivisions are uneven; e.g., mS comprises 13 Niggli characters whereas there is one character per each oP, cP, cI and cF. The Bravais types corresponding to Niggli classes and Delaunay sorts are listed in Tables 1.9 and 1.10, respectively.

Formally, knowing an exact basis of a lattice, one can reduce it, get the Niggli character of the lattice from Table 1.3 and then identify the Bravais type using Table 1.9. Similarly, the Bravais type can be determined using Delaunay reduction and Tables 1.4 and 1.10.

Since the reduction requires testing for exact equalities, practical computing based directly on the algorithm of Sect. 1.2.2 or Table 1.4 fails due to rounding errors [30] or

Table 1.10 Bravais type of a lattice of a given Delaunay sort with the sorts numbered as in Table 1.4.

1	aP	7	hR	13	tI	19	mP
2	mS	8	cI	14	aP	20	oS
3	mS	9	aP	15	mS	21	hP
4	oI	10	mS	16	oI	22	oP
5	oF	11	mS	17	hR	23	tP
6	tI	12	oI	18	cF	24	cP

experimental uncertainties [31], and a tolerance must be added. Methods for dealing with inaccurate data are described in Sect. 1.7 below.

1.4 Conventional Crystallographic Settings

Primitive lattice bases (including reduced bases) are not always convenient. Such bases frequently obscure symmetries. Bases with vectors along main crystal symmetry axes make the symmetries more transparent. Therefore, as was already mentioned, some crystal structures are described using non-primitive unit cells conveniently based on vectors along the highest symmetry axes. Additionally, constraints on lattice metrics are imposed. The standardly used bases and constraints are known as 'conventional settings'. They constitute a framework for communicating and cataloging crystallographic data.

In the standard notation, instead of $\boldsymbol{a}_i$ and $\boldsymbol{a}^i$, vectors of direct lattice bases are denoted by symbols

$$\mathbf{a} = \boldsymbol{a}_1 , \quad \mathbf{b} = \boldsymbol{a}_2 , \quad \mathbf{c} = \boldsymbol{a}_3 , \quad \mathbf{a}^* = \boldsymbol{a}^1 , \quad \mathbf{b}^* = \boldsymbol{a}^2 , \quad \mathbf{c}^* = \boldsymbol{a}^3 .$$

Moreover, the magnitudes of vectors of bases and angles between the vectors are usually denoted by

$$a = |\mathbf{a}| , \quad b = |\mathbf{b}| , \quad c = |\mathbf{c}| , \quad \alpha = \angle(\mathbf{b}, \mathbf{c}) , \quad \beta = \angle(\mathbf{c}, \mathbf{a}) , \quad \gamma = \angle(\mathbf{a}, \mathbf{b}) ,$$

plus analogous relations for $a^*, b^*, c^*, \alpha^*, \beta^*$ and γ^*; e.g., $a^* = |\mathbf{a}^*|, \alpha^* = \angle(\mathbf{b}^*, \mathbf{c}^*)$. The lengths of the edges of a unit cell and the angles between them are referred to as parameters of the lattice cell or briefly as lattice parameters (Fig. 1.15).

Details of the conventional lattice bases for particular crystal systems are listed in [*ITC-A*]; see also, e.g., [32].

⊡ Very briefly, they can be summarized as follows. All bases are right handed. Conventional lattice bases for crystals of the triclinic system are Buerger reduced. Monoclinic crystals have the conventional lattice bases with **b** along the symmetry axis, and the shortest possible **a** and **c** such that $\alpha = \pi/2 = \gamma$ and $\beta \geq \pi/2$ ('second setting').

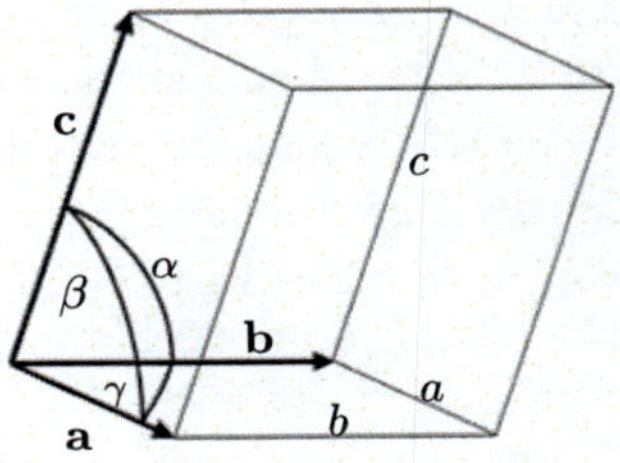

Fig. 1.15 Conventional lattice parameters: $\mathbf{a} = \boldsymbol{a}_1$, $\mathbf{b} = \boldsymbol{a}_2$, $\mathbf{c} = \boldsymbol{a}_3$, $a = |\mathbf{a}|$, $b = |\mathbf{b}|$, $c = |\mathbf{c}|$, $\alpha = \angle(\mathbf{b}, \mathbf{c})$, $\beta = \angle(\mathbf{c}, \mathbf{a})$, $\gamma = \angle(\mathbf{a}, \mathbf{b})$.

In the case of orthorhomic crystal symmetry, the lattice is conventionally based on vectors along two-fold axes; if the lattice is base-centered, **c** is the unique vector perpendicular to the centered face; with the above satisfied, the metric condition $a \leq b \leq c$ is used. If the symmetry is tetragonal, **c** is the unique vector along the four-fold axis and **a** and **b** are the shorter vectors along two-fold axes. In the cubic case, the basis vectors are along the four-fold axes.

In the conventional setting of the hexagonal system, the third basis vector **c** has the direction of the unique highest symmetry axis, and **a** and **b** are the shortest vectors along two-fold axes such that $\gamma = 120°$. If the hexagonal setting is applied to crystals of the trigonal crystal system, it is additionally required that the point $[1\,2\,2]/3$ is a lattice node. Alternatively, for trigonal crystal system, if the rhombohedral lattice setting is used, **a**, **b** and **c** constitute the shortest possible edges of a rhombohedron distributed symmetrically around the rhombohedron's three-fold axis.

The conventional unit cells of Bravais lattices are shown in Fig. 1.16. These unit cells must be distinguished from the crystal asymmetric unit. It is the smallest (simply connected) part of the crystal which is sufficient to fill the whole crystal by application of symmetry operations of the crystal space group.

1.5 Indices of Directions and Planes

1.5.1 Direction and Miller Indices

Given a non-zero vector $\check{\boldsymbol{m}} = m^i \boldsymbol{a}_i$ of a three-dimensional lattice based on $\boldsymbol{a}_i$, one may consider the one-dimensional sublattice $l\check{\boldsymbol{m}}$, where l is an arbitrary integer. With the function GCD denoting the greatest common divisor of its arguments, the vector $\check{\boldsymbol{t}} = (m^i/GCD(m^1, m^2, m^3))\boldsymbol{a}_i$ also belongs to the initial three-dimensional lattice. Depending on context, the *lattice direction* is the one-dimensional sublattice $l\check{\boldsymbol{t}}$ or the line $\alpha\check{\boldsymbol{m}}$, where α is a real number. The coefficients $m^1(= u), m^2(= v)$ and $m^3(= w)$ are *direction indices*, and the direction is denoted by the symbol $[u\ v\ w]$.

Related to direction indices are Miller[21] indices of lattice planes. A lattice plane is a plane through three non-collinear lattice nodes. Clearly, all nodes in the lattice plane constitute a two-dimensional sublattice of the lattice. A non-zero vector $\check{\boldsymbol{h}}$ of the reciprocal lattice is orthogonal to a two-dimensional sublattice of the direct lattice. With $\boldsymbol{x} = x^i \boldsymbol{a}_i$ and $\check{\boldsymbol{h}} = h_i \boldsymbol{a}^i$, the equation $\boldsymbol{x} \cdot \check{\boldsymbol{h}} = x^i h_i = 0$ represents a plane perpendicular to $\check{\boldsymbol{h}}$ through the origin. For given integer components h_i, there exist solutions of $x^i h_i = 0$ with integer values of x^i. It is easy to construct two independent direct lattice vectors satisfying that equation; their integer linear combinations also solve the equation and constitute a two-dimensional sublattice of the direct lattice.

With N_s taking integer values, equations $\boldsymbol{x} \cdot \check{\boldsymbol{h}} = x^i h_i = N_s$ determine a stack of equally spaced planes perpendicular to $\check{\boldsymbol{h}}$. If h_i are relatively prime, the planes $\boldsymbol{x} \cdot \check{\boldsymbol{h}} = x^i h_i = N_s$ are lattice planes (and there are no other lattice planes perpendicular to $\check{\boldsymbol{h}}$). The relatively prime coefficients $h_1(= h)$, $h_2(= k)$ and $h_3(= l)$ are Miller indices

[21] William Hallowes Miller (1801–1880).

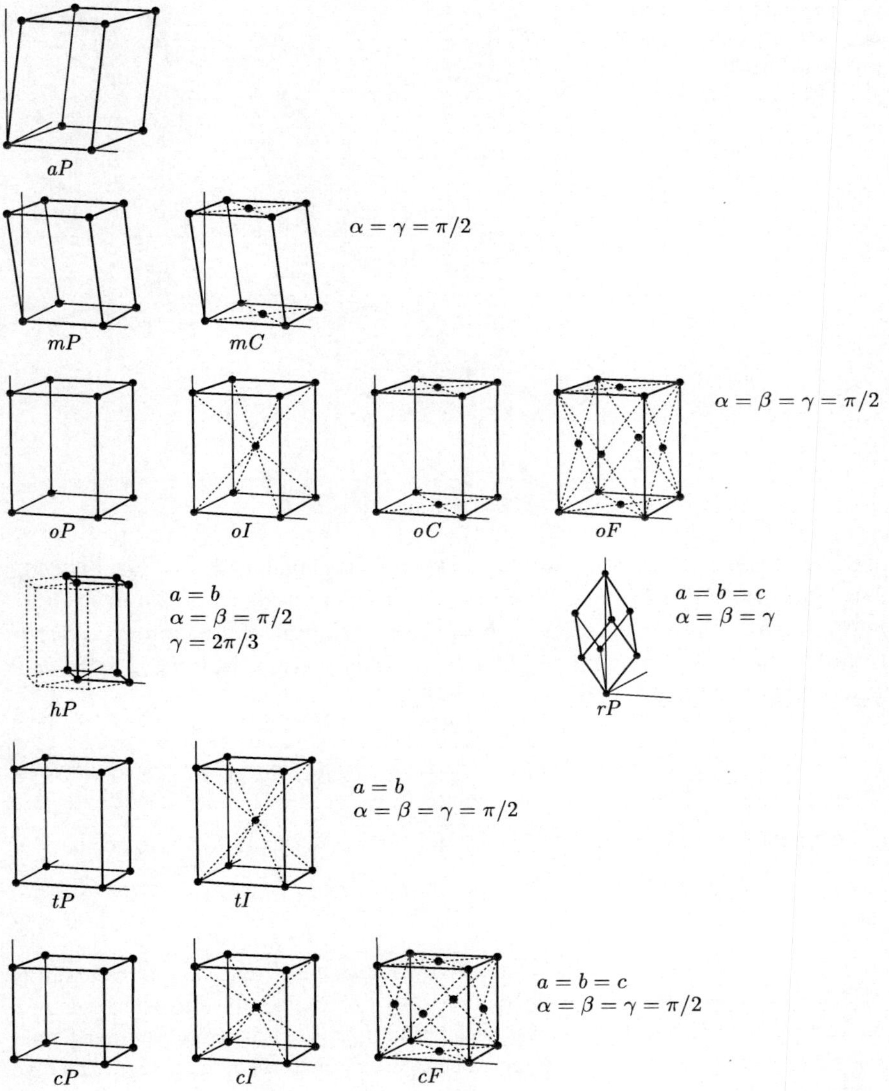

Fig. 1.16 Conventional unit cells of Bravais lattices. Orthogonal bases linked to the lattices are also sketched.

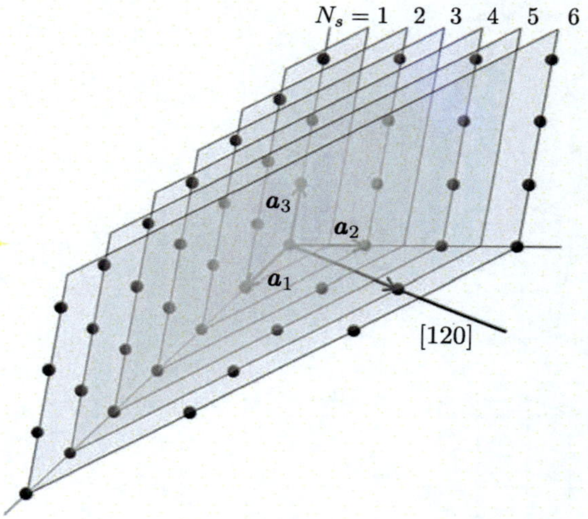

Fig. 1.17 Illustration of the stack of (120) lattice planes and the [120] lattice direction in direct space.

of the (crystallographic) planes of the stack, and the symbol $(h\,k\,l)$ is used to denote the planes. See Fig. 1.17.

Since the distance of the plane $\boldsymbol{x} \cdot \check{\boldsymbol{h}} = 1$ from the origin of the coordinate system is equal to the reciprocal of the length of $\check{\boldsymbol{h}}$, the (inter-planar) distance $d_{(h\,k\,l)}$ between nearest planes of the stack $(h\,k\,l) = (h_1\,h_2\,h_3)$ is

$$d_{(h\,k\,l)} = d_{\check{\boldsymbol{h}}} = 1/|\check{\boldsymbol{h}}| = \left(\check{\boldsymbol{h}} \cdot \check{\boldsymbol{h}}\right)^{-1/2} = (g^{ij} h_i h_j)^{-1/2} \,. \qquad (1.48)$$

If $\check{\boldsymbol{h}}$ is replaced by $n\check{\boldsymbol{h}}$, where $n > 1$ is an integer, based on (1.48), there occurs

$$d_{n\check{\boldsymbol{h}}} = d_{\check{\boldsymbol{h}}}/|n| \,.$$

Thus, the stack $\boldsymbol{x} \cdot (n\check{\boldsymbol{h}}) = nx^i h_i = N_s$ is $|n|$ times denser than that corresponding to $(h_1\,h_2\,h_3) = (h\,k\,l)$. Crystal diffraction data are frequently characterized by a parameter called resolution. The resolution limit d_{ResLim} is the smallest inter-planar spacing $d_{\check{\boldsymbol{h}}}$ contained in the data. It represents a rough measure of the power to resolve atomic positions.

It must be stressed that some planes of the denser stack do not contain any direct lattice points. As will be discussed below, plane stacks and reciprocal lattice vectors correspond to diffraction reflections. A symbolic distinction is made between the set of parallel equidistant (direct) lattice planes on one hand, and a reciprocal lattice nodes and the corresponding Bragg reflections on the other; the former are denoted by (hkl) and the latter are denoted by hkl (and sometimes referred to as Laue indices). The reflection indices can take any values including $hkl = 000$, whereas, in principle,

Miller indices given with respect to a lattice basis should be relatively prime.[22] However, these rules are sometimes violated, as some texts are clearer when the symbol (*hkl*) is used instead of hkl. Also indices of reflections are frequently referred to as Miller indices.

Summarizing, direction indices and indices of stacks of planes in the direct space are contra-variant and co-variant components of vectors, respectively. With this, the classical crystallographic calculations of an angle between two crystal directions, an angle between two crystal planes, or an angle between a plane and a direction, become trivial applications of (1) $\cos\theta = \boldsymbol{v} \cdot \boldsymbol{w}/(|\boldsymbol{v}|\,|\boldsymbol{w}|)$.

Zones and Laue Zones

A non-zero vector $\check{\boldsymbol{m}}$ of the direct lattice is orthogonal to a two-dimensional sublattice of the reciprocal lattice. With $\boldsymbol{x} = x_i \boldsymbol{a}^i$ and $\check{\boldsymbol{m}} = m^i \boldsymbol{a}_i$, the equation $\boldsymbol{x} \cdot \check{\boldsymbol{m}} = x_i m^i = 0$ represents a plane perpendicular to $\check{\boldsymbol{m}}$ through the origin. The nodes of the reciprocal lattice on this plane are said to constitute a *zone*, $\check{\boldsymbol{m}}$ is the *zone axis*, and $[u\,v\,w] = [m^1\,m^2\,m^3]$ $(\sim \check{\boldsymbol{m}})$ is the zone symbol. A plane in the direct space perpendicular to the reciprocal lattice vector $\check{\boldsymbol{h}}$ of the zone is parallel to the zone axis. The condition for a reciprocal lattice node $\check{\boldsymbol{h}} \sim (h\,k\,l) = (h_1\,h_2\,h_3)$ to be in to the zone $[u\,v\,w] = [m^1\,m^2\,m^3] \sim \check{\boldsymbol{m}}$

$$\check{\boldsymbol{h}} \cdot \check{\boldsymbol{m}} = h_i m^i = h\,u + k\,v + l\,w = 0$$

is known as the zone (or Weiss) law.

The axis of the zone with two non-parallel planes $\check{\boldsymbol{h}} \sim (h_1\,h_2\,h_3) = (h\,k\,l)$ and $\check{\boldsymbol{h}}' \sim (h'_1\,h'_2\,h'_3) = (h'\,k'\,l')$ has the indices $[u\,v\,w]/GCD(u\,v\,w)$, where $u = kl' - k'l$, $v = lh' - l'h$ and $w = hk' - h'k$. In shorter notation, $[u\,v\,w] = [m^1\,m^2\,m^3] \sim \check{\boldsymbol{m}}$, where

$$\check{\boldsymbol{m}} \propto \check{\boldsymbol{h}} \times \check{\boldsymbol{h}}' \quad \text{or} \quad m^i \propto \varepsilon^{ijk} h_j h'_k \,. \tag{1.49}$$

If a plane $\check{\boldsymbol{h}}$ belongs to two distinct zones $\check{\boldsymbol{m}}$ and $\check{\boldsymbol{m}}'$ then

$$\check{\boldsymbol{h}} \propto \check{\boldsymbol{m}} \times \check{\boldsymbol{m}}' \quad \text{or} \quad h_i \propto \varepsilon_{ijk} m^j m'^k \,.$$

These relationships follow directly from (1.13).

Planes of the same zone are called tautozonal. Normals to three tautozonal planes are linearly dependent. Thus, using (1.49), one can see that, with the indices of the i-th ($i = 1, 2, 3$) plane denoted by $(h_{i1}\,h_{i2}\,h_{i3})$, these three planes are tautozonal if and only if the determinant of the matrix with the entries h_{ij} equals zero.

---

[22] The situation is additionally complicated by the use of conventional non-primitive cells. The indices of some lattice planes in conventional bases linked to such cells are not relatively prime. For instance, it is easy to see, that in the conventional basis of face- or body centered cubic lattice, indices (200) represent lattice planes.

The zone can be referred to as the zeroth-order zone. The n-th order zone is the n-th closest 'plane' of reciprocal lattice points and parallel to the zeroth-order zone, i.e., with $\check{\boldsymbol{m}}$ being the axis of a zone, the n-th order zone comprises of reciprocal lattice vectors $\check{\boldsymbol{h}}$ such that $\check{\boldsymbol{h}} \cdot \check{\boldsymbol{m}} = n$. The reason for having this construction comes from transmission electron microscopy. In certain conditions (short camera length, optical axis parallel to low index zone axis) reflections from the zones of low orders are visible on electron diffraction patterns in the form of concentric rings and a net of spots in the center. The net is referred to as zeroth-order Laue zone, and the rings are referred to as higher-order Laue zones (HOLZ). In microscopists' nomenclature, the order of a Laue zone is the consecutive number of the *visible* ring which may differ from the integer n corresponding to this ring.

1.5.2 Generalized Indices of Directions and Planes

The algebraic approach with three-axis indices is suitable for computations for arbitrary lattices. However, for human analysis of the hexagonal and rhombohedral lattices, it is more convenient to use 'four-axis indices' reflecting their symmetry. The four-index system (with Bravais–Miller [33] plane indices and Weber [34] direction indices) has the advantage of an inbuilt symmetry check. The Bravais–Miller and Weber indexing is a particular case of a more general indexing scheme involving frames [2].

Lattice and a Frame

The formalism of Sect. 1.1.2 relying on N vectors $\boldsymbol{a}_\mu$ was introduced without referring to a lattice. Not every set of vectors $\boldsymbol{a}_\mu$ naturally induces a lattice; for some sets, integer combinations of vectors $\boldsymbol{a}_\mu$ can be arbitrarily close to each other. Even if the latter is not the case, the relationship between the vectors $\boldsymbol{a}_\mu$ and the lattice is not unique and needs to be explicitly specified.

Assuming a *canonical* case, in which all integer combinations $\tilde{k}^\mu \boldsymbol{a}_\mu$ of vectors $\boldsymbol{a}_\mu$ constitute a lattice, there is a question about requirements which must be satisfied by $\boldsymbol{a}_\mu$ for this lattice to be based on certain three linearly independent vectors $\boldsymbol{a}_i$? A tilde is added to indicate that $\tilde{k}^\mu$ generally do not satisfy the conditions (1.26). Each $\boldsymbol{a}_\mu$ must be an integer combination of vectors $\boldsymbol{a}_i$, i.e., $\boldsymbol{a}_\mu = T_\mu{}^i \boldsymbol{a}_i$, with $T_\mu{}^i = \boldsymbol{a}_\mu \cdot \boldsymbol{a}^i$ being integers. This implies that an arbitrary integer combination of the vectors $\boldsymbol{a}_\mu$ is an integer combination of the vectors $\boldsymbol{a}_i$. On the other hand, also the vector $k^i \boldsymbol{a}_i$ with integer k^i, is expressible as an integer combination of the vectors $\boldsymbol{a}_\mu$. Thus, the system of three linear Diophantine equations $T_\mu{}^i \tilde{x}^\mu = k^i$ must be solvable with respect to integer $\tilde{x}^\mu$ for each triplet of integer k^i. This means that all three elementary divisors of the integer matrix $T_\mu{}^i$ must be equal to 1. Moreover, let $T^\mu_{\ i}$ be the transposed pseudoinverse of $T_\mu{}^i$. With integer $T_\mu{}^i$, the matrix $T^\mu_{\ i}$ is rational. With the full rank

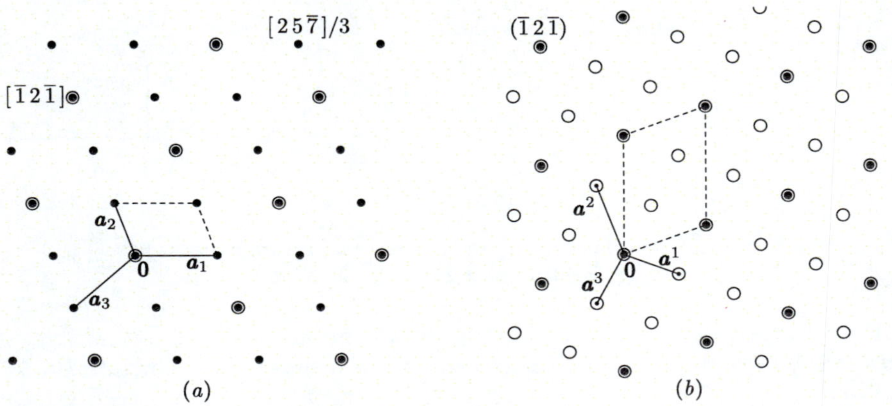

Fig. 1.18 Schematic two-dimensional illustration of canonical configuration for the vectors $\boldsymbol{a}_\mu$ shown in Fig. 1.1. In both (*a*) and (*b*), the actual lattices are depicted using disks, and their primitive cells are sketched using dashed lines. (*a*) Direct lattice comprised of all integer combinations of $\boldsymbol{a}_\mu$. The particular nodes $k^\mu \boldsymbol{a}_\mu$ with integer k^μ satisfying $k^\mu = g^\mu_{\ \nu} k^\nu$ are marked by circles. (*b*) Reciprocal lattice built of the combinations $l_\mu \boldsymbol{a}^\mu$ with integer l_μ satisfying $l_\mu = g_\mu^{\ \nu} l_\nu$. Circles mark points corresponding to all integer combinations of $\boldsymbol{a}^\mu$. (In both (*a*) and (*b*), the circles indicate nodes of lattices constructed using the 'dual-to-canonical' scheme.)

of $T_\mu^{\ i}$, one has $T_\mu^{\ i} T^\mu_{\ j} = \delta^i_{\ j}$, and hence, $\boldsymbol{a}_i = \boldsymbol{a}_\mu T^\mu_{\ i}$. Thus, some lattice vectors with integer components k^i in the basis $\boldsymbol{a}_i$ have rational (i.e., not necessarily integer) components k^μ $(= g^\mu_{\ \nu} k^\nu)$ when expressed as $k^\mu \boldsymbol{a}_\mu$. By definition of the canonical case, these vectors are linear combinations of $\boldsymbol{a}_\mu$ with integer coefficients, but such coefficients may not satisfy the conditions (1.26). Since $g_\mu^{\ \nu} = \boldsymbol{a}_\mu \cdot \boldsymbol{a}^\nu = T_\mu^{\ i} T^\nu_{\ i}$, for $\boldsymbol{a}_\mu$ to determine a lattice in the above sense, the matrix $g_\mu^{\ \nu}$ must be rational.

Other consequences of assuming the canonical scheme concern the $\boldsymbol{a}^i$-based reciprocal lattice. It is easy to see that the vector $\boldsymbol{a}^\mu = g^{\mu\nu} \boldsymbol{a}_\nu$ may not necessarily be a vector of the reciprocal lattice. (See Fig. 1.18.) The key observation is that the reciprocal lattice based on vectors $\boldsymbol{a}^i$ is identical with the lattice $k_\mu \boldsymbol{a}^\mu$, where k_μ are integer vector components satisfying $k_\mu = g_\mu^{\ \nu} k_\nu$. More explicitly, the following occurs:
– with integer k_i, the vector $k_i \boldsymbol{a}^i$ of the reciprocal lattice can be represented as $k_\mu \boldsymbol{a}^\mu$ with integer components k_μ satisfying $k_\mu = g_\mu^{\ \nu} k_\nu$,
– if k_μ are integers satisfying $k_\mu = g_\mu^{\ \nu} k_\nu$ then $k_\mu \boldsymbol{a}^\mu$ is a vector of the lattice based on $\boldsymbol{a}^i$, i.e., it is equal to $k_i\, \boldsymbol{a}^i$ with integer k_i.
The first part follows immediately. The second one can be proved using the Smith normal form of the matrix $T_\mu^{\ i}$, by taking into account that its elementary divisors are all equal to 1 [2].

Summarizing, in the canonical scheme, the direct lattice based on the vectors $\boldsymbol{a}_i$ is identical to the lattice constructed of *all* integer combinations of $\boldsymbol{a}_\mu$, and the reciprocal lattice based on the vectors $\boldsymbol{a}^i$ is reproduced by linear combinations of $\boldsymbol{a}^\mu$ with integer coefficients satisfying the conditions (1.26). The above can be expressed in a slightly oversimplified form as

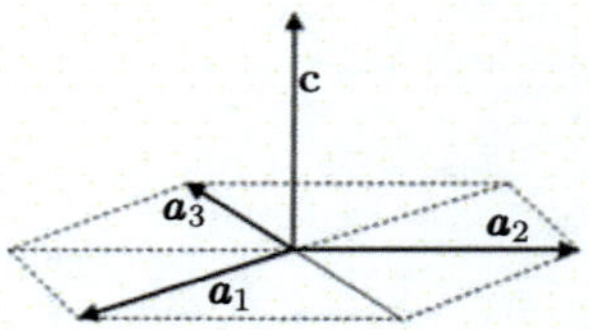

Fig. 1.19 Symmetric 'basis' of hexagonal lattice.

$$\text{canonical scheme} : \begin{cases} \text{direct lattice} & : \tilde{k}^\mu \boldsymbol{a}_\mu, \text{ where } \tilde{k}^\mu \text{ are arbitrary integers },\\ \text{reciprocal lattice} & : k_\mu \boldsymbol{a}^\mu, \text{ where } k_\mu \text{ are integers and } k_\mu = g_\mu^{\ \nu} k_\nu \ . \end{cases}$$

Direct lattice nodes can also be identified as $k^\mu \boldsymbol{a}_\mu$ with unique sets of *rational* indices k^μ satisfying $k^\mu = g^\mu_{\ \nu} k^\nu$. The integers proportional to the unique coefficients k^μ can be seen as generalized indices of lattice direction, and the coefficients k_μ can be seen as generalized indices of lattice plane.

⊡ Clearly, one may consider another natural relationship between the set of vectors $\boldsymbol{a}_\mu$ and a lattice which is in a sense dual to the canonical scheme, with the roles of lattices reversed: The direct lattice based on vectors $\boldsymbol{a}_i$ is identical to the linear combinations of $\boldsymbol{a}_\mu$ with integer coefficients satisfying the conditions (1.26), and the reciprocal lattice based on vectors $\boldsymbol{a}^i$ overlaps with linear combinations of $\boldsymbol{a}^\mu$ with arbitrary integer coefficients. It is evident that if $N = 3$, there is no difference between the canonical scheme and its dual.

With the generalized indexing scheme, the crystallographic formulas for inter-planar distances or angles (between two directions, two crystal planes, or between a plane and a direction) are analogous to those in the three-index case except that the metrics now are $g_{\mu\nu}$ and $g^{\mu\nu}$.

In the above approach, all vectors are in the three-dimensional physical space. The described formalism can be compared to that with the physical space embedded in an abstract space of higher dimension N (with $g_\mu^{\ \nu}$ and $g^\mu_{\ \nu}$ as projection matrices). Such a description of Bravais–Miller and Weber indexing was given in [35].[23]

The choice of reference axes in the hexagonal lattice below is imposed by the lattice symmetry, but one may have other motives for having redundant axes. They can be used to facilitate handling of arbitrary symmetries, e.g., symmetries of processes. The formalism is very flexible: the restrictions put on the choice of the vectors $\boldsymbol{a}_\mu$ are weak, the array of these vectors can be highly redundant, and the vectors can fit various geometries [3].

Symmetric Frame of Hexagonal Lattice

The crystallographic lattice of our interest is based on vectors $\boldsymbol{a}_1, \boldsymbol{a}_2$ and $\mathbf{c}$ satisfying the conditions

[23] A similar method is used for depiction of quasicrystals with three-dimensional quasilattices considered to be projections of subsets of lattices of higher dimensions; see Chap. 13.

$$\boldsymbol{a}_1 \cdot \boldsymbol{a}_1 = a^2 = \boldsymbol{a}_2 \cdot \boldsymbol{a}_2\,, \quad c^2 = \mathbf{c} \cdot \mathbf{c}\,, \quad \boldsymbol{a}_i \cdot \mathbf{c} = 0\,, \quad \boldsymbol{a}_1 \cdot \boldsymbol{a}_2 = -a^2/2\,. \qquad (1.50)$$

The basis is shown schematically in Fig. 1.19. For simplicity, we consider the lattice scaled by $1/a$ and the four-vector frame $\boldsymbol{a}_\mu$ ($\mu = 1, 2, 3, 4$) consisting of unit vectors $\boldsymbol{a}_1, \boldsymbol{a}_2$ (such that $\boldsymbol{a}_1 \cdot \boldsymbol{a}_2 = -1/2$),

$$\boldsymbol{a}_3 = -\boldsymbol{a}_1 - \boldsymbol{a}_2 \quad \text{and} \quad \boldsymbol{a}_4 \text{ such that } |\boldsymbol{a}_4| = c/a,\ \ \boldsymbol{a}_4 \cdot \boldsymbol{a}_\mu = 0 \text{ for } \mu = 1, 2, 3\,.$$

With the Cartesian basis $\boldsymbol{e}_i$ such that $\boldsymbol{e}_1 = \boldsymbol{a}_1, \boldsymbol{e}_2 \cdot \boldsymbol{a}_2 > 0$ and $\boldsymbol{e}_3$ along the principal axis $\boldsymbol{a}_4 = \mathbf{c}$, one has

$$\left[\mathcal{T}_\mu{}^j\right] = \begin{bmatrix} A/2 & 0 \\ 0 & c/a \end{bmatrix}, \quad \left[\mathcal{T}^\mu{}_j\right] = \begin{bmatrix} A/3 & 0 \\ 0 & a/c \end{bmatrix},$$

and

$$\left[g_{\mu\nu}\right] = \begin{bmatrix} B/2 & 0 \\ 0 & (c/a)^2 \end{bmatrix}, \quad \left[g_\mu{}^\nu\right] = \begin{bmatrix} B/3 & 0 \\ 0 & 1 \end{bmatrix}, \quad \left[g^{\mu\nu}\right] = \begin{bmatrix} 2B/9 & 0 \\ 0 & (a/c)^2 \end{bmatrix},$$

$$\text{where } A = \begin{bmatrix} 2 & 0 \\ -1 & \sqrt{3} \\ -1 & -\sqrt{3} \end{bmatrix} \text{ and } B = \begin{bmatrix} 2 & -1 & -1 \\ -1 & 2 & -1 \\ -1 & -1 & 2 \end{bmatrix}.$$

Based on the above $g_{\mu\nu}$, the vectors of the frame dual to $\boldsymbol{a}_\mu$ are

$$\boldsymbol{a}^\mu = (2/3)\,\boldsymbol{a}_\mu \text{ for } \mu = 1, 2, 3 \quad \text{and} \quad \boldsymbol{a}^4 = (a/c)^2\,\boldsymbol{a}_4\,.$$

The explicit form of the conditions (1.26) for $\boldsymbol{v} = v_\mu \boldsymbol{a}^\mu = v^\mu \boldsymbol{a}_\mu$ is $v^1 + v^2 + v^3 = 0 = v_1 + v_2 + v_3$.

Let the hexagonal lattice be linked to the vectors $\boldsymbol{a}_\mu$ via the canonical scheme. Weber[24] indices $[u\ v\ t\ w]$ of lattice directions [34] and Bravais–Miller indices[25] $(h\ k\ i\ l)$ of lattice planes [33] are, respectively, contra- and co-covariant components of vectors.

The direct lattice based on three vectors $\boldsymbol{a}_1, \boldsymbol{a}_2$ and $\boldsymbol{a}_4$, i.e., the integer combinations $k\boldsymbol{a}_1 + l\boldsymbol{a}_2 + m\boldsymbol{a}_4$, is represented by the combinations $k^\mu \boldsymbol{a}_\mu$ of all four vectors $\boldsymbol{a}_\mu$ with the coefficients

$$\left[k^1\ k^2\ k^3\ k^4\right] = [u\ v\ t\ w]\,/3 = [2k - l\ \ 2l - k\ \ -k - l\ \ 3m]\,/3\,,$$

which satisfy $k^\mu = g^\mu{}_\nu k^\nu$ (i.e., $u + v + t = 0$), and k^μ are multiples of $1/3$.

---

[24] Leonhard Weber (1883–1968).

[25] This term is used in [*ITC*]. In the field of materials science, they are usually called Miller–Bravais indices.

To get the complete reciprocal lattice using four vectors $\boldsymbol{a}^{\mu}$, one needs all nodes $k_{\mu}\boldsymbol{a}^{\mu}$, where

$$(k_1\, k_2\, k_3\, k_4) = (h\, k\, i\, l)$$

are integers and satisfy $k_{\mu} = g_{\mu}^{\ \nu} k_{\nu}$ or $h + k + i = 0$. With the three–vector base $\boldsymbol{a}_1$, $\boldsymbol{a}_2$ and $\boldsymbol{a}_4$, the conventional Miller indices of the same plane would be $(h\, k\, l)$. These indices are obtained by taking the scalar products of $\boldsymbol{h} = k_{\mu}\boldsymbol{a}^{\mu}$ with the vectors $\boldsymbol{a}_1$, $\boldsymbol{a}_2$ and $\boldsymbol{a}_4$.

1.6 Families of Equivalent Stacks of Planes

As was already mentioned in Sect. 1.5.1, stacks of planes in direct space correspond to diffraction reflections. Equivalence of stacks of planes due to a crystal symmetry implies the equality of intensities of the respective reflections. Therefore, families of equivalent stacks of planes play a role in determination of crystal symmetries. Stacks contributing to diffraction correspond to reciprocal lattice vectors and are identified by their covariant coordinates—the indices. Given indices of a stack, it is important to be able to find the indices of all stacks symmetrically equivalent to it.

Let the indices of the original stack be $(h\, k\, l) = (h_1\, h_2\, h_3)$. The corresponding reciprocal lattice vector is $\check{\boldsymbol{h}} = h_i \boldsymbol{a}^i$, where $\boldsymbol{a}^i$ $(i = 1, 2, 3)$ is the basis of the reciprocal lattice. With the symmetry S, the stack of planes represented by the vector $\check{\boldsymbol{h}}$ and the indices h_i is equivalent to the stack represented by the vector $\check{\boldsymbol{h}}' = S\check{\boldsymbol{h}}$ and the indices $h'_j = \check{\boldsymbol{h}}' \cdot \boldsymbol{a}_j = (S\check{\boldsymbol{h}}) \cdot \boldsymbol{a}_j = h_i\, (S\boldsymbol{a}^i) \cdot \boldsymbol{a}_j$, where the (direct lattice) basis $\boldsymbol{a}_i$ is dual to $\boldsymbol{a}^i$. Similar relationships are applicable if a frame is used instead of a basis.

If the used basis (or frame) reflects the symmetry of the crystal lattice, the relationships between the indices of equivalent stacks are simple and easy to deduce. As an example, the indices of stacks of planes equivalent with respect to the point groups of the trigonal crystal system are listed in Table 1.11. The family of all stacks of planes equivalent to $(h\, k\, l)$ with respect to symmetry operations of a given point group is denoted by $\{h\, k\, l\}$.

Analogous considerations apply to vectors of the direct space representing crystal directions. The set of all direct space directions equivalent to $[u\, v\, w]$ with respect to elements of a point group is denoted by $\langle u\, v\, w \rangle$.

1.7 Comparison of Lattices and Bravais-class Determination

It is one of the basic crystallographic problems to verify whether two given lattice bases represent the same lattice apart from orientation (cf. Sect. 1.2.1). More

Table 1.11 Indices of symmetrically equivalent stacks of planes for the groups C_3, C_{3v} and D_3 in rhombohedral basis (left) and in hexagonal frame with $i = -h - k$ (right). By adding inversion (i.e., each $(h\,k\,l)$ or $(h\,k\,i\,l)$ is accompanied by $(\bar{h}\,\bar{k}\,\bar{l})$ or $(\bar{h}\,\bar{k}\,\bar{i}\,\bar{l})$, respectively) to C_3 and D_3, one obtains equivalences for the groups S_6 and D_{3d}, respectively.

C_3	$(h\,k\,l)$ $(l\,h\,k)$ $(k\,l\,h)$
C_{3v}	$(h\,k\,l)$ $(l\,h\,k)$ $(k\,l\,h)$
	$(h\,l\,k)$ $(l\,k\,h)$ $(k\,h\,l)$
D_3	$(h\,k\,l)$ $(l\,h\,k)$ $(k\,l\,h)$
	$(\bar{h}\,\bar{l}\,\bar{k})$ $(\bar{l}\,\bar{k}\,\bar{h})$ $(\bar{k}\,\bar{h}\,\bar{l})$

C_3	$(h\,k\,i\,l)$ $(i\,h\,k\,l)$ $(k\,i\,h\,l)$
C_{3v}	$(h\,k\,i\,l$ $(i\,h\,k\,l)$ $(k\,i\,h\,l)$
	$(h\,i\,k\,l)$ $(i\,k\,h\,l)$ $(k\,h\,i\,l)$
D_3	$(h\,k\,i\,l)$ $(i\,h\,k\,l)$ $(k\,i\,h\,l)$
	$(\bar{h}\,\bar{i}\,\bar{k}\,\bar{l})$ $(\bar{i}\,\bar{k}\,\bar{h}\,\bar{l})$ $(\bar{k}\,\bar{h}\,\bar{i}\,\bar{l})$

generally, the problem concerns comparison and identification of lattices. Are two given lattices identical or different? This question needs to be answered to classify lattices based on comparison to hypothetical symmetric models. The usual goal is to determine the symmetry of the lattice, i.e., to ascribe a Bravais type to the lattice, and to select an appropriate conventional basis.

The determination of lattice symmetry, Bravais class and the conventional basis is straightforward in ideal error-free cases, but experimental data are affected by uncertainties, and the latter complicate lattice classification. In practice, the goal is to indicate the 'best' lattice which fits the data with only a small discrepancy.

Formally, an exactly known lattice can be ascribed to a Bravais class by reduction of the basis [36, 37]. The first step is to calculate the Niggli-reduced basis [14, 15]; see Sect. 1.2.1. Then, one needs to identify the Bravais class containing the obtained Niggli 'character'. This can be done by inspecting the tables in [38] or [39]. These tables also give the transformation from Niggli-reduced bases to standard crystallographic settings. (For a description of an implementation, see [40]. A similar method of ascribing a lattice to a Bravais class can be based on Delaunay classification [31, 41].) Theoretically, this algorithm can be carried out exactly using arbitrary precision integer arithmetic but this approach is of little use with experimental data. In the presence of experimental errors, the reduction algorithm is not stable in the sense that a small change of a parameter may lead to an assignment to a different class. With uncertain data, (in)equalities of the reduction algorithm must be replaced by softer conditions with tolerances allowing for the uncertainties and rounding errors. In this situation, the reduction process is erratic. A minor distortion of a basis may lead to a different Bravais class.

In view of the above, other methods of determination of the Bravais class in the presence of data uncertainties have been devised. Some of them are briefly described below. For an approach combining a number of basis reduction schemes, see [42].

1.7.1 Lattice Symmetry from Distribution of Two-fold Axes

The instability of the reduction affects only the angles between basis vectors; depending on decisions taken at the 'special' conditions for the reduction (see Sect. 1.2.1), different angles are obtained, but the magnitudes of the vectors of the Buerger-reduced basis are stable. Therefore, comparison of lattices via Buerger reduction and matching the vector magnitudes (and some other parameters) have been proposed [43, 44]. Clegg [45] suggested to list low-integer linear combinations of the vectors of the Niggli-reduced basis. Their magnitudes and scalar products are then used to identify the Bravais class; see also [43, 46].

More efficient, i.e., tolerant of relatively large inaccuracies in the input parameters, is the method of Le Page [47] (*CREDUC*, also implemented as *LEPAGE* [48]; see also [49, 50]). In its simplest version, the conventional unit cell and the lattice symmetry are obtained from the spatial distribution of two-fold axes.

⊡ The spatial distribution of two-fold axes unequivocally determines the geometric holohedry and the conventional cell. To determine the cell, one needs to identify the axes from error affected cell parameters. Le Page [47] observed that with Buerger-reduced basis
(a) the indices of the shortest reciprocal lattice vectors along two-fold symmetry axes have values between -2 and 2, and
(b) a direction of the shortest direct lattice vector $\check{\boldsymbol{u}}$ coincides with a two-fold axis if and only if there exists a reciprocal lattice vector $\check{\boldsymbol{h}}$ parallel to $\check{\boldsymbol{u}}$ and satisfying $\check{\boldsymbol{u}} \cdot \check{\boldsymbol{h}} \approx 1$ or 2.
Thus, the first step of Le Page's approach is the Buerger-reduction of the experimental basis. Then, following the observation (a), a direct lattice vector $\check{\boldsymbol{u}}$ is generated as an integer combination of the basis vectors with absolute values of the coefficients not exceeding 2. The vector $\check{\boldsymbol{u}}$ is considered to be along two-fold axis if it approximately satisfies the condition (b), i.e., if there exists a reciprocal lattice vector $\check{\boldsymbol{h}}$ such that $\check{\boldsymbol{u}} \cdot \check{\boldsymbol{h}} = 1$ or 2, and the angle between $\check{\boldsymbol{u}}$ and $\check{\boldsymbol{h}}$ is below a certain limit. The spatial distribution of the vectors along the two-fold axes is then compared to the distribution of axes in the seven lattice systems; the conventional cell is determined based on the number of the identified two-fold axes and the similarity of angles between their directions. With known conventional unit cell, the data are checked for centering and the Bravais class is found.

The search for lattice symmetry is usually a part of general computer programs determining symmetries of crystal structures. See, e.g. [51] and references therein.

1.7.2 Method Based on Metric Tensor

Formally, the problem of comparing lattices can be expressed in the following way: Two bases $\boldsymbol{a}_i$ and $\boldsymbol{a}'_i$ represent the same lattice if the vectors $\boldsymbol{a}'_i$ after an unknown rotation R can be expressed as a $T_i{}^j \boldsymbol{a}_j$, where $T_i{}^j$ are integers. With $R\boldsymbol{a}'_i$ denoting the rotated $\boldsymbol{a}'_i$, one has

$$R\,\boldsymbol{a}'_i = T_i{}^j \boldsymbol{a}_j \ .$$

Since both bases span primitive cells, the matrix T must be unimodular. Moreover, since rotations preserve scalar products, one has $g'_{ij} = \boldsymbol{a}'_i \cdot \boldsymbol{a}'_j = (R\boldsymbol{a}'_i) \cdot (R\boldsymbol{a}'_j) =$

$T_i{}^k T_j{}^l \boldsymbol{a}_k \cdot \boldsymbol{a}_l = T_i{}^k T_j{}^l g_{kl}$ or briefly,

$$g' = TgT^T \; ; \tag{1.51}$$

cf. (1.4). The conclusion follows that two lattices L and L' with metrics g and g', respectively, are similar if there exists an integer unimodular matrix T such that g' is close to TgT^T. The distance can be quantified by the squared Frobenius norm $\|\Delta\|^2 = \mathrm{tr}(\Delta^T\Delta)$ of the difference $\Delta = g' - TgT^T$. Therefore, T can be determined by the minimization of $\|g' - TgT^T\|^2$; cf. [52]. Other objective functions have also been tried, e.g, $\sum_{ij} |\Delta_{ij}|$ in [53].

The relationship (1.51) can be used for direct determination of lattice symmetry. The set of all integer matrices T satisfying $TgT^T = g$ is a representation of the lattice point group in the basis having the metric g. I.e., it is a representation of one of the holohedries. Since any lattice is symmetric with respect to inversion, one can limit the considerations to matrices satisfying $\det(T) = +1$, which constitute subgroups of proper rotations (hemihedries) of particular holohedral point groups. According to [53], for the Niggli-reduced basis, the entries of T have absolute values not exceeding 1. With the additional restrictions $\det(T) = +1$ (hemihedries) and $\mathrm{tr}(T) \geq -1$ (equivalence to a rotation and invariance of trace), there are only 3048 distinct matrices of this kind. Since these matrices are elements of point groups, their product must also belong to the set with the entries -1, 0 and 1 [54]. Thus, a matrix T is allowed if also T^n ($n = 2, 3, 4, 5$) have the entries -1, 0 and 1. This reduces the number of the matrices to 480. The hemihedry of the lattice can be obtained by testing all these matrices whether they satisfy $TgT^T \approx g$. The resulting group determines the lattice system; if the order of the group is 1, 2, 4, 6, 8, 12 or 24, the corresponding system is a, m, o, r, t, h or c, respectively. Simple extra tests of centering are needed to get the Bravais class within the system. Again, a tolerance must be specified to examine the approximate equality between TgT^T and g. For more details, see [53–56].

⊡ By calculating traces of the above described 480 matrices, one can easily check that, besides the identity matrix, twelve of the matrices represent rotations by $\pi/3$, 126 represent rotations by $\pi/2$, 260 represent rotations by $2\pi/3$ and 81 represent rotations by π. The last class, i.e., the class of half-turns, is important in the context of the algorithm of Le Page, as instead of looking for two-fold axes, one can directly test the 81 matrices. A matrix representing a half-turn has the form $T = M$ or $T = M^T$, where M is among

$$\begin{bmatrix} 1 & \iota_1 & \iota_2 \\ 0 & -1 & 0 \\ 0 & 0 & -1 \end{bmatrix}, \quad \begin{bmatrix} -1 & 0 & 0 \\ \iota_1 & 1 & \iota_2 \\ 0 & 0 & -1 \end{bmatrix}, \quad \begin{bmatrix} -1 & 0 & 0 \\ 0 & -1 & 0 \\ \iota_1 & \iota_2 & 1 \end{bmatrix},$$

$$\begin{bmatrix} -1 & \iota_1 & \iota_1\iota \\ 0 & 0 & \iota \\ 0 & \iota & 0 \end{bmatrix}, \quad \begin{bmatrix} 0 & 0 & \iota \\ \iota_1 & -1 & \iota_1\iota \\ \iota & 0 & 0 \end{bmatrix}, \quad \begin{bmatrix} 0 & \iota & 0 \\ \iota & 0 & 0 \\ \iota_1 & \iota_1\iota & -1 \end{bmatrix},$$

with ι_1 and ι_2 equal -1, 0 or $+1$, and $\iota = \pm 1$. After taking into account that some of the matrices are symmetric and some appear twice, one obtains 81 distinct matrices.

The ε-Similarity of Approximate Bases of a Lattice

Closely related to the above considerations is the notion of ε-similarity of two approximate bases of the same lattice. Let $\boldsymbol{a}'^i$ denote vectors of the basis reciprocal to $\boldsymbol{a}'_i$. Roughly, bases $\boldsymbol{a}_i$ and $\boldsymbol{a}'_i$ are similar if the matrix $T_i^{\ j} = \boldsymbol{a}'_i \cdot \boldsymbol{a}^j$ transforming $\boldsymbol{a}_i$ to $\boldsymbol{a}'_i$ ($T_i^{\ j}\boldsymbol{a}_j = \boldsymbol{a}'_i$) is nearly uni-modular (with determinant close to ± 1) and nearly integer. Formally, bases $\boldsymbol{a}_i$ and $\boldsymbol{a}'_i$ are ε-similar if $|| \det(T)| - 1| < \varepsilon$ and $\sum_{i,j} \left| T_i^{\ j} - \lfloor T_i^{\ j} \rceil \right| < 3\varepsilon$.

1.8 Crystal Orientation

Orientations of crystals or crystallites in polycrystalline materials are given with respect to external, usually Cartesian, reference systems. In the analysis of crystallographic textures, the external coordinate system is attached to the polycrystalline sample [57]. In diffraction experiments, since the sample may undergo various rotations, it is usually easier to consider the orientation of the crystal with respect to the experimental equipment (and, if needed, use the orientation of the sample with respect to the equipment to get the orientation of the crystal in the sample). For instance, if a diffraction pattern is acquired by transmission electron microscopy, it is usually assumed that the external Cartesian coordinate system has the '3' axis along the optical axis of the microscope, '1' is along the tilt axis, and '2' is determined by the right-hand rule; see Sect. 4.2. Knowing the orientation of the crystal in such a 'microscope' coordinate system, the orientation in the sample can be calculated based on the known tilt angles. Details on such calculations are described in Chap. 4.

Besides the basis of the crystal lattice, one attaches an auxiliary *orthonormal* basis to the crystal, and the matrix $\mathcal{T}$ links the two *crystal* bases. With this, the orientation of the crystal is specified by the relative rotation between the orthonormal basis of the crystal and the external basis (Fig. 1.20). Since both bases are orthonormal, the rotation matrices are orthogonal. One can take advantage of this and use one of numerous ways of parameterizing such rotations (Euler angles, Rodrigues parameters, quaternions, et cetera) [27].

Numerous conventions for the matrix $\mathcal{T}$ relating the lattice basis to the orthonormal basis of the crystal have been proposed.[26] In the field of four-circle diffractometry, most popular is the one introduced in [59] with $\mathcal{T} = \mathbf{B}^{-1}$, where

$$\mathbf{B} = \begin{bmatrix} a^* & b^* \cos(\gamma^*) & c^* \cos(\beta^*) \\ 0 & b^* \sin(\gamma^*) & -c^* \sin(\beta^*) \cos(\alpha) \\ 0 & 0 & 1/c \end{bmatrix},$$

[26] The first one seems to be described in 1861 by C. Neumann [58].

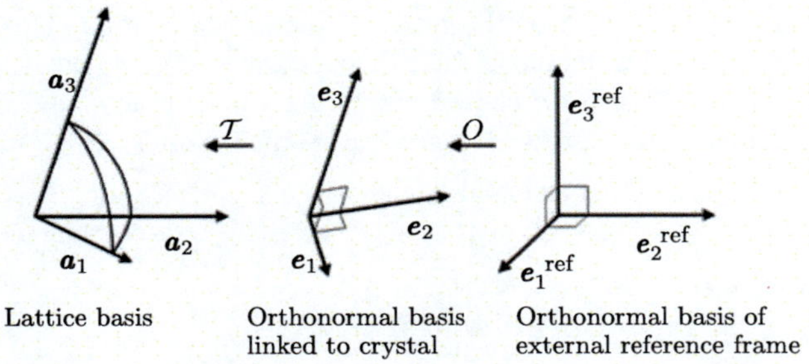

Fig. 1.20 Schematic illustration of the relationship between crystal and sample coordinate systems. One has $\boldsymbol{a}_i = \mathcal{T}_i^{\ j} \boldsymbol{e}_j$ and $\boldsymbol{e}_i = O_{ij} \boldsymbol{e}_j^{\text{ref}}$, where O is an orthogonal matrix (i.e., $O_{ij} O_{ik} = \delta_{ik}$).

and the symbols on the right-hand side denote conventional parameters of the direct and reciprocal lattices. With this convention, the direction of the vector $\boldsymbol{e}^3 = \boldsymbol{e}_3$ coincides with the direction of $\boldsymbol{a}_3$ basis vector of the direct lattice, $\boldsymbol{e}_1 = \boldsymbol{e}^1$ is in the plane containing $\boldsymbol{a}_1$ and $\boldsymbol{a}_2$ (i.e., the direction of $\boldsymbol{e}_1 = \boldsymbol{e}^1$ coincides with the direction of $\boldsymbol{a}^1 = \mathbf{a}^*$ of the reciprocal lattice), and the direction of $\boldsymbol{e}_2 = \boldsymbol{e}^2$ is determined by the right-handedness of the system; briefly, one has

$$\boldsymbol{e}_1 = \boldsymbol{a}^1/|\boldsymbol{a}^1| \,, \quad \boldsymbol{e}_3 = \boldsymbol{a}_3/|\boldsymbol{a}_3| \,, \quad \boldsymbol{e}_2 = \boldsymbol{e}_3 \times \boldsymbol{e}_1 \,.$$

In the practice of crystallographic computing, it is frequently more convenient to put the conventions aside, and to rely on a user-specified $\mathcal{T}$ matrix.

Finally, it must be stressed that for a symmetric crystal, the parameters of its orientation are not unique. This has an impact on many aspects of analysis of crystal orientations. For a detailed description of quantitative methods of such analysis, the reader is referred to [57] and [27].

1.9 Homogeneous Strain

The final chapter of this book concerns diffraction-based refinement of lattice parameters and determination of elastic strains in crystals. A general deformation comprises changes in shape and size (strains or stretches) and rigid body motions (rotations and translations). When a body is deformed, distances between points in the undeformed and deformed configurations are identical if only a rigid-body displacement has occurred. Therefore, strains are measured by the difference between metrics in the deformed and undeformed configurations.

With the interest in local analysis, only homogeneous deformation of the crystal lattice needs to be considered. Two equivalent descriptions of homogeneous deformation are given below. The first one is natural from the viewpoint of lattice specification. Elastic strain in a crystal causes changes of lattice parameters, and deformation is related to a transformation of the basis of the crystal lattice. This means that deformation is related to a change of the reference frame. On the other hand, in continuum mechanics, deformation-caused displacements of points, and thus changes of associated vectors, are expressed in given reference frames.

1.9.1 Change of Lattice Metric

The elastic strain of a crystal corresponds to a change of the parameters of its lattice. More generally, with a crystal deformation, one has a transformation of the basis $\boldsymbol{a}_i$ of the undeformed crystal to a new one denoted by $\boldsymbol{a}'_i$. As in (1.20–1.22) of Sect. 1.1, this can be formally described as

$$\boldsymbol{T}\boldsymbol{a}_i = \boldsymbol{a}'_i \ ,$$

where the transformation tensor T has the form $T = \boldsymbol{a}'_i \otimes \boldsymbol{a}^i$. The transformation comprises of a 'stretch' and a 'rigid-body' rotation. Thus, T is a composition of two tensors: one representing the stretch and denoted by U, and R representing the rotation. Formally, when applied to the matrix of components of T, this step is known as the polar decomposition. It is a factorization

$$T = RU \ ,$$

where U is symmetric (semi)positive definite and R represents an orthogonal transformation ($R^{-1} = R^T$). For physical reasons, U is positive definite and R is proper orthogonal. The quantity

$$T^T T = U^2$$

is not affected by the 'rigid-body' rotation, i.e., it is a certain measure of strain. The formula $T^T T = U^2$ corresponds to the transformation rule for the entries of the metric tensor; in components, it has the form of (1.4). With $T = \boldsymbol{a}'_i \otimes \boldsymbol{a}^i$, one has

$$U^2 = T^T T = (\boldsymbol{a}'_i \otimes \boldsymbol{a}^i)^T (\boldsymbol{a}'_j \otimes \boldsymbol{a}^j) = (\boldsymbol{a}^i \otimes \boldsymbol{a}'_i)(\boldsymbol{a}'_j \otimes \boldsymbol{a}^j) = (\boldsymbol{a}'_i \cdot \boldsymbol{a}'_j)(\boldsymbol{a}^i \otimes \boldsymbol{a}^j) = g'_{ij}\, \boldsymbol{a}^i \otimes \boldsymbol{a}^j \ ,$$

i.e., the components of the tensor U^2 in the basis $\boldsymbol{a}^i \otimes \boldsymbol{a}^j$ are $g'_{ij} = \boldsymbol{a}'_i \cdot \boldsymbol{a}'_j$. As was mentioned above, lattice strain can be measured by the difference between the lattice metric in the deformed configuration and that in the undeformed configuration g. The quantity

$$\gamma = \frac{1}{2}\left(U^2 - g\right) = \frac{1}{2}\left(g'_{ij} - g_{ij}\right) \boldsymbol{a}^i \otimes \boldsymbol{a}^j = \gamma_{ij} \boldsymbol{a}^i \otimes \boldsymbol{a}^j \ .$$

is known as the Green-Lagrange *strain tensor*. The difference $2\boldsymbol{u} \cdot \gamma \boldsymbol{u} = 2\gamma_{ij} u^i u^j$ between the squared magnitudes $g'_{ij} u^i u^j$ and $g_{ij} u^i u^j$ of the displacement vector

$\boldsymbol{u} = u^i \boldsymbol{a}_i$ in the deformed and undeformed configurations is a measure of deformation; it vanishes if the deformation is a rigid-body displacement.

1.9.2 Effect of Lattice Transformation on Its Reciprocal Lattice

With the transformation of the direct lattice by T, the reciprocal lattice transforms by $T^{-T} = (T^{-1})^T$, i.e.,

$$T^{-T}\boldsymbol{a}^i = \boldsymbol{a}'^i \ . \tag{1.52}$$

This follows already from (1.11). Alternatively, in the component-free tensor notation of Sect. 1.1.1, the tensor transforming the dual basis vectors $\boldsymbol{a}^i$ to $\boldsymbol{a}'^i$ has the form $\boldsymbol{a}'^i \otimes \boldsymbol{a}_i$, the composition of its transpose with $T = \boldsymbol{a}'_i \otimes \boldsymbol{a}^i$ gives

$$(\boldsymbol{a}'^i \otimes \boldsymbol{a}_i)^T T = (\boldsymbol{a}_i \otimes \boldsymbol{a}'^i)(\boldsymbol{a}'_j \otimes \boldsymbol{a}^j) = (\boldsymbol{a}'^i \cdot \boldsymbol{a}'_j)(\boldsymbol{a}_i \otimes \boldsymbol{a}^j) = \delta^i_{\ j}\, \boldsymbol{a}_i \otimes \boldsymbol{a}^j = \boldsymbol{a}_i \otimes \boldsymbol{a}^i = I \ ,$$

and hence $\boldsymbol{a}'^i \otimes \boldsymbol{a}_i = T^{-T}$.

The relationship (1.52) is needed to obtain the lattice deformation from diffraction experiments. The reference parameters $\boldsymbol{a}_i$ (and hence also $\boldsymbol{a}^i$) are known a priori, and the actual (experimentally measured) scattering vector $\check{\boldsymbol{h}}'$ is equal to $h_i \boldsymbol{a}'^i$ with integer h_i. The problem of diffraction-based deformation measurement can be formulated as the determination of an unknown tensor T leading to the best match between experimental vectors $\check{\boldsymbol{h}}'$ and the corresponding vectors $h_i T^{-T} \boldsymbol{a}^i$.

1.9.3 Strain Tensor in the Crystal Reference System

The formal description of deformation of a body belongs to the field of continuum mechanics. It is based on a fundamental notion of deformation gradient. In general descriptions of deformation, the deformation gradient is a quite complicated entity (a two-point tensor). However, in the considered case of homogeneous deformation, it is directly linked to the above considered tensor T.

Let $\boldsymbol{x}$ and $\boldsymbol{x}'$ be vectors to positions of a particle in the undeformed configuration and deformed configuration, respectively. In the Lagrange formulation (i.e., with $\boldsymbol{x}' = \boldsymbol{x}'(\boldsymbol{x})$), the deformation gradient is defined as $F(\boldsymbol{x}) = \nabla \boldsymbol{x}'(\boldsymbol{x})$. In the case of homogeneous deformation, the deformation gradient F is constant, and (with the translation term set to zero) the relationship $\boldsymbol{x}' = \boldsymbol{x}'(\boldsymbol{x})$ is linear, i.e., $\boldsymbol{x}' = F\boldsymbol{x}$. With the vectors $\boldsymbol{x}$ and $\boldsymbol{x}'$ specified in a single reference frame $\boldsymbol{a}_i$, and $F^i_{\ j}$ denoting the components of the deformation gradient in this frame, one has

$$x'^i = F^i_{\ j} x^j \ . \tag{1.53}$$

One can easily get the relationship between ${F^i}_j$ and components ${T_i}^j$ of T representing the same deformation in the frame $\boldsymbol{a}_i$. The deformation moves $\boldsymbol{a}_k$ to $\boldsymbol{a}'_k$. In the reference frame $\boldsymbol{a}_i$, the j-th component x^j of the vector $\boldsymbol{a}_k$ is $\boldsymbol{a}_k \cdot \boldsymbol{a}^j = {\delta_k}^j$. The i-th component x'^i of the vector $\boldsymbol{a}'_k$ in the same reference frame is $\boldsymbol{a}'_k \cdot \boldsymbol{a}^i$. Therefore $\boldsymbol{a}'_k \cdot \boldsymbol{a}^i = x'^i = {F^i}_j x^j = {F^i}_j {\delta^j}_k$ or ${F^i}_j = \boldsymbol{a}^i \cdot \boldsymbol{a}'_j$. On the other hand, ${T_i}^j = \boldsymbol{a}'_i \cdot \boldsymbol{a}^j$; see (1.22). Hence, one obtains the relationship

$$
{F^j}_i = {T_i}^j ,
$$

i.e., when expressed in the matrix form, F is the transpose of T.

Since the ij-th component of $T^T T = U^2$ in the basis $\boldsymbol{a}^i \otimes \boldsymbol{a}^j$ is $\boldsymbol{a}_i \cdot (T^T T \boldsymbol{a}_j) = {T_i}^k {T_j}^l g_{kl}$, the components $\gamma_{ij} = \boldsymbol{a}_i \cdot (\gamma \boldsymbol{a}_j)$ of the Green-Lagrange strain tensor are

$$
\gamma_{ij} = \frac{1}{2}\left(g'_{ij} - g_{ij}\right) = \frac{1}{2}\left({T_i}^k {T_j}^l g_{kl} - g_{ij}\right) = \frac{1}{2}\left(g_{kl} {F^k}_i {F^l}_j - g_{ij}\right) .
$$

Using the tensor $E = F - I_3$ which links $\boldsymbol{x}$ to the displacement $\boldsymbol{x}' - \boldsymbol{x} = F\boldsymbol{x} - \boldsymbol{x} = E\boldsymbol{x}$, one has

$$
\gamma_{ij} = \frac{1}{2}\left(g_{kj} {E^k}_i + g_{ik} {E^k}_j + g_{nl} {E^n}_i {E^l}_j\right) .
$$

For small strains, the entries ${F^i}_j$ are close to ${\delta^i}_j$, ${E^i}_j$ are small, and with quadratic terms ignored

$$
\gamma_{ij} \approx \frac{1}{2}\left(g_{kj} {E^k}_i + g_{ik} {E^k}_j\right) ,
$$

i.e., γ_{ij} is approximated by the symmetric part of $E_{ij} = g_{ik} {E^k}_j = g_{ik}({F^k}_j - {\delta^k}_j)$.

1.9.4 Strain Tensor in Cartesian Reference System

The conventional theory of elasticity is formulated using tensors given with respect to Cartesian reference systems. Due to their tensorial character, the above relationships can be easily rewritten in such systems. In the already explained notation, $\boldsymbol{a}_i = {\mathcal{T}_i}^j \boldsymbol{e}_j$, i.e., ${\mathcal{T}_i}^j$ is the j-th component of $\boldsymbol{a}_i$ in the orthonormal basis $\boldsymbol{e}_i$, and the metric is given by $g_{ij} = \mathcal{T}_i^k \mathcal{T}_j^l \, \delta_{kl}$. Similarly, the basis vectors of the deformed lattice can also be expressed in the reference basis $\boldsymbol{e}_i$ as $\boldsymbol{a}'_i = {\mathcal{T}'_i}^j \, \boldsymbol{e}_j$, and the metric of the deformed lattice is $g'_{ij} = {\mathcal{T}'_i}^k {\mathcal{T}'_j}^l \, \delta_{kl}$. Since $\boldsymbol{a}^i = {(\mathcal{T}^{-1})_k}^i \, \boldsymbol{e}^k$, the components of F in the basis $\boldsymbol{a}_i$ can be expressed as

$$
{F^i}_j = \boldsymbol{a}^i \cdot \boldsymbol{a}'_j = {\mathcal{T}'_j}^k {(\mathcal{T}^{-1})_k}^i = {\mathcal{T}'_j}^k {\mathcal{T}^i}_k . \tag{1.54}
$$

The Cartesian components of the strain tensor γ are is usually denoted by η_{ij}. One has

$$\eta_{ij} = (\mathcal{T}^{-1})_i^{\ k}(\mathcal{T}^{-1})_j^{\ l}\,\gamma_{kl} = \frac{1}{2}\left(\delta_{kl}\,\bar{F}^k_{\ i}\bar{F}^l_{\ j} - \delta_{ij}\right) \ , \tag{1.55}$$

where

$$\bar{F}^i_{\ j} = (\mathcal{T}^{-1})_j^{\ l}\,\mathcal{T}_k^{\ i}\,F^k_{\ l} = (\mathcal{T}^{-1})_j^{\ l}\,\mathcal{T}'^{\ i}_l = \mathcal{T}^l_{\ j}\,\mathcal{T}'^{\ i}_l$$

are Cartesian components of the deformation gradient. The substitution of F^k_l in the last relationship by (1.54) gives

$$\mathcal{T}'^{\ j}_i = \bar{F}^j_{\ k}\mathcal{T}_i^{\ k} \ ,$$

i.e., in agreement with $\boldsymbol{x}' = \boldsymbol{F}\boldsymbol{x}$ and (1.53), the Cartesian components $\mathcal{T}'_i{}^{j}$ of $\boldsymbol{a}'_i$ are expressed via the Cartesian components $\bar{F}^i_{\ j}$ of F and the Cartesian components $\mathcal{T}_i^{\ j}$ of $\boldsymbol{a}_i$.

As in the case of general reference frames, when the displacements are small, the entries $\bar{E}^i_{\ j} = \bar{F}^i_{\ j} - \delta^i_{\ j}$ are small, the terms quadratic in $\bar{E}^i_{\ j}$ can be ignored, and

$$\eta_{ij} \approx \left(\delta_{ik}\bar{E}^k_{\ j} + \delta_{jk}\bar{E}^k_{\ i}\right)/2 = \left(\bar{E}_{ij} + \bar{E}_{ji}\right)/2 = \epsilon_{ij} \ ,$$

i.e., ϵ_{ij} are Cartesian components of the infinitesimal strain tensor.

For the interpretation of the impact of lattice deformation on diffraction patterns, it is convenient to know how the deformation affects the Cartesian coordinates of reciprocal lattice vectors. The coordinates h_i of $\check{\boldsymbol{h}}$ in the basis $\boldsymbol{a}^i$ are linked to those in the Cartesian system, say, y_i, via (1.2), i.e., $h_i = \mathcal{T}_i^{\ j}\,y_j$. The deformation displaces the point indicated by $\check{\boldsymbol{h}} = y_i\boldsymbol{e}^i$ to $\check{\boldsymbol{h}}' = y'_i\boldsymbol{e}^i$, but the coordinates in the basis of the crystal reciprocal lattice remain unchanged, i.e., with $h'_i = \mathcal{T}'^{\ j}_i y'_j$, one has $h'_i = h_i$. Thus, $\mathcal{T}'^{\ j}_i y'_j = \mathcal{T}_i^{\ j}\,y_j$ or $y_i = (\mathcal{T}^{-1})_i^{\ k}\mathcal{T}'^{\ j}_k y'_j = y'_j\bar{F}^j_{\ i}$, and the Cartesian coordinates of the deformed reciprocal lattice vector are

$$y'_i = y_j(\bar{F}^{-1})^j_{\ i} = \bar{F}_i^{\ j}\,y_j \ .$$

In particular, if only a small symmetric strain ϵ and a small rotation represented by the antisymmetric r ($r_{ij} = -r_{ji}$) are involved, and $\bar{F}^{-1} = (I_3 + \bar{E})^{-1} \approx (I_3 + \epsilon + r)^{-1} \approx I_3 - \epsilon - r$, the Cartesian coordinates of the deformed reciprocal lattice vector are

$$y'_i = (\delta_{ij} - \epsilon_{ij} - r_{ij})\,y_j \ . \tag{1.56}$$

This simple relationship is key for determination of small lattice strains from displacements of diffraction peaks.

1.10 Lattice and Fourier Transformation

Crystal diffraction leads to a network of peaks; see Fig. 2.10 below. It is known that in far-field (Fraunhofer) diffraction, the distribution of amplitudes of waves scattered by an object is the Fourier transform of the characteristic function of the object; see Sect. 2.8. Therefore, it is instructive to consider the Fourier transform of a characteristic function of a lattice. Appendix A is devoted to obtaining appropriate expressions in suitably general forms. Briefly, the point is that the Fourier transform of a periodic function consists of weighted Dirac-delta peaks, and the Fourier transform of a function characterizing an arbitrary lattice is equal to a function characterizing its reciprocal.

As was mentioned before, an ideal unbounded crystal can be seen as a (triply) periodic arrangement of a motif (of atoms). Thus, it has two constituents: a periodic lattice and the content of the primitive cell of the lattice. Formally, the function ρ characterizing the crystal as the scattering object can be represented by the convolution of a 'function' $Ш_\Lambda$ characterizing the crystal lattice Λ and $\rho_{|\mathrm{uc}}$, i.e., ρ with the support restricted to the unit cell of the lattice. By further formal arguments, this implies that the Fourier transform of ρ is a product of the Fourier transform of $\rho_{|\mathrm{uc}}$ and $Ш_{\Lambda^*}$ characterizing the reciprocal lattice Λ^* of the crystal. At this stage, the above statement may be incomprehensible, but it indicates the direction taken in Appendix A. It explains the link between crystal diffraction patterns and the reciprocal lattice of the crystal. The understanding of the statement does not come easily. One needs to invest a bit in the mathematical formalism. The main point is to introduce the distribution $Ш_\Lambda$ and to show that the Fourier transform $\mathcal{F}$ of $Ш_\Lambda$ is directly related to $Ш_{\Lambda^*}$. More precisely, one has

$$\mathcal{F} Ш_\Lambda = V^{-1}\, Ш_{\Lambda^*} \ ,$$

i.e., the Fourier transform of $Ш_\Lambda$ characterizing the lattice Λ is equal to $Ш_{\Lambda^*}$ characterizing the reciprocal Λ^* times the volume V^{-1} of the primitive cell of Λ^*. The sections below sketch the path one needs to follow to derive this relationship.

1.11 Appendix: Fourier Transformation

1.11.1 Fourier Series and Fourier Transformation

The most basic introduction to Fourier series and Fourier transformation concerns real functions of real variables. Let f be a periodic function with the period of 1. Under some integrability conditions, which are assumed to be satisfied, the function f can be represented in the interval $[0, 1]$ by the series

$$f(x) = \sum_{n=-\infty}^{\infty} F(n) \exp(2\pi i n x) , \tag{1.57}$$

where $F(n)$ is the n-th Fourier coefficient. The (generally complex) coefficients are given by

$$F(n) = \int_0^1 f(x) \exp(-2\pi i n x) \, dx . \tag{1.58}$$

In a slightly different and more general formulation, the function f having the period of λ_p (> 0) can be represented in the interval $[-\lambda_p/2, \lambda_p/2]$ by the infinite series $f(x) = \sum_{n=-\infty}^{\infty} F(n) \exp(2\pi i n x/\lambda_p)$, where the coefficients $F(n)$ are given by

$$\lambda_p F(n) = \int_{-\lambda_p/2}^{\lambda_p/2} f(x) \exp(-2\pi i x \xi_n) \, dx ,$$

and $\xi_n = n/\lambda_p$. Linked to this representation of Fourier coefficients is the Fourier transformation. When the period λ_p becomes large, the spacing between subsequent factors ξ_n becomes small, and the discrete variable ξ_n tends to continuous variable ξ when $\lambda_p \to \infty$. With the period of the function f being infinity, the ξ-dependent Fourier transform $\mathcal{F}f$ of f is defined as

$$\mathcal{F}f\,(\xi) = \int_{-\infty}^{\infty} f(x) \exp(-2\pi i x \xi) \, dx , \tag{1.59}$$

if the integral converges. The two-character symbol $\mathcal{F}f$ stands for a function whose argument is ξ. Below, if there is a risk of confusion, extra elements describing the transform will be added; e.g., with f being a function of x, instead of the brief $\mathcal{F}f\,(\xi)$, forms of the extensive (and abusing notation rules) symbol $\mathcal{F}_x[f(x)](\xi)$ will be used. The inverse Fourier transform of f is

$$\mathcal{F}^{-1}f\,(x) = \int_{-\infty}^{\infty} f(\xi) \exp(2\pi i x \xi) \, d\xi ,$$

if the integral converges.

The Fourier series and Fourier transformation differ by the type of the support of frequencies, i.e. the spectrum of f: it is discrete in the first case (n or ξ_n) and continuous in the second one (ξ).

There are a number of generalizations of Fourier series expansions and Fourier transformations. The most basic states that the Fourier series expansions and Fourier transformations are also applicable to complex-valued functions of real variables. The second issue, considered in Sect. 1.11.2 below, is more intricate.

Clearly, both Fourier transformation and inverse Fourier transformation are linear. Under some assumptions (see 1.11.2), the transformations $\mathcal{F}$ and $\mathcal{F}^{-1}$ have the property

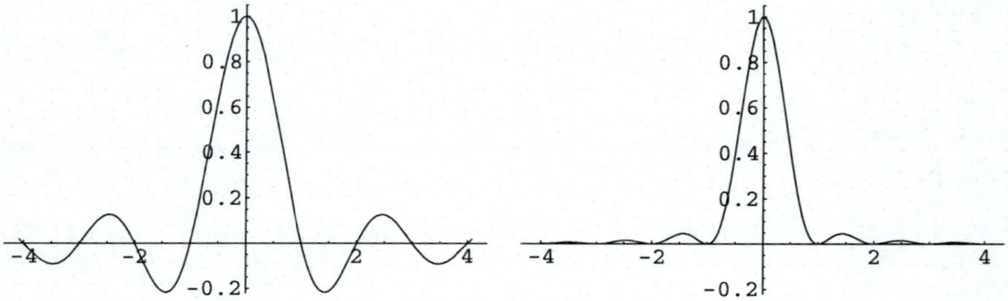

Fig. 1.21 The functions $\text{sinc}(x)$ and $\text{sinc}^2(x)$ in the vicinity of $x = 0$. As the Fourier transform of the rectangular function Π, the function $\text{sinc}(x)$ appears frequently in the analysis of diffraction phenomena. It has zeros at non-zero integers and extrema at the roots of $\tan(\pi x) = \pi x$, i.e., at 0 and close to $\pm(n + 1/2)$ for integer $n > 0$. In effect, $\text{sinc}^2(x)$ contains a strong central peak with successions of weak secondary maxima on both sides.

$$\mathcal{F}^{-1}\mathcal{F}f = f \quad \text{and} \quad \mathcal{F}\mathcal{F}^{-1}f = f \tag{1.60}$$

(at points of continuity of f). When applied to complex functions of real variable, the transformation has the property $(\mathcal{F}_x[f(x)])^* = \mathcal{F}_x[f^*(-x)]$, where $*$ denotes complex conjugate. The Fourier transform of a real-valued function f satisfies $\mathcal{F}f(-\xi) = (\mathcal{F}f(\xi))^*$. Quite obvious is the 'shift' property:

$$\mathcal{F}_x\left[f(x-a)\right](\xi) = \mathcal{F}_x\left[f(x)\right](\xi)\,\exp(-2\pi i \xi \cdot a)\,. \tag{1.61}$$

We also need to mention the 'scaling' property: the transform of f with the variable x scaled by a non-zero factor a is

$$\mathcal{F}_x[f(ax)](\xi) = |a|^{-1}\mathcal{F}_x[f(x)](\xi/a)\,. \tag{1.62}$$

Thus, stretching (squeezing) the function squeezes (stretches) its Fourier transform by the same factor.

As an important example, one needs to mention the Fourier transform of the rectangular function Π; $\Pi(x)$ equals 1 if $|x| < 1/2$, 1/2 if $|x| = 1/2$, and 0 otherwise. Its transform is known as sinc function

$$\mathcal{F}\Pi = \text{sinc}\,; \tag{1.63}$$

the latter is given by $\text{sinc}(x) = \sin(\pi x)/(\pi x)$ and $\text{sinc}(0) = 1$. See Fig. 1.21.

Example Fourier Series: Jacobi-Anger Expansion

An important example of the Fourier series is the expansion of the function $\exp(iz\sin\theta)$ periodic with respect to the angle θ; z is an arbitrary complex number.

With $\theta = 2\pi x$, the period of the x-dependent $\exp(iz\sin(2\pi x))$ is 1. One can write

$$\exp(iz\sin(2\pi x)) = \sum_{n=-\infty}^{\infty} J_n(z)\exp(2\pi inx)\ ,$$

and based on (1.58), the coefficients are $J_n(z) = \int_0^1 \exp(iz\sin 2\pi x)\exp(-2\pi inx)\,\mathrm{d}x = \int_0^1 \exp(i(z\sin(2\pi x) - 2\pi nx))\,\mathrm{d}x$ or

$$J_n(z) = \int_0^1 \cos(z\sin(2\pi x) - 2\pi nx)\,\mathrm{d}x + i\int_0^1 \sin(z\sin(2\pi x) - n2\pi x)\,\mathrm{d}x\ .$$

The term $i\int_0^1 \sin(z\sin(2\pi x) - n2\pi x)\,\mathrm{d}x = i\int_{-1/2}^{+1/2} \sin(z\sin(2\pi x) - n2\pi x)\,\mathrm{d}x$ vanishes because the integrand is an odd function of x. The expansion coefficient

$$J_n(z) = \int_0^1 \cos(z\sin(2\pi x) - 2\pi nx)\,\mathrm{d}x = \frac{1}{2\pi}\int_0^{2\pi} \cos(z\sin\theta - n\theta)\,\mathrm{d}\theta$$

is the n-th Bessel function of the first kind. The identity

$$\exp(iz\sin\theta) = \sum_{n=-\infty}^{\infty} J_n(z)\ \exp(in\theta)\ , \tag{1.64}$$

or the equivalent expression

$$\exp(iz\cos\theta) = \sum_{n=-\infty}^{\infty} i^n J_n(z)\ \exp(in\theta)\ , \tag{1.65}$$

which follows from (1.64) and $\cos\theta = \sin(\theta + \pi/2)$, is known as the Jacobi-Anger expansion.

1.11.2 Distributions

The Fourier transformation as defined above exists (and is continuous) for a function f such that $\int |f|$ is finite. However, the resulting function $\mathcal{F}f$ may not be in this set, and the inverse transform of $\mathcal{F}f$ may not exist. This complicates the use of the key property (1.60). The above condition for the existence of $\mathcal{F}f$ also means that strictly periodic functions of our interest, e.g., the electron density or the electrostatic potential in an ideal unbounded crystal, do not have the classical Fourier transform. These issues can be resolved by using a different (and more natural) domain of the Fourier transformation: generalized functions or distributions.

Both the traditional (function-based) and the distribution-based formalisms are used below. Staying within the former leads to formally invalid statements, and using only the latter would make the text too abstract. It is believed that showing them both will not be confusing and will help in transitioning from the traditional to a slightly more modern approach.

The distributions can be defined using infinitely differentiable rapidly decreasing test functions. These can be functions of compact support, but we will focus on less restricted Schwartz functions. By definition, a Schwartz function and all its derivatives tend to zero at infinity faster than any power function. A Schwartz function has a Fourier transform in the sense of the integral (1.59), and the transform is a Schwartz function. A (tempered) distribution is a continuous linear functional on Schwartz functions: a distribution T on the space of test functions assigns scalar values to functions f, f_1 and f_2 and a convergent sequence f_n in such a way that

$$\mathsf{T}(f_1 + f_2) = \mathsf{T} f_1 + \mathsf{T} f_2 \,, \quad \mathsf{T}(af) = a\mathsf{T} f \ \text{ and } \ \lim_{n\to\infty} \mathsf{T} f_n = \mathsf{T}\left(\lim_{n\to\infty} f_n\right) ,$$

where a is a constant. The product $g\mathsf{T}$ of a function g and a distribution T is defined if g is such that its product with every test function f is a test function; $g\mathsf{T}$ is the distribution such that $g\mathsf{T}(f) = \mathsf{T}(gf)$. The distribution T is determined by a function h if the result $\mathsf{T} f$ of its action on an arbitrary test function f can be expressed as

$$\mathsf{T} f = \int_{-\infty}^{\infty} h(x)\, f(x)\, \mathrm{d}x \,,$$

but clearly, not every distribution has such a representation. Distributions are sometimes defined using sequences $f_s(x)$ of test functions such that there exist the limits

$$\lim_{s\to 0} \int_{-\infty}^{\infty} f_s(x)\, f(x) \mathrm{d}x \,,$$

where f is an arbitrary test function. A distribution is a class of sequences leading to the same limits for all test functions f.

Dirac Distribution

As a functional, the Dirac distribution δ is defined simply by

$$\delta f = f(0) \,.$$

The functional is clearly linear (as $\delta(af_1 + f_2) = af_1(0) + f_2(0) = a\delta f_1 + \delta f_2$) and continuous (as for the functions t_n converging to t, the sequence $\delta f_n = f_n(0)$ converges to $f(0) = \delta f$).

The distribution δ is frequently introduced using the sequence-based definition. The sequence of Gaussian functions $(\sqrt{\pi}\, s)^{-1} \exp(-(x/s)^2) = f_s(x)$ determines a distribution because these are Schwartz functions and $\lim_{s\to 0} \int_{-\infty}^{\infty} f_s(x)\, f(x)\, \mathrm{d}x$ exists; as the limit can be shown to be equal to $f(0)$, that distribution is the δ distribution.

In conventional notation, the δ distribution is represented by the Dirac δ 'function'. The 'function' $\delta(x)$ is zero everywhere except at $x = 0$. It is defined by its integral property: assuming $a < b$,

$$\int_a^b \delta(x - x_0)\, f(x)\, \mathrm{d}x = \begin{cases} f(x_0) & \text{if } a < x_0 < b\,, \\ f(x_0)/2 & \text{if } x_0 = a \ \text{ or } x_0 = b\,, \\ 0 & \text{otherwise.} \end{cases}$$

With $f(x) = 1$, one gets $\int_{-\infty}^{\infty} \delta(x)\, \mathrm{d}x = 1$. Moreover, in this notation, for a non-zero a, one has

$$\delta(ax - x_0) = |a|^{-1} \delta(x - x_0/a)\,, \tag{1.66}$$

and in particular, $\delta(-x) = \delta(x)$.

Dirac Comb

The Dirac comb Ш is defined by

$$Ш f = \sum_{n=-\infty}^{\infty} f(n)\,, \tag{1.67}$$

where n denotes an integer. By arguments analogous to those used in the case of the δ distribution, the functional Ш is linear and continuous, i.e., the Dirac comb is a distribution. In the conventional notation, one has the Shah 'function'

$$Ш(x) = \sum_{n=-\infty}^{\infty} \delta(x - n)\,. \tag{1.68}$$

Clearly, the Shah 'function' is periodic: $Ш(x + 1) = Ш(x)$. Moreover, similarly to (1.66), it can be shown that

$$Ш(ax) = |a|^{-1} \sum_{n=-\infty}^{\infty} \delta(x - n/a)\,. \tag{1.69}$$

In particular, $Ш(-x) = Ш(x)$.

The goal is to define a Dirac comb based on a general three-dimensional lattice. It is revealing to consider first the one-dimensional Dirac comb with the period $\lambda_p > 0$.

Such a Dirac comb $Ш_{\lambda_p}$ is defined as

$$Ш_{\lambda_p} f = \sum_{n=-\infty}^{\infty} f(n\lambda_p) \ .$$

Clearly, the above considered III equals $Ш_1$. In the conventional notation, one has the Shah 'function'

$$Ш_{\lambda_p}(x) = \sum_{n=-\infty}^{\infty} \delta(x - n\lambda_p)$$

satisfying $Ш_{\lambda_p}(x + \lambda_p) = Ш_{\lambda_p}(x)$. Using the scaling property (1.69) of III, one can write

$$Ш_{\lambda_p}(x) = \lambda_p^{-1}\, \mathsf{III}(x/\lambda_p) \ . \tag{1.70}$$

The above relationship expresses the general $Ш_{\lambda_p}$ via the particular $Ш_1 = \mathsf{III}$.

Fourier Transform of a Distribution

Calculus of distributions is in a sense simpler than that of functions of real variables.[27] A test function f has a Fourier transform $\mathcal{F}f$ in the conventional sense (1.59), and the inverse Fourier transform of $\mathcal{F}f$ is equal to f. The Fourier transform of the distribution T is defined as the distribution $\mathcal{F}\mathsf{T}$ such that

$$\mathcal{F}\mathsf{T}(f) = \mathsf{T}(\mathcal{F}f) \ , \tag{1.71}$$

for arbitrary test function f. An analogous definition applies to the inverse Fourier transform of the distribution: $\mathcal{F}^{-1}\mathsf{T}(f) = \mathsf{T}(\mathcal{F}^{-1}f)$. The transformations $\mathcal{F}$ and $\mathcal{F}^{-1}$ of a distribution always exist and are mutually inverse (as $\mathcal{F}^{-1}(\mathcal{F}(\mathsf{T}f)) = \mathcal{F}^{-1}\mathsf{T}(\mathcal{F}f) = \mathsf{T}(\mathcal{F}^{-1}\mathcal{F}f) = \mathsf{T}f$).

It is worth noting that, if the test function f is even, then $\mathcal{F}f = \mathcal{F}^{-1}f$ and $\mathcal{F}\mathcal{F}f = f$. One can define even and odd distributions, and analogous relationships are applicable to distributions. In particular, in the case of the rectangular function and the Dirac distribution

$$\mathcal{F}\Pi = \mathrm{sinc} = \mathcal{F}^{-1}\Pi \ , \quad \mathcal{F}\mathrm{sinc} = \Pi = \mathcal{F}^{-1}\mathrm{sinc} \ ,$$

---

[27] Although the following note on the derivative of a distribution is not essential for our purposes, it is worth adding it to give a better picture of calculus involving distributions: Classically, for the test functions t and f and their derivatives t' and f', one has $\int_{-\infty}^{\infty} t'(x) f(x) \mathrm{d}x = -\int_{-\infty}^{\infty} t(x) f'(x) \mathrm{d}x$. With this, the sequence-based definition of distributions implies the way of defining the derivatives of distributions. The first derivative of the distribution T is defined as the distribution T' acting on test functions via $\mathsf{T}'f = -\mathsf{T}f'$. Clearly, with this definition, a distribution has derivatives of all orders. The n-th derivative $\mathsf{T}^{(n)}$ of T is given by $\mathsf{T}^{(n)} f = (-1)^n\, \mathsf{T} f^{(n)}$. For instance, the n-th derivative $\delta^{(n)}$ of Dirac distribution acts on f via $\delta^{(n)} f = (-1)^n f^{(n)}(0)$.

$$\mathcal{F}\delta = 1 = \mathcal{F}^{-1}\delta\,, \qquad \mathcal{F}1 = \delta = \mathcal{F}^{-1}1\,.$$

In relation to the latter expressions, allowing for a non-zero x_0, the shift property for the Dirac δ 'function' gives

$$\mathcal{F}_x[\delta(x-x_0)](\xi) = \exp(-2\pi i\,\xi x_0) \quad \text{and} \quad \mathcal{F}_\xi^{-1}[\exp(-2\pi i\,\xi x_0)](x) = \delta(x-x_0)\,; \tag{1.72}$$

the first relationship is based on the definition of the δ 'function', and the second one follows from (1.60).

Using (1.72), one can obtain the Fourier transform of the Fourier series (1.57); it is given by

$$\mathcal{F}_x\left[\sum_{n=-\infty}^{\infty} F(n)\exp(2\pi i n x)\right](\xi) = \sum_{n=-\infty}^{\infty} F(n)\,\delta(\xi - n)\,, \tag{1.73}$$

Thus, the Fourier transform of a periodic function is a sum of δ distributions weighted by the Fourier coefficients of the function; cf. definition (1.68) of Ш.

1.11.3 Convolution

The convolution $f_1 * f_2$ of test functions f_1 and f_2 is defined as

$$(f_1 * f_2)(x) = \int_{-\infty}^{\infty} f_1(y)\, f_2(x-y)\,dy\,. \tag{1.74}$$

The operation '$*$' is associative and commutative. By the convolution theorem, the Fourier transform of the convolution of functions is equal to the product of their Fourier transforms, and the Fourier transform of the product of functions is equal to the convolution of their Fourier transforms

$$\mathcal{F}(f_1 * f_2) = (\mathcal{F}f_1)\,(\mathcal{F}f_2)\,, \qquad \mathcal{F}(f_1 f_2) = \mathcal{F}f_1 * \mathcal{F}f_2\,.$$

The class of test functions is closed under convolution. Again, the operation (1.74) can be generalized to distributions by using the sequence-based definition of distributions. As a functional, the convolution of a test function f and a distribution T is a distribution denoted by $f * \mathsf{T}$ acting on a test function g via $(f * \mathsf{T})g = \mathsf{T}(f^- * g)$, where $f^-(x) = f(-x)$. To give examples, let us note that

$$(\delta * f)(x) = f(x) \quad \text{and} \quad (Ш * f)(x) = \sum_{n=-\infty}^{\infty} f(x-n)\,. \tag{1.75}$$

These relationships can be easily verified using the definitions of the Dirac and Shah 'functions'. The first one is just a different form of the classic $\int \delta(x-y)f(y)\mathrm{d}y = f(x)$. By the convolution theorem, the Fourier transform of $f = \delta * f$ gives $\mathcal{F}f = \mathcal{F}(\delta * f) = (\mathcal{F}\delta)\,(\mathcal{F}f)$, which is in agreement with $\mathcal{F}\delta = 1$; cf. (1.73). The second of (1.75) has an important interpretation: The convolution of f with Ш makes copies of f translated by an integer and adds them up, and this clearly results in a periodic function.

In relation to the convolution, we also need to mention the cross-correlation of functions of a real variable. It is defined as

$$(f_1 \star f_2)(x) = \int_{-\infty}^{\infty} f_1^*(y)\, f_2(x+y)\, dy\,.$$

The operation $\star$ is associative but not commutative. It is easy to verify that cross-correlation can be expressed via convolution $f_1 \star f_2 = (f_1^-)^* * f_2$. Based on this relationship, the convolution theorem and $(\mathcal{F}_x[f(x)])^* = \mathcal{F}_x[f^*(-x)]$, one has

$$|\mathcal{F}f|^2 = (\mathcal{F}f)^*\,(\mathcal{F}f) = \mathcal{F}(f \star f)\,, \tag{1.76}$$

i.e., the squared magnitude of the Fourier transform of f is equal to the Fourier transform of its autocorrelation.

1.11.4 Fourier Transform of Dirac Comb

It turns out that the Fourier transform of the Dirac comb is the Dirac comb

$$\mathcal{F}Ш = Ш\,. \tag{1.77}$$

This result follows from the Poisson summation formula: for a test function f

$$\sum_{n=-\infty}^{\infty} \mathcal{F}f(n) = \sum_{n=-\infty}^{\infty} f(n)\,. \tag{1.78}$$

⊡ Here is a sketch of proofs of these relationships. Given the test function f, the function $(Ш * f)(x) = \sum_{n=-\infty}^{\infty} f(x-n)$ is smooth and periodic with the period of 1, and it can be expanded in the Fourier series $(Ш * f)(x) = \sum_{n=-\infty}^{\infty} F(n)\exp(2\pi i n x)$ with coefficients

$$F(n) = \int_0^1 (Ш * f)(x)\,\exp(-2\pi i n x)\mathrm{d}x = \int_0^1 \left(\sum_{m=-\infty}^{\infty} f(x-m)\right)\exp(-2\pi i n x)\,\mathrm{d}x = \tag{1.79}$$

$$\sum_{m=-\infty}^{\infty} \int_{-m}^{-m+1} f(y) \exp(-2\pi i n(y+m))\, \mathrm{d}y = \int_{-\infty}^{\infty} f(y) \exp(-2\pi i n y)\, \mathrm{d}y = \mathcal{F}f(n) .$$

At $x = 0$, the function $\text{III} * f$ equals $(\text{III} * f)(0) = \sum_n f(-n) = \sum_n f(n)$, whereas its Fourier series at the same point gives $(\text{III} * f)(0) = \sum_n F(n) = \sum_n \mathcal{F}f(n)$. Hence, one gets (1.78).

Now, going back to the Fourier transform of the Dirac comb, using (1.67), (1.71) and (1.78), one has

$$(\mathcal{F}\text{III})f = \text{III}(\mathcal{F}f) = \sum_n \mathcal{F}f(n) = \sum_n f(n) = \text{III}f ,$$

and this occurs for arbitrary test function f. Hence, one has (1.77).

Since $\mathcal{F}\text{III}(\xi) = \sum_{n=-\infty}^{\infty} \int \delta(x-n) \exp(-2\pi i \xi x)\, \mathrm{d}x = \sum_{n=-\infty}^{\infty} \exp(-2\pi i n \xi)$, the identity (1.77) and the definition (1.68) of the Shah 'function' lead to

$$\sum_{n=-\infty}^{\infty} \exp(-2\pi i n \xi) = \text{III}(\xi) = \sum_{m=-\infty}^{\infty} \delta(\xi - m) . \tag{1.80}$$

It is easy to see that also $\sum_n \exp(+2\pi i n \xi)$ equals $\text{III}(\xi)$.

The next goal is to show how the relationship $\mathcal{F}\text{III} = \text{III}$ is generalized to the distribution III_{λ_p}. Application of the scaling property (1.62) of the Fourier transformation to (1.70) or $\mathit{III}_{\lambda_p}(x) = \lambda_p^{-1}\, \text{III}(x/\lambda_p)$ results in $\mathcal{F}_x[\mathit{III}_{\lambda_p}(x)](\xi) = \lambda_p^{-1} \mathcal{F}_x[\text{III}(x/\lambda_p)](\xi) = \mathcal{F}_x[\text{III}(x)](\xi\lambda_p) = \text{III}(\xi\lambda_p) = \lambda_p^{-1} \mathit{III}_{\lambda_p^{-1}}(\xi)$, or briefly

$$\mathcal{F}\mathit{III}_{\lambda_p} = \lambda_p^{-1} \mathit{III}_{\lambda_p^{-1}} , \tag{1.81}$$

i.e., III_{λ_p} characterizing a one-dimensional lattice with the period of λ_p has the Fourier transform equal to $\lambda_p^{-1} \mathit{III}_{\lambda_p^{-1}}$ characterizing the lattice with the period of $1/\lambda_p$. Already this simple formula indicates the reciprocity between lattices linked to III_{λ_p} and $\mathcal{F}\mathit{III}_{\lambda_p}$. The next step is to generalize this observation to three-dimensional lattices.

Extension to Functions of Several Real Variables

Both the Fourier series and the Fourier transformation can be extended to functions of several real variables $(x^1, x^2, ..., x^N)$, and the case with $N = 3$ is considered here. Multiple integrations and summations are used, and the products nx and ξx are replaced by $n_i x^i$ and $\xi_i x^i$, respectively. This means that with spatial positions represented by vectors $(\boldsymbol{x})$ given in contravariant components (x^i), the (spatial) frequencies become vectors $(\boldsymbol{\xi})$ given above in covariant components (ξ_i). Analogously to the case of a single variable, also the distributions are defined as linear functionals of infinitely differentiable rapidly decreasing functions of several variables.

Let $f = f(x^1, x^2, x^3) = f(\boldsymbol{x})$ be a function of the variables x^1, x^2, x^3, or $\boldsymbol{x} = x^i \boldsymbol{e}_i$ in a Cartesian frame based on vectors $\boldsymbol{e}_i = \boldsymbol{e}^i$. The definitions of distributions δ and Ш can be generalized naturally to the three-dimensional space via $\delta f = f(\boldsymbol{0})$ and $\text{Ш} f = \sum_{\check{\boldsymbol{m}}} f(\check{\boldsymbol{m}})$, where $\check{\boldsymbol{m}}$ are nodes of the lattice based on the vectors $\boldsymbol{e}_i$. Allowing for some abuse of notation, the Dirac and Shah 'functions' of $\boldsymbol{x}$ are defined as

$$\delta(x^1, x^2, x^3) = \delta(\boldsymbol{x}) = \prod_i \delta(x^i)\,, \qquad \text{Ш}(x^1, x^2, x^3) = \text{Ш}(\boldsymbol{x}) = \prod_i \text{Ш}(x^i)\,.$$

The expressions for multidimensional Fourier transforms have forms analogous to those of the one-dimensional case, e.g., in three dimensions (1.59) is replaced by

$$\mathcal{F} f(\boldsymbol{\xi}) = \int f(\boldsymbol{x})\ \exp(-2\pi i\, \boldsymbol{x} \cdot \boldsymbol{\xi})\, \mathrm{d}^3\boldsymbol{x}\,,$$

where $\boldsymbol{\xi} = \xi_i \boldsymbol{e}^i$. Analogously to the case of one variable, the multivariable functions δ and 1 are related by the Fourier transforms

$$\mathcal{F}_{\boldsymbol{x}}[\delta(\boldsymbol{x})](\boldsymbol{\xi}) = 1\,, \quad \mathcal{F}_{\boldsymbol{x}}[1](\boldsymbol{\xi}) = \delta(\boldsymbol{\xi})\,.$$

With $\Pi(\boldsymbol{x}) = \prod_i \Pi(x^i)$ and $\text{sinc}(\boldsymbol{x}) = \prod_i \text{sinc}(x^i)$, one has[28]

$$\mathcal{F}_{\boldsymbol{x}}[\Pi(\boldsymbol{x})](\boldsymbol{\xi}) = \text{sinc}(\boldsymbol{\xi})\,, \quad \mathcal{F}_{\boldsymbol{x}}[\text{sinc}(\boldsymbol{x})](\boldsymbol{\xi}) = \Pi(\boldsymbol{\xi})\,.$$

What matters below is that the three-dimensional Fourier transform of the three-dimensional Shah 'function' $\text{Ш}(\boldsymbol{x})$ related to an integer lattice (based on orthonormal vectors $\boldsymbol{e}_i$) is the Shah 'function'

$$\mathcal{F}\text{Ш} = \text{Ш}\,. \tag{1.82}$$

This multidimensional relationship follows directly from the one-dimensional (1.77). Similarly, the second of relationships (1.75) is generalized to

$$(\text{Ш} * f)(\boldsymbol{x}) = \sum_{\check{\boldsymbol{m}}} f(\boldsymbol{x} - \check{\boldsymbol{m}})\,, \tag{1.83}$$

where $\check{\boldsymbol{m}}$ are nodes of the integer lattice based on $\boldsymbol{e}_i$, i.e., a primitive cubic lattice with the primitive cell of volume 1.

---

[28] It is worth noting here the relationship

$$\mathcal{F}_{\boldsymbol{x}}[(1/|\boldsymbol{x}|)](\boldsymbol{\xi}) = 1/(\pi \xi^2)$$

used below to derive (2.32). It follows from $\nabla^2(1/|\boldsymbol{x}|) = -4\pi\delta(\boldsymbol{x})$ and $\mathcal{F}_{\boldsymbol{x}}[\partial_k f(\boldsymbol{x})](\boldsymbol{\xi}) = -2\pi i \xi_k \mathcal{F}_{\boldsymbol{x}}[f(\boldsymbol{x})](\boldsymbol{\xi})$. The first of these equalities leads to $\mathcal{F}_{\boldsymbol{x}}[\nabla^2(1/|\boldsymbol{x}|)](\boldsymbol{\xi}) = -4\pi \mathcal{F}_{\boldsymbol{x}}[\delta(\boldsymbol{x})](\boldsymbol{\xi}) = -4\pi$, and the second one gives $\mathcal{F}_{\boldsymbol{x}}[\nabla^2(1/|\boldsymbol{x}|)](\boldsymbol{\xi}) = -4\pi^2 \xi^2 \mathcal{F}_{\boldsymbol{x}}[(1/|\boldsymbol{x}|)](\boldsymbol{\xi})$.

Fourier Transform of $f(X\boldsymbol{x})$

The above properties need to be applied to general (not necessarily cubic) three-dimensional lattices. The generalization is based on the multidimensional scaling property; cf. (1.66) and (1.69). Let X be an invertible matrix and $y^i = X^i{}_j x^j$. By the change of variables, one obtains the scaling relationship $\int \delta(X^1{}_j x^j, X^2{}_j x^j, X^3{}_j x^j) f(x^i)\, \mathrm{d}x^1 \mathrm{d}x^2 \mathrm{d}x^3 = \int \delta(y^1, y^2, y^3) f((X^{-1})^i{}_j y^j)\, |\det(X^{-1})|\, \mathrm{d}y^1 \mathrm{d}y^2 \mathrm{d}y^3$, and hence, the multivariable Dirac 'function' satisfies $\delta(X^1{}_j x^j, X^2{}_j x^j, X^3{}_j x^j) = |\det(X)|^{-1}\, \delta(x^1, x^2, x^3)$, or briefly,

$$\delta(X\boldsymbol{x}) = |\det(X)|^{-1} \delta(\boldsymbol{x}) \,.$$

Similarly, by definition of the Fourier transform, one has the relationship $\mathcal{F}_{\boldsymbol{x}}\left[f(X^i{}_j x^j)\right](\boldsymbol{\xi}) = \int f(X^i{}_j x^j) \exp(-2\pi i x^k \xi_k)\, \mathrm{d}^3\boldsymbol{x} = \int f(y^i) \exp(-2\pi i (X^{-1})^k{}_j y^j \xi_k)\, |\det(X)|^{-1}\, \mathrm{d}^3\boldsymbol{y}$, and it leads to

$$\mathcal{F}_{\boldsymbol{x}}\left[f(X\boldsymbol{x})\right](\boldsymbol{\xi}) = |\det(X)|^{-1}\, \mathcal{F}_{\boldsymbol{x}}\left[f(\boldsymbol{x})\right](X^{-T}\boldsymbol{\xi}) \,, \tag{1.84}$$

which is a generalized version of the scaling formula (1.62). It is worth noting that if X is an orthogonal matrix ($X^{-1} = X^T$) then $\mathcal{F}_{\boldsymbol{x}}\left[f(X\boldsymbol{x})\right](\boldsymbol{\xi}) = \mathcal{F}_{\boldsymbol{x}}\left[f(\boldsymbol{x})\right](X\boldsymbol{\xi})$, i.e., a rotation (represented by X in a Cartesian frame) in the spatial domain implies the same rotation in the domain of frequencies.[29]

Fourier Transform of the Characteristic 'Function' of a Lattice

The rule that the Fourier transform of the Dirac comb III characterizing the lattice based on the orthonormal vectors $\boldsymbol{e}_i$ is given by the Dirac comb, can be generalized using (1.84) to lattices with arbitrary bases. Let III_Λ denote the distribution acting via

$$III_\Lambda f = \sum_{\check{\boldsymbol{m}}} f(\check{\boldsymbol{m}}) \,,$$

where $\check{\boldsymbol{m}} = m^i \boldsymbol{a}_i$ are nodes of lattice Λ based on vectors $\boldsymbol{a}_i$. The lattice is characterized by

$$III_\Lambda(\boldsymbol{x}) = \sum_{\check{\boldsymbol{m}}} \delta(\boldsymbol{x} - \check{\boldsymbol{m}}) \,,$$

where the sum is over all lattice points $\check{\boldsymbol{m}}$. Convolution of the above III_Λ with a function f results in

[29] This is linked to the physical facts that a rotation of a direct lattice implies the same rotation of the reciprocal lattice, and the rotation of a crystal implies a corresponding rotation of its diffraction pattern.

$$(Ш_\Lambda * f)(\boldsymbol{x}) = \sum_{\check{\boldsymbol{m}}} f(\boldsymbol{x} - \check{\boldsymbol{m}})\ ; \tag{1.85}$$

cf. (1.75) and (1.83). Clearly, the function $Ш_\Lambda * f$ is periodic with the periodicity of Λ: $(Ш_\Lambda * f)(\boldsymbol{x}) = (Ш_\Lambda * f)(\boldsymbol{x} + \check{\boldsymbol{m}})$ for $\check{\boldsymbol{m}}$ in Λ.

In a Cartesian system, the function $Ш_\Lambda$ has the form

$$\sum_{\check{\boldsymbol{m}}} \delta(\boldsymbol{x} - \check{\boldsymbol{m}}) = \sum_{m^1, m^2, m^3} \prod_j \delta((\boldsymbol{x} - m^i \boldsymbol{a}_i) \cdot \boldsymbol{e}_j) = \sum_{m^1, m^2, m^3} \prod_j \delta(x^j - m^i \mathcal{T}_i{}^j) = \sum_{\check{\mathrm{m}}} \delta(\boldsymbol{x} - \mathcal{T}^T \check{\mathrm{m}}),$$

where $\check{\mathrm{m}}$ denotes the triplet of integers $(m^1, m^2, m^3)^T$, and $\mathcal{T}$ links the lattice basis $\boldsymbol{a}_i$ to the Cartesian basis $\boldsymbol{e}_i$: $\boldsymbol{a}_i = \mathcal{T}_i{}^j \boldsymbol{e}_j$. Since $\delta(\boldsymbol{x} - \mathcal{T}^T \check{\mathrm{m}}) = \delta(\mathcal{T}^T(\mathcal{T}^{-T}\boldsymbol{x} - \check{\mathrm{m}})) = |\det(\mathcal{T})|^{-1}\,\delta(\mathcal{T}^{-T}\boldsymbol{x} - \check{\mathrm{m}})$, one can write a generalized version of (1.70)

$$Ш_\Lambda(\boldsymbol{x}) = |\det(\mathcal{T})|^{-1} \sum_{\check{\mathrm{m}}} \delta(\mathcal{T}^{-T}\boldsymbol{x} - \check{\mathrm{m}}) = |\det(\mathcal{T})|^{-1}\, Ш(\mathcal{T}^{-T}\boldsymbol{x})\ , \tag{1.86}$$

and by using the generalized scaling formula (1.84), one gets

$$\mathcal{F}_{\boldsymbol{x}}\left[Ш_\Lambda(\boldsymbol{x})\right](\boldsymbol{\xi}) = |\det(\mathcal{T})|^{-1}\, \mathcal{F}_{\boldsymbol{x}}\left[Ш(\mathcal{T}^{-T}\boldsymbol{x})\right](\boldsymbol{\xi}) = \mathcal{F}_{\boldsymbol{x}}\left[Ш(\boldsymbol{x})\right](\mathcal{T}\boldsymbol{\xi})\ .$$

Since $\mathcal{F}Ш = Ш$ (1.82), one obtains

$$\mathcal{F}Ш_\Lambda(\boldsymbol{\xi}) = Ш(\mathcal{T}\boldsymbol{\xi}) = \sum_{\check{\mathrm{h}}} \delta(\mathcal{T}\boldsymbol{\xi} - \check{\mathrm{h}})\ ,$$

where $\check{\mathrm{h}}$ denotes an *integer* combination of vectors $\boldsymbol{e}^i$ ($\check{\mathrm{h}} = h_i \boldsymbol{e}^i$). Finally, based on $\delta(\mathcal{T}\boldsymbol{\xi} - \check{\mathrm{h}}) = |\det(\mathcal{T})|^{-1}\delta(\boldsymbol{\xi} - (\mathcal{T}^{-1})\check{\mathrm{h}})$, $(\mathcal{T}^{-1})\check{\mathrm{h}} = h_i (\mathcal{T}^{-1})_j{}^i \boldsymbol{e}^j = h_i \boldsymbol{a}^i = \check{\boldsymbol{h}}$ and $|\det(\mathcal{T})| = V$, one gets the relationship

$$\mathcal{F}Ш_\Lambda(\boldsymbol{\xi}) = \frac{1}{V} \sum_{\check{\boldsymbol{h}}} \delta(\boldsymbol{\xi} - \check{\boldsymbol{h}})$$

with the summation over all points $\check{\boldsymbol{h}}$ of the reciprocal lattice Λ^*. The above formula can be written in the brief form

$$\mathcal{F}Ш_\Lambda = V^{-1}\, Ш_{\Lambda^*}\ ; \tag{1.87}$$

cf. (1.81). It follows then, that the Fourier transform of $Ш_\Lambda$ characterizing an arbitrary lattice Λ is equal to $Ш_{\Lambda^*}$ characterizing the reciprocal Λ^* times the volume V^{-1} of the primitive cell of Λ^*.

By arguments analogous to those given in the one-dimensional case of (1.79), one has

$$\iiint_0^1 (\text{Ш}_\Lambda * f)(\boldsymbol{x}) \exp(-2\pi i h_i x^i)\, \mathrm{d}x^1 \mathrm{d}x^2 \mathrm{d}x^3 = \frac{1}{V} \int_{\text{cell}} \left(\sum_{\check{\boldsymbol{m}}} f(\boldsymbol{x} - \check{\boldsymbol{m}}) \right) \exp(-2\pi i \check{\boldsymbol{h}} \cdot \boldsymbol{x})\, \mathrm{d}^3\boldsymbol{x}$$

$$= \frac{1}{V} \sum_{\check{\boldsymbol{m}}} \int_{\text{cell}-\check{\boldsymbol{m}}} f(\boldsymbol{y}) \exp(-2\pi i \check{\boldsymbol{h}} \cdot (\boldsymbol{y} + \check{\boldsymbol{m}}))\, \mathrm{d}^3\boldsymbol{y} = \frac{1}{V} \int f(\boldsymbol{y}) \exp(-2\pi i \check{\boldsymbol{h}} \cdot \boldsymbol{y})\, \mathrm{d}^3\boldsymbol{y} = V^{-1} \mathcal{F} f(\check{\boldsymbol{h}}) ,$$

i.e., the periodic function $\text{Ш}_\Lambda * f$ can be expressed as the Fourier series

$$(\text{Ш}_\Lambda * f)(\boldsymbol{x}) = V^{-1} \sum_{\check{\boldsymbol{h}}} (\mathcal{F} f)(\check{\boldsymbol{h}}) \exp(2\pi i \check{\boldsymbol{h}} \cdot \boldsymbol{x})$$

with Fourier coefficients given by (V^{-1}-scaled) Fourier transform of f at nodes $\check{\boldsymbol{h}}$ of the lattice Λ^* reciprocal to Λ. The above relationship arises in consideration of wave diffraction by an unbounded crystal; see (2.35) and (2.36) in Sect. 2.5.1 below.

1.11.5 Projection-Slice Theorem

Later on, in Chap. 13, a simple form of the projection-slice theorem will be needed. Based on the definition (1.59) of the Fourier transform $\mathcal{F}_x[f](\xi)$ of f, it is easy to see that $\mathcal{F}_x[f](0) = \int f(x)\, \mathrm{d}x$. Similarly, in two dimensions, with $p_2(x^1) = \int f(x^1, x^2)\, \mathrm{d}x_2$, one has

$$\mathcal{F}_{(x^1,x^2)}[f](\xi_1, 0) = \int_{-\infty}^{\infty} \mathrm{d}x_1 \int_{-\infty}^{\infty} \mathrm{d}x_2\, f(x^1, x^2) \exp\left(-2\pi i x^1 \xi_1\right) = \mathcal{F}_{x^1}[p_2](\xi_1) ,$$

i.e., the Fourier transform of the projection p_2 of the function $f(x^1, x^2)$ along the second axis, is equal to the 'slice' through the Fourier transform $\mathcal{F}[f](\xi_1, \xi_2)$ of f at $\xi_2 = 0$. The theorem can be generalized to arbitrary direction of projection, to higher dimensions and to non-Cartesian bases.

References

1. T. Hahn (ed.), *International Tables for Crystallography*, vol. A (Springer Verlag, Berlin, 2005)
2. A. Morawiec, On representing rotations by Rodrigues parameters in non-orthonormal reference systems. Acta Cryst. A **72**, 548–556 (2016)
3. A. Morawiec, Parameterization of rotations in reference frames with redundant crystallographic axes. IOP Conf. Ser.: Mater. Sci. Eng. **375**, 012027 (2018)
4. A. Ben-Israel, T.N.E. Greville, *Generalized Inverses: Theory and Applications* (Springer, New York, 2003)
5. J.W. Gibbs, E.B. Wilson, *Vector Analysis* (Charles Scribner's Sons, New York, 1901)
6. D.W.J. Cruickshank, H.J. Juretschke, N. Kato (ed.), *P.P. Ewald and his Dynamical Theory of X-ray Diffraction*, (IUCr/Oxford University Press, Oxford, 1992)

7. P.P. Ewald, Das "reziproke Gitter" in der Strukturtheorie. Z. Kristallogr. **56**, 129–156 (1921)
8. M. Laue, Les phénomènes d'interférence des rayons de Röntgen, produits par le réseau tridimensional des cristaux, in *La structure de la matière*. Rapports et discussions du Conseil de physique tenu à Bruxelles du 27 au 31 octobre 1913, sous les auspices de l'Institut international de physique Solvay. (Gauthier-Villars et cie, Paris, 1921), pp. 75–112
9. M. Laue. Die Interferenzerscheinungen an Röntgenstrahlen hervorgerufen durch das Raumgitter der Kristalle. In *Gesammelte Schriften und Vorträge. Bd. 1*. (Vieweg und Sohn, Braunschweig, 1961), pp. 285–325
10. M. Pohst, A modification of the LLL reduction algorithm. J. Symb. Comput. **4**, 123–127 (1987)
11. J. Buchmann, M. Pohst. Computing a lattice basis from a system of generating vectors, in *Proceedings of EUROCAL87, Lecture Notes in Computer Science 378* (Springer, Berlin, 1989), pp. 54–63
12. B.N. Delone, R.V. Galiulin, M.I. Shtogrin, On the Bravais types of lattices. Itogi Mauki i Tekhniki, Sovremennye Problemy Matematiki **2**, 119–254 (1973)
13. B. Gruber, The relationship between reduced cells in a general Bravais lattice. Acta Cryst. A **29**, 433–440 (1973)
14. I. Křivý, B. Gruber, A unified algorithm for determining the reduced (Niggli) cells. Acta Cryst. A **32**, 297–298 (1976)
15. A. Santoro, A.D. Mighell, Determination of reduced cells. Acta Cryst. A **26**, 124–127 (1970)
16. B. Gruber, Reduced cells based on extremal principles. Acta Cryst. A **45**, 123–131 (1989)
17. B. Gruber, Topological approach to the Niggli lattice characters. Acta Cryst. A **48**, 461–470 (1992)
18. B. Delaunay, Neue Darstellung der geometrischen Kristallographie. Z. Kristallogr. **84**, 109–149 (1933)
19. A.L. Patterson, E. Love, Remarks on the Delaunay reduction. Acta Cryst. **10**, 111–116 (1957)
20. B.N. Delone, R.V. Galiulin, N.P. Dolbilin, V.A. Zalgaller, M.I. Shtogrin, On three successive minima of a three-dimensional lattice. Dokl. Akad. Nauk SSSR **209**, 309–313 (1973)
21. H. Burzlaff, H. Zimmermann, On the metrical properties of lattices. Z. Kristallogr. **170**, 247–262 (1985)
22. B. Gruber, Classification of lattices: a new step. Acta Cryst. A **53**, 505–521 (1997)
23. A. Santoro, A.D. Mighell, Coincidence-site lattices. Acta Cryst. A **29**, 169–175 (1973)
24. Y. Billiet and M. Rolley Le Coz. Le groupe *P*1 et ses sous-groupes. II. Tables de sous-groupes. *Acta Cryst. A*, **36**, 242–248 (1980)
25. B. Gruber. Further properties of lattices. In T. Hahn, editor, *International Tables for Crystallography, Vol. A, Section 9.3* (Springer, Dodrecht, 2005) pp. 756–760
26. N.J.A. Sloane. The on-line encyclopedia of integer sequences. http://oeis.org/. Accessed Aug 2022
27. A. Morawiec, *Orientations and Rotations. Computations in Crystallographic Textures* (Springer-Verlag, Berlin, 2004)
28. H. Weyl, *Symmetry* (Princeton University Press, Princeton, 1952)
29. J.D.H. Donnay, The structural classification of crystal point symmetries. Acta Cryst. A **33**, 979–984 (1977)
30. R.W. Grosse-Kunstleve, N.K. Sauter, P.D. Adams, Numerically stable algorithms for the computation of reduced unit cells. Acta Cryst. A **60**, 1–6 (2004)
31. H. Zimmermann, H. Burzlaff, *DELOS* - a computer program for the determination of a unique conventional cell. Z. Kristallogr. **170**, 241–246 (1985)
32. E. Parthé, L.M. Gelato, The standardization of inorganic crystal-structure data. Acta Cryst. A **40**, 169–183 (1984)
33. M.A. Bravais, *Études Cristallographiques* (Gauthier-Villars, Paris, 1866)
34. L. Weber, Das viergliedrige Zonensymbol des hexagonalen Systems. Z. Kristallogr. **57**, 200–203 (1922)
35. F.C. Frank, On Miller-Bravais indices and four-dimensional vectors. Acta Cryst. **18**, 862–866 (1965)

36. R.B. Roof Jr. A theoretical extension of the reduced-cell concept in crystallography. *Los Alamos Report*, LA–4038, 1967
37. A.D. Mighell, J.R. Rodgers, Lattice symmetry determination. Acta Cryst. A **36**, 321–326 (1980)
38. A.D. Mighell, Lattice symmetry and identification - the fundamental role of reduced cells in materials characterization. J. Res. Natl. Inst. Sand. Technol. **106**, 983–995 (2001)
39. P.M. de Wolff, Reduced bases. In T. Hahn, editor, *International Tables for Crystallography, Vol. A, Section 9.2* (Springer, Dodrecht, 2005) pp. 750–755
40. T. Higgins, R. Dark, P. McArdle, J. Simmie, *BRVCEL* - a computer program for cell reduction and Bravais lattice determination. Computers Chem. **14**, 33–36 (1990)
41. C. Katayama, Bravais-lattice determination for automatic data collection. J. Appl. Cryst. **19**, 69–72 (1986)
42. R. Oishi-Tomiyasu, Rapid Bravais-lattice determination algorithm for lattice parameters containing large observation errors. Acta Cryst. A **68**, 525–535 (2012)
43. L.C. Andrews, H.J. Bernstein, G.A. Pelletier, A perturbation stable cell comparison technique. Acta Cryst. A **36**, 248–252 (1980)
44. A.D. Mighell, V.L. Himes, Compound identification and characterization using lattice-formula matching techniques. Acta Cryst. A **42**, 101–105 (1986)
45. W. Clegg, Cell reduction and lattice symmetry determination. Acta Cryst. A **37**, 913–915 (1981)
46. G. Ferraris, G. Ivaldi, A simple method for Bravais lattice determination. Acta Cryst. A **39**, 565–569 (1983)
47. Y. Le Page, The derivation of the axes of the conventional unit cell from the dimensions of the Buerger-reduced cell. J. Appl. Cryst. **15**, 255–259 (1982)
48. A.L. Spek, *LEPAGE* - an MS-DOS program for the determination of the metrical symmetry of a translation lattice. J. Appl. Cryst. **21**, 578–579 (1988)
49. Y. Le Page, Computer derivation of the symmetry elements implied in a structure description. J. Appl. Cryst. **20**, 264–269 (1987)
50. Y. Le Page. *MISSYM 1.1*—a flexible new release. J. Appl. Cryst. **21**, 983–984 (1988)
51. A. Togo, I. Tanaka. *Spglib*: a software library for crystal symmetry search, 2018. https://arxiv.org/pdf/1808.01590.pdf. Accessed Aug 2022
52. A. Santoro, A.D. Mighell, J.R. Rodgers, The determination of the relationship between derivative lattices. Acta Cryst. A **36**, 796–800 (1980)
53. J. Macíček and A. Yordanov. *BLAF*—a robust program for tracking out admittable Bravais lattice(s) from the experimental unit-cell data. J. Appl. Cryst. **25** 73–80 (1992)
54. A.A. Lebedev, A.A. Vagin, G.N. Murshudov, Intensity statistics in twinned crystals with examples from the PDB. Acta Cryst. D **62**, 83–95 (2006)
55. V.L. Himes, A.D. Mighell, A matrix method for lattice symmetry determination. Acta Cryst. A **38**, 748–749 (1982)
56. V.L. Karen, A.D. Mighell, Converse-transformation analysis. J. Appl. Cryst. **24**, 1076–1078 (1991)
57. U.F. Kocks, C.N. Tomé, and H.R. Wenk. *Texture and Anisotropy. Preferred Orientations in Polycrystals and their Effect on Materials Properties*. (Cambridge University Press, Cambridge, 1998)
58. C. Neumann, Ueber die thermischen Axen der Krystalle des ein- und eingliedrigen Systems. Ann. Phys. Chem. **114**, 492–504 (1861)
59. W.R. Busing, H.A. Levy, Angle calculations for 3- and 4-circle X-ray and neutron diffractometers. Acta Cryst. **22**, 457–464 (1967)

Chapter 2
Basic Aspects of Crystal Diffraction

2.1 Scattering of Waves in Solids

To be able to determine lattice parameters or crystal orientations, it is essential to know how features in diffraction patterns are related to crystal structures. In general, the term *diffraction* refers to physical phenomena occurring when waves encounter obstacles, and the directions of wave propagation are altered. When the dimensions of the obstacles are of the order of the length of waves, the diffraction results in a specific superposition (interference) of waves. This leads to the formation of patterns on wave detectors.[1]

We are interested in diffraction patterns formed by waves scattered by periodically arranged atoms in crystals. When the radiation penetrates a crystal, the waves scattered by the atoms interfere with each other. Assuming coherent scattering preserving the relative phases of waves emanating from different atoms, waves propagating in some directions add to each other (constructive interference), whereas waves propagating in other directions subtract from each other (destructive interference). In effect, the intensities of the scattered radiation vary leading to a *crystal* diffraction pattern. For the interference to occur, the wavelength must be of the order of the inter-atomic distances. Hence, only relatively short waves are applicable. In practice, X-rays, electrons and neutrons are used. X-ray wavelengths applied in crystallography range from 0.5 to 2.5Å. The wavelengths of electrons in conventional electron microscopy are usually in the range from 0.02 to 0.10Å. Also neutrons are slowed down so the corresponding wavelength is of the order of 1Å. Below are basic facts about scattering of these three types of radiation.

[1] For an extensive introduction to diffraction and the early history of research on this phenomenon, see [1].

A. Morawiec, *Indexing of Crystal Diffraction Patterns*, Springer Series in Materials Science 326, https://doi.org/10.1007/978-3-031-11077-1_2

X-ray Scattering

An electromagnetic wave incident on a charged particle sets it into periodic motion, and the oscillating particle, in turn, emits electromagnetic radiation with the same frequency as the incoming beam (Thomson scattering). Since the nuclei are heavy, the interaction of X-rays with them is weak and can be neglected, and the considerations can be focused on X-rays scattered by electrons. Thus, the description of X-ray scattering is based on the spatial distribution of electrons in a crystal.[2] In general, a wave is scattered either without any loss of energy (elastic scattering) or with a loss of energy and increased wavelength (inelastic or Compton scattering). Our interest is in elastically scattered waves. The complete description of the X-ray scattering—the dynamical theory of X-ray diffraction taking into account multiple scattering events—relies on the Maxwell equations of the classical electromagnetic theory.

Scattering of High-Energy Electrons

Electrons, as charged particles, interact with matter through the Coulomb forces. Therefore, the basic description of electron scattering is based on the spatial distribution of the Coulomb potential. The incident electrons are influenced by both the nuclei and the electrons in the material. Due to the strength of the interaction, electrons penetrate solids only for short distances. As in the case of X-rays, both elastic and inelastic components play a role. Some patterns are dominated by elastically scattered electrons, whereas in other cases (e.g., Kikuchi patterns) inelastic scattering is essential. A physically correct approach to electron diffraction takes into account the fact that interference effects follow from the wave-like nature of electrons, and the dynamical theory of electron diffraction relies on equations of quantum mechanics.

Non-Magnetic Neutron Scattering

Free neutrons are scattered by interactions with nuclei. With this process, the interaction of neutrons with solids is much weaker than that of X-rays, and large penetration depths are possible. Differently than for X-rays or electrons, the scattering power of an atom does not decrease with growing scattering angle, and moreover, light atoms diffract strongly; in fact, for neutrons, there is no clear correlation between the atomic number and the scattering power, and diffracted intensities can be different for different isotopes. Nuclear scattering of neutrons provides information about atomic positions only indirectly affected by electron density and atomic bonding.

[2] One needs to be aware that, due to atomic bonding, the distribution of electrons in the crystal differs from a theoretical sum of spherically symmetric distributions of isolated atoms.

2.1.1 Coherence

Crucial for diffraction is the aforementioned assumption of coherent scattering. Briefly, two waves are fully coherent if they have the same wavelength and are in phase with each other. The temporal coherence describes the bandwidth (range of frequencies) of the wave; a single frequency (monochromatic) wave has a perfect temporal coherence. Second, with a wave extended over an area, one needs to consider the spatial coherence, i.e., the constancy of the relationship between phases at different points in space. Spatial coherence of radiation is related to the size of the radiation source. The types of coherence are illustrated in Fig. 2.1.

Real radiation sources are partially coherent. They have finite (non-zero) physical extent, and they radiate in a finite part of spectrum. Clearly, use of filtering devices and pinholes is always at the loss of intensity. Key for many experiments is that at large distances from a small incoherent source the waves appear nearly coherent. In other words, a finite nearly monochromatic incoherent radiation source produces a partially spatially coherent wave front as the radiation propagates.

As the diffraction is based on interference of waves influenced by different parts of the diffracting object, incoherence obscures diffraction patterns and hides the structure of the object.

2.1.2 Diffraction Theories

In-depth theories of diffraction are rather involved. There are, however, a number of approximations. This introduction to diffraction begins with an elementary geometric description. Then, basic information on the kinematical theory is given. Some aspects

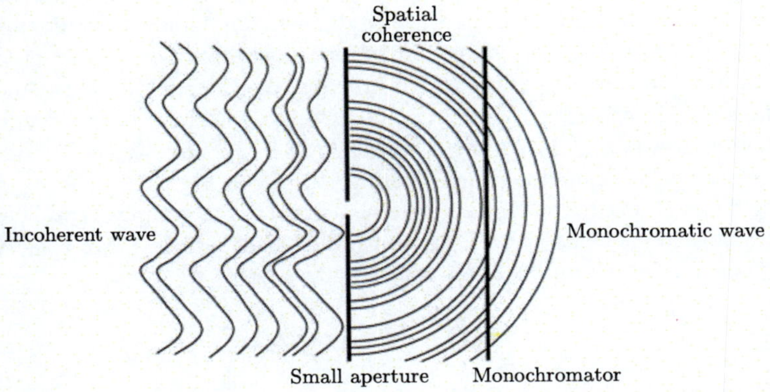

Fig. 2.1 Schematic illustration of (in)coherence [2]. Incoherent wave becomes spatially coherent after passing through a pinhole and temporally coherent after passing through a monochromator.

of the dynamical theory of electron diffraction will be presented in Chap. 3. In the considerations below, it will be assumed that a crystal is illuminated by monochromatic plane wave of radiation. If polychromatic radiation is considered, this will be explicitly stated. Moreover, the energy of the scattered wave is assumed to be equal to that of the incident wave (elastic scattering).

The foundations of the description of the phenomenon of diffraction were laid in the field of light optics. The key conclusion is that in the far-field diffraction (also known as Fraunhofer diffraction) the distribution of the amplitudes of waves scattered by an object is the Fourier transform of the transfer function of the object. See Appendix 2.8. In cases considered below, the waves may be of different kinds but the conditions for Fraunhofer diffraction are satisfied, and the amplitudes of scattered waves are Fourier transforms of functions characterizing the scattering objects.

2.2 Geometry of Crystal Diffraction

2.2.1 Laue Equation

For a given radiation wavelength, the directions of the elastically scattered high intensity rays are determined by the direction of the beam of incident radiation and the crystal orientation. The simplest description of crystal diffraction is analogous to that of diffraction of light on a diffraction grating—equidistant and parallel slits separated by a distance comparable to the light wavelength.

Diffraction by a Grating

With Fraunhofer diffraction, i.e., with the screen located at a distance large compared to the distance between slits and the wavelength, coherent monochromatic radiation passing through a grating creates a pattern of repeated (diffraction) peaks. The peaks appear at locations corresponding to constructive interference of waves emerging from the slits. Constructive interference takes place when the path difference of the radiation rays is a multiple of the wavelength. It is instructive to consider the simple planar case (one-dimensional grating) shown in Fig. 2.2. With α_0 and α denoting the angles between the rays and the normal to the grating and a being the distance between the slits, the path difference is $a(\sin\alpha - \sin\alpha_0)$. For constructive interference, it must a multiple of the wavelength λ, i.e.,

$$a(\sin\alpha - \sin\alpha_0) = n\lambda \, .$$

This is the well known (one-dimensional) diffraction grating equation. Clearly, the longest wave which the grating can diffract has the length $\lambda = 2a$. Let $\boldsymbol{k}_0$ and $\boldsymbol{k}$ be the wave vectors of the incident and diffracted waves respectively ($|\boldsymbol{k}_0| = 1/\lambda = |\boldsymbol{k}|$),

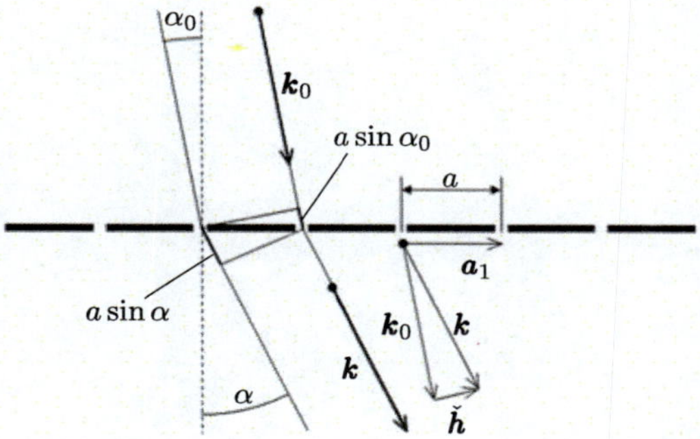

Fig. 2.2 Schematic illustration of diffraction on one-dimensional grating. With α_0 and α denoting the angles between the rays and the normal to the grating and a being the distance between the slits, the path difference for rays through neighboring slits is $a(\sin \alpha - \sin \alpha_0)$.

and let $\boldsymbol{a}_1$ be the vector between neighboring slits. With $\check{\boldsymbol{h}} = \boldsymbol{k} - \boldsymbol{k}_0$ and $n = h_1$, the diffraction grating equation can be written as

$$\boldsymbol{a}_1 \cdot \check{\boldsymbol{h}} = h_1 \ .$$

This formula can be generalized to higher dimensions. Two superimposed diffraction gratings form a two-dimensional periodic array of apertures[3] (Fig. 2.3). The constructive interference occurs when the diffraction grating equation is satisfied for each of the two component gratings

$$\boldsymbol{a}_1 \cdot \check{\boldsymbol{h}} = h_1 \ , \quad \boldsymbol{a}_2 \cdot \check{\boldsymbol{h}} = h_2 \ , \tag{2.1}$$

where $\check{\boldsymbol{h}} = \boldsymbol{k} - \boldsymbol{k}_0, \boldsymbol{a}_i$ is the vector linking nearest slits of the i-th grating, and h_i are integers. It turns out that also three-dimensional structures may serve as diffraction gratings.

Crystal as a Diffraction Grating

A 'crystal' contains an $\boldsymbol{a}_i$-based ($i = 1, 2, 3$) lattice of diffracting centers, and such a lattice constitutes a three-dimensional grating. For the crystal, the Fraunhofer diffraction of radiation having wavelength comparable to $|\boldsymbol{a}_i|$ is governed by relationships analogous to (2.1)

$$\boldsymbol{a}_i \cdot \check{\boldsymbol{h}} = h_i \ ,$$

where $i = 1, 2, 3$, and h_i are integers. These equations are solved with respect to $\check{\boldsymbol{h}}$ by

$$\check{\boldsymbol{h}} = h_i \boldsymbol{a}^i \ , \tag{2.2}$$

[3] The shapes of the apertures and of the grating are ignored here. They, respectively, determine the large and fine structures of the pattern.

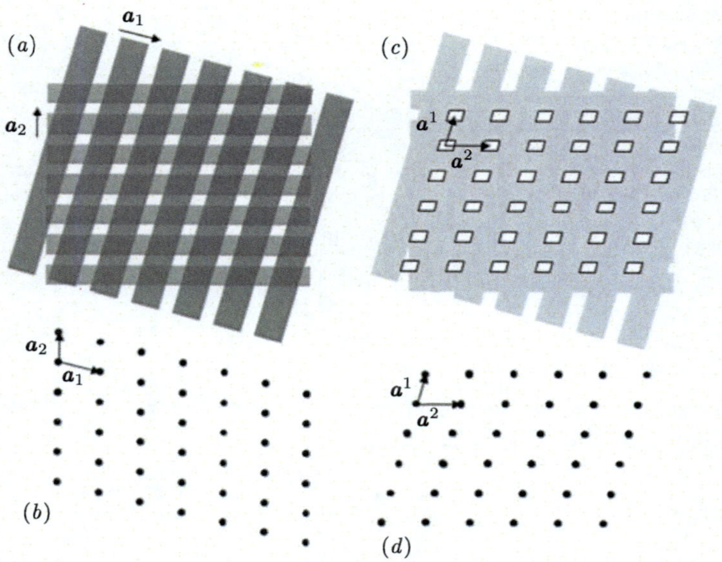

Fig. 2.3 (*a*) Two superimposed diffraction gratings with slits of the i-th grating separated by vector $\boldsymbol{a}_i$. (*b*) Schematic of the lattice based on vectors $\boldsymbol{a}_1$ and $\boldsymbol{a}_2$. (*c*) Two-dimensional array of apertures; apertures along the slit of the i-th grating are separated by vector $\boldsymbol{a}^i$. (*d*) Schematic of the lattice based on vectors $\boldsymbol{a}_1$ and $\boldsymbol{a}_2$. This lattice is a scaled reciprocal of the lattice shown in (*b*). (In the realm of the diffraction optics, real gratings do not form a point-like image; image details depend on the extent of the gratings and shapes of the apertures. Cf. footnote 3.)

where the vectors $\boldsymbol{a}^i$ constitute a basis reciprocal to the basis $\boldsymbol{a}_i$. Hence, the diffraction peaks are enumerated by vectors of the crystal reciprocal lattice, and the wave vector $\boldsymbol{k} = \boldsymbol{k}_{\check{\boldsymbol{h}}}$ corresponding to the diffraction peak $\check{\boldsymbol{h}}$ is determined by

$$\boldsymbol{k}_{\check{\boldsymbol{h}}} - \boldsymbol{k}_0 = \check{\boldsymbol{h}} . \tag{2.3}$$

This diffraction condition is known as the *Laue equation*[4]: the difference between wave vectors of the incident and diffracted waves (referred to as the *scattering vector*) is equal to a reciprocal lattice vector. Together with the energy conservation expressed via the wave vectors

$$\boldsymbol{k}_{\check{\boldsymbol{h}}}^2 = \boldsymbol{k}_0^2 , \tag{2.4}$$

the Laue equation governs the simple geometric description of crystal diffraction. In particular, they imply that diffraction peaks may arise only for reciprocal lattice vectors satisfying the condition $|\check{\boldsymbol{h}}| \leq |\boldsymbol{k}_{\check{\boldsymbol{h}}}| + |\boldsymbol{k}_0| = 2/\lambda$.

[4] Max Theodor Felix von Laue (1879–1960), co-discoverer (with Paul Knipping and Walter Friedrich) of crystal diffraction.

The vector $\check{\boldsymbol{h}}$ is perpendicular to $\boldsymbol{k}_{\check{h}} + \boldsymbol{k}_0$; one has $(\boldsymbol{k}_{\check{h}} + \boldsymbol{k}_0) \cdot \check{\boldsymbol{h}} = (\boldsymbol{k}_{\check{h}} + \boldsymbol{k}_0) \cdot (\boldsymbol{k}_{\check{h}} - \boldsymbol{k}_0) = \boldsymbol{k}_{\check{h}}^2 - \boldsymbol{k}_0^2 = 0$. This relationship and the Laue equation lead to[5]

$$\check{\boldsymbol{h}} \cdot (\check{\boldsymbol{h}} - 2\boldsymbol{k}_{\check{h}}) = 0 \tag{2.5}$$

and

$$\check{\boldsymbol{h}} \cdot (\check{\boldsymbol{h}} + 2\boldsymbol{k}_0) = 0\,. \tag{2.6}$$

Each of these two formulas links the reciprocal lattice vector to just one of wave vectors $\boldsymbol{k}_0$ and $\boldsymbol{k}_{\check{h}}$. The formulas implicate that with diffraction occurring for a non-zero $\check{\boldsymbol{h}}$, the inequalities $\check{\boldsymbol{h}} \cdot \boldsymbol{k}_{\check{h}} > 0 > \check{\boldsymbol{h}} \cdot \boldsymbol{k}_0$ are satisfied.

For a non-zero vector $\check{\boldsymbol{h}} = |\check{\boldsymbol{h}}|\widehat{\boldsymbol{n}}_{\check{h}}$, the Laue (2.3) and (2.6) lead to $\boldsymbol{k}_{\check{h}} = \boldsymbol{k}_0 + \check{\boldsymbol{h}} = \boldsymbol{k}_0 - 2\check{\boldsymbol{h}}\,(\check{\boldsymbol{h}} \cdot \boldsymbol{k}_0)/(\check{\boldsymbol{h}} \cdot \check{\boldsymbol{h}})$. Hence, one has

$$\boldsymbol{k}_{\check{h}} = \left(I - 2\widehat{\boldsymbol{n}}_{\check{h}} \otimes \widehat{\boldsymbol{n}}_{\check{h}}\right)\boldsymbol{k}_0\,, \tag{2.7}$$

i.e., the wave vector $\boldsymbol{k}_{\check{h}}$ is obtained by reflection of $\boldsymbol{k}_0$ with respect to the plane normal to $\widehat{\boldsymbol{n}}_{\check{h}}$ and $\check{\boldsymbol{h}}$; cf. (1.41). Although a beam diffracted by a crystal differs from a light-ray reflected by a mirror by its discrete character (see the next paragraph), because of (2.7), it is still described as a 'reflection'.

It needs to be stressed that there is a fundamental difference between diffraction from a two-dimensional grating and from a three-dimensional crystal structure: a two-dimensional grating diffracts for a continuous range of directions of the incident beam, whereas a three-dimensional structure diffracts only for certain incident beam directions. Why? A three-dimensional 'grating' can also be seen as a periodic sequence of identical two-dimensional gratings (crossed gratings). The beam diffracted by the first grating of the sequence is incident on the second one. It will be diffracted again by this (and subsequent) gratings only in certain circumstances: roughly, that will occur if the small 'apertures' of the second grating match peaks of the diffraction pattern of the first grating, or formally, if the Laue condition is satisfied.[6] With fixed $\boldsymbol{k}_0$ and $\check{\boldsymbol{h}}$ being a reciprocal lattice vector of (2.2), there may be no vectors $\boldsymbol{k}$ satisfying conditions (2.3) and (2.4). On the other hand, for the two-dimensional grating, (2.1) are solved by $\check{\boldsymbol{h}} = h_1\boldsymbol{a}^1 + h_2\boldsymbol{a}^2 + x\boldsymbol{a}^3$, where x is an arbitrary real number, and the conditions for diffraction become less stringent. The selectiveness of crossed gratings is best explained by Ewald construction.

---

[5] Equation (2.5) is frequently written as $(\check{\boldsymbol{h}}/2)^2 = \boldsymbol{k}_{\check{h}} \cdot (\check{\boldsymbol{h}}/2)$; for readers familiar with the notion of Brillouin zone, this relationship simply means that the wave vectors $\boldsymbol{k}_{\check{h}}$ of diffracted beams have terminal points on the planes bounding the zones.

[6] If the crystal is sufficiently thin, it becomes similar to a two-dimensional grating, and (as in the case of electron spot patterns) diffraction spots appear even if the Laue condition is only approximately satisfied.

2.2.2 Ewald Construction

The Ewald[7] construction is a graphic visualization of the Laue equation. The construction involves points of the crystal reciprocal lattice and the (Ewald) sphere (of reflection) through the origin **0** of the coordinate system linked to the lattice, with the sphere center at $-\boldsymbol{k}_0$. A vector $\boldsymbol{k}_{\check{h}}$ having its initial point at the center and such that $|\boldsymbol{k}_{\check{h}}| = |\boldsymbol{k}_0|$ has the terminal point on the sphere. See Fig. 2.4. In other words, there are two characteristic points in the Ewald construction: the first one is the center of the sphere, which is also the origin of the secondary wave vectors, and thus can be seen as a point were diffraction takes place, and the second point is the origin of reciprocal lattice (which is also marked as the terminal point of $\boldsymbol{k}_0$).

According to the Laue equation, diffraction will occur when $\boldsymbol{k}_{\check{h}} - \boldsymbol{k}_0$ equals $\check{\boldsymbol{h}}$, i.e., when the sphere passes through the reciprocal lattice point indicated by $\check{\boldsymbol{h}}$. The diffracted beam will not be observed if the Ewald sphere does not pass through a reciprocal lattice point. Therefore, in order to observe diffracted beams, some of the quantities involved must vary. This could be achieved by a rotation of the crystal with respect to the incident beam, or by the use of polychromatic radiation with λ covering a certain spectrum. Both will lead to diffraction spots on two-dimensional detectors. If two parameters are varied, a planar detector will register lines. For instance, by illuminating many crystallites with various orientations (as in methods based on crystalline powders or polycrystalline solids) one obtains diffraction rings. Another geometric possibility is to have a convergent incident beam with $\boldsymbol{k}_0$ vectors covering certain solid angle; this leads to conic lines in (K-line) diffraction patterns. Section 2.3 contains more detailed analysis of the geometries of some diffraction techniques.

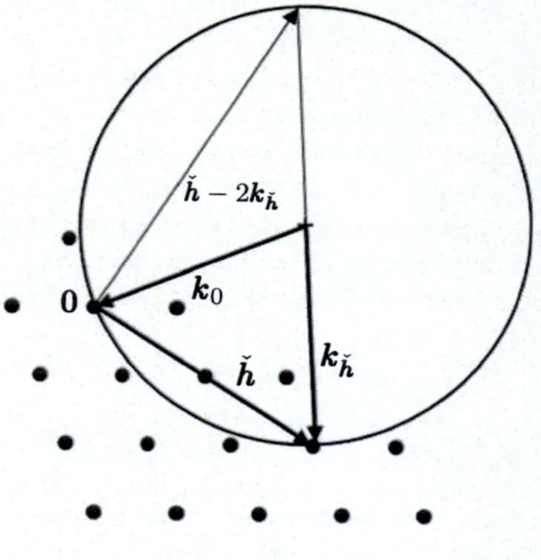

Fig. 2.4 Schematic of Ewald sphere of reflection. The figure also illustrates (2.5); since the third side of the triangle with two sides given by the vectors $\check{\boldsymbol{h}}$ and $\check{\boldsymbol{h}} - 2\boldsymbol{k}_{\check{h}}$ is a diameter of the circle, the triangle is right-angled, i.e. the vectors are perpendicular.

[7] Paul Peter Ewald (1888–1985).

2.2.3 Bragg's Law

Let $\check{\boldsymbol{h}}$ be a vector of the crystal reciprocal lattice determining a stack of crystal planes with indices $h\,k\,l$ and inter-planar distance $d_{(h\,k\,l)} = d_{\check{h}} = 1/|\check{\boldsymbol{h}}|$; cf. (1.48). As above, $\boldsymbol{k}_0$ and $\boldsymbol{k}_{\check{h}}$ denote the wave vectors indicating the directions of the incident beam and the diffracted beam, respectively, and having equal magnitudes $|\boldsymbol{k}_0| = 1/\lambda = |\boldsymbol{k}_{\check{h}}|$. The (Bragg) angle θ between the plane and the incident ray (Fig. 2.5) satisfies the condition

$$2\sin\theta = \cos(\pi/2-\theta) - \cos(\pi/2+\theta) = \widehat{\boldsymbol{k}}_{\check{h}}\cdot\widehat{\boldsymbol{h}} - \widehat{\boldsymbol{k}}_0\cdot\widehat{\boldsymbol{h}} = \lambda(\boldsymbol{k}_{\check{h}} - \boldsymbol{k}_0)\cdot\check{\boldsymbol{h}}/|\check{\boldsymbol{h}}| = \lambda\,|\check{\boldsymbol{h}}|\,, \quad (2.8)$$

where $\widehat{\boldsymbol{h}} = \check{\boldsymbol{h}}/|\check{\boldsymbol{h}}|, \widehat{\boldsymbol{k}}_0 = \boldsymbol{k}_0/|\boldsymbol{k}_0| = \lambda\boldsymbol{k}_0$, and the unit vector $\widehat{\boldsymbol{k}}_{\check{h}}$ is defined in analogous way. Hence, one obtains Bragg's law [3]: for the constructive interference to occur, there must exist a stack of planes $h\,k\,l$ such that[8]

$$\lambda = 2d_{(h\,k\,l)}\sin\theta\,; \quad (2.9)$$

moreover, the direction of the scattered beam is at the same angle with respect to the crystal plane as that of the incident beam, and both beams are in the plane (of incidence) perpendicular to the crystal plane. The important point is that not only does Bragg's law follow from the Laue equation, but also the Laue equation can be derived from Bragg's law[9]. Thus, the two descriptions of the geometry of crystal diffraction are equivalent. It is worth noting that

$$\lambda^2\,\boldsymbol{k}_0\cdot\boldsymbol{k}_{\check{h}} = \widehat{\boldsymbol{k}}_0\cdot\widehat{\boldsymbol{k}}_{\check{h}} = \cos(2\theta)\,; \quad (2.10)$$

thus, the Bragg angle is half of the angle between the wave vectors $\boldsymbol{k}_0$ and $\boldsymbol{k}_{\check{h}}$.

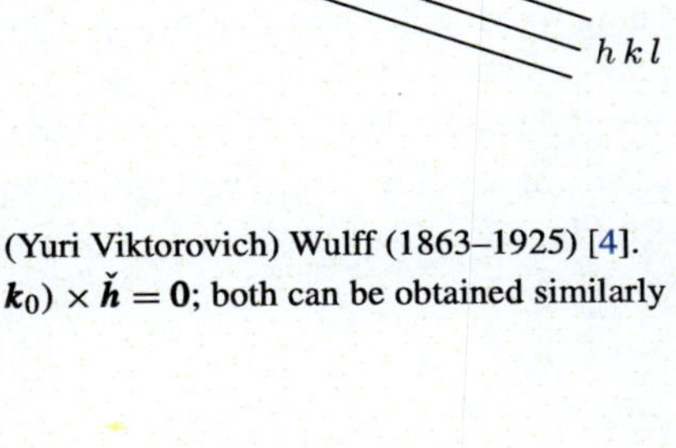

Fig. 2.5 Illustration of Bragg's law and its link to wave vectors $\boldsymbol{k}_{\check{h}}$ and $\boldsymbol{k}_0$ used in Laue equation.

[8] The law was independently discovered by Georg V. (Yuri Viktorovich) Wulff (1863–1925) [4].

[9] For that, one needs $(\boldsymbol{k}_{\check{h}} - \boldsymbol{k}_0)\cdot\check{\boldsymbol{h}} = \check{\boldsymbol{h}}\cdot\check{\boldsymbol{h}}$ and $(\boldsymbol{k}_{\check{h}} - \boldsymbol{k}_0)\times\check{\boldsymbol{h}} = \boldsymbol{0}$; both can be obtained similarly as (2.8).

Let n be a small positive integer. The formula (2.9) applies to all planes including $nh\,nk\,nl$, i.e., a diffraction peak is expected for θ satisfying $\lambda = 2d_{(nh\,nk\,nl)} \sin\theta$. Using the relationship $d_{(nh\,nk\,nl)} = d_{(h\,k\,l)}/n$ for the inter-planar distance in the stack of planes (see, Sect. 1.5.1), one obtains Bragg's law in the conventional form $n\lambda = 2d \sin\theta$. It is customary to consider first-order scattering on $nh\,nk\,nl$ instead of the scattering of n-th order on $h\,k\,l$.

2.3 Geometries of Selected Diffraction Techniques

There are numerous diffraction techniques for investigating crystalline materials. This section outlines the geometries of those that are referenced later in the book.

2.3.1 *X-ray Diffractometry*

One of the most commonly used experimental setups for investigating crystal diffraction—the X-ray diffractometer – consists of a fixed X-ray source, a monochromator limiting the wavelength range, and a detector registering intensities at varying scattering angles. Nowadays, the detectors are usually two-dimensional for concurrent detection of reflections at various angles. They replace linear and point detectors (moving on a circle around the diffracting material).

If a single crystal is examined, it is rotated about some axes depending on the geometry of the instrument. This could be a 'four circle diffractometer' with three variable angles to orient the crystal with respect to the incident beam, and the fourth θ angle for positioning the detector; see Fig. 4.1. The point of interaction of the beam with the crystal coincides with the unique point of intersection of the rotation axes. At some crystal orientations, the Laue conditions are satisfied and diffraction spots appear on the diffraction pattern.

In the *powder method*, the radiation is scattered by a collection of crystals, and the intensities are registered for varying scattering angle. This can be done with a cylindrical (Debye-Scherrer type) camera or a linear detector. With a point-like detector, in order to maintain a simple geometry, the so-called '$\theta - 2\theta$' scan is used, in which both the detector and the material rotate with synchronized rates, the rate of the detector being twice faster than that of the material (Bragg-Brentano technique). In terms of the wave vectors, this approach corresponds to the direction of $\boldsymbol{k} - \boldsymbol{k}_0$ being fixed with respect to the specimen, and to a varying magnitude of this vector. The scattering angle is directly linked to the inter-planar spacing (via Bragg's law), and the positions of diffraction peaks are drawn versus the spacing. With a given metric of the lattice, the crude geometry of such pattern follows from $d_{\check{h}} = 1/\sqrt{\check{\boldsymbol{h}} \cdot \check{\boldsymbol{h}}}$; see Fig. 2.6.

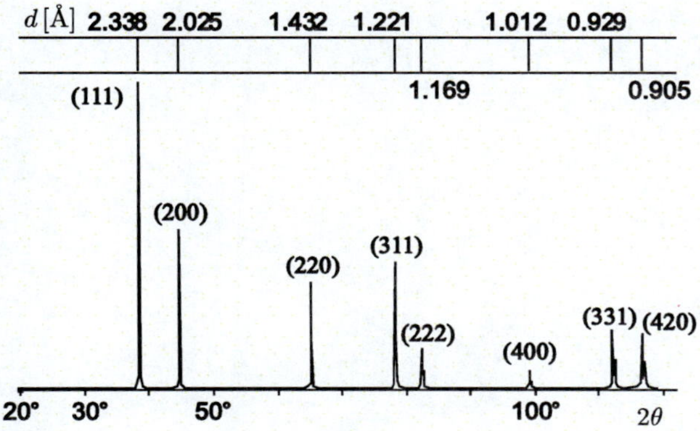

Fig. 2.6 Example X-ray powder diffractogram of fcc metal (Al). (Cu-K_α radiation with $\lambda \approx$ 1.540Å.)

The positions of peaks on a detector are determined by the directions of propagation of the diffracted waves. A diffraction pattern can be seen as intensity I being a function on (a part of) the unit sphere, i.e., $I = I(\widehat{\boldsymbol{k}})$, where $\widehat{\boldsymbol{k}} = \boldsymbol{k}/|\boldsymbol{k}|$, and $\boldsymbol{k}$ is a possible wave vector. In experimental situations, what matters are vectors $\widehat{\boldsymbol{k}}$ pointing towards maxima of the function $I = I(\widehat{\boldsymbol{k}})$. To determine these vectors from the positions of peaks, one needs the relationship between the detector coordinates and the (spherical) coordinates of $\widehat{\boldsymbol{k}}$ vectors.

2.3.2 *Planar Detector*

Most of the diffraction patterns are recorded on planar (position sensitive) detectors. The plane of a detector is determined by a vector $\boldsymbol{L}$ perpendicular to it with the magnitude equal to the distance of the detector to the point of scattering $\boldsymbol{0}$; see Fig. 2.7. With initial point of $\boldsymbol{L}$ at the point of scattering, its terminal point on the detector plane is called the *pattern center*. Every point $\boldsymbol{x}$ satisfying $(\boldsymbol{x} - \boldsymbol{L}) \cdot \boldsymbol{L} = 0$ is on the plane of the detector. Let $\boldsymbol{y} = \boldsymbol{x} - \boldsymbol{L}$ be a vector in the detector plane with the origin at the pattern center. The expression for the position of the trace of the reflection corresponding to the wave vector $\boldsymbol{k}$ such that $\boldsymbol{k} \cdot \boldsymbol{L} > 0$ is

$$\boldsymbol{y} = \frac{L^2}{\widehat{\boldsymbol{k}} \cdot \boldsymbol{L}} \widehat{\boldsymbol{k}} - \boldsymbol{L} = \frac{\boldsymbol{L} \times (\widehat{\boldsymbol{k}} \times \boldsymbol{L})}{\widehat{\boldsymbol{k}} \cdot \boldsymbol{L}} = \frac{\boldsymbol{L} \times (\boldsymbol{k} \times \boldsymbol{L})}{\boldsymbol{k} \cdot \boldsymbol{L}}, \tag{2.11}$$

where $\widehat{\boldsymbol{k}} = \lambda \boldsymbol{k}$, $|\widehat{\boldsymbol{k}}| = 1$. The point $\boldsymbol{y}$ is the gnomonic projection of points along $\widehat{\boldsymbol{k}}$ on the plane determined by $\boldsymbol{L}$. Knowing the position $\boldsymbol{y}$ of a peak on the detector, one can get the direction of the corresponding wave vector $\widehat{\boldsymbol{k}} = (\boldsymbol{y} + \boldsymbol{L})/|\boldsymbol{y} + \boldsymbol{L}|$.

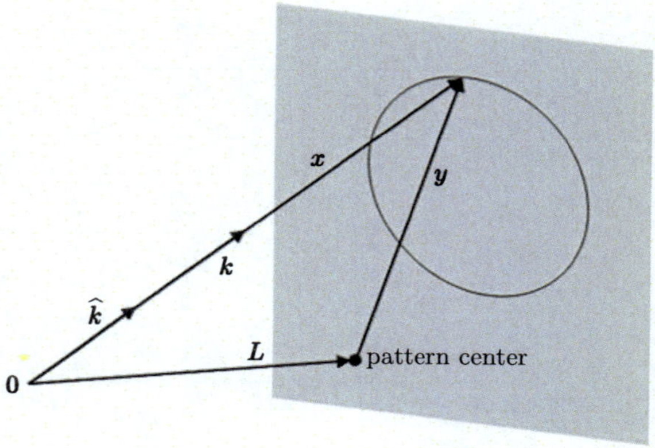

Fig. 2.7 Schematic illustration of (2.11) for a planar detector. The plane of the detector is determined by $\boldsymbol{L}$ via $(\boldsymbol{x} - \boldsymbol{L}) \cdot \boldsymbol{L} = 0$. The pattern center is the terminal point of $\boldsymbol{L}$. The vector $\boldsymbol{y} = \boldsymbol{x} - \boldsymbol{L}$ in the detector plane marks a position on the detector.

2.3.3 Geometry of K-lines

The so-called K-line diffraction patterns are obtained using monochromatic radiation with convergent incident beam. Thus, the geometry of K-line patterns is based on the assumption that the direction of the vector $\boldsymbol{k}_0$ is arbitrary, but its magnitude is fixed; possible directions of $\boldsymbol{k}_{\check{h}}$ follow from (2.5). Diffracted beams have directions given by all vectors $\boldsymbol{k}_{\check{h}}$ satisfying the relationships $\check{\boldsymbol{h}} \cdot (\check{\boldsymbol{h}} - 2\boldsymbol{k}_{\check{h}}) = 0$ and $|\boldsymbol{k}_{\check{h}}| = 1/\lambda$. Since $\boldsymbol{k}_0$ is not involved in these equations, the geometry of a K-line diffraction pattern is determined by the radiation wavelength and the orientation of the crystal. For a given reciprocal lattice vector $\check{\boldsymbol{h}}$, the equations are satisfied by

$$\boldsymbol{k}_{\check{h}} = \check{\boldsymbol{h}}/2 + \boldsymbol{l} \, , \tag{2.12}$$

where $\boldsymbol{l}$ is a vector perpendicular to $\check{\boldsymbol{h}}$ (i.e., $\boldsymbol{l} \cdot \check{\boldsymbol{h}} = 0$) of magnitude $|\boldsymbol{l}| = \sqrt{(1/\lambda^2 - (\check{\boldsymbol{h}}/2)^2)}$. In relation to this expression, it is worth recalling that a reflection can arise only if $|\check{\boldsymbol{h}}| < 2/\lambda$; cf. Sect. 2.2.1. The directions of vectors $\boldsymbol{k}_{\check{h}}$ of (2.12) cover a (Kossel) cone around the axis coinciding with $\check{\boldsymbol{h}}$. The angle between these directions and the plane perpendicular to $\check{\boldsymbol{h}}$ is equal to the Bragg angle θ (Fig. 2.8). An intersection of such cones by a detector leads to (Kossel) conics on a diffraction pattern (Fig. 2.9). The position of the conics on a planar detector is governed by (2.11).

The family of K-line diffraction patterns includes several pattern types. The best known are X-ray Kossel patterns (e.g., [5–7]) and electron Kikuchi patterns; closely related are divergent-beam X-ray patterns [8], diffuse multiple scattering patterns generated using synchrotron radiation [9], electron channeling patterns (e.g., [10]), convergent beam electron diffraction (CBED) patterns [11], electron back-scattered diffraction (EBSD) patterns [12, 13] or more exotic protonograms obtained by scattering of protons.

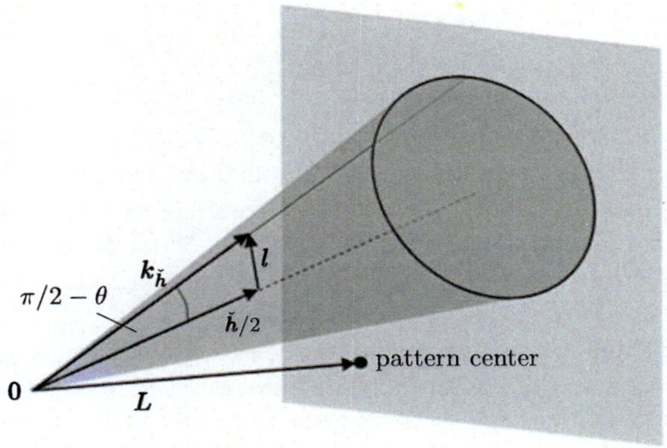

Fig. 2.8 Kossel cone and Kossel line corresponding the reciprocal lattice vector $\check{\boldsymbol{h}}$. The axis of the cone is parallel to $\check{\boldsymbol{h}}$. Cone generating lines are along the wave vectors of diffracted beams $\boldsymbol{k}_{\check{h}} = \check{\boldsymbol{h}}/2 + \boldsymbol{l}$, where $\boldsymbol{l}$ is perpendicular to $\check{\boldsymbol{h}}$ and $|\boldsymbol{l}| = \sqrt{(1/\lambda^2 - (\check{\boldsymbol{h}}/2)^2)}$. The angle between $\boldsymbol{k}_{\check{h}}$ and $\check{\boldsymbol{h}}$ is $\pi/2 - \theta$, where θ is the Bragg angle. The Kossel line is the intersection of the cone and the detector plane.

Fig. 2.9 Experimental Kossel pattern of CuAlBe (shape memory) alloy. Courtesy of D. Bouscaud. Original Kossel patterns were obtained by X-ray fluorescence. (In a sample exposed to X-rays, photons knock electrons out of their orbits. With electrons falling from outer shells onto the incomplete inner shells, the excess energy is emitted as fluorescence radiation.) The pattern shown above was generated by SEM with fluorescence radiation excited by electron beam. Cf. Sect. 14.3.

2.3.4 Electron Spot Patterns

Besides those listed above, there is a variety of other electron diffraction techniques. In high energy transmission electron microscopy (TEM), it is common to observe spot diffraction patterns. We will use the informal term *spot* patterns in reference to selected area diffraction patterns obtained with parallel beam, and patterns obtained with small-angle convergence of the incident beam. If the direct beam direction is along a low-index zone axis of the crystal, the pattern is similar to that of crossed gratings, and it is referred to as a zone axis pattern. Cf. Sect. 2.2.1.

With fixed $\boldsymbol{k}_0$, the diffraction condition (2.3) is satisfied only for some crystal orientations. Contrary to this assertion, TEM *spot* diffraction patterns are generally visible for arbitrary orientations. The reason lies in the thinness of the specimen prepared for transmission of electrons.

If a crystal is limited in size, its periodicity is disturbed and, as will be shown below (Sects. 2.5 and 3.4), the diffraction condition becomes 'stretched'; cf. footnote 6. In particular, if the crystal is a thin foil perpendicular to $\widehat{\boldsymbol{n}}$, diffraction occurs if

$$\boldsymbol{k}_{\check{h}} - \boldsymbol{k}_0 = \check{\boldsymbol{h}} + s_{\check{h}}\widehat{\boldsymbol{n}} ,$$

(cf. 2.3), where the *excitation error* $s_{\check{h}}$ is a parameter such that $|s_{\check{h}}|\, t < 1$, where t is the foil thickness. To take account of the excitation error, reciprocal lattice points are considered to be streaked out in the direction perpendicular to the foil. Such *re*ciprocal *l*attice *rods* are referred to as relrods (Fig. 2.11).

2.3.5 Geometry of Laue Patterns

Laue diffraction patterns are obtained with fixed incident beam direction and *polychromatic* radiation, i.e., the unit vector $\widehat{\boldsymbol{k}}_0 = \lambda \boldsymbol{k}_0$ is fixed, but the wavelength λ is in a certain interval $[\lambda_{min}, \lambda_{max}]$ [14]. Based on the Laue equation (2.3), a high intensity spot may appear if there exist a reciprocal lattice vector $\check{\boldsymbol{h}}$ and $\lambda_{\check{h}}$ in $[\lambda_{min}, \lambda_{max}]$ such that (2.6) or $\lambda_{\check{h}}\, \check{\boldsymbol{h}} \cdot \check{\boldsymbol{h}} + 2\widehat{\boldsymbol{k}}_0 \cdot \check{\boldsymbol{h}} = 0$ is satisfied. Thus, the incident beam direction $\widehat{\boldsymbol{k}}_0$ and the reciprocal lattice vector $\check{\boldsymbol{h}}$ ($\neq \boldsymbol{0}$) may lead to a spot if

$$-\frac{2\widehat{\boldsymbol{k}}_0 \cdot \check{\boldsymbol{h}}}{\check{\boldsymbol{h}} \cdot \check{\boldsymbol{h}}} = \lambda_{\check{h}} \tag{2.13}$$

is in the interval $[\lambda_{min}, \lambda_{max}]$. With this condition satisfied, the direction $\widehat{\boldsymbol{k}}_{\check{h}}$ of the diffracted beam follows from the Laue equation

$$\widehat{\boldsymbol{k}}_{\check{h}} = \widehat{\boldsymbol{k}}_0 + \lambda_{\check{h}} \check{\boldsymbol{h}} . \tag{2.14}$$

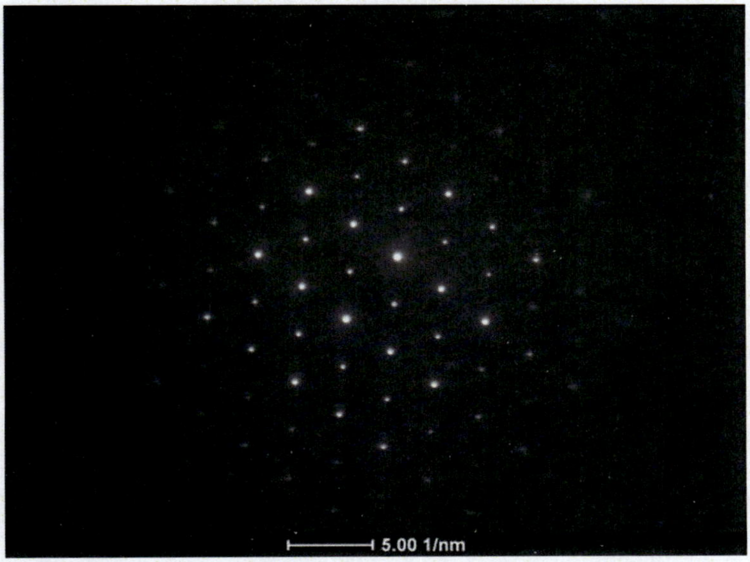

Fig. 2.10 An off-axis (selected area) electron diffraction *spot* pattern of austenite phase of Ni-Mn-Sn-In Heusler-type alloy. Courtesy of W. Maziarz.

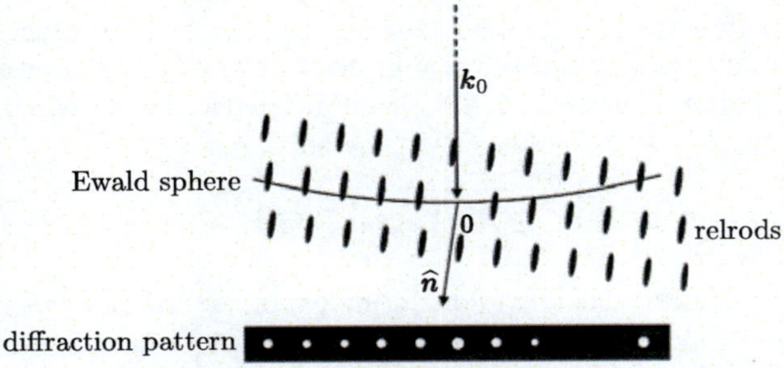

Fig. 2.11 Two-dimensional section through Ewald sphere and relrods – reciprocal lattice nodes streaked out along direction of vector $\widehat{\boldsymbol{n}}$ perpendicular to the foil. Spots in the (one-dimensional) diffraction pattern correspond to relrods intersecting the Ewald sphere.

Laue patterns are usually recorded on planar detectors perpendicular to the direction of the incident beam (in transmission, with $\widehat{\boldsymbol{L}} = +\widehat{\boldsymbol{k}}_0$, or in back-reflection, with $\widehat{\boldsymbol{L}} = -\widehat{\boldsymbol{k}}_0$).[10] Reflective geometry with specimen, detector and incident beam at some angles is less common; see, e.g., [15]. Again, the location of the spot on a planar detector is obtained by using (2.11).

[10] Reflective (transmission) geometry is typically used in the case of crystals with small (large) unit cells.

Let $\check{t}$ be a direct lattice vector. Based on (2.14), an arbitrary reciprocal-lattice point $\check{h}$ belonging to the zone $\check{t}$ (i.e., such that $\check{h} \cdot \check{t} = 0$) satisfies

$$\widehat{k}_{\check{h}} \cdot \check{t} = \widehat{k}_0 \cdot \check{t} \ .$$

Thus, the diffracted beams corresponding to vectors of one zone are all at the same angle to $\check{t}$, i.e., they lie on a cone, and—with a planar detector – the diffraction spots produced by these beams are located on a conic section.

2.4 Structure Factor

2.4.1 Introduction

Besides the diffraction geometry, which is fundamental for the indexing of diffraction patterns, also important are the intensities of particular diffraction peaks. They are necessary for structural analyses. The wave energy is proportional to the squared wave amplitude; one needs the amplitudes of particular waves for estimating the intensities of corresponding reflections and vice versa. The waves scattered by atoms interfere with each other, and the amplitude of the superposition of the scattered waves depends on the phase difference $\Delta\varphi$ via the factor $\exp(i\Delta\varphi)$. Let us consider interference of the waves scattered at points $\mathbf{0}$ and $\boldsymbol{x}$ in the same direction $\boldsymbol{k}$; see Fig. 2.12. As was already mentioned in Sect. 2.2, the path difference for the two waves equals $\Delta p_0 - \Delta p = \widehat{\boldsymbol{k}}_0 \cdot \boldsymbol{x} - \widehat{\boldsymbol{k}} \cdot \boldsymbol{x}$ and the difference of phases is

$$\Delta\varphi = 2\pi(\Delta p_0 - \Delta p)/\lambda = 2\pi(\boldsymbol{k}_0 - \boldsymbol{k}) \cdot \boldsymbol{x} \ .$$

Constructive interference leading to the formation of the diffracted wave will occur if $\Delta\varphi$ is close to a multiple of 2π, or if $(\boldsymbol{k}_0 - \boldsymbol{k}) \cdot \boldsymbol{x}$ is close to an integer.

In a more complete description, the incident harmonic plane wave of frequency ν, which reaches a scattering center (an atom) located at the point $\boldsymbol{x}_n$ at the time t, is described by the function $\psi_0 \propto \exp(2\pi i(\boldsymbol{k}_0 \cdot \boldsymbol{x}_n - \nu t))$. The actual disturbance is the real part of ψ_0. Let f_n denote the scattering power of the center (atom). With some simplification, the center at $\boldsymbol{x}_n$ can be considered as a point source for a spherical wave $f_n \psi_0 \exp(2\pi i(\boldsymbol{k} \cdot (\boldsymbol{x} - \boldsymbol{x}_n) - \nu t'))/|\boldsymbol{x} - \boldsymbol{x}_n|$, and with $\boldsymbol{x}$ far from $\boldsymbol{x}_n$, it is close

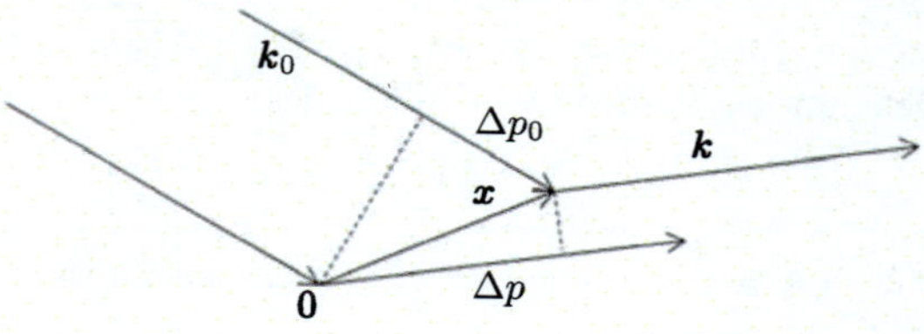

Fig. 2.12 Schematic for calculation of phase difference $\Delta\varphi = 2\pi(\Delta p_0 - \Delta p)/\lambda$. With $|\boldsymbol{k}_0| = 1/\lambda = |\boldsymbol{k}|$, it is given by $\Delta\varphi = 2\pi(\boldsymbol{k}_0 - \boldsymbol{k}) \cdot \boldsymbol{x}$.

to the plane wave

$$f_n \psi_0 \exp(2\pi i (\boldsymbol{k} \cdot (\boldsymbol{x} - \boldsymbol{x}_n) - \nu t')) = f_n \exp(-2\pi i (\boldsymbol{k} - \boldsymbol{k}_0) \cdot \boldsymbol{x}_n) \exp(i\alpha) ,$$

where $\alpha = 2\pi(\boldsymbol{k} \cdot \boldsymbol{x} - \nu(t + t'))$ does not depend on the location of the scattering centers. What matters for the relationship between structure and intensity is the factor $f_n \exp(-2\pi i (\boldsymbol{k} - \boldsymbol{k}_0) \cdot \boldsymbol{x}_n)$. With multiple centers (atoms) involved, the total undulation is

$$\sum_n f_n \exp(-2\pi i (\boldsymbol{k} - \boldsymbol{k}_0) \cdot \boldsymbol{x}_n) , \tag{2.15}$$

where the summation is over the scattering atoms. This is the fundamental equation of the kinematical theory of diffraction. According to this formula, the amplitude of the entire scattered wave is a weighted sum of separate waves scattered from atomic sites.

One needs to note that the scattering at an atom may involve a phase change, and the change can be different for atoms of different kind. This would lead to an additional n-dependent term in the argument of the exponent in (2.15). One can formally take this into account by allowing complex atomic scattering powers f_n. In most cases of our concern, however, the diversity of phase changes can be ignored and f_n are assumed to be real.

2.4.2 X-ray Form Factors

Let us focus for a while on the scattering of X-rays. Its description is based on the spatial distribution of electrons. The electron density function ρ is periodic with the periodicity of the crystal lattice. Thus, it can be expressed in the form of a Fourier series. With the contravariant coordinates x^1, x^2, x^3 of the vector $\boldsymbol{x}$ given in the basis of the lattice determining the crystal unit cell, the dimensionless function $\rho' = V_{\text{uc}}\, \rho$ can be written as

$$\rho'(\boldsymbol{x}) = V_{\text{uc}}\, \rho(\boldsymbol{x}) = \sum_{h_1,h_2,h_3}^{\infty} F(h_1, h_2, h_3) \exp(2\pi i (h_1 x^1 + h_2 x^2 + h_3 x^3)) , \tag{2.16}$$

where h_1, h_2, h_3 are integers and V_{uc} denotes the volume of the unit cell. With h_1, h_2, h_3 interpreted as the covariant components of vector $\check{\boldsymbol{h}}$, one has

$$\rho'(\boldsymbol{x}) = \sum_{\check{\boldsymbol{h}}} F(\check{\boldsymbol{h}}) \exp(2\pi i \check{\boldsymbol{h}} \cdot \boldsymbol{x}) , \tag{2.17}$$

where the summation is over all vectors $\check{\boldsymbol{h}}$ of the reciprocal lattice. Physically, this can be seen as expressing the density function in terms of plane waves $\exp(2\pi i \check{\boldsymbol{h}} \cdot \boldsymbol{x})$

propagating along reciprocal lattice vectors $\check{\boldsymbol{h}}$, i.e., waves with planes parallel to direct lattice planes; cf. Sect. 1.5.1. The coefficients $F(\check{\boldsymbol{h}})$ are known as *structure factors*.

Following (1.58) for Fourier coefficients, the structure factor for a given reciprocal lattice vector $\check{\boldsymbol{h}}$ is given by

$$F(\check{\boldsymbol{h}}) = \int_{\mathrm{uc}} \rho'(\boldsymbol{x}) \exp(-2\pi i \check{\boldsymbol{h}} \cdot \boldsymbol{x}) \, \mathrm{d}^3\boldsymbol{x} \,, \tag{2.18}$$

where uc denotes the unit cell, the integration is carried out using the coordinates in the lattice basis, $\mathrm{d}^3\boldsymbol{x} = \mathrm{d}x^1\mathrm{d}x^2\mathrm{d}x^3$, and the integration limits are from 0 to 1. To integrate in the physical space using Cartesian coordinates, one needs a transformation between the coordinate systems. The Jacobian matrix of the transformation from the coordinates in the lattice basis to Cartesian coordinates is $(\mathcal{T}^{-1})$ with $|\det(\mathcal{T})| = V_{\mathrm{uc}}$. Thus, using (1.7), the structure factor can be expressed as

$$F(\check{\boldsymbol{h}}) = \frac{1}{V_{\mathrm{uc}}} \int_{\mathrm{uc}} \rho'(\boldsymbol{x}) \exp(-2\pi i \check{\boldsymbol{h}} \cdot \boldsymbol{x}) \, \mathrm{d}^3\boldsymbol{x} = \int_{\mathrm{uc}} \rho(\boldsymbol{x}) \exp(-2\pi i \check{\boldsymbol{h}} \cdot \boldsymbol{x}) \, \mathrm{d}^3\boldsymbol{x} \,, \tag{2.19}$$

where the integration is carried out in the Cartesian system linked to the lattice. Based on the definition of the Fourier transformation (1.59), one can write

$$F(\check{\boldsymbol{h}}) = \mathcal{F}_{\boldsymbol{x}}[\rho_{|\mathrm{uc}}(\boldsymbol{x})](\check{\boldsymbol{h}}) \,, \tag{2.20}$$

where $\rho_{|\mathrm{uc}}$ denotes the function ρ with the support restricted to the unit cell uc, i.e.,

$$\rho_{|\mathrm{uc}}(\boldsymbol{x}) = 1_{|\mathrm{uc}}(\boldsymbol{x}) \, \rho(\boldsymbol{x}) = \begin{cases} \rho(\boldsymbol{x}) & \text{if } \boldsymbol{x} \text{ is in the unit cell,} \\ 0 & \text{otherwise.} \end{cases}$$

It is worth noting that the factor $F(\boldsymbol{0}) = \int_{\mathrm{uc}} \rho(\boldsymbol{x}) \mathrm{d}^3\boldsymbol{x}$ is equal to the number of electrons per unit cell.

The scattering factor for the unit cell can be expressed via scattering amplitudes of individual atoms. Let $\boldsymbol{p}_n$ be the location of the nucleus of the atom n, and let $\boldsymbol{r}_n$ be a vector from the nucleus to a given point so $\boldsymbol{x} = \boldsymbol{p}_n + \boldsymbol{r}_n$ (Fig. 2.13). Let $\rho_n(\boldsymbol{r})$ be the density of electrons of the n-th atom at the point separated by $\boldsymbol{r}$ from the atom nucleus. In the *independent atom approximation*, the density ρ is a sum of contributions $\rho_n(\boldsymbol{r}_n)$ from individual atoms

$$\rho(\boldsymbol{x}) = \sum_{n}^{\infty} \rho_n(\boldsymbol{r}_n) = \sum_{n}^{\infty} \rho_n(\boldsymbol{x} - \boldsymbol{p}_n) \,, \tag{2.21}$$

with the summation over all atoms of the crystal. The vector $\boldsymbol{p}_n$ indicating a position of the n-th atom can be decomposed into $\check{\boldsymbol{m}} + \boldsymbol{x}_n = \boldsymbol{p}_n$, where $\boldsymbol{x}_n$ is in the unit cell and $\check{\boldsymbol{m}}$ is a direct lattice vector. Thus, the density $\rho(\boldsymbol{x})$ can be expressed as a

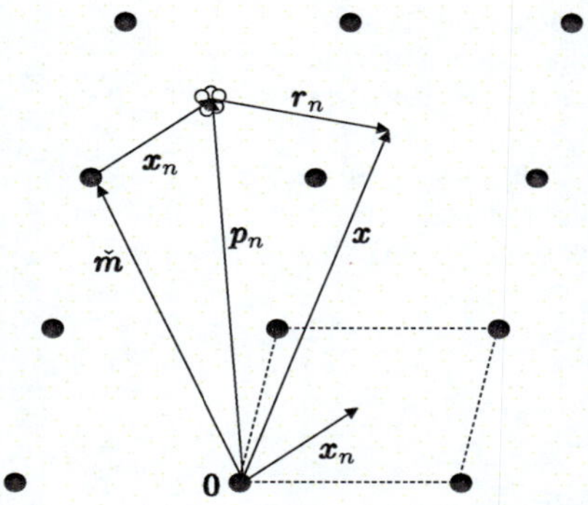

Fig. 2.13 Illustration of vectors used in this section: $\boldsymbol{x}$ – location at which the density is estimated, $\boldsymbol{p}_n$ – location of the n-th nucleus, $\boldsymbol{x}_n$ – position of the n-th nucleus in the unit cell, $\boldsymbol{r}_n$ – deviation of $\boldsymbol{x}$ from the nucleus and $\check{\boldsymbol{m}}$ – direct lattice vector.

sum over all N_{uc} atoms of the unit cell (index n) and over the unit cells (index $\check{\boldsymbol{m}}$) $\rho(\boldsymbol{x}) = \sum_{n=1}^{N_{\text{uc}}} \sum_{\check{\boldsymbol{m}}} \rho_n(\boldsymbol{x} - (\check{\boldsymbol{m}} + \boldsymbol{x}_n))$. Substitution of ρ in (2.19) gives $F(\check{\boldsymbol{h}}) = \int_{\text{uc}} \sum_{n=1}^{N_{\text{uc}}} \sum_{\check{\boldsymbol{m}}} \rho_n(\boldsymbol{x} - (\check{\boldsymbol{m}} + \boldsymbol{x}_n)) \exp(-2\pi i \check{\boldsymbol{h}} \cdot \boldsymbol{x})\, \mathrm{d}^3\boldsymbol{x}$, or with $\boldsymbol{r}_n = \boldsymbol{x} - (\check{\boldsymbol{m}} + \boldsymbol{x}_n)$,

$$F(\check{\boldsymbol{h}}) = \sum_{n=1}^{N_{\text{uc}}} \exp(-2\pi i \check{\boldsymbol{h}} \cdot \boldsymbol{x}_n) \sum_{\check{\boldsymbol{m}}} \exp(-2\pi i \check{\boldsymbol{h}} \cdot \check{\boldsymbol{m}}) \int_{\text{uc}-(\check{\boldsymbol{m}}+\boldsymbol{x}_n)} \rho_n(\boldsymbol{r}_n) \exp(-2\pi i \check{\boldsymbol{h}} \cdot \boldsymbol{r}_n)\, \mathrm{d}^3\boldsymbol{r}_n\,,$$

where $\sum_{\check{\boldsymbol{m}}}$ denotes summation over all cells, and uc $- (\check{\boldsymbol{m}} + \boldsymbol{x}_n)$ represents the unit cell uc shifted by $(\check{\boldsymbol{m}} + \boldsymbol{x}_n)$. Since $\check{\boldsymbol{h}}$ and $\check{\boldsymbol{m}}$ are vectors of mutually reciprocal lattices, the product $\check{\boldsymbol{h}} \cdot \check{\boldsymbol{m}}$ is an integer, and the factor $\exp(-2\pi i \check{\boldsymbol{h}} \cdot \check{\boldsymbol{m}})$ equals 1. Moreover

$$\sum_{\check{\boldsymbol{m}}} \int_{\text{uc}-(\check{\boldsymbol{m}}+\boldsymbol{x}_n)} \rho_n(\boldsymbol{r}_n) \exp(-2\pi i \check{\boldsymbol{h}} \cdot \boldsymbol{r}_n)\, \mathrm{d}^3\boldsymbol{r}_n = \int_{\infty} \rho_n(\boldsymbol{r}_n) \exp(-2\pi i \check{\boldsymbol{h}} \cdot \boldsymbol{r}_n)\, \mathrm{d}^3\boldsymbol{r}_n = f_n^X(\check{\boldsymbol{h}})\,,$$

where f_n^X is the Fourier transform of ρ_n

$$f_n^X(\check{\boldsymbol{h}}) = \mathcal{F}_{\boldsymbol{r}_n}[\rho_n(\boldsymbol{r}_n)](\check{\boldsymbol{h}})\,; \tag{2.22}$$

cf. (1.59). Hence, one has

$$F(\check{\boldsymbol{h}}) = \sum_{n}^{N_{\text{uc}}} f_n^X(\check{\boldsymbol{h}}) \exp(-2\pi i \check{\boldsymbol{h}} \cdot \boldsymbol{x}_n)\,, \tag{2.23}$$

where the summation is over atoms of the unit cell (Fig. 2.14).

The structure factor can also be expressed using the function

$$\rho_0(\boldsymbol{x}) = \sum_{n=1}^{N_{\text{uc}}} \rho_n(\boldsymbol{x} - \boldsymbol{x}_n)$$

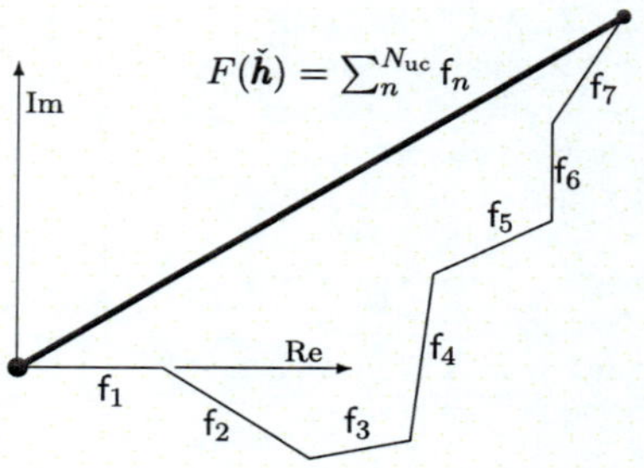

Fig. 2.14 Schematic illustration of the sum (2.23) on the complex plane for a fixed $\check{\boldsymbol{h}}$. The term $\mathrm{f}_n = f_n^X(\check{\boldsymbol{h}}) \exp(-2\pi i \check{\boldsymbol{h}} \cdot \boldsymbol{x}_n)$ is determined by n-th atom position $\boldsymbol{x}_n$ and its scattering power $f_n^X(\check{\boldsymbol{h}})$.

with summation over atoms of the unit cell. It represents the contribution of a single cell to the periodic $\rho(\boldsymbol{x}) = \sum_{\check{\boldsymbol{m}}} \rho_0(\boldsymbol{x} - \check{\boldsymbol{m}})$. Fourier transform of ρ_0 can be written as

$$\mathcal{F}_x[\rho_0(\boldsymbol{x})](\check{\boldsymbol{h}}) = \sum_{n=1}^{N_{uc}} \mathcal{F}_{\boldsymbol{x}}[\rho_n(\boldsymbol{x} - \boldsymbol{x}_n)](\check{\boldsymbol{h}}) = \sum_{n=1}^{N_{uc}} \mathcal{F}_{\boldsymbol{r}_n}[\rho_n(\boldsymbol{r}_n)](\check{\boldsymbol{h}}) \exp(-2\pi i \check{\boldsymbol{h}} \cdot \boldsymbol{x}_n) \, ,$$

where the last step is simply the shift property (1.61). By using the definition (2.22) and (2.23), one obtains

$$\mathcal{F}_{\boldsymbol{x}}[\rho_0(\boldsymbol{x})](\check{\boldsymbol{h}}) = \sum_{n=1}^{N_{uc}} f_n^X(\check{\boldsymbol{h}}) \exp(-2\pi i \check{\boldsymbol{h}} \cdot \boldsymbol{x}_n) = F(\check{\boldsymbol{h}}) \, . \tag{2.24}$$

Comparison of this expression with (2.20) leads to

$$\mathcal{F}\rho_0(\check{\boldsymbol{h}}) = \mathcal{F}\rho_{|\mathrm{uc}}(\check{\boldsymbol{h}}) \, .$$

This equality holds at the nodes $\check{\boldsymbol{h}}$ of the reciprocal lattice. Clearly, there is no ground for equality of $\mathcal{F}\rho_0$ and $\mathcal{F}\rho_{|\mathrm{uc}}$ at points other than the lattice nodes. The above relationship matters because, by definition (2.20), the structure factors are the values of the Fourier transform of $\rho_{|\mathrm{uc}}$ with the support restricted to a lattice cell, whereas in practice, they are computed via Fourier transform of the function ρ_0 with unrestricted support (Fig. 2.15).

With the Laue equation $\check{\boldsymbol{h}} = \boldsymbol{k}_{\check{\boldsymbol{h}}} - \boldsymbol{k}_0$, there is a clear analogy between relations (2.15) and (2.23) or (2.24). With the latter relationships, the structure factor $F(\check{\boldsymbol{h}})$ can be interpreted as the scattering power of the unit cell. The factor $f_n^X(\boldsymbol{h})$ is the scattering amplitude of an individual atom, also known as the *atomic scattering factor* (or atomic form factor). For $\boldsymbol{h} = \boldsymbol{0}$, the scattering factor of a neutral atom is equal to its atomic number Z_n: $f_n^X(\boldsymbol{0}) = \int \rho_n(\boldsymbol{r}_n) \, \mathrm{d}^3\boldsymbol{r}_n = Z_n$. For non-zero $\boldsymbol{h}$, destructive interferences take place, and the scattering amplitude decreases with growing $|\boldsymbol{h}|$ (Fig. 2.16). The difference in form factors between light and heavy atoms means that, if both types are present, the contributions of the former are less 'visible' in diffraction patterns.

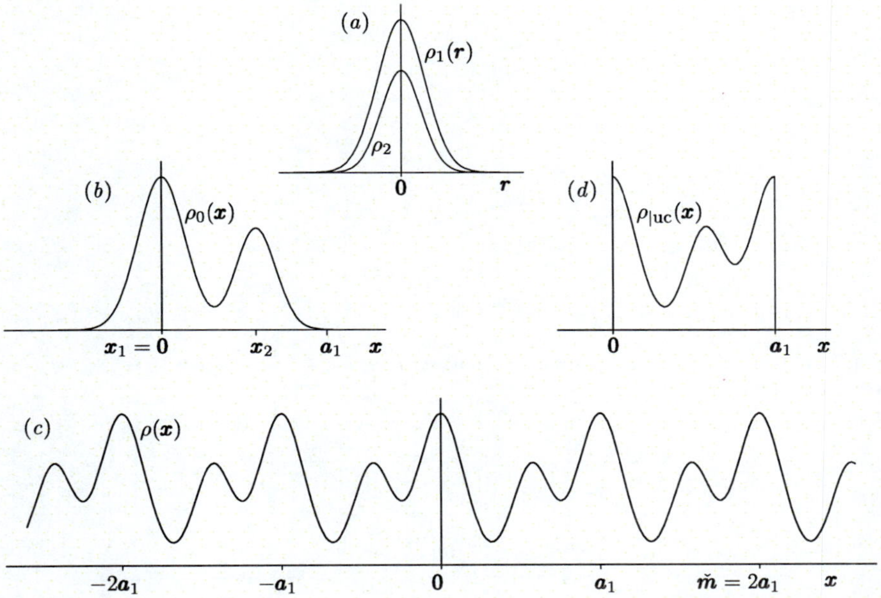

Fig. 2.15 One-dimensional example of the electron density function for $N_{\mathrm{uc}} = 2$. (*a*) Densities ρ_n $(n = 1, 2)$ for individual atoms versus atom radius. (*b*) Contribution of all atoms of the unit cell $\rho_0(\boldsymbol{x}) = \rho_1(\boldsymbol{x} - \boldsymbol{x}_1) + \rho_2(\boldsymbol{x} - \boldsymbol{x}_2)$. (*c*) The periodic electron density function $\rho(\boldsymbol{x}) = \sum_{\check{\boldsymbol{m}}} \rho_0(\boldsymbol{x} - \check{\boldsymbol{m}})$ in the crystal. (*d*) The function $\rho_{|\mathrm{uc}}(\boldsymbol{x}) = 1_{|\mathrm{uc}}(\boldsymbol{x})\, \rho(\boldsymbol{x})$; it equals $\rho(\boldsymbol{x})$ when $\boldsymbol{x}$ is in the unit cell and it is zero otherwise.

In the independent atom model, atomic scattering factors are assumed to be isotropic.[11] If the density $\rho_n(\boldsymbol{r}_n)$ depends only on the magnitude $r = |\boldsymbol{r}_n|$, i.e., $\rho_n(\boldsymbol{r}_n) = \rho_n(r)$, then one can integrate (2.22) over spherical angles, and this results in the isotropic factor

$$f_n^X(\boldsymbol{h}) = f_n^X(|\boldsymbol{h}|) = 4\pi \int_0^\infty \rho_n(r)\, \mathrm{sinc}(2r|\boldsymbol{h}|)\, r^2\, \mathrm{d}r\,, \tag{2.25}$$

where $\mathrm{sinc}(x) = \sin(\pi x)/(\pi x)$; see Sect. 1.11.1. Of no practical importance, but convenient for illustration is the analytically solvable case of a free hydrogen atom with electron density in the ground state decaying exponentially with the distance r

[11] Which means that the distributions of electrons round atomic nuclei are assumed to be spherically symmetric. Anisotropic charge distribution is accounted for in *multipole density formalism* of Hansen-Coppens [16] with the direction dependent part of $\rho_n(\boldsymbol{r}_n)$ expressed via spherical harmonics. In that case, each atom needs to be considered in a specific orientation with respect to its surrounding.

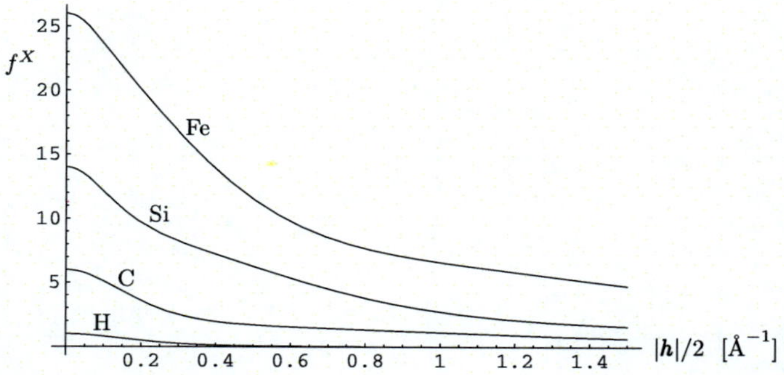

Fig. 2.16 Dependence of X-ray isotropic atomic form factors f_n^X on Z and $|\boldsymbol{h}|/2 = (\sin\theta)/\lambda$ for some elements after [19].

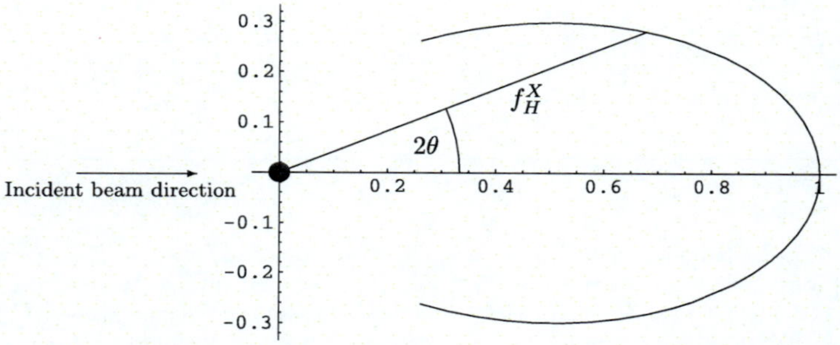

Fig. 2.17 Parametric plot of $f_H^X(|\boldsymbol{h}|) = f_H^X(2(\sin\theta)/\lambda)$ given by (2.27) versus 2θ for $\lambda = 3a_0$.

$$\rho_H(r) = \frac{e^{-2r/a_0}}{\pi a_0^3}\,, \tag{2.26}$$

where $a_0 \equiv \varepsilon_0 h^2/(\pi m_0 e^2) \approx 0.529177$Å is the Bohr radius; m_0 is the electron mass, e is the elementary charge, ε_0 is the electric permittivity of free space and h is the Planck constant. The integration of (2.25) with $\rho_n(r) = \rho_H(r)$ gives (Fig. 2.17)

$$f_H^X(|\boldsymbol{h}|) = \left(1 + (\pi a_0|\boldsymbol{h}|)^2\right)^{-2}\,. \tag{2.27}$$

Isotropic atomic scattering factors for other elements and ions are tabulated (or stored in software) as functions of the parameter $s = (\sin\theta)/\lambda = |\boldsymbol{h}|/2$; see e.g., [17–19].

2.4.3 Electron Atomic Scattering Factors

Electrons are scattered by the electrostatic potentials of atoms. The expressions for structure factor and atomic scattering factor for electrons are analogous to those for X-rays except that the density ρ_n is replaced by the atomic electrostatic Coulomb potential V_n of the n-th atom (times the constant factor $1/a_1$, where $a_1 = h^2/(2\pi m_0 e) \approx 0.47877643 \times 10^{-18}\,\mathrm{V\,m^2}$).[12] Thus, analogously to (2.22), the atomic scattering factor for electron diffraction is

$$f_n^e(\check{\boldsymbol{h}}) = \frac{1}{a_1}\,\mathcal{F}_{\boldsymbol{r}_n}[V_n(\boldsymbol{r}_n)](\check{\boldsymbol{h}})\;. \tag{2.28}$$

The scattering factors f_n^e are in units of length [m] (while f_n^X is a dimensionless 'number of electrons'). Within the standard independent atom model, they are tabulated as spherically symmetric functions of $s = |\check{\boldsymbol{h}}|/2$; see, e.g., [18, 20]. As in the case of X-rays, the structure factor $F(\check{\boldsymbol{h}})$ for electron scattering is

$$F(\check{\boldsymbol{h}}) = \sum_n^{N_{\mathrm{uc}}} f_n^e(\check{\boldsymbol{h}})\exp(-2\pi i\check{\boldsymbol{h}}\cdot\boldsymbol{x}_n)\;, \tag{2.29}$$

i.e., it is given by expression similar to (2.23) but with f_n^X replaced by f_n^e.

The electrostatic potential at $\boldsymbol{x}$ denoted by $V(\boldsymbol{x})$ is a sum of contributions V_n from particular atoms, i.e., analogously to (2.21), one has $V(\boldsymbol{x}) = \sum_n V_n(\boldsymbol{x} - \boldsymbol{x}_n)$. Based on the analogy between (2.28, 2.29) and (2.22, 2.23) it is clear that, in the same manner as in (2.16, 2.17), one obtains

$$(V_{\mathrm{uc}}/a_1)\,V(\boldsymbol{x}) = \sum_{\check{\boldsymbol{h}}} F(\check{\boldsymbol{h}})\exp(2\pi i\check{\boldsymbol{h}}\cdot\boldsymbol{x})\;, \tag{2.30}$$

i.e., the structure factors for electron scattering are Fourier expansion coefficients of the periodic function $V(\boldsymbol{x})$ (times the constant V_{uc}/a_1).

An approximation of the amplitude for electrons f_n^e can be obtained from the atomic scattering amplitude for X-rays f_n^X given by (2.22). The Coulomb potential is the sum of the (attractive) interaction with the nucleus of charge $Z_n e$ and the (repulsive) interaction with electrons, i.e., with the nucleus at $\mathbf{0}$

$$V_n(\boldsymbol{r}_n) = \frac{e}{4\pi\varepsilon_0}\left(\frac{Z_n}{|\boldsymbol{r}_n|} - \int_\infty \frac{\rho_n(\boldsymbol{y})}{|\boldsymbol{r}_n - \boldsymbol{y}|}\,\mathrm{d}^3\boldsymbol{y}\right) = \frac{e}{4\pi\varepsilon_0}\left(Z_n g(\boldsymbol{r}_n) - (\rho_n * g)(\boldsymbol{r}_n)\right)\;, \tag{2.31}$$

where $g(\boldsymbol{x}) = |\boldsymbol{x}|^{-1}$. Hence, with a_0 denoting the Bohr radius, using the relationship $\mathcal{F}[g](\boldsymbol{h}) = 1/(\pi\boldsymbol{h}^2)$ (see footnote 28), one obtains the Mott formula

[12] The factor a_1 can be expressed via the Bohr radius a_0 as $a_1 = a_0 e/(2\varepsilon_0)$.

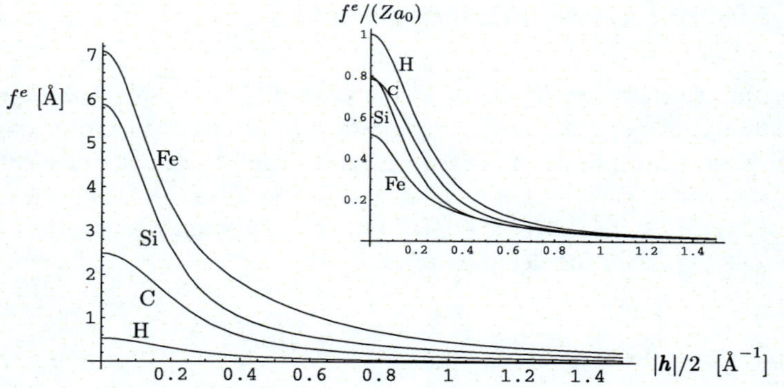

Fig. 2.18 Dependence of (elastic, isotropic) electron atomic scattering factors f^e on Z and $|\boldsymbol{h}|/2 = (\sin\theta)/\lambda$ for H based on (2.34), and for C, Si and Fe after [20]. The inset shows the dimensionless function $f^e/(Za_0)$; it illustrates the non-linear dependence of f^e on Z at small $|\boldsymbol{h}|$.

$$f_n^e(\boldsymbol{h}) = \frac{1}{8\pi^2 a_0}\,\frac{Z_n - f_n^X(\boldsymbol{h})}{(\boldsymbol{h}/2)^2} = 2.3934 \times 10^8[\mathrm{m}^{-1}] \times \frac{Z_n - f_n^X(\boldsymbol{h})}{(\boldsymbol{h}/2)^2} \tag{2.32}$$

relating the electron atomic scattering factors to X-ray form factors. It needs to be noted that the Mott formula is singular at $\boldsymbol{h} = \boldsymbol{0}$, and another approximation is needed in this case. Using the approximation $\mathrm{sinc}(x) \approx 1 - \pi^2 x^2/6$ for x near zero, and substituting $f_n^X(\boldsymbol{h})$ in the Mott formula by the isotropic (2.25), one obtains

$$f_n^e(\boldsymbol{0}) = Z_n\,\langle r^2\rangle/(3a_0) = 6.2991 \times 10^9[\mathrm{m}^{-1}] \times Z_n\langle r^2\rangle\,, \tag{2.33}$$

where $\langle r^2\rangle$ is the atom mean-square radius defined as $\langle r^2\rangle = \int_0^\infty \rho_n(r)\,r^4\mathrm{d}r/\int_0^\infty \rho_n(r)\,r^2\mathrm{d}r$ [21]. With growing $|\boldsymbol{h}|$, the term $f_n^X(\boldsymbol{h})$ decreases (Fig. 2.16), and the first (Z_n) term in (2.32) dominates, i.e., the larger the scattering angles, the larger the effect of the electron interaction with nuclei (Fig. 2.18).

For the hydrogen atom, the substitution of f_n^X in the Mott formula by the structure factor f_H^X given by (2.27) leads to

$$f_H^e(\boldsymbol{h}) = f_H^e(|\boldsymbol{h}|) = \frac{a_0}{2}\left(\frac{1}{\zeta} + \frac{1}{\zeta^2}\right) \quad \text{where } \zeta = \zeta(|\boldsymbol{h}|) = 1 + (\pi a_0|\boldsymbol{h}|)^2\,. \tag{2.34}$$

Accordingly, with the electron density $\rho_n(r) = \rho_H(r)$ given by (2.26), the mean-square radius of the atom is $\langle r^2\rangle = 3a_0^2$ and (2.33) results in $f_H^e(\boldsymbol{0}) = a_0$.

2.5 Formal Approach to Crystal Diffraction

One may consider diffraction from a more formal point of view based on general principles of optics. As was indicated in Sect. 2.8, in the Fraunhofer (far-field) diffraction, the observed interference pattern is the Fourier transform of an 'aperture' or 'transmittance' function of spatial coordinates, representing the amplitude of the wave originating from particular points of the diffracting object. This principle was already applied in the form of (2.22) and (2.20) for an atom and the unit cell, respectively. With a proper 'transmittance' function, one can use the same formalism to obtain directly the amplitudes of diffracted rays for an unbounded crystal.

In the text below, to the end of this chapter, the notation of X-ray diffraction is used, but similar considerations apply to electron diffraction. In the case of X-ray diffracting crystals, the 'transmittance' is the periodic electron density. With properly assigned boundaries of the unit cell, the periodicity of the electron density is incorporated by convolving the density restricted to the unit cell with a function characterizing the lattice based on the vectors spanning the unit cell.

2.5.1 *Fourier Transform of the Transfer Function of an Unbounded Crystal*

Let ρ denote the electron density with the domain over an unbounded crystal, and let $\rho_{|\mathrm{uc}}(\boldsymbol{x}) = 1_{|\mathrm{uc}}(\boldsymbol{x})\,\rho(\boldsymbol{x})$ be the restriction of ρ to the unit cell of the crystal lattice Λ. The function ρ is the sum of $\rho_{|\mathrm{uc}}$ translated by lattice vectors, i.e.,

$$\rho(\boldsymbol{x}) = \sum_{\check{\boldsymbol{m}}} \rho_{|\mathrm{uc}}(\boldsymbol{x} - \check{\boldsymbol{m}}) ,$$

where the summation is over translation vectors $\check{\boldsymbol{m}}$ of Λ. Based on (1.85), ρ can be expressed as the convolution of Ш_Λ characterizing the direct lattice Λ and the function $\rho_{|\mathrm{uc}}$ supported by the cell of the lattice

$$\rho = \text{Ш}_\Lambda * \rho_{|\mathrm{uc}} . \tag{2.35}$$

By the convolution theorem, and taking into account the definition (2.20) of the structure factor ($F = \mathcal{F}\rho_{|\mathrm{uc}}$), the Fourier transform of the electron density ρ in the whole crystal is

$$\mathcal{F}\rho(\boldsymbol{h}) = \mathcal{F}\rho_{|\mathrm{uc}}(\boldsymbol{h})\,\mathcal{F}\text{Ш}_\Lambda(\boldsymbol{h}) = \frac{F(\boldsymbol{h})}{V_{\mathrm{uc}}}\,\text{Ш}_{\Lambda^*}(\boldsymbol{h}) . \tag{2.36}$$

Thus, the Fourier transform of the electron density ρ – the transfer function of the unbounded crystal – is a weighted Dirac comb characterizing the *reciprocal* lattice of the crystal. Hence, the amplitudes of waves diffracted by an unbounded crystal are non-zero at nodes of the its reciprocal lattice with the values at nodes weighted by structure factors [22, 23]. Since the intensities of reflections are determined by the amplitudes, they are increased at the nodes of the reciprocal lattice. Modification of the atomic content of the primitive cell of a given lattice changes intensities of diffraction peaks without moving the peaks, as the peak positions are determined by the reciprocal lattice. Depending on the acquisition geometry, experimental diffraction patterns are specific sections through the 'reciprocal space' or its projections. The geometry of the reciprocal lattice is directly observable on some patterns. A prominent example of this effect are electron spot patterns (Fig. 2.10).

Equation (2.35) has an analogue which, instead of $\rho_{|\mathrm{uc}}$, involves the sum ρ_0 of electron densities of atoms in the unit cell. As was already noted, there is an important difference between $\rho_{|\mathrm{uc}}$ and ρ_0: The support of $\rho_{|\mathrm{uc}}$ is the unit cell of the lattice, whereas ρ_0 – as the sum of potentials with unbounded domains – has the support reaching beyond the cell. As in the case of $\rho_{|\mathrm{uc}}$, based on $\rho(\boldsymbol{x}) = \sum_{\check{\boldsymbol{m}}} \rho_0(\boldsymbol{x} - \check{\boldsymbol{m}})$ and (1.85), the density ρ can be expressed as the convolution of $Ш_\Lambda$ and ρ_0

$$\rho = Ш_\Lambda * \rho_0 \,.$$

Similarly to (2.36), the Fourier transform of ρ can be expressed as

$$\mathcal{F}\rho(\boldsymbol{h}) = \mathcal{F}\rho_0(\boldsymbol{h})\; \mathcal{F}Ш_\Lambda(\boldsymbol{h}) = \mathcal{F}\rho_0(\boldsymbol{h})\; Ш_{\Lambda^\star}(\boldsymbol{h}) \,, \tag{2.37}$$

i.e., it is the Dirac comb characterizing the *reciprocal* lattice of the crystal weighted by $\mathcal{F}\rho_0(\boldsymbol{h})$. By comparing (2.36) and (2.37) at the nodes $\check{\boldsymbol{h}}$ of the reciprocal lattice, one obtains the equality $\mathcal{F}\rho_0(\check{\boldsymbol{h}}) = \mathcal{F}\rho_{|\mathrm{uc}}(\check{\boldsymbol{h}}) = F(\boldsymbol{h})$ already derived in Sect. 2.4.2.

2.5.2 *Crystal of Finite Dimensions*

Let a crystal have finite dimensions, and let it be extended over the region called Cryst. Instead of $Ш_\Lambda$, one needs to use $Ш_{\Lambda|\,\mathrm{Cryst}}$ which is $Ш_\Lambda$ in Cryst and zero outside this region. With the characteristic function $1_{|\,\mathrm{Cryst}}$ describing the crystal shape and $Ш_\Lambda$ corresponding to the lattice on which the unit cell is based, the electron density ρ is given by

$$\rho = \left(Ш_\Lambda * \rho_{|\mathrm{uc}}\right) \times 1_{|\,\mathrm{Cryst}} \tag{2.38}$$

Hence, the Fourier transform of ρ is

$$\mathcal{F}\rho = V_{\mathrm{uc}}^{-1}\,(F \times Ш_{\Lambda^\star}) * \mathcal{F}1_{|\,\mathrm{Cryst}} \,. \tag{2.39}$$

Thus, the amplitude of waves diffracted by the finite crystal is a lattice reciprocal to the lattice of the crystal with structure-factor-weighted nodes spread by the Fourier

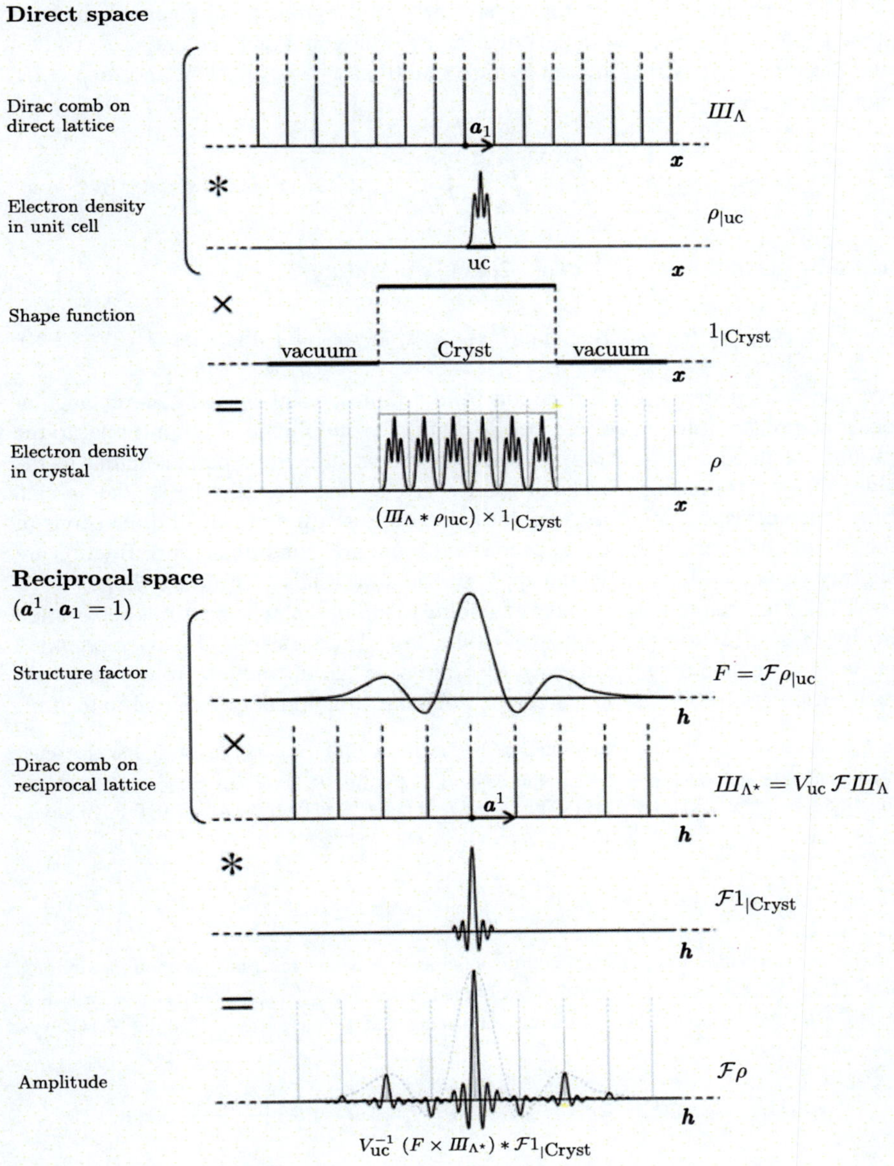

Fig. 2.19 Schematic one-dimensional illustration of (2.38) and (2.39). See text for details.

transform of the characteristic function $1_{|\,\mathrm{Cryst}}$. In other words, with a slight simplification, the transform of the shape is repeated around each node of the reciprocal lattice and scaled by the structure factor; see Fig. 2.19.

Let the external shape of the crystal be a rectangular parallelepiped (box) of dimensions t^1, t^2 and t^3, with the edges along orthogonal unit vectors $\boldsymbol{e}_1$, $\boldsymbol{e}_2$ and $\boldsymbol{e}_3$, and the vector $\boldsymbol{x}$ in Cryst limited by the conditions $\boldsymbol{x} \cdot \boldsymbol{e}_i \leq t^i/2$. With $\boldsymbol{x}$ expressed as $\boldsymbol{x} = x^i \boldsymbol{e}_i$, one gets

$$\mathcal{F}_{\boldsymbol{x}}[1_{|\,\mathrm{Cryst}}(\boldsymbol{x})](\boldsymbol{h}) = \int_{-t^1/2}^{t^1/2} \int_{-t^2/2}^{t^2/2} \int_{-t^3/2}^{t^3/2} \exp(-2\pi i x^i \boldsymbol{e}_i \cdot \boldsymbol{h})\, \mathrm{d}x^1 \mathrm{d}x^2 \mathrm{d}x^3$$

and subsequently, since $\int_{-t/2}^{t/2} \exp(-2aix)\mathrm{d}x = \sin(at)/a$, one has

$$\mathcal{F}_{\boldsymbol{x}}[1_{|\,\mathrm{Cryst}}(\boldsymbol{x})](\boldsymbol{h}) = t^1 t^2 t^3 \operatorname{sinc}(t^1 \boldsymbol{e}_1 \cdot \boldsymbol{h}) \operatorname{sinc}(t^2 \boldsymbol{e}_2 \cdot \boldsymbol{h}) \operatorname{sinc}(t^3 \boldsymbol{e}_3 \cdot \boldsymbol{h})\,; \qquad (2.40)$$

see Fig. 1.21. Thus, in the diffraction from a finite crystal shaped as a rectangular box, reciprocal lattice nodes are smeared out. The amplitude is proportional to the product of three sinc functions with arguments in directions perpendicular to the sides of the parallelepiped. The smaller a given dimension, the larger the spread. Since the corresponding peaks are not indefinitely sharp, with small dimensions of the crystal, diffraction will occur even if the geometric conditions for diffraction are slightly violated. This conclusion applies to small crystals of arbitrary shape.

A TEM foil can be seen as a slab extending to infinity in two directions and limited in the third direction perpendicular to the foil. Thus, as was already mentioned in Sect. 2.3 concerning electron *spot* patterns, in the diffraction from a thin foil, reciprocal lattice nodes are smeared out along the direction perpendicular to the foil.

⊡ For crystals of reasonably simple shapes, the functions $\mathcal{F}_{\boldsymbol{x}}[1_{|\,\mathrm{Cryst}}(\boldsymbol{x})]$ have similar character. As an example, one may consider a cylindrical wire of radius R. If its axis is along $\boldsymbol{e}_3 = \boldsymbol{e}^3$, the function $1_{|\,\mathrm{Cryst}}(\boldsymbol{x})$ has the value of 1 if $\boldsymbol{x}^2 - (\boldsymbol{x} \cdot \boldsymbol{e}^3)^2 \leq R^2$ and 0 otherwise. With the cylindrical coordinates (r_c, ϕ, x^3) of $\boldsymbol{x}$ and (h_c, ϕ_h, h_3) of $\boldsymbol{h}$, one has

$$\mathcal{F}_{\boldsymbol{x}}[1_{|\,\mathrm{Cryst}}(\boldsymbol{x})](\boldsymbol{h}) = \int_0^R r_c \mathrm{d}r_c \int_0^{2\pi} \mathrm{d}\phi \int_{-\infty}^{\infty} \mathrm{d}x^3 \exp(-2\pi i (r_c h_c \cos(\phi - \phi_h) + x^3 h_3))\,.$$

Fourier transformation in cylindrical coordinates involves Bessel functions of the first kind; cf. Sect. 12.3.3. With J_n denoting the n-th Bessel function, one has $\int_0^{2\pi} \exp(ai \cos x)\, \mathrm{d}x = 2\pi J_0(|a|)$ and $\int_0^a x J_0(x)\, \mathrm{d}x = a J_1(a)$. Using these relationships and $\int_{-\infty}^{\infty} \exp(-2\pi i x \xi))\mathrm{d}x = \mathcal{F}_x[1](\xi) = \delta(\xi)$, one obtains

$$\mathcal{F}_{\boldsymbol{x}}[1_{|\,\mathrm{Cryst}}(\boldsymbol{x})](\boldsymbol{h}) = 2\pi\delta(h_3) \int_0^R r_c J_0(2\pi r_c h_c)\, \mathrm{d}r_c = R\, \frac{J_1(2\pi R h_c)}{h_c}\, \delta(h_3)\,,$$

where $h_c = \sqrt{\boldsymbol{h}^2 - (\boldsymbol{h} \cdot \boldsymbol{e}_3)^2}$ and $h_3 = \boldsymbol{h} \cdot \boldsymbol{e}_3$. The case of a cylindrical wire combines features of a crystal infinite in one direction (which leads to $\delta(h_3)$) and finite in two remaining directions (with smearing corresponding to $J_1(2\pi R h_c)/h_c$). The function $J_1(\pi x)/(\pi x)$ is shown in Fig. 2.20.

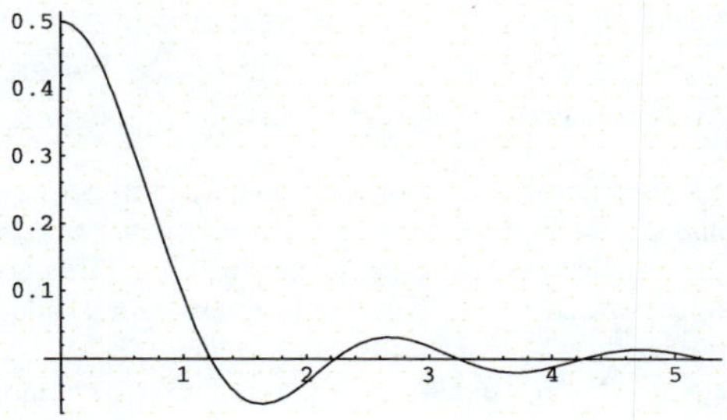

Fig. 2.20 The function $J_1(\pi x)/(\pi x)$ in the vicinity of $x = 0$. Like sinc, it has a strong peak at $x = 0$ and a wavy tail. Cf. Fig. 1.21.

2.6 Intensities of Reflections

The energy of a wave is proportional to square of its amplitude. Ordinarily, the intensity $I(\check{\boldsymbol{h}})$ of the peak corresponding to $\check{\boldsymbol{h}}$ is taken to be proportional to the squared magnitude of $F(\check{\boldsymbol{h}})$. In the general case, with complex structure factors, the intensity is expressed in the form

$$I(\check{\boldsymbol{h}}) \propto |F(\check{\boldsymbol{h}})|^2 = F^*(\check{\boldsymbol{h}})\,F(\check{\boldsymbol{h}}) = \sum_{m=1}^{N_{\mathrm{uc}}}\sum_{n=1}^{N_{\mathrm{uc}}} f_m^*(\check{\boldsymbol{h}})\,f_n(\check{\boldsymbol{h}})\,\exp(2\pi i\check{\boldsymbol{h}}\cdot(\boldsymbol{x}_m - \boldsymbol{x}_n))\,, \tag{2.41}$$

where the asterisk $*$ denotes the complex conjugate, and the summations are over atoms of the unit cell. Since the Fourier transformation has the property $(\mathcal{F}_{\boldsymbol{x}}[f(\boldsymbol{x})])^* = \mathcal{F}_{\boldsymbol{x}}[f^*(-\boldsymbol{x})]$, the relationship (2.20) results in

$$F^*(\check{\boldsymbol{h}})\,F(\check{\boldsymbol{h}}) \propto \mathcal{F}_{\boldsymbol{x}}[P(\boldsymbol{x})](\check{\boldsymbol{h}})\,,$$

where the convolution $P(\boldsymbol{x}) = \int \rho_{|\mathrm{uc}}(\boldsymbol{y})\,\rho_{|\mathrm{uc}}(\boldsymbol{y}+\boldsymbol{x})\,\mathrm{d}_3\boldsymbol{y}$ is known as the Patterson[13] function (e.g., [23]). P is the autocorrelation of $\rho_{|\mathrm{uc}}$; see Sect. 1.11.3. Formally, the inverse Fourier transform $\mathcal{F}_{\boldsymbol{h}}^{-1}[I(\boldsymbol{h})]$ at $\boldsymbol{x}$ is proportional to the value of the Patterson function at that point. With $F(\boldsymbol{h}) = \sum_{n=1}^{N_{\mathrm{uc}}} f_n(\boldsymbol{h})\,\exp(-2\pi i\boldsymbol{h}\cdot\boldsymbol{x}_n)$ (c.f. (2.23)), the function can be expressed as

$$P(\boldsymbol{x}) = \mathcal{F}_{\boldsymbol{h}}^{-1}[F^*(\boldsymbol{h})\,F(\boldsymbol{h})](\boldsymbol{x}) = \mathcal{F}_{\boldsymbol{h}}^{-1}\left[\sum_{m,n}^{N_{\mathrm{uc}}} f_m^*(\boldsymbol{h})\,f_n(\boldsymbol{h})\,\exp(2\pi i\boldsymbol{h}\cdot(\boldsymbol{x}_m - \boldsymbol{x}_n))\right](\boldsymbol{x})$$

i.e., it has peaks located at $\boldsymbol{x} = \boldsymbol{x}_m - \boldsymbol{x}_n$ and weighted by $f_m^*(\boldsymbol{h})\,f_n(\boldsymbol{h})$. In other words, P has peaks at locations given by vectors separating maxima of the electron density function. In particular, with each maximum separated from itself by $\mathbf{0}$, there is a high peak at the origin $\boldsymbol{x} = \mathbf{0}$. As the products of values of the density function at the ends of the separating vectors $\boldsymbol{x}$ and $-\boldsymbol{x}$ are equal, the Patterson function is always

[13] Arthur Lindo Patterson (1902–1966).

centrosymmetric with respect to $\boldsymbol{x} = \boldsymbol{0}$: $P(\boldsymbol{x}) = P(-\boldsymbol{x})$. Experimental diffraction data are sufficient for computing the Patterson functions, but the Fourier formalism does not provide direct access to the structure determining density $\rho_{|\mathrm{uc}}$.

⊡ The intensity function is the squared Fourier transform of the function describing the structure, and it is the Fourier transform of the structure autocorrelation function. The structure determination problem is to extract the structure from the diffraction data, i.e., from the squared Fourier transform of the structure. The solution is generally not unique; e.g., enantiomorphs have the same squared transform. There exist structures which are neither congruent nor enantiomorphic and have the same squared Fourier transform. They are known as homomorphs or homometric structures. Homometric structures, indistinguishable by conventional diffraction, have the same auto-correlation functions. From the formal viewpoint, the problem is: under what conditions does an auto-correlation determine the underlying structure in a unique way? It is easy to present the problem in the case of point sets. For a point set, autocorrelation reduces to a set of vectors representing relative positions of the points. Homometric point sets have identical sets of such vectors. Even in this geometrical setting, obtaining meaningful results is not trivial [24].

The fact that detectable intensities depend on the magnitudes $|F(\check{\boldsymbol{h}})|$ but not on the phases of structure factors significantly complicates structure determination. The issue is referred to as the phase problem. Interpretation of Patterson functions is a classical technique of resolving it. More sophisticated theoretical approaches based on some dependences between magnitudes and phases of structure factors are known as direct methods. For detailed account on these subjects the reader is refereed to other texts, e.g. [25].

It must be noted that analysis of intensities cannot be limited to expressions like (2.41). Within the geometrical approximation, coherent scattering occurs exactly in directions satisfying the Laue equation. Clearly, this cannot be physically realistic. A need for adding a tolerance follows already from the considerations for finite size crystals (Sect. 2.5) but there are a number of other reasons for that. Experimentally accessible are the so-called integrated intensities obtained by integration of 'counts' over scattering directions near maxima; they correspond to theoretical integrals of $|F(\check{\boldsymbol{h}})|^2$ over variable parameters of a given experiment.

Practical expressions for integrated intensities involve additional factors based on geometry and physics of scattering of a particular type of radiation. Among the aspects which need to be considered are the angular divergence of the scattered beams, the length of the path of the radiation in the medium combined with absorption and extinction, crystallite size, presence of strain, polarization in the case of X-rays, preferred orientations if polycrystals are examined, multiplicity (i.e., the number of symmetrically equivalent planes that contribute to the same point of measurement). Most of these factors depend only on the scattering angle. There are numerous sources on estimating intensities in various experimental conditions. For illustration, two examples of such factors are given, one for X-ray diffraction and one for electron diffraction.

Intensities of Spots in X-ray Laue Diffraction

With $S(\lambda)$ describing the spectrum of the incident beam and detector efficiency, the intensities in a Laue pattern can be estimated using (cf. [26])

$$I(\check{\boldsymbol{h}}) \propto S(\lambda_{\check{h}})\, |F(\check{\boldsymbol{h}})|^2 \, \frac{\lambda_{\check{h}}^4}{\sin^2\theta} \left(1+\cos^2(2\theta)\right) ,$$

where $\check{\boldsymbol{h}}$ is the reciprocal lattice vector, $\lambda_{\check{h}}$ is given by (2.13), and the value of θ follows from $\cos(2\theta) = \widehat{\boldsymbol{k}}_{\check{h}} \cdot \widehat{\boldsymbol{k}}_0 = 1 + \lambda_{\check{h}} \check{\boldsymbol{h}} \cdot \widehat{\boldsymbol{k}}_0 = 1 - 2\,(\widehat{\boldsymbol{k}}_0 \cdot \widehat{\boldsymbol{h}})^2$, which is based on (2.14). The geometric (Lorentz) factor $\lambda^4/\sin^2\theta$ appears because of the need to integrate over the angular divergence of the beams; in general, its form depends on experimental arrangements. In Thomson scattering of X-rays, re-emission of radiation in various directions has a non-uniform distribution; assuming unpolarized incident radiation, the intensity depends on the scattering angle via the polarization factor $(1+\cos^2(2\theta))/2$. The intensity at a given Laue spot is a sum of intensities of different orders of diffraction coinciding in the spot. For other factors influencing the intensities in Laue patterns see, e.g., [27].

Intensities in Electron Spot Diffraction Patterns

With regards to electron diffraction, worth noting is a formula frequently used in the case of TEM spot patterns. Based on (2.39) and (2.40), the intensity of the spot corresponding to the reciprocal lattice vector $\check{\boldsymbol{h}}$ is

$$I(\check{\boldsymbol{h}}) \propto \left|\mathcal{F}[\rho_{|\mathrm{uc}}](\check{\boldsymbol{h}})\right|^2 = \frac{|F(\check{\boldsymbol{h}})|^2\, t^2}{V_{\mathrm{uc}}^2}\, \mathrm{sinc}^2(t\,\widehat{\boldsymbol{n}} \cdot \check{\boldsymbol{h}}) , \tag{2.42}$$

i.e., besides the squared modulus of the structure factor, the geometric factor involving the foil thickness t and its inclination $\widehat{\boldsymbol{n}}$ is taken into account.

Resolution

Determined crystal structures are characterized by their resolution, i.e., the smallest distance between distinguishable points. For a given radiation wavelength, the resolution is determined by the largest scattering angle for which reflections can be detected. (Cf. Sect. 1.5.1.) Therefore, it is sometimes limited by the angular extent of detectors, but the key factor really determining the resolution is the 'quality' of the crystal. The better the quality, the more peaks appear on diffraction patterns, the higher the resolution. Resolution is relative to the size of the unit cell; e.g., the resolution of, say, 1.5Å will be low in the case of a small-cell inorganic crystal and high in the case of of a large-cell macromolecular crystal.

Table 2.1 Character of structure factors $F(\check{\boldsymbol{h}})$ for centrosymmetric crystals (with suitably chosen origin) and non-centrosymmetric crystals (i.e., without the inversion in the point group) for cases with and without absorption.

	No absorption (real ρ, V)	With absorption (complex ρ, V)
Centrosymmetric	real, even: $F(-\check{\boldsymbol{h}}) = F(\check{\boldsymbol{h}}) = F^*(\check{\boldsymbol{h}})$	complex, even: $F(-\check{\boldsymbol{h}}) = F(\check{\boldsymbol{h}})$
Non-centrosymmetric	complex: $F(-\check{\boldsymbol{h}}) = F^*(\check{\boldsymbol{h}})$	complex

2.6.1 Systematic Absences

For some reciprocal lattice vectors of some crystals, the structure factor is zero irrespective of the values of the atomic scattering factors. This causes absences of some reflections in diffraction patterns. The absences (extinctions) are called *systematic*.

The systematic absences may appear due to microsymmetry operations. Glide planes and screw axes introduce extra planes between those determined by the unit cell dimensions. This elongates the reciprocal lattice vector corresponding to the stack of these planes, and has an impact on the diffraction pattern. The effect can be explained in formal terms. For the element $(R, \boldsymbol{t})$ of the crystal space group, one has $\rho(\boldsymbol{x}) = \rho((R, \boldsymbol{t})\boldsymbol{x})$ (and consequently $\rho_{|\mathrm{uc}}(\boldsymbol{x}) = \rho_{|\mathrm{uc}}((R, \boldsymbol{t})\boldsymbol{x})$). From the definition of the structure factor, one obtains

$$F(\check{\boldsymbol{h}}) = \exp(2\pi i (R\check{\boldsymbol{h}}) \cdot \boldsymbol{t}) \; F(R\check{\boldsymbol{h}}) .$$

When the reciprocal lattice vector $\check{\boldsymbol{h}}$ happens to be invariant with respect to the rotation R (i.e., $R\check{\boldsymbol{h}} = \check{\boldsymbol{h}}$) and $\check{\boldsymbol{h}} \cdot \boldsymbol{t}$ is not an integer, then $F(\check{\boldsymbol{h}}) = \exp(2\pi i\check{\boldsymbol{h}} \cdot \boldsymbol{t}) \; F(\check{\boldsymbol{h}})$ with $\exp(2\pi i\check{\boldsymbol{h}} \cdot \boldsymbol{t}) \neq 1$; hence, one has $F(\check{\boldsymbol{h}}) = 0$, and the reflection $\check{\boldsymbol{h}}$ is absent.

Another reason for the absence of reflections is the use of non-primitive cells; see, Sect. 1.2.5. This second type of absences affects a large fraction of potential reflections, whereas only few are affected by the presence of microsymmetry operations. It is illustrative to consider a crystal with the A2 structure (one atom at each node of cI lattice) given in a unit cell based on mutually perpendicular vectors of equal magnitude directed along four-fold symmetry axes. With two atoms of the same kind located at $\boldsymbol{x}_1 = (0, 0, 0)$ and $\boldsymbol{x}_2 = (1/2, 1/2, 1/2)$ having the same scattering factors $f(h\,k\,l)$ (e.g., as in the case of Ferrite), (2.23) and (2.29) lead to $F(\check{\boldsymbol{h}}) = f(h\,k\,l) \exp(-2\pi i 0) + f(h\,k\,l) \exp(-\pi i(h + k + l))$ or

$$F(\check{\boldsymbol{h}}) = \begin{cases} 2f(h\,k\,l) & \text{if } h + k + l \text{ is even,} \\ 0 & \text{if } h + k + l \text{ is odd.} \end{cases} \tag{2.43}$$

Thus, there are no reflections with odd $h + k + l$. In fact, as we noted in Sect. 1.2.5, the corresponding vectors do not belong to the reciprocal of the primitive lattice. (If the atoms are of two different kinds, say, A and B with different scattering factors f_A and f_B, the above unit cell is primitive,

$$F(\check{\boldsymbol{h}}) = \begin{cases} f_A(h\,k\,l) + f_B(h\,k\,l) \text{ if } h+k+l \text{ is even,} \\ f_A(h\,k\,l) - f_B(h\,k\,l) \text{ if } h+k+l \text{ is odd,} \end{cases} \tag{2.44}$$

and there are no systematic absences for odd $h+k+l$.)

It is easy to see that a reflection condition analogous to (2.43) is applicable to all crystals with body-centered lattices and with arbitrary content of the unit cell. Due to the lattice translation $\boldsymbol{t} = (1/2, 1/2, 1/2)$, the positions $\boldsymbol{x}_n$ and $\boldsymbol{x}_n + \boldsymbol{t}$ are equivalent. Half of N_{uc} atoms of the unit cell having the positions $\boldsymbol{x}_n$ $(n = 1, \ldots N_{\text{uc}}/2)$ can be ascribed to $\boldsymbol{0} = (0, 0, 0)$ in such a way that the positions of the other half are $\boldsymbol{x}_{n+N_{\text{uc}}/2} = \boldsymbol{x}_n + \boldsymbol{t} + k_n^i \boldsymbol{a}_i$, where k_n^i are integers and the vectors $\boldsymbol{a}_i$ span the cell. Thus, based on (2.23) and (2.29),

$$F(\check{\boldsymbol{h}}) = \sum_{n=1}^{N_{\text{uc}}/2} \left(f_n \exp(-2\pi i \check{\boldsymbol{h}} \cdot \boldsymbol{x}_n) + f_n \exp(-2\pi i \check{\boldsymbol{h}} \cdot (\boldsymbol{x}_n + \boldsymbol{t} + k_n^i \boldsymbol{a}_i)) \right) .$$

Since $\exp(-2\pi i \check{\boldsymbol{h}} \cdot (k_n^i \boldsymbol{a}_i)) = 1$, one obtains

$$F(\check{\boldsymbol{h}}) = \left(1 + \exp(-2\pi i \check{\boldsymbol{h}} \cdot \boldsymbol{t})\right) \sum_{n=1}^{N_{\text{uc}}/2} f_n \exp(-2\pi i \check{\boldsymbol{h}} \cdot \boldsymbol{x}_n) .$$

The factor $1 + \exp(-2\pi i \check{\boldsymbol{h}} \cdot \boldsymbol{t}) = 1 + \exp(-\pi i (h+k+l))$ is non-zero only if $h + k + l$ is even. In summary, the reflection condition for all body-centered cells has the form $h + k + l = 2n$, where n is an integer.

By similar arguments, the reflection condition for centered on the base C is that the sums $h + k$ are even, and for face-centered cells $h + k$ and $k + l$ are even.

⊡ It is also worth considering the case of crystals with rhombohedral lattices described using hexagonal axes. Coordinates of nodes of an R-centered cell of hexagonal lattice (obverse setting) expressed with respect to in the hexagonal axes are at $\boldsymbol{x}_1 = (0, 0, 0)$, $\boldsymbol{x}_2 = (2/3, 1/3, 1/3)$ and $\boldsymbol{x}_2 = (1/3, 2/3, 2/3)$; see Fig. 1.14*b*. Hence, (2.23) and (2.29) lead to $F(\check{\boldsymbol{h}}) \propto \exp(-2\pi i 0) + \exp(-2\pi i (2h + k + l)/3) + \exp(-2\pi i (h + 2k + 2l)/3) = 1 + 2\cos(2\pi(-h + k + l)/3)$. Since $1 + 2\cos(2\pi(3n \pm 1)/3)$ equals zero for integer n, if $-h + k + l = 3n \pm 1$, one has $F(\check{\boldsymbol{h}}) = 0$, and the *hkl* reflection is absent. The reflection may be present only if

$$-h + k + l = 3n .$$

This is the reflection condition for crystals with rhombohedral lattices $rP = hR$ and indices specified with respect to hexagonal axes.

2.6.2 *Friedel's Law*

Based on Neumann's principle (see Sect. 1.3.4), the symmetry of the complete set of diffraction patterns of a given crystal must be isomorphic to or higher than the crystal point group. A well know example of the latter case concerns non-centrosymmetric crystals. With real atomic scattering factors, the complex conjugate to the structure factor (2.23 and 2.29) satisfies $F^*(\check{\boldsymbol{h}}) = F(-\check{\boldsymbol{h}})$ (Table 2.1). Hence, one has $F(\check{\boldsymbol{h}})\, F^*(\check{\boldsymbol{h}}) = F^*(-\check{\boldsymbol{h}})\, F(-\check{\boldsymbol{h}})$, and based on (2.41), the intensities of the reflections $\check{\boldsymbol{h}}$ and $-\check{\boldsymbol{h}}$ are equal

$$I(\check{\boldsymbol{h}}) = I(-\check{\boldsymbol{h}})\ . \tag{2.45}$$

This property is known as Friedel's law. Taking into account that the positions of diffraction peaks are based on the reciprocal lattice (which is centrosymmetric), Friedel's law implicates that diffraction patterns exhibit a symmetry with inversion,[14] and diffraction cannot be used for detecting the absence of the inversion center in the crystal point group. Only the Laue group can be deduced. The reciprocal lattice vectors $\check{\boldsymbol{h}}$ and $-\check{\boldsymbol{h}}$ are called Friedel pair. They are called Bijvoet pair if the vectors are not equivalent with respect to point symmetries of the crystal.

Friedel's law may affect the orientation determination. If a crystal is not centrosymmetric, two non-equivalent orientations correspond to the same diffraction pattern. Formally, the law breaks down if atomic scattering factors are allowed to be complex (see Sect. 2.7.1), but the complexity is not sufficient for that. For instance, if ρ in (2.19) is an even function ($\rho(\boldsymbol{x}) = \rho(-\boldsymbol{x})$), then also F is even ($F(\check{\boldsymbol{h}}) = F(-\check{\boldsymbol{h}})$), and consequently the equality (2.45) holds even if ρ is complex. The symmetry of a crystal diffraction pattern may be higher than the crystal point group for reasons other than Friedel's law; see Sect. 10.1.

2.7 Other Factors Affecting Intensities

Intensities on diffraction patterns are affected by a number of other factors. First, there is absorption. Second, they are influenced by atomic disorder and thermal vibrations of atoms. Both subjects are discussed very briefly below. Moreover, energy is shuffled between incident and diffracted beams (extinction). One may also observe exchange of energy between diffracted beams (multiple diffraction), when more than one plane happens to be in a reflection position. Extinction becomes of importance in crystal domains with dimensions exceeding an extinction length. The extinction lengths for 1Å long waves of X-rays, 1Å waves of neutrons and 0.05Å waves of electrons are of the order of 10^4Å, 10^5Å and 10^2–10^3Å, respectively [28]. Extinction effects are naturally accounted for in dynamical theory; see, Chap. 3.

[14] Some authors call it the Friedel symmetry; see, e.g., [14].

2.7.1 Absorption

As radiation interacts with matter, it looses intensity. Only a part of the attenuation of the incident beam is caused by the scattering. The incident and diffracted beams also loose intensity due to other absorption mechanisms. Not all waves are scattered elastically. Inelastic scattering leads to diffuse intensities in diffraction patterns. The processes of inelastic scattering are physically complicated, and frequently the description of absorption is phenomenological, i.e., consequences are considered without analyzing the physical reasons. If a planar wave $A\exp(2\pi i\boldsymbol{k}\cdot\boldsymbol{x})$ decays as it propagates, its amplitude decreases and the decrease can be assumed to be exponential, i.e. $A=\exp(-2\pi\boldsymbol{q}\cdot\boldsymbol{x})$. Hence, the decaying wave is $\exp(2\pi i(\boldsymbol{k}+i\boldsymbol{q})\cdot\boldsymbol{x})$, or equivalently, $\exp(2\pi i\boldsymbol{k}'\cdot\boldsymbol{x})$ with the complex wave vector $\boldsymbol{k}'=\boldsymbol{k}+i\boldsymbol{q}$. The complexity of wave vectors can be introduced by assuming that the densities ρ_n or potentials V_n are not real but complex (Table 2.1). To take absorption into account, atomic scattering amplitudes f_n of particular atoms in (2.23) and (2.29) are completed with imaginary components representing their particular absorptive properties; see, e.g., [20].

⊡ The simplest description of Thompson scattering assumes free electrons. However, with electrons bound to atoms, if the energy of the incident wave is close to the energy bringing the atom into an excited state, anomalous (resonant) scattering occurs. (The corresponding wavelength is called an absorption edge.) The phenomenon has an impact on the intensity of scattered waves, i.e., on the atomic scattering factor. Again, it is accounted for by allowing atomic scattering factors to be complex numbers. Since resonant scattering makes a difference between intensities at $\check{\boldsymbol{h}}$ and $-\check{\boldsymbol{h}}$, its principal application is the determination of absolute structures of chiral crystals; see Sect. 10.1.

2.7.2 Occupancy and Thermal Vibrations

The most essential step towards description of diffraction in real crystals is to account for basic atomic disorders and thermal displacement disorder. If a perfect crystal with its ordered structure consists of more than one element, it is a chemical compound. In solid-state solutions, atoms of the solute may randomly substitute some atoms of the solvent in such a way that the crystal as a whole is homogeneous, and the crystalline structure of the solvent is unchanged. Such a *substitutionally* disordered structure is frequent in metal alloys and usually concerns elements close on the periodic table. Disorder is *positional* when atoms of a given kind may occupy different sites in different cells. Materials with disordered structures are not compounds because of the random distribution of atoms. The probability P_n^ν that atoms of type ν occupy the site $\boldsymbol{x}_n$ is referred to as the *occupancy*. If all atoms of a given kind occupy some sites in the unit cell, or each site is occupied by an atom of a certain kind, the occupancies satisfy the conditions $\sum_n P_n^\nu=1$ and $\sum_\nu P_n^\nu=1$, respectively.

As was already mentioned, atoms vibrate about their equilibrium positions. With the occupancy taken into account, similarly to (2.21), the density of electrons can be

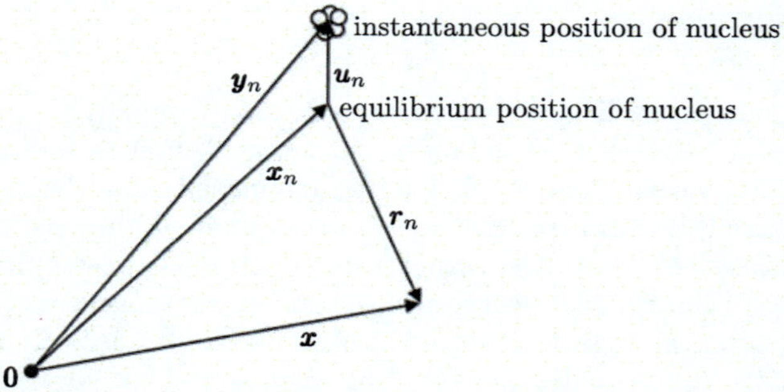

Fig. 2.21 Vectors used for describing the electronic density with atom nucleus displaced from its equilibrium position $\boldsymbol{x}_n$ to $\boldsymbol{y}_n = \boldsymbol{x}_n + \boldsymbol{u}_n$.

expressed as

$$\rho(\boldsymbol{x}) = \sum_n \sum_\nu P_n^\nu \int_\infty \rho_\nu(\boldsymbol{x} - \boldsymbol{y}_n)\, p_\nu(\boldsymbol{y}_n - \boldsymbol{x}_n)\, \mathrm{d}^3\boldsymbol{y}_n$$

where $\rho_\nu(\boldsymbol{x} - \boldsymbol{y}_n)$ is the electron density at $\boldsymbol{x}$ contributed by the atom of type ν having the nucleus at $\boldsymbol{y}_n$, $p_\nu(\boldsymbol{y}_n - \boldsymbol{x}_n)$ is the probability of the nucleus to be displaced from the equilibrium position $\boldsymbol{x}_n$ to point $\boldsymbol{y}_n$ (Fig. 2.21), and the summation is over all sites (n) in the unbounded crystal and all types of atoms (ν). The probability function p_ν 'smears out' the contribution of the static density ρ_ν. Substitution of $\rho(\boldsymbol{x})$ in (2.19) results in a formula slightly more complicated but similar to (2.23). With the substitutions $\boldsymbol{r}_n = \boldsymbol{x} - \boldsymbol{x}_n$ and $\boldsymbol{u}_n = \boldsymbol{y}_n - \boldsymbol{x}_n$, one has

$$F(\check{\boldsymbol{h}}) = \sum_n^{N_{\mathrm{uc}}} \sum_\nu P_n^\nu \exp(-2\pi i \check{\boldsymbol{h}} \cdot \boldsymbol{x}_n) \int_\infty \left(\int_\infty \rho_\nu(\boldsymbol{r}_n - \boldsymbol{u}_n)\, p_\nu(\boldsymbol{u}_n)\, \mathrm{d}^3\boldsymbol{u}_n \right) \exp(-2\pi i \check{\boldsymbol{h}} \cdot \boldsymbol{r}_n)\, \mathrm{d}^3\boldsymbol{r}_n\,.$$

or $F(\check{\boldsymbol{h}}) = \sum_{n,\nu} \exp(-2\pi i \check{\boldsymbol{h}} \cdot \boldsymbol{x}_n)\, \mathcal{F}\left[(\rho_\nu * p_\nu)(\boldsymbol{r}_n)\right](\check{\boldsymbol{h}})$. By the convolution theorem, this leads to

$$F(\check{\boldsymbol{h}}) = \sum_n^{N_{\mathrm{uc}}} \sum_\nu P_n^\nu\, D_{n,\nu}(\check{\boldsymbol{h}})\, f_{n,\nu}^X(\check{\boldsymbol{h}}) \exp(-2\pi i \check{\boldsymbol{h}} \cdot \boldsymbol{x}_n)\,, \tag{2.46}$$

where $f_{n,\nu}^X(\check{\boldsymbol{h}})$ is given by (2.22), and $D_{n,\nu}(\check{\boldsymbol{h}}) = \mathcal{F}_{\boldsymbol{u}_n}\left[p_\nu(\boldsymbol{u}_n)\right](\check{\boldsymbol{h}})$ is the *Debye-Waller temperature factor*[15,16]. The Debye-Waller factor describes the decrease of intensity of Bragg peaks due to thermal lattice vibrations.

---

[15] Peter Joseph William Debye (1884–1966).

[16] Ivar Waller (1898–1991).

Hereafter, for simplicity, the indices n and ν identifying atom position and type will be omitted. Assuming that the anisotropic smearing function p is a three-dimensional normalized Gaussian function[17] characterized by a symmetric positive definite matrix V,

$$p(\boldsymbol{u}) = \sqrt{\frac{\det(V)}{(2\pi)^3}} \exp(-V_{ij}u^i u^j/2) ,$$

the Fourier transform $D(\check{\boldsymbol{h}})$ of Gaussian p is Gaussian

$$D(\check{\boldsymbol{h}}) = \mathcal{F}[p](\check{\boldsymbol{h}}) = \exp(-2\pi^2 U^{ij} h_i h_j) ,$$

where U is the moment matrix $U^{ij} = \int u^i u^j p(\boldsymbol{u})\, \mathrm{d}^3\boldsymbol{u}$ inverse to V: $V_{ij}U^{jk} = \delta_i^{\ k}$.[18] The components of U are referred to as anisotropic displacement parameters. With $\check{\boldsymbol{h}} = 2s\,\widehat{\boldsymbol{n}}$ ($\widehat{\boldsymbol{n}}^2 = 1$) and n_i denoting components of $\widehat{\boldsymbol{n}}$, one has

$$D = \exp(-8\pi^2 U^{ij} n_i n_j s^2) = \exp(-8\pi^2 \langle \boldsymbol{u}^2 \rangle s^2) ,$$

where $\langle \boldsymbol{u}^2 \rangle = U^{ij} n_i n_j$ is known as the mean square amplitude of the vibration in the direction $\widehat{\boldsymbol{n}}$. In the isotropic case, with $8\pi^2 U^{ij} n_i n_j = B$, one obtains the well-known expression $D = \exp(-Bs^2)$. Clearly, the reduction of intensity caused by thermal vibrations is larger for reflections corresponding to vectors of larger magnitude $|\check{\boldsymbol{h}}| = 2s$.

⊡ There are numerous experimental papers providing values of (harmonic isotropic Debye-Waller) B factor for various materials and temperatures. These values can also be calculated under various assumptions. For instance, the dependence of X-ray B on absolute temperature T for an elemental crystal with atomic mass m and (mean) Debye temperature Θ can be estimated using

$$B = B(T) = c_B \phi(\Theta/T) , \text{ where } \phi(x) = \frac{1}{4} + \frac{1}{x}\left(\frac{1}{x}\int_0^x \frac{\xi}{\exp \xi - 1} \mathrm{d}\xi\right) \text{ and } c_B = \frac{6h^2}{m k_B \Theta} ,$$

where h and k_B are Planck and Boltzman constants, respectively [29]. For small T/Θ, one has $B(T) \approx c_B \left(1/4 + (\pi^2/6)(T/\Theta)^2\right)$, whereas for large T/Θ the relationship is linear $B(T) \approx c_B T/\Theta$. The function is shown in Fig. 2.22. For polyatomic crystals, overall (i.e., averaged over atomic sites) Debye-Waller factors are available.

Diffuse Scattering

The Debye-Waller factor is the simplest (kinematical) way of accounting for the temperature dependence of intensities of diffraction peaks. The proper description

---

[17] This occurs for an atom vibrating in harmonic potential field.

[18] To obtain the transform, one can express V as $V = O\Lambda^2 O^T = O\Lambda(O\Lambda)^T$ with positive definite diagonal matrix Λ and orthogonal O, and use (1.84).

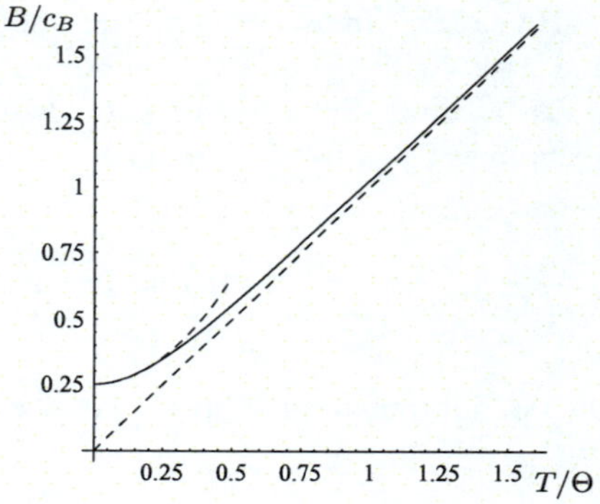

Fig. 2.22 Debye-Waller B factor versus absolute temperature for monoatomic crystals. Dashed lines represent two approximations mentioned in the text.

of thermal oscillations is more complex than that given above. It was ignored that the displacements of atoms are coupled: equilibrium position $\boldsymbol{x}_n$ of a given atom is not fixed but it is relative to positions of its neighbors. The coupling of vibrating objects is expressed via waves which can be decomposed in normal modes of vibrations characterized by vectors of reciprocal space. In a quantum-mechanical description, the thermal effect in crystals is seen as scattering by phonons – quanta of lattice vibrations. Scattering by phonons is also known as thermal diffuse scattering (TDS). Clearly, TDS increases as crystal temperature increases. In a formal approach, the function describing scattering intensities under the assumption of harmonic vibrations, is expanded into a series. Its zeroth order term contains the above described Debye-Waller factor decreasing intensity at Bragg peaks, i.e., at $\check{\boldsymbol{h}}$ representing vectors of the reciprocal lattice. The remaining terms involve arbitrary reciprocal space vectors. TDS effects are linked to thermal motions as they spoil the translational symmetry, which is the basis of the discrete character of 'kinematical' crystal diffraction patterns.

Diffuse scattering effects (diffuse background, unexpected streaks or peaks) arise also due to partial disorder or the presence of correlations in the crystal structure. The correlations may have different ranges, from short (e.g., when two near-neighbor atomic sites are more frequently occupied by the same pair of chemical species) through medium (e.g., with the presence of different repeat motifs spanning multiple unit cells) to long range (e.g., superstructures superimposed on the crystal structure). Diffuse scattering caused by correlations and structural imperfections of all kinds is sometimes referred to as *structural diffuse scattering*.

Ordering

Clearly, ordering and disorder affect the intensities of diffraction peaks. Ordering enhances reflections which are faint or absent in the disordered state. For illustration, it is convenient to use the example considered in the last paragraph of Sect. 2.6.1 with a cubic crystal lattice, two atoms A and B located at $\boldsymbol{x}_1 = (0, 0, 0)$ and $\boldsymbol{x}_2 = (1/2, 1/2, 1/2)$, and equal fractions of A and B. Based on (2.46) with Debye-Waller factor ignored, $f_{1,A}(\check{\boldsymbol{h}}) = f_{2,A}(\check{\boldsymbol{h}}) = f_A(\check{\boldsymbol{h}})$, $f_{1,B}(\check{\boldsymbol{h}}) = f_{2,B}(\check{\boldsymbol{h}}) = f_B(\check{\boldsymbol{h}})$ and atoms randomly distributed at sites 1 and 2, i.e., with the occupancies $P_1^A = P_2^B = P_1^B = P_2^A = 1/2$, one has $F(\check{\boldsymbol{h}}) = (f_A(\check{\boldsymbol{h}}) + f_B(\check{\boldsymbol{h}}))/2 \left(\exp(-2\pi i \check{\boldsymbol{h}} \cdot \boldsymbol{x}_1) + \exp(-2\pi i \check{\boldsymbol{h}} \cdot \boldsymbol{x}_2)\right)$. In terms of indices hkl after substitution of $\boldsymbol{x}_1$ and $\boldsymbol{x}_2$, one obtains

$$F(\check{\boldsymbol{h}}) = \begin{cases} f_A(h\,k\,l) + f_B(h\,k\,l) & \text{if } h+k+l \text{ is even,} \\ 0 & \text{if } h+k+l \text{ is odd,} \end{cases}$$

i.e., reflections with odd $h+k+l$ will be absent; cf. (2.43). Thus, as expected, the disordered crystal diffracts like a crystal with the same 'average' atom at all nodes of a body centered lattice. Any ordering will cause forbidden reflections to appear. They will be faint in the case of partial ordering, and they will reach intensities determined by the structure factors (2.44) for the ordered alloy with the occupancies $P_1^A = 1 = P_2^B$ and $P_1^B = 0 = P_2^A$.

2.8 Appendix: A Note on the Diffraction of Light

Diffraction optics, unlike geometrical (ray) optics, takes into account that light is a wave. With the simplest scalar wave theory,[19] a phenomenological description of wave propagation in general, and wave diffraction in particular, are based on the principles of Huygens and Fresnel: every point of a wave-front is a source of a spherical wavelet, the wavelets interfere, and later wave-fronts are envelopes of the wavelets. Formally, wave propagation is governed by the wave equation. With the separation of spatial and temporal variables, the wave equation leads to the time-independent Helmholtz equation

$$\nabla^2 \Psi(\boldsymbol{x}) + 4\pi^2 k^2 \Psi(\boldsymbol{x}) = 0\,, \tag{2.47}$$

where $\Psi = \Psi(\boldsymbol{x})$ is the wave amplitude at $\boldsymbol{x}$, and k is the inverse of the wavelength $\lambda = 1/k$.

The simplest description of diffraction based on the above-mentioned principles is known as Kirchhoff theory [30]. It gives a formula for the amplitude $\Psi(\boldsymbol{x})$ of a monochromatic wave encountering a planar barrier with an aperture. The formula is a solution to the Helmholtz equation (2.47) under boundary conditions determined by this obstacle.[20] With the point source of light at $\boldsymbol{x}_0$ and the aperture A in a planar opaque barrier normal to the unit vector $\widehat{\boldsymbol{n}}$ (Fig. 2.23), under the assumption that

---

[19] Not adequate for describing polarized radiation.

[20] The disturbance Ψ and its first derivatives in the direction normal to the barrier vanish at the blocking parts of the barrier.

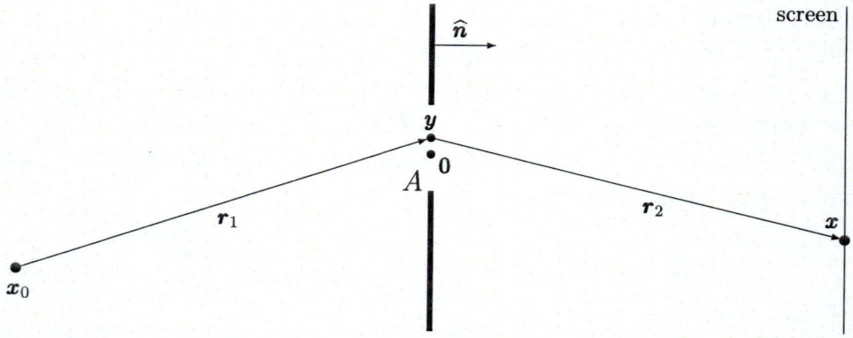

Fig. 2.23 Schematic illustration of the construction used to justify (2.48). The source of the wave is at $\boldsymbol{x}_0$. A is an aperture in an opaque barrier normal to $\widehat{\boldsymbol{n}}$. One looks for the amplitude of the wave at $\boldsymbol{x}$ under the assumption that the distances of the aperture to the source and the screen are large compared to the wavelength. The point $\boldsymbol{y}$ is in the aperture.

the distances of the aperture to the source and the screen are large compared to the wavelength, the disturbance Ψ at point $\boldsymbol{x}$ (after the aperture) is given by[21]

$$\Psi(\boldsymbol{x}) \propto \int_A \frac{\exp(2\pi i k r_1)}{r_1} \frac{\exp(2\pi i k r_2)}{r_2} (\widehat{\boldsymbol{n}} \cdot \widehat{\boldsymbol{r}}_1 - \widehat{\boldsymbol{n}} \cdot \widehat{\boldsymbol{r}}_2)\, \mathrm{d}^2\boldsymbol{y} ,$$

where the integration is over points $\boldsymbol{y}$ of the aperture A, $\boldsymbol{r}_1 = \boldsymbol{y} - \boldsymbol{x}_0$, $\boldsymbol{r}_2 = \boldsymbol{x} - \boldsymbol{y}$ and $r_i = |\boldsymbol{r}_i|$. The first two factors under the integral follow directly from the Huygens principle for the wave propagating in free space.[22] The factor $\exp(2\pi i k r_1)/r_1$ represents the amplitude at $\boldsymbol{y}$ of the wave originating at $\boldsymbol{x}_0$, whereas $\exp(2\pi i k r_2)/r_2$ represents the amplitude at $\boldsymbol{x}$ of the wave originating from the fictitious point source at $\boldsymbol{y}$. The inclination factors $\widehat{\boldsymbol{n}} \cdot \widehat{\boldsymbol{r}}_1$ and $\widehat{\boldsymbol{n}} \cdot \widehat{\boldsymbol{r}}_2$ are cosines of illumination and observation angles, respectively.

In Fraunhofer (far-field) diffraction, with r_1 and r_2 very large compared to the dimensions of the aperture and small angles between $\widehat{\boldsymbol{n}}$ and vectors $\boldsymbol{r}_1$ and $\boldsymbol{r}_2$, the factors $(\widehat{\boldsymbol{n}} \cdot \boldsymbol{r}_1/r_1 - \widehat{\boldsymbol{n}} \cdot \boldsymbol{r}_2/r_2)$ and $1/(r_1 r_2)$ are almost constant over the aperture; they vary slowly compared to the exponential factor $\exp(2\pi i k(r_1 + r_2))$. Therefore, they can be taken outside the integral and

$$\Psi(\boldsymbol{x}) \propto \int_A \exp(2\pi i k(r_1 + r_2))\, \mathrm{d}^2\boldsymbol{y} .$$

With the origin of the coordinate system located at the centroid of the aperture (i.e., with $|\boldsymbol{y}|$ small compared to $|\boldsymbol{x}|$ and $|\boldsymbol{x}_0|$), one has $r_2 = \sqrt{(\boldsymbol{x} - \boldsymbol{y})^2} \approx \sqrt{\boldsymbol{x}^2 - 2\boldsymbol{x} \cdot \boldsymbol{y}} \approx$

[21] The physically observed disturbance is the real part of the integral.

[22] The factors $1/r_i$ appear because the wave amplitude decreases like the inverse of the distance from the source (as the surface of the wave front grows with r_i^2, and thus the wave energy $\propto$ amplitude^2 decreases with r_i^{-2}).

$|\boldsymbol{x}| - \boldsymbol{x} \cdot \boldsymbol{y}/|\boldsymbol{x}| = |\boldsymbol{x}| - \widehat{\boldsymbol{x}} \cdot \boldsymbol{y}$, and analogously $r_1 \approx |\boldsymbol{x}_0| - \widehat{\boldsymbol{x}}_0 \cdot \boldsymbol{y}$. Hence, with the $\boldsymbol{y}$-independent factors omitted

$$\Psi(\boldsymbol{x}) \propto \int_A \exp(-2\pi i\, k(\widehat{\boldsymbol{x}}_0 + \widehat{\boldsymbol{x}}) \cdot \boldsymbol{y})\, \mathrm{d}^2\boldsymbol{y} = \int_A \exp(-2\pi i\, (\boldsymbol{k} - \boldsymbol{k}_0) \cdot \boldsymbol{y})\, \mathrm{d}^2\boldsymbol{y}\,,$$

where the constant $\boldsymbol{k}_0 = -k\widehat{\boldsymbol{x}}_0$ and $\boldsymbol{x}$-dependent $\boldsymbol{k} = k\widehat{\boldsymbol{x}}$ are the wave vectors before and after the aperture. Thus, in the far-field diffraction, the amplitudes $\Psi(\boldsymbol{x})$ of waves scattered in the direction $\widehat{\boldsymbol{k}} = \widehat{\boldsymbol{x}}$ are proportional to the two-dimensional Fourier transform of the characteristic function $1_{|A}$ of the aperture at $\boldsymbol{k} - \boldsymbol{k}_0$

$$\Psi(\boldsymbol{x}) \;\propto\; \mathcal{F}1_{|A}(\boldsymbol{k} - \boldsymbol{k}_0)\,. \tag{2.48}$$

Generally, in the far-field diffraction—a part of the so-called Fourier optics—the distribution of the amplitudes of waves scattered by an object is the Fourier transform of the transfer function of the object.

2.8.1 *Pattern at the Focal Plane of a Converging Lens*

What matters in far-field diffraction is the direction of wave propagation. Waves (originating form various points of the object) propagating in the same direction contribute to the same point of the diffraction pattern. It is easy to see that this is also a property of a thin converging lens. A plane wave after passing through a lens converges to a spot (focus). Planes propagating at different directions converge to different spots (Fig. 2.24), all located on the (back focal) plane normal to the axis of a lens. With coherent monochromatic light passing through an object (a transparency seen as a light source) and a subsequent converging lens, the wave amplitudes in the focal plane are proportional to the Fourier transform of the wave field in the source plane. One can say that the pattern at the focal plane of a lens is the Fraunhofer diffraction pattern of the transparency. This property is a basis for observation of some diffraction patterns; see Sect. 3.4 below.

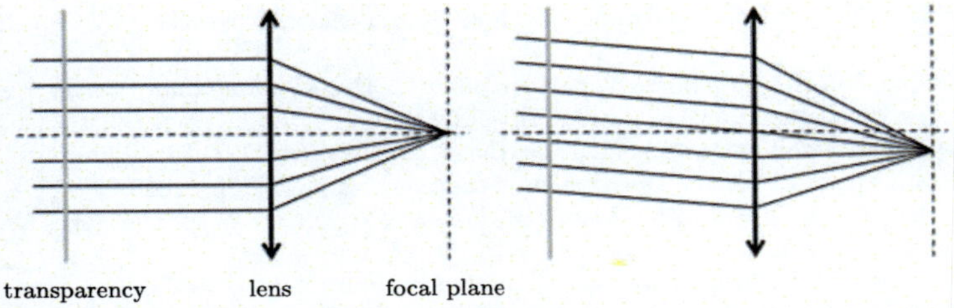

Fig. 2.24 Waves propagating at different directions after passing through a lens converge to different spots.

References

1. C.F. Meyer, *The Diffraction of Light, X-rays, and Material Particles* (The University of Chicago Press, Chicago, 1934)
2. A. L. Schawlow, Laser light. Sci. Am. **219**(3), 120–136 (1968)
3. W.L. Bragg, The diffraction of short electromagnetic waves by a crystal. P. Camb. Philos. Soc. **17**, 43–57 (1913)
4. G. Wulff, Über die Kristallröntgenogramme. Phys. Z. **14**, 217–222 (1913)
5. W. Kossel, V. Loeck, H. Voges, Die Richtungsverteilung der in einem Kristall entstandenen charakteristischen Röntgenstrahlung. Z. Phys. **94**, 139–144 (1935)
6. R. Tixier, C. Wache, Kossel patterns. J. Appl. Cryst. **3**, 466–485 (1970)
7. V.V. Lider, X-ray divergent beam (Kossel) technique: A review. Crystallogr. Rep. **56**, 169–189 (2011)
8. K. Lonsdale, Divergent-beam X-ray photography of crystals. Philos. Trans. R. Soc. A **240**, 219–250 (1947)
9. A.G.A. Nisbet, G. Beutier, F. Fabrizi, B. Moser, S.P. Collins, Diffuse multiple scattering. Acta Cryst. A **71**, 20–25 (2015)
10. E.M. Schulson, Electron channelling patterns in scanning electron microscopy. J. Mater. Sci. **12**, 1071–1087 (1977)
11. J.P. Morniroli, *Large-Angle Convergent-Beam Electron Diffraction. Applications to Crystal Defects* (SFμ, Paris, 2002)
12. M.N. Alam, M. Blackman, D.W. Pashley, High-angle Kikuchi patterns. *Proc. Roy. Soc. (London) A*, **221**, 224–242 (1954)
13. S. Zaefferer, The electron backscatter diffraction technique – a powerful tool to study microstructures by SEM. JEOL News **39**, 10–15 (2004)
14. J.L. Amorós, M.J. Buerger, M.C. de Amorós, *The Laue Method* (Academic Press, New York, 1975)
15. N. Tamura, *XMAS*: a versatile tool for analyzing synchrotron X-ray microdiffraction data, in *Strain and Dislocation Gradients from Diffraction*. ed. by R. Barabash, G. Ice (Imperial College Press, London, 2014), pp.125–155
16. N.K. Hansen, P. Coppens, Testing aspherical atom refinements on small-molecule data sets. Acta Cryst. A **34**, 909–921 (1978)
17. D.T. Cromer, J.B. Mann, X-ray scattering factors computed from numerical Hartree-Fock wave functions. Acta Cryst. A **24**, 321–324 (1968)
18. P.A. Doyle, P.S. Turner, Relativistic Hartree-Fock X-ray and electron scattering factors. Acta Cryst. A **24**, 390–397 (1968)
19. D. Waasmaier, A. Kirfel, New analytical scattering-factor functions for free atoms and ions. Acta Cryst. A **51**, 416–431 (1995)
20. A. Weickenmeier, H. Kohl, Computation of absorptive form factors for high-energy electron diffraction. Acta Cryst. A **47**, 590–597 (1991)
21. J.A. Ibers, Atomic scattering amplitudes for electrons. Acta Cryst. **11**, 178–183 (1958)
22. P.P. Ewald, X-ray diffraction by finite and imperfect crystal lattices. Proc. Phys. Soc. **52**, 167–174 (1940)
23. A. Guinier, *X-Ray Diffraction: In crystals, Imperfect Crystals, and Amorphous Bodies* (Freeman, San Francisco, 1963)
24. M. Senechal, A point set puzzle revisited. Eur. J. Combin. **29**, 1933–1944 (2008)
25. W. Massa, *Crystal Structure Determination* (Springer Verlag, Berlin, 2000)
26. W.H. Zachariasen, *Theory of X-Ray Diffraction in Crystals* (J. Wiley and Sons, New York, 1945)
27. Z. Ren, K. Moffat, Quantitative analysis of synchrotron Laue diffraction patterns in macromolecular crystallography. J. Appl. Cryst. **28**, 461–481 (1995)
28. P.H. Dederichs, Dynamical diffraction theory. Technical report, KFA—JUL–797–FF, Jülich, Germany, 1971

29. R.W. James, *The Optical Principles of the Diffraction of X-Rays* (G. Bell & Sons Ltd., London, 1954)
30. M. Born, E. Wolf, *Principles of Optics: Electromagnetic Theory of Propagation, Interference, and Diffraction of Light* (Macmillan, New York, 1964)

Chapter 3
Diffraction of High Energy Electrons

3.1 Introduction to Dynamical Diffraction

Rough simulations of diffraction patterns are usually based on the geometry of diffraction and on the kinematical approach to estimate intensities. In the kinematical approximation (Sects. 2.4 and 2.6), it is assumed that an atom of the scattering crystal is acted upon only by the incident wave, and the incident wave is diffracted only once. However, atoms interact with both the incident wave and the diffracted waves, and a diffracted wave may be scattered again and again. Multiple scattering is manifested by the presence of forbidden reflections. Let $\check{\boldsymbol{h}}$ and $\check{\boldsymbol{g}}$ be reciprocal lattice points corresponding to two allowed reflections. With double diffraction, when the diffracted beam $\check{\boldsymbol{h}}$ is diffracted again by $\check{\boldsymbol{g}}$, then the reflection corresponding to the vector $\check{\boldsymbol{h}} + \check{\boldsymbol{g}}$ will appear even if it is forbidden by the kinematic (single-diffraction) theory.[1] Violation of kinematic reflection conditions or change of reflection intensities due to this mechanism is known as the Renninger effect or Umweganregung.

If small dimensions of the crystal prevent multiple scattering events, the kinematical approximation may be close to reality, but in the general case, to obtain simulated diffraction patterns resembling experimental patterns with accurate estimation of intensities, it is necessary to use more sophisticated methods based on the dynamical theory of diffraction. This is particularly important in electron diffraction due to the strong interaction of electrons with matter.

Different from kinematical theories with the incident beam and diffracted beams seen as separate waves, dynamical theories consider a more general object—the wave field inside the crystal, at its boundaries, and in vacuo. The wave field can be seen as a single entity consisting of coupled beams continuously exchanging their energy. Thus, dynamical theories take into account multiple scattering or wave interactions.

There are a number of alternative approaches to the description of dynamical diffraction. These approaches can be seen as methods of solving a wave equation

[1] Clearly, this mechanism will not produce reflections which are forbidden because of a non-primitive unit cell.

A. Morawiec, *Indexing of Crystal Diffraction Patterns*, Springer Series in Materials Science 326, https://doi.org/10.1007/978-3-031-11077-1_3

under given boundary conditions. Details of the interaction of the radiation with matter need to be considered, and they are specific to the type of radiation. Calculations of intensities are based on models of the propagation of radiation in a crystalline solid. For a comprehensive treatment of dynamical X-ray diffraction, the interested reader is referred to [1]. There are numerous articles on dynamical electron diffraction; e.g., one can consult Cowley's overview [2] or the article [3] for a brief review of computational aspects of particular methods.

The field of dynamical diffraction is broad, issues considered there are sometimes slightly 'theoretical', and application of the theories is not trivial. In light of these complications, the practical use of dynamical diffraction (e.g. for the determination of crystal orientations or lattice parameters) is still limited. Our account on the subject will be rather superficial. To give the flavor of the dynamical theory, we will briefly describe the (so-called Bloch wave) method of Bethe[2] [4] concerning diffraction of elastically scattered electrons. Approaches with similar features exist for X-rays and neutrons. In all cases, derivations are based on the underlying differential wave equations for a periodic medium. For electron diffraction, it is a quantum mechanical wave equation with periodic Coulomb potential. In the case of neutrons, Fermi's pseudo-potential is used. For the description of X-rays, one applies the wave equation following from Maxwell's equations with dielectric 'constant' varying periodically in the crystal.[3]

Below, after a general part concerning propagation of high energy electrons in crystals, more and more limiting assumptions will be taken. At the end, we will focus on the dynamical simulation of spot patterns and convergent beam electron diffraction (CBED) patterns. Typically, these diffraction patterns are formed using transmission electron microscopy (TEM) with electrons having energies in the range of 100–300 keV.

3.1.1 Bloch Waves

Before getting to the heart of this chapter, one needs to mention the Floquet-Bloch theorem. The theorem is applicable to elliptic partial differential equations with smooth periodic coefficients. With $\check{\boldsymbol{m}}$ being a vector of their periodicity lattice, the coefficients at point $\boldsymbol{x}$ take the same values as at $\boldsymbol{x} + \check{\boldsymbol{m}}$. According to the theorem, the equation is satisfied by the function $\psi_{\boldsymbol{k}}$ being the product of the plane wave-like factor $\exp(2\pi i\, \boldsymbol{k} \cdot \boldsymbol{x})$ and the amplitude $u_{\boldsymbol{k}}(\boldsymbol{x})$ periodic in the direct space, i.e.,

$$\psi_{\boldsymbol{k}}(\boldsymbol{x}) \equiv u_{\boldsymbol{k}}(\boldsymbol{x}) \exp(2\pi i\, \boldsymbol{k} \cdot \boldsymbol{x}) \;, \tag{3.1}$$

[2] Hans Albrecht Bethe (1906–2005).

[3] The 'Bloch wave' based theory of electron diffraction has its origins in the works of Ewald on X-ray diffraction [5–7].

where $u_{\boldsymbol{k}}(\boldsymbol{x}) = u_{\boldsymbol{k}}(\boldsymbol{x} + \check{\boldsymbol{m}})$, and $\boldsymbol{k}$ is a vector in the reciprocal space. The (Bloch) wave described by $\psi_{\boldsymbol{k}}$ can be seen as a plane wave modulated by the (Bloch) function $u_{\boldsymbol{k}}(\boldsymbol{x})$ having the periodicity of the lattice. Since $\psi_{\boldsymbol{k}}(\boldsymbol{x} + \check{\boldsymbol{m}}) = \psi_{\boldsymbol{k}}(\boldsymbol{x}) \exp(2\pi i \boldsymbol{k} \cdot \check{\boldsymbol{m}})$, the function $\psi_{\boldsymbol{k}}$ differs between equivalent lattice positions by the phase factor $\exp(2\pi i \boldsymbol{k} \cdot \check{\boldsymbol{m}})$. Due to their periodicity, the Bloch functions $u_{\boldsymbol{k}}(\boldsymbol{x})$ can be expanded in the basis of the lattice as the discrete sums

$$u_{\boldsymbol{k}}(\boldsymbol{x}) = \sum_{\check{\boldsymbol{h}}} C_{\check{\boldsymbol{h}}}(\boldsymbol{k}) \exp(2\pi i\, \check{\boldsymbol{h}} \cdot \boldsymbol{x}) \,, \tag{3.2}$$

where $\check{\boldsymbol{h}}$ is a reciprocal lattice vector; cf. Sect. 2.4.2. Thus, the Bloch waves have the form

$$\psi_{\boldsymbol{k}}(\boldsymbol{x}) = \sum_{\check{\boldsymbol{h}}} C_{\check{\boldsymbol{h}}}(\boldsymbol{k}) \exp(2\pi i\, (\boldsymbol{k} + \check{\boldsymbol{h}}) \cdot \boldsymbol{x}) \,. \tag{3.3}$$

A general solution $\Psi(\boldsymbol{x})$ to the elliptic equation is a linear combination of Bloch waves

$$\Psi(\boldsymbol{x}) = \sum_{\boldsymbol{k}} \alpha_{\boldsymbol{k}} \psi_{\boldsymbol{k}}(\boldsymbol{x}) \,. \tag{3.4}$$

The differential equation and boundary conditions determine which Bloch waves actually arise, and what are their particular forms, i.e., what are the $\boldsymbol{k}$ vectors, and the pairs $\check{\boldsymbol{h}}$ and $C_{\check{\boldsymbol{h}}}$, respectively.[4]

With $\check{\boldsymbol{h}}$ and $\check{\boldsymbol{g}}$ being reciprocal lattice vectors, the terms $C_{\check{\boldsymbol{h}}} \exp(2\pi i\, (\boldsymbol{k} + \check{\boldsymbol{h}}) \cdot \boldsymbol{x})$ and $C_{\check{\boldsymbol{h}}+\check{\boldsymbol{g}}} \exp(2\pi i\, (\boldsymbol{k} + \check{\boldsymbol{h}} + \check{\boldsymbol{g}}) \cdot \boldsymbol{x})$ of the Bloch waves (3.3) represent plane waves with wave vectors related by the Laue equation if the magnitudes of $\boldsymbol{k} + \check{\boldsymbol{h}}$ and $\boldsymbol{k} + \check{\boldsymbol{h}} + \check{\boldsymbol{g}}$ are the same. In such a case, an exchange of energy between these waves is expected. The key point of the dynamical approach to diffraction is to determine the possible $\boldsymbol{k}$ vectors and the amplitudes $C_{\check{\boldsymbol{h}}}$ of excited waves based on the wave equation and boundary conditions applicable to a particular experimental situation.

The substitution of Bloch waves into the differential equation describing the wave propagation leads to a homogeneous system of linear algebraic equations with respect to $C_{\check{\boldsymbol{h}}}$. The condition for the existence of non-zero solutions of the system (vanishing of its determinant) limits the spectrum of waves (i.e., admissible $\boldsymbol{k}$ vectors) which can potentially be excited. Which waves are actually excited depends on the cause of the excitation; in practice, these waves are determined based on boundary conditions. Thus, elements used for constructing the dynamical theory are as follows:

— homogeneous linear wave equation,

— series expansion of the wave function into plane-wave solutions,

— replacing the wave equation by a homogeneous system of linear algebraic equations,

— determinant of the system must be zero for the existence of solutions

[4] The generality of the expansion (3.3) needs to be compared to that of (2.17).

(this imposes conditions on admissible wave vectors),
— boundary conditions determine waves which are actually excited.
Subsequent sections contain details of this scheme for high energy electron diffraction.

3.2 Wave Equation for a Single Electron in an Electrostatic Potential

The speeds attained by electrons in transmission electron microscopes are known to be comparable to the speed of light in vacuum (see below), and a proper description of their propagation must be based on relativistic relationships. Formally, such propagation is described by Dirac's equation. However, it turns out that electron spin has negligible effect [8]. Therefore, relativistic but spinless electrons are considered, and it is sufficient to use the relativistic Schrödinger equation for a charged particle in an electric field.

There is a question about mutual interaction of the high-energy electrons of the beam in the specimen. In a typical setup (200 kV scope, average electron flux and foil thickness of the order of 10^2 nm), there is just one electron at a time inside the specimen. Therefore, the task is to describe a single-electron scattering.

Let the electron velocity with respect to the laboratory reference system be denoted by $\boldsymbol{v}$. In the conventional notation, one has

$$\beta = |\boldsymbol{v}|/c\,, \quad \gamma = 1/\sqrt{1-\beta^2}\,, \quad m = \gamma m_0\,, \quad \boldsymbol{p} = m\boldsymbol{v}\,,$$

where c and m_0 denote the speed of light in vacuum and the electron (rest) mass, respectively, and $\boldsymbol{p}$ is the momentum of the electron. The kinetic energy of a free electron E_k equals the difference between its total energy mc^2 and the rest energy m_0c^2, i.e., $E_k = mc^2 - m_0c^2$. Moreover, one has $m^2c^4 - m_0^2c^4 = m_0^2c^4(\gamma^2-1) = m_0^2c^4\beta^2\gamma^2 = \boldsymbol{p}^2c^2$. Hence, the magnitude of momentum and the kinetic energy are linked via $\boldsymbol{p}^2c^2 = (E_k + m_0c^2)^2 - m_0^2c^4 = 2m_0c^2E_k + E_k^2$ or

$$\boldsymbol{p}^2 = 2m_0E_k\left(1 + \frac{E_k}{2m_0c^2}\right)\,. \tag{3.5}$$

The de Broglie wavelength λ of the electron is given by

$$\lambda = h/|\boldsymbol{p}|\,, \tag{3.6}$$

where h $(= 2\pi\hbar)$ is the Planck's constant. In the practice of electron microscopy, the accelerating voltage V_{acc} is used instead of the wavelength. The kinetic energy acquired by an electron in the potential difference V_{acc} is

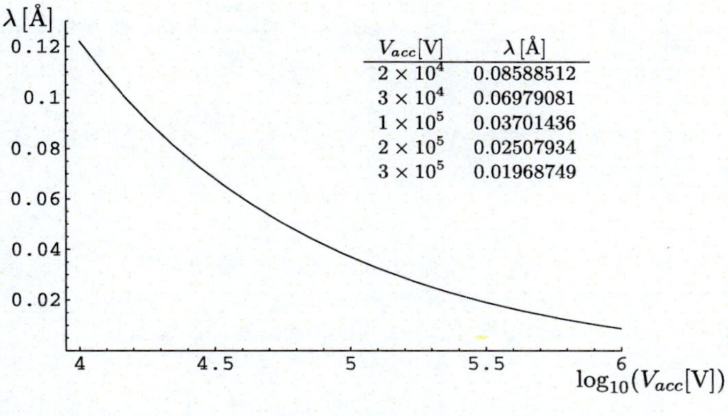

Fig. 3.1 De Broglie wavelength of electrons versus logarithm of accelerating voltage.

$$E_k = eV_{acc} \ . \tag{3.7}$$

Hence, the factor γ expressed via the voltage equals

$$\gamma = \frac{mc^2}{m_0c^2} = \frac{m_0c^2 + E_k}{m_0c^2} = 1 + \frac{eV_{acc}}{m_0c^2} \ . \tag{3.8}$$

E.g., for $V_{acc} = 200$ kV, one has $\gamma \approx 1.39$, i.e., the speed of electrons is about $0.7c$; this confirms the need for the relativistic approach.

⊡ It is worth noting here that, based on (3.5–3.8), the de Broglie wavelength is

$$\lambda = h \left(2m_0eV_{acc} \left(1 + \frac{eV_{acc}}{2m_0c^2} \right) \right)^{-1/2} , \tag{3.9}$$

and the inverse relationship for the voltage reads

$$V_{acc} = \frac{m_0c^2}{e} \left(\left(1 + \left(\frac{h}{m_0c} \frac{1}{\lambda} \right)^2 \right)^{1/2} - 1 \right) .$$

For λ in angstroms and V_{acc} in volts, one has $\lambda = 12.264259/\sqrt{(1. + 9.7847567 \times 10^{-7}\, V_{acc})V_{acc}}$ (Fig. 3.1) and $V_{acc} = -510998.91 + 12398.419\sqrt{1698.6635 + 1/\lambda^2}$.

It follows from (3.5–3.7) that the squared magnitude of the wave vector $\boldsymbol{k}_0$ of the incident beam of electrons accelerated in vacuum by V_{acc} is given by

$$\boldsymbol{k}_0^2 = \frac{1}{\lambda^2} = \frac{V_{acc}^{rel}}{\pi a_1} ,$$

where V_{acc}^{rel} and a_1 are abbreviations for

$$V_{acc}^{rel} = \left(1 + \frac{eV_{acc}}{2m_0c^2}\right) V_{acc} = \frac{1+\gamma}{2} V_{acc} \text{ and } a_1 = \frac{h^2}{2\pi m_0 e} \approx 0.47877643 \times 10^{-18}\text{Vm}^2 ;$$

cf. Sect. 2.4.3.

In the crystal, the kinetic energy E_k of (3.7) is modified by the inner electrostatic potential V, i.e., it equals

$$e(V_{acc} + V) ,$$

where $V = V(\boldsymbol{x}) = \sum_n V_n(\boldsymbol{x} - \boldsymbol{x}_n)$ is the sum of potentials V_n of particular atoms, and V_n is approximated by (2.31). By substituting E_k in (3.5) by $e(V_{acc} + V)$, and with the small quadratic term $V^2/(2m_0c^2)$ neglected, one obtains

$$\boldsymbol{p}^2 = 2m_0e\left(V_{acc}\left(1 + \frac{eV_{acc}}{2m_0c^2}\right) + V\left(1 + \frac{eV_{acc}}{m_0c^2}\right)\right)$$

or briefly

$$\boldsymbol{p}^2 = 2m_0e\left(V_{acc}^{rel} + \gamma V\right) .$$

With constant total energy, the above relationship is turned into the quantum mechanical differential wave equation by the substitution $\boldsymbol{p} \to -i\hbar\nabla$. Hence, with $\Psi = \Psi(\boldsymbol{x})$ denoting a single electron solution, one finally has

$$\frac{a_1}{4\pi}\nabla^2\Psi + \left(V_{acc}^{rel} + \gamma V\right)\Psi = 0 . \tag{3.10}$$

The same equation can be obtained directly based on the relativistic Schrödinger equation for a charged particle in electromagnetic field.

⊡ For the electrostatic potential V (and magnetic interaction terms omitted), the equation is derived from $(E_{tot} - (-e)V)^2 = c^2\boldsymbol{p}^2 + m_0^2c^4$ by the substitutions $E_{tot} \to i\hbar\partial/\partial t$ and $\boldsymbol{p} \to -i\hbar\nabla$. With time dependent $\psi = \psi(\boldsymbol{x}, t)$, the equation reads (see, e.g., [9], p. 308)

$$-\hbar^2\frac{\partial^2\psi}{\partial t^2} + 2eVi\hbar\frac{\partial\psi}{\partial t} + ie\psi\hbar\frac{\partial V}{\partial t} + e^2V^2\psi = -c^2\hbar^2\nabla^2\psi + m_0^2c^4\psi .$$

For static V (i.e., $\partial V/\partial t = 0$) and constant energy E_{tot} (i.e., $\psi(\boldsymbol{x}, t) \to \Psi(\boldsymbol{x})\exp(-2\pi i(E_{tot}/h)t)$), the above equation is transformed to $E_{tot}^2\Psi + 2eVE_{tot}\Psi + e^2V^2\Psi = -c^2\hbar^2\nabla^2\Psi + m_0^2c^4\Psi$. By neglecting the term with V^2, and taking into account that $E_{tot} = eV_{acc} + m_0c^2$ one obtains (3.10).

Besides the Bloch-wave method which is discussed below, there are a number of other approaches to solving (3.10) for simulation of TEM images and diffraction patterns of various types [10–13]. As will be shown below, the Bloch-wave method is particularly convenient for the simulation of convergent beam diffraction patterns of perfect crystals. It is also used for computing (diffraction-based) contrast images of structures.

3.2.1 Solutions for an Unbounded Crystal

In vacuum, outside the crystal, where $V = 0$, (3.10) is the classic Helmholtz equation $(\nabla^2 + 4\pi^2 \boldsymbol{k}_0^2)\Psi = 0$, satisfied (in general, by non-dispersive waves, and in particular) by the plane wave with the wave vector $\boldsymbol{k}_0$, i.e., $\Psi(\boldsymbol{x}) = \exp(2\pi i\, \boldsymbol{k}_0 \cdot \boldsymbol{x})$. The point is to solve (3.10) inside the crystal for non-zero V. The potential $V(\boldsymbol{x})$ has the periodicity of the direct lattice, and therefore it is expressible as a Fourier series

$$V(\boldsymbol{x}) = \sum_{\check{\boldsymbol{h}}} V_{\check{\boldsymbol{h}}} \exp(2\pi i \check{\boldsymbol{h}} \cdot \boldsymbol{x}) \; ; \tag{3.11}$$

cf. Sect. 2.4.3. It follows from comparison with (2.30) that the expansion coefficients $V_{\check{\boldsymbol{h}}}$ are linked to structure factors via

$$V_{\check{\boldsymbol{h}}} = \frac{h^2}{2\pi m_0 e} \frac{F(\check{\boldsymbol{h}})}{V_{\mathrm{uc}}} = a_1 \frac{F(\check{\boldsymbol{h}})}{V_{\mathrm{uc}}} \, .$$

Since (3.10) is elliptic with periodic coefficients, the Bloch theorem is applicable. Substitution of $\Psi(\boldsymbol{x})$ by the Bloch wave (3.3) and $V(\boldsymbol{x})$ by the expansion (3.11) gives

$$\sum_{\check{\boldsymbol{h}}} \left((V_{acc}^{rel} - \pi a_1 (\boldsymbol{k} + \check{\boldsymbol{h}})^2) C_{\check{\boldsymbol{h}}} + \gamma \sum_{\check{\boldsymbol{l}}} C_{\check{\boldsymbol{h}}-\check{\boldsymbol{l}}} V_{\check{\boldsymbol{l}}} \right) \exp(2\pi i\, (\boldsymbol{k} + \check{\boldsymbol{h}}) \cdot \boldsymbol{x}) = 0 \, .$$

The equality occurs for arbitrary $\boldsymbol{x}$. Hence, the coefficients in the parentheses must be zero. This condition can be expressed as $(V_{acc}^{rel} - \pi a_1 (\boldsymbol{k} + \check{\boldsymbol{h}})^2 + \gamma V_0) C_{\check{\boldsymbol{h}}} + \gamma \sum_{\check{\boldsymbol{l}},\, \check{\boldsymbol{l}} \neq \check{\boldsymbol{h}}} C_{\check{\boldsymbol{l}}} V_{\check{\boldsymbol{h}}-\check{\boldsymbol{l}}} = 0$, or equivalently, for each $\check{\boldsymbol{h}}$, one has[5]

$$b_{\check{\boldsymbol{h}}} C_{\check{\boldsymbol{h}}} + \sum_{\check{\boldsymbol{l}},\, \check{\boldsymbol{l}} \neq \check{\boldsymbol{h}}} C_{\check{\boldsymbol{l}}} U_{\check{\boldsymbol{h}}-\check{\boldsymbol{l}}} = 0 \tag{3.12}$$

where $U_{\check{\boldsymbol{g}}} \equiv \gamma V_{\check{\boldsymbol{g}}}/(\pi a_1) = \gamma F(\check{\boldsymbol{g}})/(\pi V_{\mathrm{uc}})$, and $b_{\check{\boldsymbol{h}}}(\boldsymbol{k}) \equiv V_{acc}^{rel}/(\pi a_1) + U_0 - (\boldsymbol{k} + \check{\boldsymbol{h}})^2$. (The units of $U_{\check{\boldsymbol{g}}}$ and $b_{\check{\boldsymbol{h}}}$ are [length^{-2}].) The constant term U_0 was isolated from the series of potentials and added to the energy of the incident beam. In high energy electron diffraction, it is small compared to $V_{acc}^{rel}/(\pi a_1) = \boldsymbol{k}_0^2$, and in most cases, it can be neglected. It is conventional to use the "mean electron wave vector" [14] denoted by $\boldsymbol{K}$ and satisfying the relationship

$$\boldsymbol{K}^2 \equiv V_{acc}^{rel}/(\pi a_1) + U_0 = \boldsymbol{k}_0^2 + U_0 \; ; \tag{3.13}$$

the mean electron wave vector can be seen as $\boldsymbol{k}_0$ corrected for refraction. With this vector, $b_{\check{\boldsymbol{h}}}(\boldsymbol{k})$ takes the form

---

[5] No summation over $\check{\boldsymbol{h}}$. The same concerns similar expressions below.

$$b_{\check{h}}(\boldsymbol{k}) = \boldsymbol{K}^2 - (\boldsymbol{k} + \check{\boldsymbol{h}})^2 \, . \tag{3.14}$$

The relationships (3.12) can be written as

$$\sum_{\check{l}} B_{\check{h}\check{l}} \, C_{\check{l}} = 0 \, , \tag{3.15}$$

where $B_{\check{h}\check{l}} = b_{\check{h}} \delta_{\check{h}\check{l}} + U_{\check{h}-\check{l}} - U_0 \delta_{\check{h}\check{l}}$. The coefficients $B_{\check{h}\check{l}}$ are entries of the matrix

$$B = B(\boldsymbol{k}) = \begin{bmatrix} b_{\check{g}_1} & U_{\check{g}_1-\check{g}_2} & U_{\check{g}_1-\check{g}_3} & U_{\check{g}_1-\check{g}_4} & \cdots \\ U_{\check{g}_2-\check{g}_1} & b_{\check{g}_2} & U_{\check{g}_2-\check{g}_3} & U_{\check{g}_2-\check{g}_4} & \cdots \\ U_{\check{g}_3-\check{g}_1} & U_{\check{g}_3-\check{g}_2} & b_{\check{g}_3} & U_{\check{g}_3-\check{g}_4} & \cdots \\ \cdots & \cdots & \cdots & \cdots & \end{bmatrix}$$

indexed by discrete reciprocal lattice vectors. Formally, the matrix B has infinite dimensions. In practical computations, the dimensions must be finite. They are $N_b \times N_b$, where N_b is the number of terms used in (3.11) and (3.2). The main criteria for including a given term (beam) are the magnitude of the corresponding reciprocal lattice vector and the proximity of its terminal point to the Ewald sphere, cf. [15, 16].

Equation (3.15) is a homogeneous system of linear equations with respect to $C_{\check{h}}$. The condition for its solvability

$$\det B(\boldsymbol{k}) = 0 \tag{3.16}$$

imposes constraints on possible $\boldsymbol{k}$ vectors. Based on (3.14), this is an algebraic equation of degree $2N_b$ in $\boldsymbol{k}$. Bloch waves have the same total energy but the balance between kinetic and potential energies depends on $\boldsymbol{k}$, i.e., it varies from wave to wave. For a given set of beams, (3.16) relates the allowed wave vectors to the total energy, i.e., it is a dispersion equation. Its solutions with respect to $\boldsymbol{k}$ determine an equi-energy (dispersion) surface in reciprocal space. If all the potentials $U_{\check{h}}$ were zero, the dispersion surface would be an array of spheres described by $b_{\check{h}}(\boldsymbol{k}) = 0$ or $(\boldsymbol{k} + \check{\boldsymbol{h}})^2 = \boldsymbol{K}^2$, i.e., each of them would have radius $|\boldsymbol{K}|$ and center determined by $\check{\boldsymbol{h}}$. With potentials deviating from zero, the dispersion surface deviates from an array of spheres. The character of the deviation can be seen by calculating the dispersion surface for the two-beam case.

3.2.2 *Two-Beam Centro-Symmetric Case*

The simplest nontrivial case of dynamic diffraction involves two beams: the 'incident' beam (corresponding to the reciprocal lattice vector $\check{\boldsymbol{g}}_1 = \boldsymbol{0}$), and the diffracted beam (corresponding to $\check{\boldsymbol{g}}_2 = \check{\boldsymbol{h}}$). The $B(\boldsymbol{k})$ matrix has the form

$$B(\boldsymbol{k}) = \begin{bmatrix} b_0(\boldsymbol{k}) & U_{-\check{\boldsymbol{h}}} \\ U_{\check{\boldsymbol{h}}} & b_{\check{\boldsymbol{h}}}(\boldsymbol{k}) \end{bmatrix} . \tag{3.17}$$

With a real potential V, one has $U_{-\check{\boldsymbol{h}}} = U^*_{\check{\boldsymbol{h}}}$ and

$$\det B(\boldsymbol{k}) = (\boldsymbol{K}^2 - \boldsymbol{k}^2)\left(\boldsymbol{K}^2 - (\boldsymbol{k} + \check{\boldsymbol{h}})^2\right) - |U_{\check{\boldsymbol{h}}}|^2 .$$

Based on the dispersion equation $\det B(\boldsymbol{k}) = 0$ for fixed $\boldsymbol{K}^2$, $\check{\boldsymbol{h}}$ and $|U_{\check{\boldsymbol{h}}}|$, the terminal points of the allowed $\boldsymbol{k}$ vectors constitute a two-dimensional surface of revolution in the reciprocal space. If $|U_{\check{\boldsymbol{h}}}|$ were zero, the dispersion surface would consist of two spheres ($\boldsymbol{k}^2 = \boldsymbol{K}^2$ and $(\boldsymbol{k} + \check{\boldsymbol{h}})^2 = \boldsymbol{K}^2$) with centers separated by $\check{\boldsymbol{h}}$. Since the potential is not zero and satisfies $|U_{\check{\boldsymbol{h}}}| \ll \boldsymbol{K}^2$, the surface consists of two branches. With the dimensionless parameter w defined as $w = \left(\boldsymbol{k}^2 - (\boldsymbol{k} + \check{\boldsymbol{h}})^2\right)/(2U_{\check{\boldsymbol{h}}})$, the relationship $\det B(\boldsymbol{k}) = 0$ can be written in the form

$$\boldsymbol{K}^2 = \boldsymbol{k}^2 - U_{\check{\boldsymbol{h}}}\left(w \pm \sqrt{w^2 + 1}\right) ; \tag{3.18}$$

the sign "+" ("−") corresponds to the outer (inner) branch of the dispersion surface in the schematic illustration of the surface shown in Fig. 3.2.

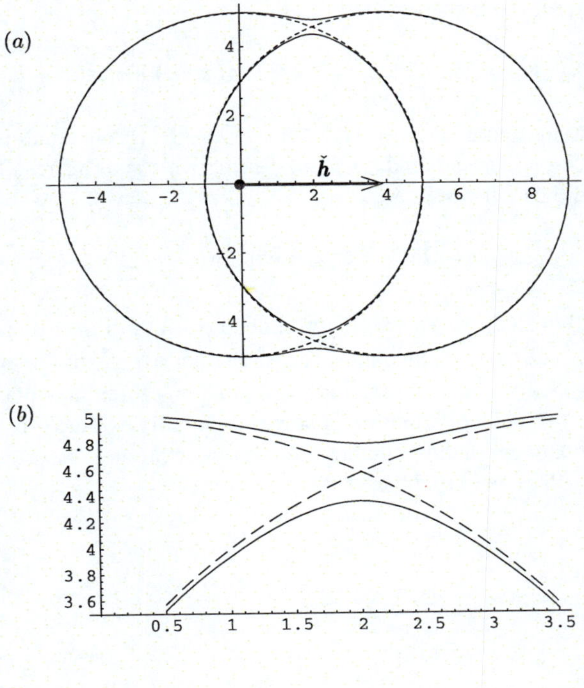

Fig. 3.2 Schematic illustration of a solution of the dispersion equation for the two-beam case. The curves show positions of all two-dimensional vectors $\boldsymbol{k}$ satisfying the equation $(\boldsymbol{K}^2 - \boldsymbol{k}^2)(\boldsymbol{K}^2 - (\boldsymbol{k} + \check{\boldsymbol{h}})^2) = u^2$ for fictitious $\boldsymbol{K}^2 = 5^2$, $\check{\boldsymbol{h}} = (4, 0)$ in the drawn Cartesian system, $u = 0$ (dashed lines) and $u = 2$ (solid lines). By convention, vectors $\boldsymbol{k}$ have terminal points at $\mathbf{0}$ with the curve representing the positions of their initial points. The complete dispersion surface is obtained by rotating the solid curves around the abscissa. Details near the 'tie-point' are shown in (*b*).

Substitution of $\boldsymbol{K}^2$ in (3.14) by (3.18) results in $b_{\mathbf{0}} = U_{\check{\boldsymbol{h}}}\left(-w \pm \sqrt{w^2+1}\right)$ and $b_{\check{\boldsymbol{h}}} = U_{\check{\boldsymbol{h}}}\left(+w \pm \sqrt{w^2+1}\right)$. One of (3.15) takes the form

$$C_{\check{\boldsymbol{h}}} = \left(w \pm \sqrt{w^2+1}\right) C_{\mathbf{0}} , \tag{3.19}$$

and the second one is linearly dependent. This relationship and the normalization condition $C_{\mathbf{0}}^2 + C_{\check{\boldsymbol{h}}}^2 = 1$ determine the coefficients $C_{\mathbf{0}}$ and $C_{\check{\boldsymbol{h}}}$.

If $|w|$ is large and the beam $\boldsymbol{k}$ is far from the Bragg condition (i.e., $(\boldsymbol{k}+\check{\boldsymbol{h}})^2$ is very different from $\boldsymbol{k}^2$), then either $C_{\mathbf{0}} \approx \pm 1$ or $C_{\check{\boldsymbol{h}}} \approx \pm 1$ with the other coefficient being close to 0.

To consider the general case, it is convenient to replace the parameter w by ω in the range $(0, \pi/2)$ such that $w = \cot(2\omega)$. Clearly, $w = 0$ corresponds to $\omega = \pi/4$. and

$$\left(w + \sqrt{w^2+1}\right) = \cot\omega \quad \text{and} \quad \left(w - \sqrt{w^2+1}\right) = -\tan\omega .$$

The relationships (3.19), i.e., $C_{\check{\boldsymbol{h}}} = \cot(\omega)\, C_{\mathbf{0}}$ and $C_{\check{\boldsymbol{h}}} = -\tan(\omega)\, C_{\mathbf{0}}$, in combination with the normalization condition lead to four solutions of the form

$$\left(C_{\mathbf{0}},\ C_{\check{\boldsymbol{h}}}\right) = (\sin(\omega + m\pi/2),\ \cos(\omega + m\pi/2)) ,$$

where $m = 0, 1, 2, 3$. With these coefficients, based on (3.3) defining the Bloch waves, one has

$$\psi_{\boldsymbol{k}}(\boldsymbol{x}) = \sin(\omega + m\pi/2)\exp(2\pi i\, \boldsymbol{k}\cdot\boldsymbol{x}) + \cos(\omega + m\pi/2)\exp(2\pi i\, (\boldsymbol{k}+\check{\boldsymbol{h}})\cdot\boldsymbol{x}) .$$

In particular, if $w = 0$ (and $\omega = \pi/4$), the Bragg condition $(\boldsymbol{k}+\check{\boldsymbol{h}})^2 - \boldsymbol{k}^2 = \check{\boldsymbol{h}}\cdot(\check{\boldsymbol{h}} + 2\boldsymbol{k}) = 0$ is satisfied (cf. (2.6)), the above equations result in $C_{\mathbf{0}} = \pm\sqrt{2}/2 = C_{\check{\boldsymbol{h}}}$, and the Bloch waves are

$$\psi_{\boldsymbol{k}}(\boldsymbol{x}) = (-1)^{\lfloor m/2 \rfloor}\, i^m \sqrt{2}\, \exp\left(\pi i (\check{\boldsymbol{h}} + 2\boldsymbol{k})\cdot\boldsymbol{x}\right)\, \cos\left(\pi(\check{\boldsymbol{h}}\cdot\boldsymbol{x} + m/2)\right) . \tag{3.20}$$

Two such waves are shown schematically in Fig. 3.3. The waves propagating along $\check{\boldsymbol{h}} + 2\boldsymbol{k}$, i.e., along planes perpendicular to $\check{\boldsymbol{h}}$, are modulated by the factors $\cos(\pi\check{\boldsymbol{h}}\cdot\boldsymbol{x})$ and $\sin(\pi\check{\boldsymbol{h}}\cdot\boldsymbol{x})$ with a half-period equal to the inter-planar spacing $1/|\check{\boldsymbol{h}}|$.

Which combination of waves is actually excited (i.e., what are the values of $\boldsymbol{k}$ and w), depends on boundary conditions involving the orientation of the incident beam with respect to the crystal.

Fig. 3.3 Schematic illustration of Bloch waves of (3.20). Drawn planes represent lattice planes corresponding to $\check{\boldsymbol{h}}$. Wave vector $\boldsymbol{k}$ is such that $\check{\boldsymbol{h}} \cdot (\check{\boldsymbol{h}} + 2\boldsymbol{k}) = 0$.

3.3 Bloch Waves in Semi-Infinite and Plate-Like Crystals

The waves in the crystal are excited by a beam coming from vacuum. The simplest approach is to assume that the vacuum and the crystal are separated by a plane, with an abrupt cut-off of crystal potentials (Fig. 3.4). Although this approach is not really physical (as the existence of the surface, with obvious asymmetry between the bulk and vacuum, causes reorganization of atoms leading to a loss of the periodicity present in the bulk), and generally, bounding a crystal violates periodicity assumptions needed for (3.11), it results in a good agreement with experiment, and it is commonly used.

As in classical light optics, with the incident beam crossing the surface, one obtains the transmitted and reflected components. Inside the crystal, the transmitted electron beam excites Bloch waves, leading to diffracted beams. The latter may have various directions of propagation, including the directions towards the bounding surface (back-diffracted waves). In conventional use of transmission microscopes, diffraction occurs under the so-called Laue (transmission) conditions[6] (or case), with the incident beam at large angle to the crystal surface. In the Laue case and with high energy of electrons, the reflected and back-diffracted beams are weak, and are neglected.[7]

The requirement of continuity of physical quantities across the surface of the crystal gives boundary conditions. Let $\widehat{\boldsymbol{n}}$ be an inward directed unit vector perpendicular to the flat entry surface of the considered semi-infinite crystal. It is convenient to assume that the surface goes through the origin of the coordinate system. The total wave function and its gradient along the direction normal to the surface are expected to be continuous on the surface. With the reflected and back-diffracted waves ignored, the first condition formally means that for all $\boldsymbol{x}_\perp$ in the bounding plane ($\widehat{\boldsymbol{n}} \cdot \boldsymbol{x}_\perp = 0$), the incident wave $\exp(2\pi i \boldsymbol{k}_0 \cdot \boldsymbol{x})$ at $\boldsymbol{x} = \boldsymbol{x}_\perp$ is equal to the total wave in the crystal Ψ at $\boldsymbol{x}_\perp$, or via (3.4), to the combination of the Bloch waves $\sum_{\boldsymbol{k}} \alpha_{\boldsymbol{k}} \psi_{\boldsymbol{k}}(\boldsymbol{x}_\perp)$, i.e.,

$$\sum_{\boldsymbol{k}} \alpha_{\boldsymbol{k}} \sum_{\check{\boldsymbol{h}}} C_{\check{\boldsymbol{h}}}(\boldsymbol{k}) \exp\left(2\pi i (\boldsymbol{k} + \check{\boldsymbol{h}}) \cdot \boldsymbol{x}_\perp\right) = \exp\left(2\pi i \boldsymbol{k}_0 \cdot \boldsymbol{x}_\perp\right) .$$

This condition is satisfied if one concurrently has, first,

$$\sum_{\boldsymbol{k}} \alpha_{\boldsymbol{k}} C_{\check{\boldsymbol{h}}}(\boldsymbol{k}) = \delta_{\check{\boldsymbol{h}}\boldsymbol{0}} , \tag{3.21}$$

and, second, $\boldsymbol{k} \cdot \boldsymbol{x}_\perp = \boldsymbol{k}_0 \cdot \boldsymbol{x}_\perp$, or equivalently[8],

$$\boldsymbol{k} = \boldsymbol{k}_0 + \tilde{\gamma} \widehat{\boldsymbol{n}} . \tag{3.22}$$

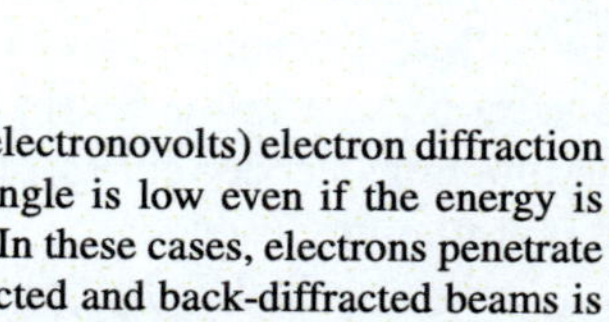

Fig. 3.4 Illustration of the non-physical assumption about the abrupt cut-off of crystal potentials. With such a model of the crystal surface some atoms are ‘incomplete’.

[6] As opposed to the Bragg case with reflection geometry.

[7] This cannot be assumed in the case of low energy (hundreds of electronovolts) electron diffraction (LEED). The approximation is also violated if the incidence angle is low even if the energy is high, as in reflection high energy electron diffraction (RHEED). In these cases, electrons penetrate only a few atomic layers, and the information contained in reflected and back-diffracted beams is affected only by these layers. Therefore, LEED and RHEED are predominantly surface characterisation methods. Besides them, the backwards traveling waves contribute to patterns obtained using other experimental techniques like the popular electron back-scatter diffraction (EBSD) or electron channeling diffraction (ECP).

[8] Conventionally, instead of $\tilde{\gamma}$, the symbol γ is used. Here, the accent $\tilde{}$ was added to distinguish the coefficient $\tilde{\gamma}$ from the relativistic Lorentz factor $\gamma = 1/\sqrt{1-\beta^2}$.

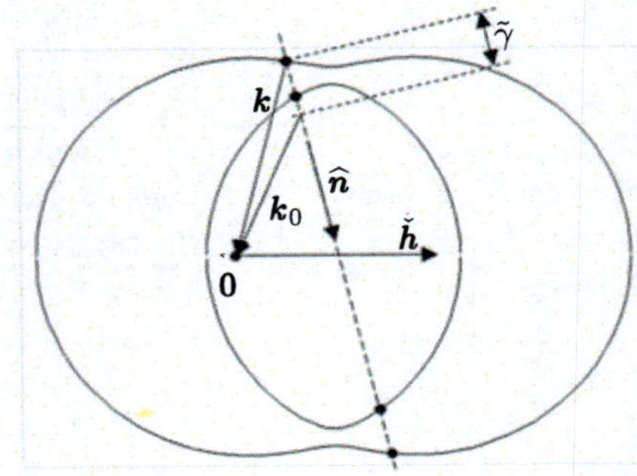

Fig. 3.5 Schematic illustration of the intersections of the line $\boldsymbol{k}_0 + \tilde{\gamma}\widehat{\boldsymbol{n}}$ with the dispersion surface. For two ($N_b = 2$) beams, the number of intersection points is at most $4\ (= 2N_b)$.

For a given incident beam direction, orientation of the crystal and orientation of the bounding plane, the vectors $\boldsymbol{k}_0$, $\widehat{\boldsymbol{n}}$ and the $\check{\boldsymbol{h}}$ vectors are fixed, and the only flexible parameter is $\tilde{\gamma}$ influencing the component of $\boldsymbol{k}$ parallel to $\widehat{\boldsymbol{n}}$. In the N_b beam case, this component meets the dispersion surface at $2N_b$ points at most. This limits the spectrum of possible $\boldsymbol{k}$ vectors to a discrete set. See Fig. 3.5.

With $\boldsymbol{k} = \boldsymbol{k}_0 + \tilde{\gamma}\widehat{\boldsymbol{n}}$, the system of (3.15) leads to the quadratic eigenvalue problem

$$\sum_{\check{l}} A_{\check{h}\check{l}}\, C_{\check{l}} = 2\tilde{\gamma}\,\widehat{\boldsymbol{n}} \cdot (\boldsymbol{k}_0 + \check{\boldsymbol{h}})\, C_{\check{h}} + \tilde{\gamma}^2\, C_{\check{h}}\ , \tag{3.23}$$

where A is defined as $A = B(\boldsymbol{k}_0)$. For a given crystal orientation, the matrix A is determined by $\boldsymbol{k}_0$ and the list of beams; it differs from the general $\boldsymbol{k}$-dependent B by fixed elements on the diagonal, which are given by $A_{\check{h}\check{h}} = \boldsymbol{K}^2 - (\boldsymbol{k}_0 + \check{\boldsymbol{h}})^2$.

⊡ The quadratic eigenvalue problem (3.23) can be linearized using standard methods [17]: Let M denote the matrix with entries $M_{\check{h}\check{l}} = 2\widehat{\boldsymbol{n}} \cdot (\boldsymbol{k}_0 + \check{\boldsymbol{h}})\, \delta_{\check{h}\check{l}}$. One has $\sum_{\check{l}} A_{\check{h}\check{l}}\, C_{\check{l}} = \tilde{\gamma} D_{\check{h}}$, where $D_{\check{h}} = \sum_{\check{l}} \left(M_{\check{h}\check{l}} + \tilde{\gamma}\delta_{\check{h}\check{l}}\right) C_{\check{l}}$. This is a linear eigenvalue problem of dimension $2N_b$, which can be briefly expressed as

$$\begin{bmatrix} -M & I \\ A & 0 \end{bmatrix} \begin{bmatrix} C \\ D \end{bmatrix} = \tilde{\gamma} \begin{bmatrix} C \\ D \end{bmatrix},$$

where C and D are arrays of $C_{\check{l}}$ and $D_{\check{l}}$, respectively, and I is the identity matrix.

In practice, various assumptions are made, and (3.23) is simplified. In the case of high energy diffraction, $\boldsymbol{k}_0^2$ is much larger than U_0, the difference between $\boldsymbol{k}_0$ and $\boldsymbol{K}$ can be ignored, and one can write $\boldsymbol{k} = \boldsymbol{K} + \tilde{\gamma}\widehat{\boldsymbol{n}}$ instead of (3.22). Moreover, the error of the same order of magnitude is introduced by neglecting the quadratic term $\tilde{\gamma}^2 C_{\check{h}}$ in (3.23) [18]. The linearization has an additional effect. Half of the eigenvalues $\tilde{\gamma}$ of (3.23) correspond to forward traveling waves, and the other half—to the backward traveling waves. Neglecting $\tilde{\gamma}^2 C_{\check{h}}$ leads to the N_b-dimensional linear problem

$$\sum_{\check{l}} A_{\check{h}\check{l}}\, C_{\check{l}} = 2\tilde{\gamma}(\widehat{\boldsymbol{n}} \cdot \boldsymbol{K}) \left(1 + \frac{\widehat{\boldsymbol{n}} \cdot \check{\boldsymbol{h}}}{\widehat{\boldsymbol{n}} \cdot \boldsymbol{K}}\right) C_{\check{h}}\ , \tag{3.24}$$

Table 3.1 The character of matrices $\mathcal{A}$ and $\mathcal{S}$ for non-centrosymmetric and centrosymmetric crystals (with suitably chosen origin) for cases with and without absorption.

Matrix $\mathcal{A}$	No absorption	With absorption
Centrosymmetric	Real, symmetric	Complex, symmetric
Non-centrosymmetric	Complex, Hermitian	Complex
Matrix $\mathcal{S}$	No absorption	With absorption
Centrosymmetric	Symmetric, unitary	Symmetric
Non-centrosymmetric	Unitary	General

with the eigenvalues $\tilde{\gamma}$ corresponding only to the forward traveling waves. By the transformation [19]

$$\widetilde{C}_{\check{h}} = \left(1 + \frac{\widehat{\boldsymbol{n}} \cdot \check{\boldsymbol{h}}}{\widehat{\boldsymbol{n}} \cdot \boldsymbol{K}}\right)^{1/2} C_{\check{h}} \,, \tag{3.25}$$

Equation (3.24) is turned into

$$\sum_{\check{h}} \mathcal{A}_{\check{g}\check{h}} \, \widetilde{C}_{\check{h}} = \tilde{\gamma} \, \widetilde{C}_{\check{g}} \,,$$

where

$$\mathcal{A}_{\check{g}\check{h}} = A_{\check{g}\check{h}} \left(4 \, (\widehat{\boldsymbol{n}} \cdot (\boldsymbol{K} + \check{\boldsymbol{g}})) \, (\widehat{\boldsymbol{n}} \cdot (\boldsymbol{K} + \check{\boldsymbol{h}}))\right)^{-1/2} .$$

If the potential V is real ($U_{\check{g}} = U^*_{-\check{g}}$), the matrix $\mathcal{A}$ is Hermitian $\mathcal{A} = \mathcal{A}^{*T}$, and its eigenvalues $\tilde{\gamma}_j$ ($j = 1, \ldots, N_b$) are real; see Table 3.1.

The eigenvector corresponding to $\tilde{\gamma}_j$ will be denoted by $\widetilde{C}^{\,j}_{\check{h}}$. With proper normalization and proper arrangement in degenerate cases, the eigenvectors satisfy the relationships

$$\sum_{\check{h}} \widetilde{C}^{\,i*}_{\check{h}} \, \widetilde{C}^{\,j}_{\check{h}} = \delta_{ij} \quad \text{and} \quad \sum_{i} \widetilde{C}^{\,i*}_{\check{g}} \, \widetilde{C}^{\,i}_{\check{h}} = \delta_{\check{g}\check{h}} \,, \tag{3.26}$$

i.e., the matrix $[\ ]_{\check{g}i} \equiv \widetilde{C}^{\,i}_{\check{g}}$ is unitary. The spectral representation of $\mathcal{A}$ has the form

$$\mathcal{A}_{\check{h}\check{g}} = \sum_{j} \widetilde{C}^{\,j}_{\check{h}} \, \tilde{\gamma}_j \, \widetilde{C}^{\,j*}_{\check{g}} \,.$$

The $\boldsymbol{k}$ vectors and $C_{\check{h}}$ coefficients corresponding to the j-th eigenvalue $\tilde{\gamma}_j$ are also provided with an index, i.e., $\boldsymbol{k}_j = \boldsymbol{K} + \tilde{\gamma}_j \widehat{\boldsymbol{n}}$. The corresponding wave functions are denoted by $\psi_{\boldsymbol{k}_j}$. As was mentioned above (see (3.4)), the total wave function inside the material is a linear combination of the Bloch waves $\psi_{\boldsymbol{k}_j}$, i.e., $\Psi(\boldsymbol{x}) = \sum_j \alpha_j \psi_{\boldsymbol{k}_j}(\boldsymbol{x})$ or

$$\Psi(\boldsymbol{x}) = \sum_j \alpha_j \sum_h C_{\check{\boldsymbol{h}}}^{\,j} \exp(2\pi i\, (\boldsymbol{k}_j + \check{\boldsymbol{h}}) \cdot \boldsymbol{x})\ . \tag{3.27}$$

Numerical values of the coefficients α_j, are determined by the conditions (3.21). Multiplication of both sides of $\sum_j \alpha_j C_{\check{\boldsymbol{h}}}^{\,j} = \delta_{\check{\boldsymbol{h}}\boldsymbol{0}}$ by $(1 + \widehat{\boldsymbol{n}} \cdot \check{\boldsymbol{h}}/(\widehat{\boldsymbol{n}} \cdot \boldsymbol{K}))C_{\check{\boldsymbol{h}}}^{\,i*}$ and summation over $\check{\boldsymbol{h}}$ give $\sum_j \alpha_j \sum_{\check{\boldsymbol{h}}} \widetilde{C}_{\check{\boldsymbol{h}}}^{\,j} \widetilde{C}_{\check{\boldsymbol{h}}}^{\,i*} = \widetilde{C}_{\boldsymbol{0}}^{\,i*}$. Based on the first of (3.26), the excitation amplitude α_i on the i-th branch of the dispersion surface is

$$\alpha_i = \widetilde{C}_{\boldsymbol{0}}^{\,i*} = C_{\boldsymbol{0}}^{\,i*}\ . \tag{3.28}$$

Thus, the solution (3.27) can be written as a sum of waves enumerated by $\check{\boldsymbol{h}}$

$$\Psi(\boldsymbol{x}) = \sum_{\check{\boldsymbol{h}}} \Psi_{\check{\boldsymbol{h}}}(\boldsymbol{x}) \exp(2\pi i\, (\boldsymbol{K} + \check{\boldsymbol{h}}) \cdot \boldsymbol{x})\ , \tag{3.29}$$

where

$$\Psi_{\check{\boldsymbol{h}}}(\boldsymbol{x}) = \sum_j C_{\boldsymbol{0}}^{\,j*} C_{\check{\boldsymbol{h}}}^{\,j} \exp(2\pi i \tilde{\gamma}_j \widehat{\boldsymbol{n}} \cdot \boldsymbol{x})\ .$$

On the plane $\boldsymbol{x} = t\widehat{\boldsymbol{n}} + \boldsymbol{x}_\perp$, where $\boldsymbol{x}_\perp$ is an arbitrary vector perpendicular to $\widehat{\boldsymbol{n}}$ and $t > 0$, one has $\widehat{\boldsymbol{n}} \cdot \boldsymbol{x} = t$, and $\Psi_{\check{\boldsymbol{h}}}(\boldsymbol{x}) = \Psi_{\check{\boldsymbol{h}}}(t) = \sum_j C_{\boldsymbol{0}}^{\,j*} C_{\check{\boldsymbol{h}}}^{\,j} \exp(2\pi i \tilde{\gamma}_j t)$. Using (3.25), one obtains

$$\Psi_{\check{\boldsymbol{h}}}(t) = \left(1 + \frac{\widehat{\boldsymbol{n}} \cdot \check{\boldsymbol{h}}}{\widehat{\boldsymbol{n}} \cdot \boldsymbol{K}}\right)^{-1/2} \sum_j \widetilde{C}_{\check{\boldsymbol{h}}}^{\,j} \exp(2\pi i \tilde{\gamma}_j t)\, \widetilde{C}_{\boldsymbol{0}}^{\,j*} = \left(1 + \frac{\widehat{\boldsymbol{n}} \cdot \check{\boldsymbol{h}}}{\widehat{\boldsymbol{n}} \cdot \boldsymbol{K}}\right)^{-1/2} [\exp(2\pi i\, \mathcal{A}t)]_{\check{\boldsymbol{h}}\boldsymbol{0}}\ , \tag{3.30}$$

where the exponential of a matrix[9] is $\exp(X) = \sum_{n=0}^{\infty} X^n/n!$. The matrix

$$\mathcal{S} = \mathcal{S}(t) = \exp(2\pi i\, \mathcal{A}\, t)\ , \tag{3.31}$$

is known as the *scattering matrix*.

With the incident electron beam nearly perpendicular to the crystal surface and the scattering of high energy electrons involving beams corresponding to reciprocal lattice vectors nearly perpendicular to the incident beam direction, the ratio $(\widehat{\boldsymbol{n}} \cdot \check{\boldsymbol{h}})/(\widehat{\boldsymbol{n}} \cdot \boldsymbol{K})$ is small, and (3.30) is further simplified by neglecting it. One obtains

$$\Psi_{\check{\boldsymbol{h}}}(t) = \mathcal{S}_{\check{\boldsymbol{h}}\boldsymbol{0}}(t)\ , \tag{3.32}$$

i.e., the wave function $\Psi_{\check{\boldsymbol{h}}}$ at t is equal to the $\check{\boldsymbol{h}}\boldsymbol{0}$ entry of the scattering matrix.

□ The approximation $(\widehat{\boldsymbol{n}} \cdot \check{\boldsymbol{h}})/(\widehat{\boldsymbol{n}} \cdot \boldsymbol{K}) \approx 0$ is related to the frequently considered 'symmetric Laue case' with $\boldsymbol{K} \approx \boldsymbol{k}_0$ parallel to $\widehat{\boldsymbol{n}}$ and all $\check{\boldsymbol{h}}$ vectors perpendicular to $\widehat{\boldsymbol{n}}$. The separate assumptions

[9] For numerical methods of calculating the matrix exponential see, e.g., [20].

$\widehat{\boldsymbol{n}} \cdot \boldsymbol{K} = |\boldsymbol{K}|$ and $\widehat{\boldsymbol{n}} \cdot \check{\boldsymbol{h}} = 0$ are usually taken at the outset. If they are applied to (3.24), it takes the form $\sum_{\check{l}} A_{\check{h}\check{l}} C_{\check{l}} = 2\tilde{\gamma}|\boldsymbol{K}|C_{\check{h}}$, and the wave function $\Psi_{\check{h}}$ at t is again given by (3.31–3.32) but with simpler $\mathcal{A} = A/(2|\boldsymbol{K}|)$. With this approach, one ignores two aspects: the inclined illumination (when the incident beam is not perpendicular to the specimen surface) and out-of-zero-Laue-zone reflections (i.e., reflections from higher order Laue zones or diffraction at arbitrary crystal orientations away from any low-index zone axis).

With (3.32), the scattering matrix has a convenient interpretation [21–23]: It follows directly from the definition (3.31) that the matrix product of the $\mathcal{S}$ matrices for t and $t' > 0$ equals to the scattering matrix for $t + t'$:

$$\mathcal{S}(t)\mathcal{S}(t') = \mathcal{S}(t + t') \,.$$

Hence, one has

$$\sum_{\check{h}} \mathcal{S}_{\check{g}\check{h}}(t)\Psi_{\check{h}}(t') = \sum_{\check{h}} \mathcal{S}_{\check{g}\check{h}}(t)\mathcal{S}_{\check{h}\mathbf{0}}(t') = \mathcal{S}_{\check{g}\mathbf{0}}(t + t') = \Psi_{\check{g}}(t + t') \,, \tag{3.33}$$

which means that the amplitudes of the waves exciting the layer at $t + t'$ are linear combinations of the amplitudes of the waves entering the layer at t', and the element $\mathcal{S}_{\check{g}\check{h}}$ can be seen as the coefficient of transfer of the wave $\Psi_{\check{h}}$ to $\Psi_{\check{g}}$.[10] In particular, the amplitudes of waves after traversing a layer of thickness t are determined by the incident beam $\Psi_{\check{h}}(0) = \delta_{\check{h}\mathbf{0}}$, i.e., $\Psi_{\check{g}}(t) = \sum_{\check{h}} \mathcal{S}_{\check{g}\check{h}}(t)\Psi_{\check{h}}(0) = \sum_{\check{h}} \mathcal{S}_{\check{g}\check{h}}(t)\delta_{\check{h}\mathbf{0}} = \mathcal{S}_{\check{g}\mathbf{0}}(t)$ in agreement with (3.32).

It is clear from (3.33) that the amplitude $\Psi_{\check{h}}$ can be non-zero even if the potential $U_{\check{h}}$ and, accordingly, the structure factor $F(\check{\boldsymbol{h}})$ are equal to zero. With the dynamical theory, a kinematically forbidden reflection may appear in a diffraction pattern due to transfers between allowed reflections. Such reflections are frequently observed in experimental electron diffraction patterns.

The crystal investigated by TEM is usually assumed to be a plate-like slab bounded by two parallel crystal–vacuum interfaces.[11] The exit surface is treated in a similar way as the entrance surface with neglect of refraction and reflection phenomena. The waves exiting the crystal form an image or a diffraction pattern. The mechanism of this formation is a basic aspect of functioning of a transmission electron microscope. As was described above, when an incident (direct) beam of electrons passes through the crystalline specimen, besides the direct beam also diffracted beams emerge. The beams pass through objective lenses. They form a diffraction pattern on the back focal plane; see Sect. 2.8.1 and Fig. 3.6. The key point is that the intensities on the back focal plane are determined by *directions* of wave vectors at the exit surface

[10] There is an analogy between $\mathcal{S}$ and the characteristic matrices of stratified media in classical optics, e.g., [24].

[11] In reality, a TEM specimen is rarely such a slab. Frequently, it has the shape of a wedge. In general, the specimen is tilted, and one should take into account inclined illumination. Convergent beam electron diffraction, by definition involves incident beams with various inclinations to vacuum–crystal interface. Nevertheless, in Bloch-wave based simulations, it is common to ignore these complications.

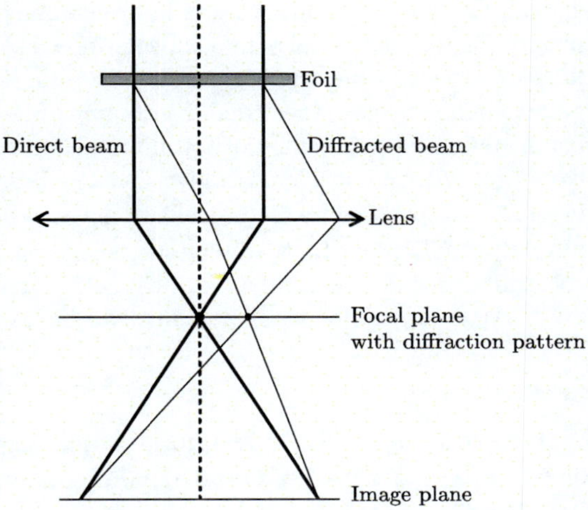

Fig. 3.6 Formation of a diffraction pattern. Waves from different points of the foil but propagating in the same direction are added at one point of the pattern.

of the crystal, and they are *not* affected by the locations of the exit points on this surface. The intensities on the image plane are distributed in the same way as at the exit surface. Either magnified image or enlarged diffraction pattern is produced on a fluorescent screen (or detector of a camera) by focusing subsequent lenses on the image plane or the back-focal plane, respectively.

It is worth adding that this arrangement allows for image filtering, i.e. processing the image by altering its Fourier transform. Just after the back-focal plane is an objective aperture used for selecting one of the beams. If the direct beam is selected, the so-called bright-field image is produced on the image plane. If one of the diffracted beams is used, one obtains a dark-field image; see, e.g., [25].

3.4 Intensities on TEM Diffraction Patterns

Formally, the intensity registered by a detector is proportional to the number of electrons crossing the detector in a unit time. A vector of magnitude representing the number of particles passing in a unit time through a surface of unit area perpendicular to the vector direction is known as the flux of particles. The quantum mechanical expression for the flux $\boldsymbol{j}$ of the wave function Ψ is $\boldsymbol{j} = -(i\hbar/(2m))\ (\Psi^*\nabla\Psi - \Psi\nabla\Psi^*) = \Psi^*\Psi\ \times$ a factor which can be interpreted as particle velocity, i.e., the intensity I is proportional to the product $\Psi^*\Psi$ representing the probability density of finding the particle in a unit volume.

The amplitudes of waves (3.29) at the exit surface depend on location $\boldsymbol{x}_\perp$ on the surface and on the beam directions. Two extreme cases can be considered. The first, with the beam directions ignored, the intensity on the exit surface, and thus on the image plane, is $\Psi^*\Psi$ as a function of $\boldsymbol{x}_\perp$. This will be a microscopic image of the crystal, On the other hand, as was indicated above, the intensities on a diffraction pattern (on the back-focal plane) are determined by the directions of wave vectors at the exit surface of the crystal, and they are *not* affected by the locations of the exit points on this surface. The actual distribution of beam directions depends on various parameters, e.g., the method of foil illumination or the selected field of view. Only two cases will be considered below. These will be spot patterns, with beam directions $\boldsymbol{K} + \check{\boldsymbol{h}}$ specified by reciprocal lattice vectors $\check{\boldsymbol{h}}$, and convergent beam electron diffraction (CBED), with the directions depending on both $\check{\boldsymbol{h}}$ and variable $\boldsymbol{K}$. The last subsection is about the formation of Kikuchi patterns; this mechanism goes beyond the above described theory since it involves inelastic scattering.

Spot patterns

With given $\boldsymbol{K}$ and a list of reciprocal lattice vectors $\check{\boldsymbol{h}}$, one has a discrete set of directions $\boldsymbol{K} + \check{\boldsymbol{h}}$ and a discrete set of corresponding diffraction spots; see, Fig. 2.10. Based on (3.32), the intensity of the beam propagating in the direction $\boldsymbol{K} + \check{\boldsymbol{h}}$ on the exit plane $\boldsymbol{x} = t\widehat{\boldsymbol{n}} + \boldsymbol{x}_{\perp}$ equals

$$I(\check{\boldsymbol{h}}) = \mathcal{S}^{*}_{\check{\boldsymbol{h}}\boldsymbol{0}}(t)\mathcal{S}_{\check{\boldsymbol{h}}\boldsymbol{0}}(t) .$$

These intensities correspond to maxima of particular reflections in spot diffraction patterns. In practice, more representative are integrated intensities, i.e., the quantities obtained by integrating local intensities over a region around the exact Bragg position.

If the total thickness t of the specimen is very small, higher order terms in the expansion of $\exp(2\pi i\,\mathcal{A}t)$ can be neglected. With only the first order term taken into account, one has $\mathcal{S} \approx I + 2\pi i\,\mathcal{A}t$, and

$$I(\check{\boldsymbol{h}}) = \mathcal{S}^{*}_{\check{\boldsymbol{h}}\boldsymbol{0}}\mathcal{S}_{\check{\boldsymbol{h}}\boldsymbol{0}} \approx (\delta_{\check{\boldsymbol{h}}\boldsymbol{0}} - 2\pi i\,\mathcal{A}^{*}_{\check{\boldsymbol{h}}\boldsymbol{0}}t)(\delta_{\check{\boldsymbol{h}}\boldsymbol{0}} + 2\pi i\,\mathcal{A}_{\check{\boldsymbol{h}}\boldsymbol{0}}t) .$$

Since $\mathcal{A}_{\boldsymbol{00}} = 0$, the intensity of the directly transmitted beam is $I(\boldsymbol{0}) = \mathcal{S}^{*}_{\boldsymbol{00}}\mathcal{S}_{\boldsymbol{00}} \approx 1$. The intensity of the diffracted beams ($\check{\boldsymbol{h}} \neq \boldsymbol{0}$ and $\delta_{\check{\boldsymbol{h}}\boldsymbol{0}} = 0$) equals

$$I(\check{\boldsymbol{h}}) \approx 4\pi^2 t^2 \mathcal{A}^{*}_{\check{\boldsymbol{h}}\boldsymbol{0}}\mathcal{A}_{\check{\boldsymbol{h}}\boldsymbol{0}} = \frac{\pi^2 t^2}{(\widehat{\boldsymbol{n}} \cdot \boldsymbol{K})^2}\, U^{*}_{\check{\boldsymbol{h}}} U_{\check{\boldsymbol{h}}} .$$

By taking into account the definition of $U_{\check{\boldsymbol{h}}}$, one gets

$$I(\check{\boldsymbol{h}}) \approx \left(\frac{\gamma\,|F(\check{\boldsymbol{h}})|\,t}{V_{\mathrm{uc}}\,\widehat{\boldsymbol{n}} \cdot \boldsymbol{K}} \right)^2 , \tag{3.34}$$

i.e., the kinematical approximation is obtained; cf. (2.42).

In the second order approximation, one has

$$\mathcal{S}_{\check{\boldsymbol{h}}\boldsymbol{0}} \approx \delta_{\check{\boldsymbol{h}}\boldsymbol{0}} + 2\pi i\,\mathcal{A}_{\check{\boldsymbol{h}}\boldsymbol{0}}t - 4\pi^2 t^2 (\mathcal{A}\mathcal{A})_{\check{\boldsymbol{h}}\boldsymbol{0}} .$$

For the diffracted beams ($\check{\boldsymbol{h}} \neq \boldsymbol{0}$ and $\delta_{\check{h}\boldsymbol{0}} = 0$), the calculation leads to

$$I(\check{\boldsymbol{h}}) = \mathcal{S}^*_{\check{h}\boldsymbol{0}} \mathcal{S}_{\check{h}\boldsymbol{0}} \approx 4\pi^2 t^2 \, |\mathcal{A}_{\check{h}\boldsymbol{0}}|^2 + 16\pi^3 \, t^3 \, \eta(\check{\boldsymbol{h}}) \, ,$$

where $\eta(\check{\boldsymbol{h}}) = i(\mathcal{A}^*_{\check{h}\boldsymbol{0}}(\mathcal{A}\mathcal{A})_{\check{h}\boldsymbol{0}} - \mathcal{A}_{\check{h}\boldsymbol{0}}(\mathcal{A}^*\mathcal{A}^*)_{\check{h}\boldsymbol{0}})/2$. For real potentials V ($U_{\check{g}} = U^*_{-\check{g}}$, Hermitian $\mathcal{A}$), the function η is odd: $\eta(-\check{\boldsymbol{h}}) = -\eta(\check{\boldsymbol{h}})$. Thus, $\eta(\check{\boldsymbol{h}})$ equals zero for centro-symmetric crystals but it is generally non-zero for non-centrosymmetric crystals. This indicates violation of Friedel's law (without involvement of absorption effects).

Two-beam centro-symmetric case

As an example, it is worth considering in detail the two-beam case under the conditions that the potential V is real ($U_{-\check{h}} = U^*_{\check{h}}$) and $\widehat{\boldsymbol{n}} \cdot \check{\boldsymbol{h}} = 0$. Based on (3.17) and definitions of A and $\mathcal{A}$, these matrices have the forms

$$A = \begin{bmatrix} 0 & U^*_{\check{h}} \\ U_{\check{h}} & \boldsymbol{K}^2 - (\boldsymbol{K} + \check{\boldsymbol{h}})^2 \end{bmatrix} \quad \text{and} \quad \mathcal{A} = \frac{1}{2} \begin{bmatrix} 0 & 1/\xi^*_{\check{h}} \\ 1/\xi_{\check{h}} & 2s_{\check{h}} \end{bmatrix} ,$$

where $1/\xi_{\check{h}} = U_{\check{h}}/(\widehat{\boldsymbol{n}} \cdot \boldsymbol{K})$ and $s_{\check{h}} = (\boldsymbol{K}^2 - (\boldsymbol{K} + \check{\boldsymbol{h}})^2)/(2\widehat{\boldsymbol{n}} \cdot \boldsymbol{K})$. The scattering matrix $\mathcal{S} = \exp(2\pi i \mathcal{A} t)$ can be expressed as

$$\mathcal{S} = \exp(i\pi s_{\check{h}} t) \left(\cos\left(\pi s_{\check{h}}^{\text{eff}} t\right) \begin{bmatrix} 1 & 0 \\ 0 & 1 \end{bmatrix} + i \sin\left(\pi s_{\check{h}}^{\text{eff}} t\right) \frac{1}{s_{\check{h}}^{\text{eff}}} \begin{bmatrix} -s_{\check{h}} & 1/\xi^*_{\check{h}} \\ 1/\xi_{\check{h}} & s_{\check{h}} \end{bmatrix} \right) ,$$

where $s_{\check{h}}^{\text{eff}} = \sqrt{1/|\xi_{\check{h}}|^2 + s^2_{\check{h}}}$. Hence, the intensity of the diffracted beam is given by

$$I(\check{\boldsymbol{h}}) = \mathcal{S}^*_{\check{h}\boldsymbol{0}} \mathcal{S}_{\check{h}\boldsymbol{0}} = \left(\frac{\pi}{|\xi_{\check{h}}|} \right)^2 \frac{\sin^2\left(\pi s_{\check{h}}^{\text{eff}} t\right)}{\left(\pi s_{\check{h}}^{\text{eff}}\right)^2} = \left(\frac{\gamma \, |F(\check{\boldsymbol{h}})| \, t}{V_{\text{uc}} \widehat{\boldsymbol{n}} \cdot \boldsymbol{K}} \right)^2 \text{sinc}^2\left(s_{\check{h}}^{\text{eff}} t\right) .$$

This is to be compared with (3.34) obtained from the truncated series $\mathcal{S} \approx I + 2\pi i \, \mathcal{A} t$ and again, with the kinematical intensity given in Sect. 2.6 (2.42). Since $\mathcal{S}$ is unitary, the intensity of the directly transmitted beam equals $I(\boldsymbol{0}) = \mathcal{S}^*_{\boldsymbol{00}} \mathcal{S}_{\boldsymbol{00}} = 1 - \mathcal{S}^*_{\check{h}\boldsymbol{0}} \mathcal{S}_{\check{h}\boldsymbol{0}} = 1 - I(\check{\boldsymbol{h}})$, i.e., the energy conservation condition is satisfied.

The parameters $s_{\check{h}}$ and $\xi_{\check{h}}$ known as the excitation error (cf. 2.3.4) and the extinction distance, respectively, have concrete interpretations. With $\boldsymbol{s}_{\check{h}} = s_{\check{h}} \widehat{\boldsymbol{n}}$, by definition of $s_{\check{h}}$, one has $\boldsymbol{K}^2 - (\boldsymbol{K} + \check{\boldsymbol{h}})^2 = 2\boldsymbol{K} \cdot \boldsymbol{s}_{\check{h}}$. For small $s_{\check{h}}$ ($\boldsymbol{s}^2_{\check{h}} \approx 0$) and $\check{\boldsymbol{h}} \cdot \widehat{\boldsymbol{n}} \approx 0$, the above relationship can be written as

$$\boldsymbol{K}^2 = (\boldsymbol{K} + \check{\boldsymbol{h}} + \boldsymbol{s}_{\check{h}})^2 .$$

Thus, $s_{\check{h}}\widehat{n}$ is the deviation of the vector $\boldsymbol{K} + \check{\boldsymbol{h}}$ from the Ewald's sphere of radius $|\boldsymbol{K}|$. If the crystal orientation deviates from the exact Bragg condition by a small rotation about an axis perpendicular to the reciprocal lattice vector $\check{\boldsymbol{h}}$, then the excitation error is linked to the magnitude of $\check{\boldsymbol{h}}$ and the rotation angle $\Delta\theta$ (in radians) by $s_{\check{h}} \approx |\check{\boldsymbol{h}}|\Delta\theta$.

⊡ The excitation error can be estimated using Kikuchi patterns. (See below.) Geometrically, when the excitation error (as a variable) equals zero, i.e., the crystal is in exact Bragg condition for the $\check{\boldsymbol{h}}$ reflection, the corresponding diffraction spot lies on the excess Kikuchi line $\check{\boldsymbol{h}}$, and the deficiency line $-\check{\boldsymbol{h}}$ goes through the spot of the direct beam. A (small) tilt of the crystal by $\Delta\theta$ about the axis parallel to the considered Kikuchi lines also rotates the attached Kikuchi cones, and displaces the Kikuchi lines, but it has little impact on the position of the diffraction spot; it only decreases the spot intensity. The rate of the decrease with growing $\Delta\theta$ is a measure of the excitation error $s_{\check{h}}$.

The interpretation of the extinction distance $\xi_{\check{h}}$ is simple. When the crystal is centrosymmetric, it is a real number. With $s_{\check{h}} = 0$ (i.e., in the exact Bragg condition), the effective excitation error is $s_{\check{h}}^{\text{eff}} = 1/\xi_{\check{h}}$, and the intensity $I(\check{\boldsymbol{h}}) \propto \text{sinc}^2(t/\xi_{\check{h}})$ oscillates with foil thickness t. Complementary intensity oscillations occur for the directly transmitted beam. The extinction distance $\xi_{\check{h}}$ is the distance between zeros (i.e., half-period) of these oscillations. On the other hand, based on its definition, the extinction distance is linked to $U_{\check{h}}$ and, in consequence, to the distance between the inner and outer branches of the dispersion surface. With the direct beam perpendicular to the specimen surface and some small terms ignored, $1/\xi_{\check{h}}$ is equal to the distance between the two branches.

Similar intensity oscillations will arise for constant foil thickness and varying excitation error (i.e., varying orientation of the beam with respect to the foil). The two-beam dynamical diffraction provides a link between fringes observed in CBED patterns and the thickness, and it is a foundation of methods for determination of extinction distance and foil thickness.

⊡ By rocking the beam about an axis perpendicular to $\check{\boldsymbol{h}}$, one changes the effective excitation error $s_{\check{h}}^{\text{eff}}$, and based on $I(\check{\boldsymbol{h}}) \propto \text{sinc}^2(s_{\check{h}}^{\text{eff}}\, t)$, intensity oscillations appear. In a CBED pattern, the zeros at the $I(\check{\boldsymbol{h}})$ disk correspond to maxima at $I(\boldsymbol{0})$, and vice versa. As was mentioned in the caption to Fig. 1.21, the function sinc^2 has minima at non-zero integers. Let the minima be numbered by $n\ (\neq 0)$, i.e., they occur for $\pi s_{\check{h}}^{\text{eff}}\, t = n$, or when $t^2\left(1/|\xi_{\check{h}}|^2 + s_{\check{h}}^2\right) = n^2$. This is a relationship between the squared excitation error $s_{\check{h}}^2$ at subsequent minima and the squared indices n^2 of these minima [26–28]. This equation makes sense for $n > t/\xi_{\check{h}}$. It is better known in the form

$$\frac{s_{\check{h}}^2}{n^2} + \frac{1}{n^2}\frac{1}{\xi_{\check{h}}^2} = \frac{1}{t^2}$$

given in [29]. The actual method is based on linear regression of $s_{\check{h}}^2/n^2$ versus $1/n^2$ resulting in the line slope representing $1/\xi_{\check{h}}^2$ and an intersection with the ordinate representing the squared inverse of the thickness $1/t^2$. It must be stressed that, in reality, other beams and absorption effects influence the patterns. For a more accurate determination of $\xi_{\check{h}}$ and t, the experimental intensity profiles need to be matched to patterns simulated with these factors taken into account [30].

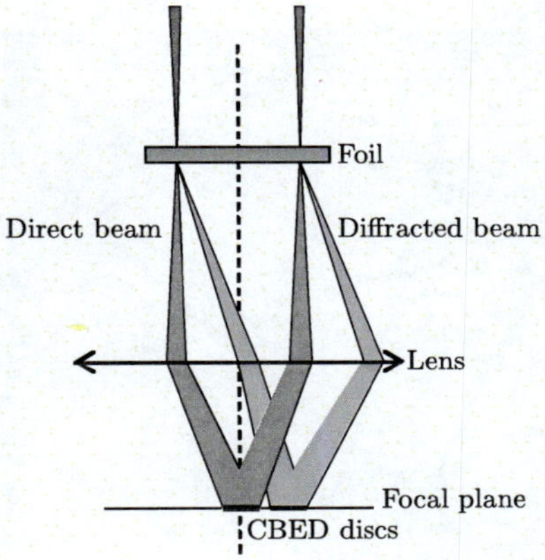

Fig. 3.7 Formation of a CBED pattern.

It is worth noting here that the two beam case and the fringes observed in CBED patterns were the basis for linking the experimental electron microscopy with the dynamic theory of electron diffraction. This important step was made by Mac Gillavry[12] [26].

CBED patterns

The formalism of Sect. 3.3 can be used for the simulation of convergent beam electron diffraction patterns. There are a number of variants of this TEM technique [32]. Their crucial common feature is the convergence of the incident beam. By use of probe-forming lenses, a specimen is illuminated by a convergent 'cone' of electrons. In other words, the directions of $\boldsymbol{k}_0 \approx \boldsymbol{K}$ vectors cover a certain (spherical) disc. In effect, each diffraction spot becomes a disk; Fig. 3.7. Its diameter is directly linked to the convergence angle—the solid angle covered by $\boldsymbol{k}_0$ vectors. With sufficiently large diameter, the disks contain excess and deficiency K-lines (Sect. 2.2). The whole disks are simulated by 'rocking' $\boldsymbol{k}_0$ within the convergence angle, and calculating intensities for particular directions of the vector, with the distribution of intensity in the disk $\check{\boldsymbol{h}}$ being simply $S^*_{\boldsymbol{0}\check{\boldsymbol{h}}} S_{\boldsymbol{0}\check{\boldsymbol{h}}}$ expressed as a function of $\boldsymbol{k}_0 \approx \boldsymbol{K}$.

The central disk corresponding to the direct beam $\check{\boldsymbol{h}} = \boldsymbol{0}$ is of particular interest. The geometry of K-lines in the central disks can be used for the refinement of lattice parameters (and, subsequently, for strain determination; see Chap. 14). An example simulated central disk is shown in Fig. 3.8.

Kikuchi Patterns

In electron microscopy, crystal orientation determination is frequently based on Kikuchi patterns (Fig. 3.9). Closely related backscatter Kikuchi (or EBSD) patterns are a basis of a prevailing orientation mapping technique.

---

[12] Carolina Henriette Mac Gillavry (1904–1993).

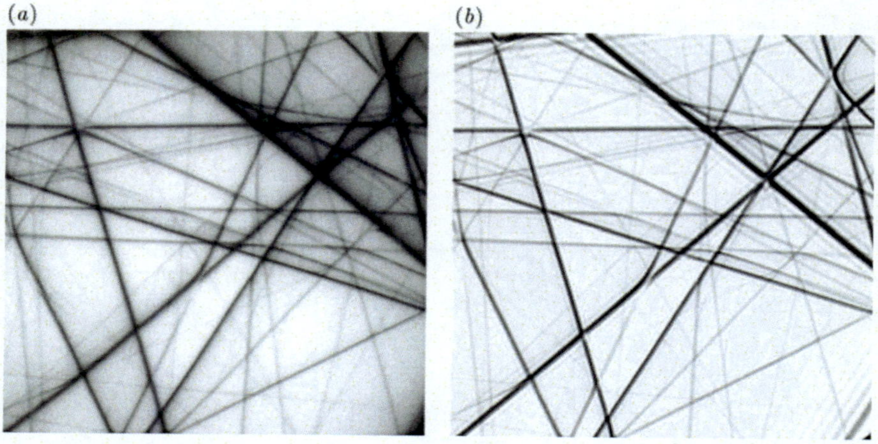

Fig. 3.8 (*a*) Experimental central disc of an off-axis (near [$\overline{16}$ 67 73]) CBED pattern of Si (courtesy of G. Brunetti [31]) and (*b*) corresponding dynamically simulated pattern. The pattern was simulated using TEM*Strain* of Sect. 14.2.

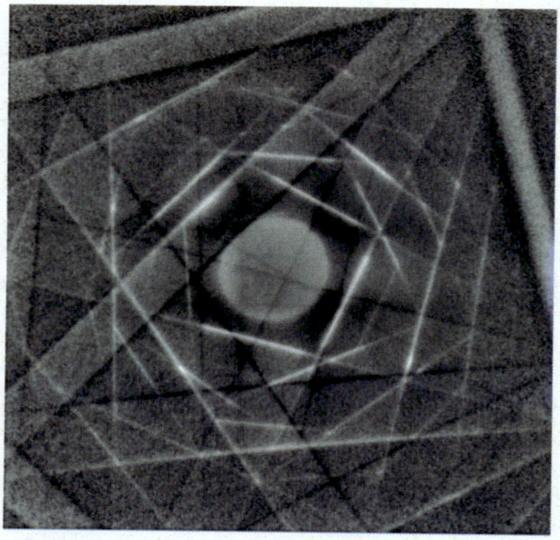

Fig. 3.9 Example experimental Kikuchi pattern of Al. Courtesy of H. Paul.

There exists a simplistic quasi-kinematical explanation of the lines visible in Kikuchi patterns: The distribution of directions of scattered electrons is highly anisotropic with most of them moving in directions close to that of the transmitted beam. Thus, the distribution of the scattered electrons on the viewing screen is centro-symmetric with a gradual decrease of intensity with a distance from the center. The diffraction deviates a given fraction of electrons. Because of the uneven distribution, this fraction of the electrons deviated from the center toward the pattern

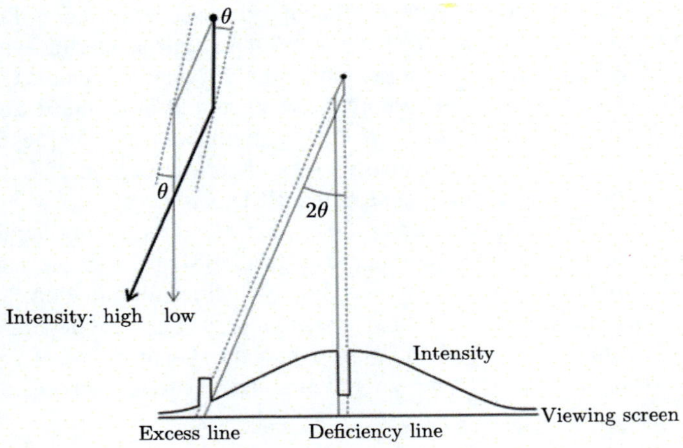

Fig. 3.10 Schematic illustration of the simplistic explanation of the formation of Kikuchi lines.

peripheries is large compared to the number of electrons directed from the peripheries toward the center; see Fig. 3.10. In effect, the pair of Kikuchi lines consists of a dark (deficiency) line near the pattern center and a bright (excess) line away from the pattern center. Clearly, this explanation does not justify the complete intensity distribution; cf. Fig. 3.9. In particular, it does not give a reason for Kikuchi bands—pairs of lines located symmetrically on both sides of the center. The general rule of thumb is that Kikuchi lines and bands with the greatest contrast correspond to the same scattering vectors as the most intense peaks in spot diffraction patterns. For accurate simulations a more realistic physical model is needed. It must take into account inelastic scattering.

Kikuchi patterns are formed via elastic scattering of inelastically scattered electrons; see, e.g., [33]. Inelastic electron scattering occurs via a number of different mechanisms. The formation of Kikuchi bands and lines is mainly attributed to phonon excitations, i.e., to thermal diffuse scattering (TDS). (Cf. Sect. 2.7.2.) TDS of electrons is quasi-elastic with energy losses of about 0.1 eV. The impact of TDS strongly increases with increasing crystal temperature. Conventional Bloch-wave simulations can include thermal diffuse intensity simply by using potentials with Debye-Waller factors. In this approach, however, the scattering is assumed to be elastic, and therefore, more sophisticated models are needed to simulate KIkuchi patterns.

⊡ There are a number of models of thermal vibrations. The simplest (Einstein) model is based on the assumption that atoms are independent (non-interacting) quantum harmonic oscillators. More sophisticated phonon models, in particular the Debye model, take into account coupling between neighboring atoms and correlations in their vibrations. It turns out that Kikuchi lines and bands can be simulated based on the Einstein model [34]. The phonon model with correlated atom vibrations is required only for detailed simulations [35].

In the simplest description, one can see Kikuchi patterns as a result of a single elastic scattering of inelastically scattered electrons. With multiple scattering events—both elastic and inelastic—Kikuchi patterns have highly dynamical character, and the corresponding theory is relatively complicated. Quantum mechanical treatment of inelastic scattering in the formation of Kikuchi patterns was given in [36, 37]. There are a number of numerical implementations these theories; see, e.g., [34, 38] (or [35, 39] for simulations of Kikuchi component in CBED patterns).

The case of EBSD is even more complex than the simulation of transmission Kikuchi patterns. In standard experimental setups, EBSD patterns are registered at lower electron energies (≤30 kV), and with highly inclined illumination.[13] The latter leads to surface effects. Since EBSD patterns are recorded in the reflection geometry (Bragg case), some solutions of (3.23) correspond to complex wave vectors and evanescent waves. These issues are ignored in practical simulations; see e.g., [40].

References

1. A. Authier, *Dynamical Theory of X-Ray Diffraction* (Oxford University Press, Oxford, 2004)
2. J.M. Cowley, P. Goodman, B.K. Vainshtein, B.B. Zvyagin, D.L. Dorset, Electron diffraction and electron microscopy in structure determination, in *International Tables for Crystallography, Vol. B, Section 2.5*, ed. by U. Shmueli (Springer, Dordrecht, 2001), pp. 276–345
3. P.G. Self, M.A. O'Keefe, P.R. Buseck, A.E.C. Spargo, Practical computation of amplitudes and phases in electron diffraction. Ultramicroscopy **11**, 35–52 (1983)
4. H.A. Bethe, Theorie der Beugung von Elektronen in Kristallen. Ann. Phys. Leipzig **87**, 55–129 (1928)
5. P.P. Ewald, On the foundation of crystal optics. English translation (by L.M. Hollingsworth) of Ewald's dissertation. Air Force Cambridge Research Laboratories Report AFCRL-70-0580, Cambridge, MA, 1970, 1912
6. P.P. Ewald, Introduction to the dynamical theory of X-ray diffraction. Acta Cryst. A **25**, 103–108 (1969)
7. D.W.J. Cruickshank, H.J. Juretschke, N. Kato (eds.), *P.P. Ewald and his Dynamical Theory of X-ray Diffraction* (IUCr/Oxford University Press, Oxford, 1992)
8. K. Fujiwara, Relativistic dynamical theory of diffraction. J. Phys. Soc. Jpn. **16**, 2226–2238 (1961)
9. L.I. Schiff, *Quantum Mechanics* (McGraw-Hill, New York, 1949)
10. J.M. Cowley, A.F. Moodie, The scattering of electrons by atoms and crystals. I. A new theoretical approach. Acta Cryst. **10**, 609–619 (1957)
11. D. van Dyck, W. Coene, The real space method for dynamical electron diffraction calculations in high resolution electron microscopy: I. Principles of the method. Ultramicroscopy **15**, 29–40 (1984)
12. P.B. Hirsch, A. Howie, R.B. Nicholson, D.W. Pashley, M.J. Whelan, *Electron Microscopy of Thin Crystals* (Krieger Publishing, New York, 1977)
13. E.J. Kirkland, *Advanced Computing in Electron Microscopy* (Springer, New York, 2010)
14. C.J. Humphreys, The scattering of fast electrons by crystals. Rep. Prog. Phys. **42**, 1825–1887 (1979)
15. J.M. Zuo, A.L. Weickenmeier, On the beam selection and convergence in the Bloch-wave method. Ultramicroscopy **57**, 375–383 (1995)

[13] Usually, the angle between the beam and the normal to the specimen surface is close to 70°.

16. C. Birkeland, R. Holmestad, K. Marthinsen, R. Høier, Efficient beam-selection criteria in quantitative convergent beam electron diffraction. Ultramicroscopy **66**, 89–99 (1996)
17. F. Tisseur, K. Meerbergen, The quadratic eigenvalue problem. SIAM Rev. **43**, 235–286 (2001)
18. A.L. Lewis, R.E. Villagrana, A.J.F. Metherell, A description of electron diffraction from higher-order Laue zones. Acta Cryst. A **34**, 138–139 (1978)
19. P.H. Dederichs, Dynamical diffraction theory. Technical report, KFA – JUL–797–FF, Jülich, Germany, 1971
20. C. Moler, C. van Loan, Nineteen dubious ways to compute the exponential of a matrix, twenty-five years later. SIAM Rev. **45**, 3–49 (2003)
21. L. Sturkey, The use of electron diffraction intensities in structure determination. Acta Cryst. **10**, 858–859 (1957)
22. L. Sturkey, The calculation of electron diffraction intensities. Proc. Phys. Soc. **80**, 321–354 (1962)
23. H. Niehrs, Formulation of electron diffraction by a scattering matrix and its practical application. Z. Naturf. **14A**, 504–511 (1959)
24. M. Born, E. Wolf, *Principles of Optics: Electromagnetic Theory of Propagation, Interference, and Diffraction of Light* (Macmillan, New York, 1964)
25. D.B. Williams, C.B. Carter, *Transmission Electron Microscopy. A Textbook for Materials Science* (Springer, New York (USA), 2009)
26. C.H. Mac Gillavry, Zur Prüfung der dynamischen Theorie der Elektronenbeugung am Kristallgitter. Physica **7**, 329–343 (1940)
27. I. Ackermann, Beobachtungen an dynamischen Interferenzerscheinungen im konvergenten Elektronenbündel I. Ann. Phys. (Leipzig) **437**, 19–40 (1948)
28. I. Ackermann, Beobachtungen an dynamischen Interferenzerscheinungen im konvergenten Elektronenbündel II. Ann. Phys. (Leipzig) **437**, 41–54 (1948)
29. P.M. Kelly, A. Jostsons, R.G. Blake, J.G. Napier, The determination of foil thickness by scanning transmission electron microscopy. Phys. Status Solidi (a) **31**, 771–780 (1975)
30. D. Delille, R. Pantel, E. Van Cappellen, Crystal thickness and extinction distance determination using energy filtered CBED pattern intensity measurement and dynamical diffraction theory fitting. Ultramicroscopy **87**, 5–18 (2001)
31. G. Brunetti, E. Bouzy, J.J. Fundenberger, A. Morawiec, A. Tidu, Determination of lattice parameters from multiple CBED patterns. A statistical approach. Ultramicroscopy **110**, 269–277 (2010)
32. J.P. Morniroli, *Large-Angle Convergent-Beam Electron Diffraction. Applications to Crystal Defects*, (SFμ, Paris, 2002)
33. L. Reimer, *Transmission Electron Microscopy: Physics of Image Formation and Microanalysis* (Springer, Berlin, 1997)
34. K. Omoto, K. Tsuda, M. Tanaka, Simulations of Kikuchi patterns due to thermal diffuse scattering on MgO crystals. J. Electron Microsc. **51**, 67–78 (2002)
35. D.A. Muller, B. Edwards, E.J. Kirkland, J. Silcox, Simulation of thermal diffuse scattering including a detailed phonon dispersion curve. Ultramicroscopy **86**, 371–380 (2001)
36. Y. Kainuma, The theory of Kikuchi patterns. Acta Cryst. **8**, 247–257 (1955)
37. H. Yoshioka, Effect of inelastic waves on electron diffraction. J. Phys. Soc. Jpn. **12**, 618–628 (1957)
38. K. Omoto, K. Tsuda, M. Tanaka, Simulations of Kikuchi patterns and comparison with experimental patterns. JEOL News **37E**(1), 14–19 (2002)
39. T. Yamazaki, M. Ohtsuka, Y. Kotaka, K. Watanabe, Bloch wave simulations in the frozen lattice approximation. Ultramicroscopy **135**, 16–23 (2013)
40. A. Winkelmann, C. Trager-Cowan, F. Sweeney, A.P. Day, P. Parbrook, Many-beam dynamical simulation of electron backscatter diffraction patterns. Ultramicroscopy **107**, 414–421 (2007)

Chapter 4
Cartesian Reference Frames in Diffractometry

Crystal or crystallite orientations are specified in various reference frames. Usually, the interest is in orientations of crystal structures with respect to the sample (as described in Sect. 1.8). These orientations are the subject of texture analysis, they are interpreted using pole figures or texture functions, shown on orientation maps, et cetera. In experiments, to determine the crystal orientation with respect to the sample, it is necessary to know the orientations of the crystal and the sample with respect to a measuring instrument.

Thus, as was mentioned in Sect. 1.8, besides the crystal and sample coordinate systems, a third reference frame linked to the instrument is needed. Various conventions are used in experimental practice. Here, all coordinate systems are assumed to be Cartesian and to have the same (right) handedness. With the same origin of coordinate systems at the point of diffraction, the aforementioned orientations can be represented by special orthogonal matrices.

The symbol $O_{c|s}$ will denote the orthogonal matrix representing the orientation of the crystal in the sample. The matrix $O_{c|m}$ will represent the orientation of the crystal in the measuring instrument; this orientation and the matrix $O_{c|m}$ are obtained directly from diffraction patterns. Finally, the orientation of the sample in the instrument is represented by $O_{s|m}$. With the conventions used in texture analysis, these matrices are related by

$$O_{c|m} = O_{c|s} O_{s|m} \ , \tag{4.1}$$

and they need to be linked to orientation data obtained from the instrument.

It is convenient to identify orientation of a given reference frame with respect to another frame with rotations relating the frames. Let $R^{\theta}_{\widehat{n}}$ denote the rotation by the angle θ about the axis indicated by $\widehat{\boldsymbol{n}}$, with entries of the corresponding orthogonal matrix given by

$$(R^{\theta}_{\widehat{n}})_{ij} = \delta_{ij} \cos\theta + n_i n_j (1 - \cos\theta) - \sin\theta \, \varepsilon_{ijk} n_k \ ;$$

see (1.18). In texture analysis, the standard way to parametrize crystal orientations is by using Euler angles φ_1, ϕ, φ_2 such that

A. Morawiec, *Indexing of Crystal Diffraction Patterns*, Springer Series in Materials Science 326, https://doi.org/10.1007/978-3-031-11077-1_4

$$R = \mathcal{R}(\varphi_1, \phi, \varphi_2) = R^{\varphi_2}_{[001]} R^{\phi}_{[100]} R^{\varphi_1}_{[001]} \ ;$$

see, e.g., [1, 2]. The inverse relationships are[1]

$$\varphi_1 = \text{atan2}(R_{31}, -R_{32}) \ , \ \ \phi = \arccos(R_{33}) \ , \ \ \varphi_2 = \text{atan2}(R_{13}, R_{23})$$

if $R_{33} \neq \pm 1$. Otherwise, if $R_{33} = +1$, then only the sum $\varphi_1 + \varphi_2 = \text{atan2}(R_{12}, R_{11})$ is determinable, and if $R_{33} = -1$, one can get only the difference $\varphi_1 - \varphi_2 = \text{atan2}(R_{12}, R_{11})$.

There are a lot of different goniometric setups. Manufacturers of instruments use different names and conventions even if some aspects of the instruments are the same. Rotation axes of instruments may be the same, but senses of positive rotations may vary. Therefore, here the senses of rotations are specified in a way which does not fit any particular instrument; to apply the formulas listed below, one needs to specify the sign in front of used angles. Since there is a large literature on goniometry used in X-ray diffractometry (e.g., [3–5] and references therein), this subject is considered only very briefly. The rest of the chapter concerns the description of orientations in electron microscopes.

4.1 X-ray Diffractometer

There is a variety of goniometric systems used in X-ray diffractometry. Consequently, various coordinate systems are used, depending on the experimental instrument, manufacturer or software. We will focus on the four-circle Eulerian-cradle diffractometer (which is also a kind of model for other techniques) and the Kappa diffractometer.

As the name indicates, the four-circle diffractometer allows for four rotations. The rotations are about three different axes through one point—the center of the diffractometer, where investigated specimens are placed, and where diffraction occurs (Fig. 4.1). The point of diffraction, the radiation source and a counter (point detector) determine the scattering plane. The axis perpendicular to the scattering plane through the center of the diffractometer is known as the principal axis of the instrument. It is a common axis of two rotations: the first moving just the counter and determining the Bragg angle θ, and the second moving the Eulerian-cradle with the specimen and known as rotation by ω angle. With the common axis, compositions of these two rotations are obtained by summing/subtracting the θ and ω angles, and one can easily calculate the orientations of the beams with respect to the cradle. The orientation of the specimen with respect to the instrument involves two other rotations. The first rotation is about the axis of the Eulerian-cradle (i.e., perpendicular the plane of the cradle and to the principal axis) by the angle χ. The second is linked to the rotating sample stage; it is a rotation about the axis perpendicular to the axis of the

[1] The function atan2 is similar to arctan but its values cover the full range of angles. With $x = a \sin \xi$, $y = a \cos \xi$ and $a > 0$, one has $\text{atan2}(x, y) = \xi$.

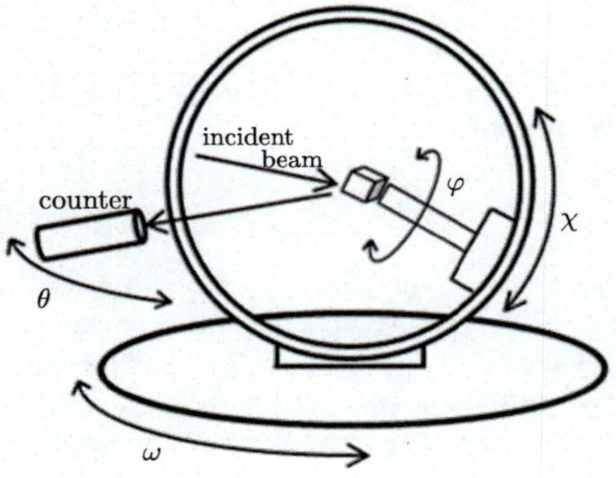

Fig. 4.1 Schematic of Eulerian four-circle diffractometer with a point detector.

Eulerian-cradle by the angle usually denoted by φ . The zeros of θ and ω are defined by the configuration with undeflected beam indicating toward the detector. The zeros of χ and φ are determined by the condition that for $\omega = \chi = \varphi = 0$ the sample coordinate system is aligned with the instrument coordinate system; with $\chi = 0$, the axis of φ coincides with the principal axis.

Let the Cartesian coordinate system of the instrument have the first axis (1) coinciding with the axis of the Eulerian-cradle at $\omega = 0$ (pointing toward the side with the source), the third axis (3) perpendicular to the scattering plane (pointing up). The second axis (2) follows from the handedness of the coordinate system. With this definition, the orientation of the sample (s) in the instrument (m) is given by

$$O_{m|s} = R^{\omega}_{[001]} R^{\chi}_{[100]} R^{\varphi}_{[001]} = \mathcal{R}(\varphi, \chi, \omega) \,. \tag{4.2}$$

Knowing $O_{m|s} = O^{-1}_{s|m}$ and the orientation $O_{c|m}$ of the crystal (c) with respect the instrument (m) (from a diffraction pattern), one can easily calculate the orientation of the crystal in the sample reference frame $O_{c|s} = O_{c|m} O_{m|s}$.

The Eulerian goniometer is conceptually simple, but it is not always convenient in practical applications as the χ-circle is an obstacle hindering access to the specimen. In single crystal diffraction more popular are Kappa goniometers. Their first and third axes are analogous to those of the Eulerian goniometer, and the angles of rotations will be denoted by, respectively, ω_κ and φ_κ. With $\omega_\kappa = 0$, the middle axis with rotation by κ is directed along $\widehat{\boldsymbol{\kappa}}_{axis} = [\sin\alpha_x \;\, 0 \;\, \cos\alpha_x]^T$. The instrumental angle α_x between the κ-axis and the principal axis of the instrument (ω_κ-axis) usually equals 50° (or 55° or 24°). More generally, one can take arbitrary α_x in the range $0 < \alpha_x < 2\pi/3$. With such arrangement, the orientation of the sample is

$$O_{m|s} = R^{\omega_\kappa}_{[001]} R^{\kappa}_{\widehat{\boldsymbol{\kappa}}_{axis}} R^{\varphi_\kappa}_{[001]} \,.$$

By comparing this expression to (4.2), one obtains the relationship between the Eulerian angles φ, χ, ω and the angles φ_κ, κ, ω_κ of Kappa goniometer

$$\varphi_\kappa = \varphi - \Delta\,, \quad \kappa = 2\,\mathrm{atan2}(p,\, q) \;\text{ and }\; \omega_\kappa = \omega - \Delta\,,$$

where $p = \sin(\chi/2)$, $q = \sqrt{s^2 - p^2}$, $s = \sin\alpha_x$ and $\Delta = \mathrm{atan2}(p\,\cos\alpha_x,\, q)$. The inverse relationships are

$$\varphi = \varphi_\kappa + \Delta\,, \quad \chi = 2\,\arcsin(s\,\sin(\kappa/2)) \;\text{ and }\; \omega = \omega_\kappa + \Delta\,,$$

where $\Delta = \mathrm{atan2}(\sin(\kappa/2)\,\cos\alpha_x,\, \cos(\kappa/2))$. (Cf. [4].)

With $\varphi_\kappa, \kappa, \omega_\kappa$ covering the complete angular domains, the Eulerian angle χ satisfies $\chi \leq 2\alpha_x$. If $|\chi| < 2\alpha_x$, and the angles φ and ω cover the complete angular domains, the angle κ satisfies $|\kappa| < 2\,\mathrm{arccot}\sqrt{2\cos\alpha_x + 1}$. Clearly, not all specimen orientations are accessible with Kappa goniometers.

4.2 Crystal Orientation in Transmission Electron Microscope

The natural way to allocate the coordinate system in a transmission electron microscope is to put one of its axes (z or 3) along the optical axis of the microscope (with the sense up); Fig. 4.2. The other axes are linked to the devices for observing or recording diffraction patterns (e.g., fluorescent screen of the microscope, CCD detector) and to the specimen holder. The first axis (1) can be conveniently chosen to be along the tilt axis of the holder (i.e., along the holder rod); in standard configurations, the axis is parallel to the horizontal line in the pattern on the fluorescent screen of the microscope. However, if the pattern is recorded by a camera it may be rotated about the optical axis (due to the use of electromagnetic lenses). This "absolute lens rotation" [6] (or 'magnetic' rotation) must be taken into account in calculations involving the pattern and microscope coordinate systems.

4.2.1 *Tilt Angles and Specimen Orientation*

The orientation of the sample in a transmission microscope $O_{s|m}$ is determined by the angles of the goniometer of the sample holder. There are a number of types of holders; the most widely used are the single-tilt holder, the double-tilt holder and the single-tilt rotation holder. The first one allows for rotations about just one axis – the holder rod, whereas the other two allow for rotations about two axes each; see Fig. 4.3.

For the double-tilt holder, with α being the angle of rotation about the principal axis of the holder (rod), β denoting the angle of rotation about the direction perpendicular the rod and to the foil normal, and γ standing for the "absolute lens rotation angle" [6] (i.e., the 'magnetic' rotation angle), one has

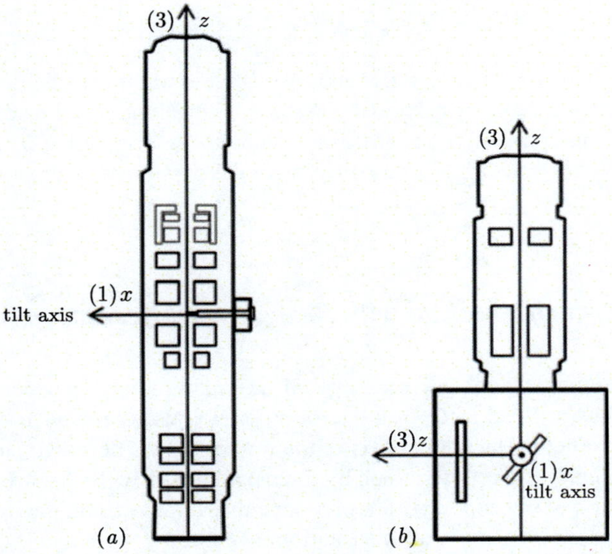

Fig. 4.2 Schematic illustration of coordinate systems linked to transmission (*a*) and scanning (*b*) electron microscopes. Two alternative *z* axes are shown in (*b*).

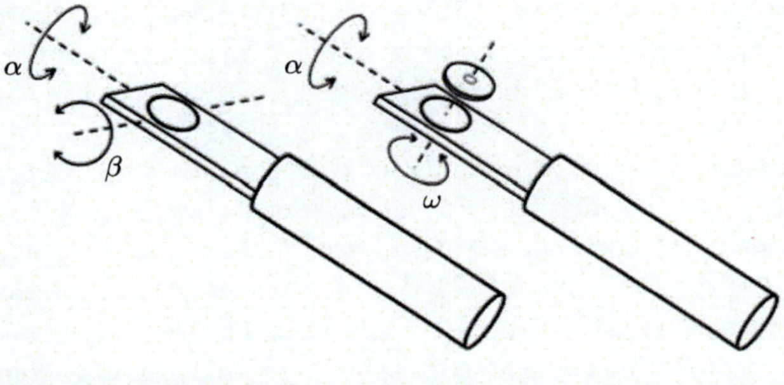

Fig. 4.3 Schematics of double-tilt holder (left) and 'single-tilt rotation' holder (right).

$$O_{s|m} = R^{\beta}_{[010]} R^{\alpha}_{[100]} R^{\gamma}_{[001]} = R^{\pi/2}_{[100]} \mathcal{R}(\gamma, \alpha - \pi/2, \beta) \,. \tag{4.3}$$

The angles α and β are changeable by a user, whereas γ is characteristic for a given camera length.

For the single-tilt rotation holder, β is fixed at zero but, instead, there is a possibility to rotate about the direction normal to the foil. Let the angle of this rotation be ω. The orthogonal matrix representing orientation of the sample in the microscope is[2]

---

[2] The composition of rotations described by (4.4) is similar to that described by (4.2) but the relationships are different. This might be seen as a lack of consequence. Clearly, (4.4) can be rewritten in the form analogous to (4.2); one could write $O_{m|s} = O_{s|m}^{-1} = R^{\gamma'}_{[001]} R^{\alpha'}_{[100]} R^{\omega'}_{[001]} = \mathcal{R}(\omega', \alpha', \gamma')$

$$O_{s|m} = R^{\omega}_{[001]} R^{\alpha}_{[100]} R^{\gamma}_{[001]} = \mathcal{R}(\gamma, \alpha, \omega) \ . \tag{4.4}$$

As in Sect. 4.1, knowing the tilt angles and absolute lens rotation angle (and thus $O_{s|m}$) and the orientation of the crystal with respect the microscope coordinate system $O_{c|m}$ (from a diffraction pattern), one can easily calculate the orientation of the crystal in the sample $O_{c|s} = O_{c|m} O^{-1}_{s|m}$. For an alternative description of orientations and tilts, see [8].

4.2.2 Crystal Orientation with Respect the Microscope Axis

TEM microscopists usually express crystal orientations with respect to the incident beam. Thus, instead of $O_{c|m}$, a microscopist will use the zone axis $[uvw]$ (see Sect. 1.5.1) parallel to the optical axis of the microscope. These two characteristics are not equivalent: $O_{c|m}$ is determined by three parameters while $[uvw]$ only by two. The specification of the zone axis does not give the rotation about the optical axis of the microscope, and the angle of this rotation is ignored.

It is easy to calculate zone axis from the known orientation matrix $O_{c|m}$. With the microscope reference frame specified above, the coordinates v^i of the unit vector along the optical axis of the microscope in the Cartesian system linked to the crystal are obtained from $\boldsymbol{v} = O_{c|m}[0\,0\,1]^T$. To get (scaled) indices $[uvw] \propto [w^1\, w^2\, w^3]$ of the zone axis, one needs to use the inverse of the $\mathcal{T}$ matrix linking the basis of the Cartesian system to the basis of the crystal lattice, i.e., $w^i = v^j (\mathcal{T}^{-1})_j^{\ i} = \mathcal{T}^i_{\ j} v^j$; cf. Sect. 1.1.

Two parameters of the orientation matrix $O_{c|m}$ can be obtained from the known zone axis $[uvw]$ in a similar way. In the Cartesian coordinate system linked to the crystal lattice, the components of the vector along the zone axis are given by $\boldsymbol{u} = \mathcal{T}^T\ [u\, v\, w]^T$ (or $u^i = \mathcal{T}_j^{\ i} w^j$). Let $\boldsymbol{v} = \widehat{\boldsymbol{v}} = \boldsymbol{u}/|\boldsymbol{u}|$. The Euler angles of $O_{c|m} = \mathcal{R}(\varphi_1', \phi', \varphi_2')$ are $\phi' = \arccos(v^3)$ and $\varphi_2' = \mathrm{atan2}(v^1, v^2)$; if both v^1 and v^2 are zero, then $\varphi_2' = 0$. The angle φ_1' corresponding to the rotation about the optical axis is not determinable from $[uvw]$.

4.2.3 Tilting a Crystal to a Given Zone Axis

Another important issue is to display a diffraction pattern corresponding to a particular zone axis. In general, the initial pattern obtained after inserting a specimen into a microscope without any tilts differs from the desired pattern corresponding to a specific zone axis. The point is to determine the tilt angles needed to

with the angles $\omega' = -\omega$, $\alpha' = -\alpha$, $\gamma' = -\gamma$. Expressions (4.3) and (4.4) are preferred because the scheme based on the angles α, β and γ (including the tilting to a given zone axis described in Sect. 4.2.3) is used in practice [7].

display the desired pattern. This can be done by taking the following steps. Based on the initial pattern, the orientation $O_{c|s}$ of the crystal in the specimen is calculated. From the target zone axis, two essential parameters of the target orientation of the crystal in the microscope ($O_{c|m}$) are obtained (Sect. 4.2.2). Thus, the needed orientation of the sample in the microscope equipped with the double-tilt holder is $O_{s|m} = R^{\beta}_{[010]} R^{\alpha}_{[100]} R^{\gamma}_{[001]} = O^{-1}_{c|s} O_{c|m}$. Hence, one has

$$R^{\beta}_{[010]} R^{\alpha}_{[100]} = O^{-1}_{c|s} O_{c|m} R^{-\gamma}_{[001]} \,. \tag{4.5}$$

Similarly, with the single-tilt rotation holder, one has $O_{s|m} = R^{\omega}_{[001]} R^{\alpha}_{[100]} R^{\gamma}_{[001]} = O^{-1}_{c|s} O_{c|m}$, and

$$R^{\omega}_{[001]} R^{\alpha}_{[100]} = O^{-1}_{c|s} O_{c|m} R^{-\gamma}_{[001]} \,. \tag{4.6}$$

The factor $R^{-\gamma}_{[001]}$ and the component $R^{\varphi'_1}_{[001]}$ of $O_{c|m} = \mathcal{R}(\varphi'_1, \phi', \varphi'_2)$ as rotations about the optical axis of the microscope have no significance for getting the needed zone axis and can be ignored. The formulas (4.5) and (4.6) allow for calculating the angle α and the angles β or ω, respectively. For alternative descriptions see, e.g., [9–12].

It is important to note that due to crystal symmetry, there are a number of symmetrically equivalent target zone axes. The crystal orientation in the specimen $O_{c|s}$ in (4.5) can be replaced by $S_i O_{c|s}$, where the proper orthogonal matrix S_i represents the i-th proper rotation of the crystal point group. Pairs of the tilt angles α and β (or the angles α and ω) can be obtained for each S_i. In practice, only some of the tilts will be within the reach of the goniometer.

4.2.4 Determination of 'Magnetic' Rotation Angle

The simplest way to determine the angle of absolute lens rotation is to use a crystal with known orientation $O_{c|s}$: collect the pattern corresponding to zero tilt angles ($\alpha = 0$ and $\beta = 0$ or $\omega = 0$), and use this pattern to get $O_{c|m}$. From (4.1), one has

$$R^{\gamma}_{[001]} = O^{-1}_{c|s} O_{c|m} \,.$$

The fact that $O^{-1}_{c|s} O_{c|m}$ is expected to represent a rotation about [001] allows for partial verification of the correctness of the result.

The angle γ can also be determined without knowing the orientation $O_{c|s}$. This can be done by solving patterns obtained from the same crystal at different tilt angles. With two such patterns (collected at $\beta = 0$ or $\omega = 0$), one has two tilts $R^{\alpha_1}_{[100]}$ and $R^{\alpha_2}_{[100]}$, and two corresponding solutions $O^1_{c|m}$ and $O^2_{c|m}$. With $\beta = 0$ or $\omega = 0$, (4.5–4.6) take the form $R^{\alpha_i}_{[100]} R^{\gamma}_{[001]} = O^{-1}_{c|s} O^i_{c|m}$, where $i = 1, 2$. By eliminating $O_{c|s}$, one has

$$R^{\gamma}_{[001]} \, R^{\xi}_{\hat{n}} \, R^{-\gamma}_{[001]} = R^{\alpha_1 - \alpha_2}_{[100]} \,,$$

where $R^{\xi}_{\hat{n}} = (O^2_{c|m})^{-1} O^1_{c|m}$. This means that the rotation axis of $R^{\xi}_{\hat{n}}$ is related to that of $R^{\alpha_1-\alpha_2}_{[100]}$ by the magnetic rotation $R^{\gamma}_{[001]}$ and their rotation angles are equal, i.e., one has $R^{\gamma}_{[001]}\hat{\boldsymbol{n}} = [100]$ and $\xi = \alpha_1 - \alpha_2$. The first relationship allows for an unambiguous determination of γ as it entails

$$[\cos\gamma \ \ \sin\gamma \ \ 0]^T = \hat{\boldsymbol{n}}\,.$$

The vanishing of the third component of $\hat{\boldsymbol{n}}$ and $\xi = \alpha_1 - \alpha_2$ can be used to confirm correctness of computations.

4.3 Crystal Orientation in Scanning Electron Microscope

With SEM, the orientation determination is usually based on EBSD with relatively simple relationships between reference frames; see Fig. 4.2*b*. Other measurements (e.g., SEM-based Kossel microdiffraction) have similar geometry. Usually, the specimen is tilted with respect to the electron beam. If the first axis (1) of the instrument coordinate system coincides with the tilt axis, it is perpendicular to the incident beam direction and it is parallel to the plane of detector and to the sample surface. The third axis (3) can be selected along the optical axis of the microscope or in the direction perpendicular to the plane of detector. In both cases, a simplified version of relationship (4.5) is applicable; with just one tilt angle and no other rotations ($\beta = 0 = \gamma$), one has $O_{c|s} R^{\alpha}_{[100]} = O_{c|m}$. One needs to note that other ways of positioning the specimen or the detector in scanning microscopes are also in use; see, e.g., [13, 14].

References

1. H.J. Bunge, *Texture Analysis in Materials Science* (Butterworth's, London, 1982)
2. A. Morawiec, *Orientations and Rotations. Computations in Crystallographic Textures* (Springer-Verlag, Berlin, 2004)
3. W.R. Busing, H.A. Levy, Angle calculations for 3- and 4-circle X-ray and neutron diffractometers. Acta Cryst. **22**, 457–464 (1967)
4. P. Dera, A. Katrusiak. Towards general diffractometry. I. Normal-beam equatorial geometry. Acta Cryst. A **54**, 653–660 (1998)
5. W.A. Paciorek, M. Meyer, G. Chapuis, On the geometry of a modern imaging diffractometer. Acta Cryst. A **55**, 543–557 (1999)
6. P. Fraundorf, Stereo analysis of single crystal electron diffraction data. Ultramicroscopy **6**, 227–236 (1981)
7. A. Morawiec, J.J. Fundenberger, E. Bouzy, J.S. Lecomte, EP - a program for determination of crystallite orientations from TEM Kikuchi and CBED diffraction patterns. J. Appl. Cryst. **35**, 287 (2002)
8. C.T. Chou, Computer software for specimen orientation adjustment using double-tilt or rotation holders. J. Electron Micr. Tech. **7**, 263–268 (1987)

9. T. Duden, A. Gautam, U. Dahmen, *KSpaceNavigator* as a tool for computer-assisted sample tilting in high-resolution imaging, tomography and defect analysis. Ultramicroscopy **111**, 1574–1580 (2011)
10. N. Cautaerts, R. Delville, D. Schryvers, *ALPHABETA*: a dedicated open-source tool for calculating TEM stage tilt angles. J. Microsc. **273**, 189–198 (2019)
11. R.X. Xie, W.Z. Zhang, $\tau ompas$: a free and integrated tool for online crystallographic analysis in transmission electron microscopy. J. Appl. Cryst. **53**, 561–568 (2020)
12. Y. Zhang, R. Yan, T. Sun, Y. Ma, A simple program for fast tilting electron-beam sensitive crystals to zone axes. Ultramicroscopy **211**, 112941 (2020)
13. R.R. Keller, R.H. Geiss, Transmission EBSD from 10 nm domains in a scanning electron microscope. J. Microsc. **245**, 245–251 (2012)
14. J.J. Fundenberger, E. Bouzy, D. Goran, J. Guyon, H. Yuan, A. Morawiec, Orientation mapping by transmission-SEM with an on-axis detector. Ultramicroscopy **161**, 17–22 (2016)

Chapter 5
Ab Initio Indexing of Single-Crystal Diffraction Patterns

5.1 Indexing in General

The process of ascribing indices to peaks on diffraction patterns is referred to as indexing. Indexing appears in various forms in most crystallographic studies, from those based on single-crystal data, through various analyses of bicrystals, multi-crystals and polycrystalline materials, to solving powder diffraction patterns, and it concerns all sorts of radiation (X-ray, electron, neutron, proton) in both mono- and polychromatic modes.

There are two significantly different types of indexing: *ab initio* indexing associated with determination of lattice parameters (which is a step of crystal structure determination [1]), and indexing of patterns originating from known crystalline structures. The latter type is commonly used for determination of orientations of crystal lattices; since it can also be a final stage of the ab initio indexing, after [2], it will be briefly referred to as *end*-indexing. Between these two limiting cases are the phase discrimination and phase identification problems: a pattern is known to originate from one of a number of admissible phases with known lattice metrics, and the point is to identify the phase. The number of admissible lattice metrics differs the indexing in phase discrimination/identification from those in structure determination or in orientation determination: this number is, respectively, finite but larger than one, infinite, and just one.

⊡ It is worth to make a terminological note. With several phases expected to be present in the material, after taking note of all possibilities, the diffracting phase is the one with best matching diffraction data. If the list of expected phases is short, the procedure is called phase *discrimination*. If a large database of phases is searched to identify the phase, the procedure is called phase *identification*. Finally, phase *determination* encompasses both discrimination and identification, and it also allows for an unknown phase with a yet undetermined structure.

The number of possible metrics has a large impact on the complexity of the indexing problem. The indexing with an a priori known metric or metrics is simpler than the indexing comprising reconstruction of the lattice. However, in all circumstances,

A. Morawiec, *Indexing of Crystal Diffraction Patterns*, Springer Series in Materials Science 326, https://doi.org/10.1007/978-3-031-11077-1_5

indexing requires the assignment of integer indices based on approximate data, and therefore, it can be seen as a problem in combinatorial optimization.

Due to relative complexity, repetitive character, and the need for speed, the indexing problems are best solved by computers. Nowadays, almost all indexing is carried out automatically. In the case of ab initio indexing, automation is needed mainly because of the complexity of the studied structures, whereas end-indexing is frequently applied to large sets of diffraction patterns.

Attempts to resolve the indexing problem resulted in very creative programming activities. Generally, an ideal indexing program is expected to be highly reliable, robust, fast, fully automatic, operating without the need for visual inspection of results and well documented. Some of the existing programs are very efficient, but not all are documented. Descriptions of proprietary software are sketchy.

In some cases, e.g., indexing of a diffraction pattern for crystal structure determination, one might be interested in a complete list of possible indexing solutions. An exhaustive indexing method (or program) would provide all solutions within rigorously defined tolerances. Standard time-efficient indexing programs are non-exhaustive. With automatic indexing of multiple patterns, one usually wants to obtain a single optimal solution per pattern.

⊡ In the pre-computer era, numerous 'manual' methods of indexing have been devised. For instance, various graphical methods (e.g., Hull-Davey, Bjurström, Harrington, Bunn, Frevel-Blumer charts [3–7]) were used for indexing powder diffraction patterns. The Bernal charts [8] were applied for solving single-crystal rotation photographs. The Leonhardt and Greninger charts [9, 10] were designed for solving transmission and back-reflection Laue patterns, respectively. One also needs to mention various atlases for indexing and crystal orientation determination; see, e.g., 'standard spot patterns' in [11] and [12] or atlases of Laue patterns [13] and Kikuchi patterns [14]. A solution to an experimental pattern was found my matching it to a pattern in the atlas. These methods were a step toward contemporary algorithms, but since they cannot compete with the current approaches, they are ignored here.

Worth mentioning are former experimental X-ray techniques intended to simplify indexing. They relied on the use of moving X-ray films. The de Jong-Bouman "retigraph" was a camera for making undistorted X-ray photographs of reciprocal lattices [15, 16]. With Weissenberg X-ray camera [17], crystal rotation is coordinated with movement of a cylindrical film along the crystal rotation axis, and with Buerger precession camera [18], crystal and film are in correlated precession motion. These techniques separate reflections, i.e., prevent superposition of diffraction spots.

5.2 Ab Initio Indexing for Structure Determination

Structural analysis is the essence of experimental crystallography, and a variety of methods are used for structure determination. They are roughly classified into single-crystal methods and powder diffraction methods. As these names indicate, the first term is applicable if diffraction patterns originate from a single crystal, whereas the second term is used when a pattern is produced by a large number of randomly oriented crystallites.

The indexing is a part of the procedure referred to as *data processing*. Other steps of this procedure include refinement of experimental parameters and crystal orientation, integration of intensities of diffraction peaks, and scaling of the intensities. After data processing and space-group assignment, at further stages of structure determination, the phase problem is solved, atomic positions are determined and the structure is refined. Indexing is a prerequisite for these stages; if indexing is incorrect, the subsequent stages will give incorrect results.

Although no prior knowledge of the lattice geometry is assumed, common are some expectations based on the type of the considered crystal. E.g., these are usually limits on the volume of the unit cell which correspond to limits on the number of atoms in the unit cell. The magnitude of the separation of peaks in the diffraction pattern indicates reasonable ranges of these parameters.

There is a huge literature on structure determination. Only a small fraction of it is devoted to indexing,[1] and only this particular aspect of structure determination is discussed below. The rest of this chapter is devoted to ab initio indexing in monochromatic single-crystal methods, the next one is about ab initio indexing of Laue patterns, and indexing of powder diffraction patterns is considered in the subsequent chapter.

5.3 Experimental Single-Crystal Techniques

Structure determination is largely based on single-crystal techniques. Most common are techniques relying on conventional X-rays. However, with growing complexity of the studied structures, high angular resolution is required, and—in effect—the number of cases resolved using synchrotron radiation is growing. There are also methods of resolving structures with high spatial resolution by 'electron crystallography'.

Large structures are often solved using the X-ray oscillation method. With this technique, the number of peaks on diffraction pattern is sufficiently large only with a finite oscillation angle. On the other hand, indexing procedures are based on the assumption that the patterns originate from a single crystal with a fixed orientation. Therefore, in practice, 'still' photographs obtained with a small range of oscillations are used.

Many aspects of indexing depend on the type of experiment by which the data are obtained (e.g., level and anisotropy of errors, type of noise, shift of the origin, et cetera), but there are some general features and only these will be considered. No matter what is the method of recording a single-crystal diffraction pattern,[2] the first step of its analysis is to search for peaks which will be used for indexing. The

[1] In structural analysis, the computer assisted determination of lattice parameters and assignment of indices is frequently referred to as *auto-indexing*.

[2] Formerly, the single-crystal techniques utilized photographic films and point detectors. With proper goniometers, numerous orientations of the specimen are reached, and the reciprocal space is uniformly searched for diffraction peaks. Nowadays, most popular are various area detectors (e.g., X-ray CCD cameras) with digital analysis of registered patterns.

positions of these peaks are then used to calculate the scattering vectors. Details of the calculation depend on the method and the experimental setup, but generally, it is based on the relative positions of the radiation source, the specimen and the detector, and on the relationships between the coordinate systems involved. (For illustration, a scheme used for getting the scattering vectors from Kossel lines is described in Appendix.)

It must be noted at the outset that indexing based on ideal data is straightforward. Problems arise when the diffraction data are not perfect. Consistent and reliable indexing based on poor-quality data or with multiple lattices contributing to a pattern can be difficult. Robustness to both noise (random errors in locations of diffraction peaks and instrument settings) and gross errors (e.g., satellite or alien reflections) is crucial for the quality of indexing software. Therefore most of this software is built of robust components (optimization under uncertainty, voting schemes, accumulators, iterative steps starting from the most reliable part of data et cetera). The correctness of indexing is assessed based on some indicators. Computer software relies on quantitative measures of how well simulated patterns match experimental patterns. Non-quantitative indicators of correctness involve visual inspection and comparison of results, and a critical analysis of patterns (e.g., comparison of solutions listed for various program settings). The procedures considered below are intended for indexing individual patterns, but generally, the reliability and accuracy of the lattice parameter determination can be improved if it is based on multiple patterns [19].

Most indexing programs are intrinsic parts of structure-determination packages, e.g., [19–25], but also a number of stand-alone indexing tools have been reported, e.g., [26–29]. There are a number of reasons for considering various approaches to indexing and different indexing programs. Indexing packages are frequently designed to particular problems, i.e., they are specialized and, in consequence, limited; a program flawlessly indexing patterns of one type may not be suitable for solving other patterns. Moreover, each indexing program has specific built-in mechanisms for controlling the extent of search for solutions. In difficult cases, the choice of the control parameters and criteria is essential for getting the right result, and it is important to have diverse tools for solving such patterns. Generally, reliability of results of indexing a diffraction pattern can be verified or improved by entering data to multiple programs or using multiple indexing methods.

5.4 The Problem of Indexing Single-Crystal Data

The input for indexing a single-crystal diffraction pattern is a set of scattering vectors obtained from positions of particular diffraction peaks via the Laue equation (2.3). These vectors approximate positions of some reciprocal lattice points. The goal of ab initio indexing is to reconstruct the geometry of the crystal lattice (including its orientation and symmetry), and to assign consistently indices to the peaks.

Let the input data be N experimental scattering vectors $\boldsymbol{h}^n$ $(n = 1, 2, \ldots, N)$ given by their components v_i^n in a Cartesian laboratory reference frame $\boldsymbol{e}_L^i$, i.e.,

$\boldsymbol{h}^n = v^n_i \boldsymbol{e}^i_L$. Legitimate scattering vectors $\boldsymbol{h}^n$ are close to vectors $\check{\boldsymbol{h}}^n$ of an unknown (reciprocal) lattice. Since an integer linear combination of lattice vectors is a lattice vector, the list of vectors $\boldsymbol{h}^n$ used in calculations can be extended. For instance, besides the experimentally measured scattering vectors, one can also use differences between these vectors; see Sect. 5.7.

Most approaches to indexing are based on *binary* information from diffraction patterns: a diffraction peak is either taken into consideration or completely ignored, but clearly, weights can be used to account for the reliability of the data. The weights may or may not be related to peak intensities. For instance, the measured scattering vectors may have larger impact than the derived difference vectors [26], or the weights of shorter vectors can be larger, e.g., they can be equal to inverses of the magnitudes of the scattering vectors [30].

5.4.1 Basics

In order to introduce basic notions, it is convenient to consider the ideal case without any experimental errors. An accurate reciprocal lattice vector $\check{\boldsymbol{h}}$ can be expressed in a yet unknown basis $\boldsymbol{a}^i$ as $h_i \boldsymbol{a}^i$, where h_i are integers. On the other hand, the experiment provides laboratory Cartesian components v_i of the $\check{\boldsymbol{h}}$ vector. With $\mathcal{T}$ transforming the basis $\boldsymbol{a}^i$ to the Cartesian crystal coordinate system $\boldsymbol{e}^i_c$ (i.e., $\mathcal{T}_j^{\ i} \boldsymbol{a}^j = \boldsymbol{e}^i_c$) and the crystal coordinate system linked to the external coordinate system $\boldsymbol{e}^i_L$ by the orientation matrix O (i.e., $\boldsymbol{e}^i_c = O^i_{\ j} \boldsymbol{e}^j_L$ and $O = O_{c|L}$ in the notation of Chap. 4), one has

$$v_i = O^k_{\ i} (\mathcal{T}^{-1})_k^{\ j} h_j = A_i^{\ j} h_j \ . \tag{5.1}$$

The matrix $A = O^T \mathcal{T}^{-1}$ with the entries $A_i^{\ j} = O^k_{\ i} (\mathcal{T}^{-1})_k^{\ j} = \mathcal{T}^j_{\ k} O^k_{\ i} = \boldsymbol{e}^L_i \cdot \boldsymbol{a}^j$ links the bases $\boldsymbol{a}^i$ and $\boldsymbol{e}^i_L$, i.e., $A_i^{\ j} \boldsymbol{e}^i_L = \boldsymbol{a}^j$. Since $A_i^{\ j} = \boldsymbol{e}^L_i \cdot \boldsymbol{a}^j$, the j-th column of the matrix A contains the components of (reciprocal lattice) basis vector $\boldsymbol{a}^j$ in the laboratory reference frame $\boldsymbol{e}^L_i$. It follows from the definition of A that $A^T A = \mathcal{T}^{-T} O O^T \mathcal{T}^{-1} = \left(\mathcal{T}\mathcal{T}^T\right)^{-1}$ is equal to the metric of the reciprocal lattice or

$$\delta^{kl} A_k^{\ i} A_l^{\ j} = g^{ij} = \boldsymbol{a}^i \cdot \boldsymbol{a}^j \ . \tag{5.2}$$

Similarly, with $X = A^{-1} = \mathcal{T} O$ (i.e., $X_i^{\ j} = \mathcal{T}_i^{\ k} O_k^{\ j}$ and $\boldsymbol{e}^i_L = X_j^{\ i} \boldsymbol{a}^j$), the relationships inverse to (5.1) are

$$h_i = X_i^{\ j} v_j \ . \tag{5.3}$$

Since $X_i^{\ j} = \boldsymbol{a}_i \cdot \boldsymbol{e}^j_L$, the i-th row of the matrix X contains the components of (direct lattice) basis vector $\boldsymbol{a}_i$ in the laboratory reference frame $\boldsymbol{e}^i_L$, and (5.3) is equivalent to $h_i = \boldsymbol{a}_i \cdot \check{\boldsymbol{h}}$. It follows from the definition of X that $X X^T = \mathcal{T} O O^T \mathcal{T}^T = \mathcal{T}\mathcal{T}^T$ is equal to the metric of the direct lattice, or $X_i^{\ k} X_j^{\ l} \delta_{kl} = g_{ij}$.

It is easy to verify that in the notation of Sect. 1.1.1, the operators $\mathcal{T}$, O and X can be expressed as $\mathcal{T} = \boldsymbol{e}_c^i \otimes \boldsymbol{a}_i = \mathcal{T}_i^{\ j} \boldsymbol{a}^i \otimes \boldsymbol{a}_j$, $O = \boldsymbol{e}_i^L \otimes \boldsymbol{e}_c^i = O^i_{\ j} \boldsymbol{e}_i^L \otimes \boldsymbol{e}_L^j$ and $X = \boldsymbol{e}_L^i \otimes \boldsymbol{a}_i = X_i^{\ j} \boldsymbol{a}^i \otimes \boldsymbol{a}_j$. In the field of X-ray diffractometry, the above scheme is conventionally presented using the matrices U and B of Busing and Levy [31].

⊡ *Busing-Levy* UB *matrix*
A reciprocal lattice vector $\check{\boldsymbol{h}} = h_i \boldsymbol{a}^i$ can be expressed as

$$\check{\boldsymbol{h}} = v_i \boldsymbol{e}_L^i = v_i^c \boldsymbol{e}_c^i ,$$

where $\boldsymbol{e}_L^i$ is an orthonormal basis of a laboratory reference frame and $\boldsymbol{e}_c^i$ is an orthonormal basis attached to the crystal. The two Cartesian bases are linked by

$$U = \boldsymbol{e}_c^i \otimes \boldsymbol{e}_i^L = U_i^{\ j} \boldsymbol{e}_L^i \otimes \boldsymbol{e}_j^L ,$$

where $U_i^{\ j} = \boldsymbol{e}_i^L \cdot \boldsymbol{e}_c^j$, and these entries constitute an orthogonal matrix. One has

$$U \boldsymbol{e}_L^j = U_i^{\ j} \boldsymbol{e}_L^i = \boldsymbol{e}_c^j \quad \text{and} \quad v_i = \check{\boldsymbol{h}} \cdot \boldsymbol{e}_i^L = v_j^c \boldsymbol{e}_c^j \cdot \boldsymbol{e}_i^L = U_i^{\ j} v_j^c . \tag{5.4}$$

The U matrix determines the crystal orientation.

The reciprocal lattice basis $\boldsymbol{a}^i$ and the crystal orthonormal basis $\boldsymbol{e}_i^c$ are related by

$$B = \boldsymbol{a}^i \otimes \boldsymbol{e}_i^c = B_i^{\ j} \boldsymbol{e}_c^i \otimes \boldsymbol{e}_j^c ,$$

where $B_i^{\ j} = \boldsymbol{e}_i^c \cdot \boldsymbol{a}^j$; one has

$$B \boldsymbol{e}_c^j = B_i^{\ j} \boldsymbol{e}_c^i = \boldsymbol{a}^j \quad \text{and} \quad v_i^c = \check{\boldsymbol{h}} \cdot \boldsymbol{e}_i^c = h_j \boldsymbol{a}^j \cdot \boldsymbol{e}_i^c = B_i^{\ j} h_j . \tag{5.5}$$

The Busing-Levy B and U matrices are linked to the above-used $\mathcal{T}$, O and A via $B = \mathcal{T}^{-1}$, $U = O^T$ and $A = UB$.

The matrix $B = \mathcal{T}^{-1}$ is determined by the six (unknown) parameters of the unit cell. There are a number of different conventions for selecting the B matrix, or in other words, for attaching an orthonormal basis to the crystal [30–34]; cf. Sect. 1.8. Knowing the matrix A, in the final stage of indexing (end-indexing), the matrices $U = O^T$ and B can be obtained via QR-decomposition of A into a product of special orthogonal U and upper triangular B with positive diagonal entries. Since A is invertible (as it links two bases), such decomposition of A into the product of U and B is unique.

The components v_i^n of the vectors $\boldsymbol{h}^n$ in the Cartesian laboratory system are known a priori. In the error-free case, the indexing problem is to determine the unknown non-singular matrix $X = A^{-1}$ such that $h_i^n = X_i^{\ j} v_j^n$ are integers. It is easy to see that with an integer matrix, say T, also $T_i^{\ j} h_j^n = T_i^{\ j} X_j^{\ k} v_k^n$ are integers. This expresses the fact that if the scattering vectors $\boldsymbol{h}^n$ point to nodes of a certain (reciprocal) lattice, they also point to nodes of its superlattices. Therefore, one needs to limit the primitive cell of the reciprocal lattice by demanding that $|\det X|$ takes the smallest possible value. Since $|\det(TX)| = |\det T||\det X|$, the distinct solutions of the problem are related by *unimodular* integer matrices T. (If $|\det T| > 1$ then $|\det(TX)| > |\det X|$, i.e., $|\det(TX)|$ is not the smallest value. If $|\det T| < 1$ then $|\det(TX)| = 0$, i.e., TX is singular.) With unimodular integer T, the solutions X and TX correspond to bases of the same lattice; see Sect. 1.2.

5.4.2 *Indexing Error-Free Data*

It was mentioned before that indexing accurate data is straightforward. Ideal data are easily indexed with a proper software, but this does not mean that manual indexing would be easy. If one goes beyond the simplest cases with a very small number of scattering vectors, manual indexing becomes a formidable task.

The problem is to compute a basis of a lattice from a set of generating lattice vectors. In general, it can be formulated as follows: Knowing N exact vectors $\boldsymbol{h}^n$ $(n = 1, 2, \ldots, N)$ pointing to nodes of a lattice, what is the basis of the lattice? In the case of indexing, the vector space is three-dimensional, the vectors $\boldsymbol{h}^n$ are given in a Cartesian coordinate system, and it is assumed that the lattice generated by the vectors is also three-dimensional.

⊡ To see that indexing is tedious even if data are exact, one may try to get a basis of the lattice generated by five vectors $\boldsymbol{h}^n$ $(n = 1, 2, \ldots, 5)$ with coordinates in a 'laboratory' Cartesian reference frame collected in the matrix

$$H = \left[\boldsymbol{h}^1\, \boldsymbol{h}^2\, \boldsymbol{h}^3\, \boldsymbol{h}^4\, \boldsymbol{h}^5\right] = \begin{bmatrix} 3/2 & 3/2 & 3/2 & 1 & 5/2 \\ \sqrt{2}/2 & \sqrt{2}/2 & \sqrt{2} & \sqrt{2}/2 & \sqrt{2} \\ 3\sqrt{2} & 2\sqrt{2} & 3\sqrt{2}/2 & 3\sqrt{2}/2 & 7\sqrt{2}/2 \end{bmatrix} . \tag{5.6}$$

These vectors and the matrix will be used below to illustrate a systematic method of indexing accurate data.

There are a number of ways to index error-free data. For instance, one can search the vectors $\boldsymbol{h}^n$ and their integer linear combinations for three shortest linearly independent vectors $\boldsymbol{h}^{n_1}, \boldsymbol{h}^{n_2}, \boldsymbol{h}^{n_3}$. With the assignment $h_k^{n_j} = \delta_k^j$, (5.1) leads to a tentative solution

$$A_i^{\ j} = A_i^{\ k}\delta_k^j = A_i^{\ k}h_k^{n_j} = v_i^{n_j} \ .$$

Such obtained $X = A^{-1}$ is tested against the remaining vectors $\boldsymbol{h}^n$: it is verified whether all products $X\boldsymbol{h}^n$ are triplets of integers. If this occurs, a solution of the indexing problem was found. If not, bases of low-index superlattices of the lattice based on the triplet $\boldsymbol{h}^{n_j}$ $(j = 1, 2, 3)$ need to be tested.

The above simple approach is related to a more elegant systematic method for computing a basis of a lattice from a set of generating lattice vectors. It consists of two steps: First, one needs to express the generating lattice vectors in a reference system in which they have integer coordinates. Then, with these coordinates collected in an integer matrix, one can determine a lattice basis using the Hermite normal form of the matrix. Details are given below.

Transforming to an integer matrix

With the assumption that the lattice generated by the vectors is three-dimensional, three of these vectors must be linearly independent. The numbering of the vectors can be changed so $\boldsymbol{h}^1$, $\boldsymbol{h}^2$ and $\boldsymbol{h}^3$ are linearly independent. Given the $3 \times N$ matrix H with coordinates of individual lattice vectors in a 'laboratory' Cartesian reference frame as its columns

$$H = \left[\boldsymbol{h}^1\, \boldsymbol{h}^2\, \ldots\, \boldsymbol{h}^N\right] \qquad (\text{or } H_{in} = v_i^n)$$

and the non-singular matrix $T_0 = \left[\boldsymbol{h}^1\, \boldsymbol{h}^2\, \boldsymbol{h}^3\right]^{-1}$, the $3 \times n$ product $T_0 H$ is the matrix of components of the vectors $\boldsymbol{h}^n$ $(n = 1, 2, \ldots, N)$ in the reference system based on $\boldsymbol{h}^1$, $\boldsymbol{h}^2$ and $\boldsymbol{h}^3$. Since the lattice generated by $\boldsymbol{h}^1$, $\boldsymbol{h}^2$ and $\boldsymbol{h}^3$ is a sublattice of that generated by all vectors $\boldsymbol{h}^n$, the entries of $T_0 H$ are rational numbers. With m_L being the least common multiple of the denominators of the entries of $T_0 H$, the matrix

$$M = m_L T_0 H$$

has integer entries.

⊡ For the matrix H given by (5.6) and $T_0 = \left[\boldsymbol{h}^1\, \boldsymbol{h}^2\, \boldsymbol{h}^3\right]^{-1}$, one has

$$T_0 H = \begin{bmatrix} 1 & 0 & 0 & 1/3 & 1/3 \\ 0 & 1 & 0 & 0 & 1 \\ 0 & 0 & 1 & 1/3 & 1/3 \end{bmatrix}.$$

The least common multiple of the denominators of the entries of the matrix $T_0 H$ is $m_L = 3$, and thus the integer matrix M has the form

$$M = m_L T_0 H = \begin{bmatrix} 3 & 0 & 0 & 1 & 1 \\ 0 & 3 & 0 & 0 & 3 \\ 0 & 0 & 3 & 1 & 1 \end{bmatrix}. \tag{5.7}$$

Hermite normal form

The definition of Hermite normal form given in Sect. 1.2.4 for square integer matrices, can be extended to rectangular $m \times N$ integer matrices. All that is needed here is the case of matrices of rank m with $N \geq m$ and matrix columns containing vector components. A matrix in Hermite normal form has the shape

$$[\mathbf{0}\mathbf{0}\ldots\mathbf{0}\, A] = \begin{bmatrix} 0 & 0 & \ldots & 0 & \star & \star & \ldots & \star \\ 0 & 0 & \ldots & 0 & 0 & \star & \ldots & \star \\ \vdots & \vdots & \ldots & \vdots & \vdots & \ddots & \ddots & \vdots \\ 0 & 0 & \ldots & 0 & 0 & 0 & 0 & \star \end{bmatrix},$$

where the square matrix A is upper triangular and its entries satisfy the conditions $A_{(i)(i)} > A_{ij} \geq 0$ for $j > i$. The columns preceding A have only zero entries. Every integer matrix of full row rank has a unique HNF. For an arbitrary $m \times N$ integer matrix M of rank m, there exists a (non-unique) unimodular integer matrix $\bar{U}$ such that the matrix $M\bar{U}$ is in Hermite normal form.[3] The Hermite normal form of M will be denoted by $\mathrm{HNF}(M) = M\bar{U}$.

Since $\bar{U}$ is integer and unimodular, $\bar{U}^{-1}$ is also an integer matrix. With given M, the relationship $\mathrm{HNF}(M)\bar{U}^{-1} = M$ is a system of linear Diophantine equations for unknown $\mathrm{HNF}(M)$ and $\bar{U}^{-1}$. The Hermite normal form can be determined by a procedure similar to Gaussian elimination used for solving systems of linear equations. Only three operations on matrix columns are needed: change of the sign of a column, swapping two columns, and change of a column by adding an integer multiple of another column; see, e.g., Chap. 2 of [35]. Such simple algorithms are easy to implement, but they frequently lead to a phenomenon known as number size explosion: some integers become large, and a library supporting infinite precision integer arithmetic is needed. For this reason, other algorithms have been invented. They not only avoid the number size explosion, but also speed up computations [35–37].

Reduction of a matrix to its Hermite normal form can be seen as the determination of a basis of the lattice generated by columns of the matrix. The formula $M = \mathrm{HNF}(M)\bar{U}^{-1}$ is nothing else but a way of expressing columns of M via linear combinations of the columns of $\mathrm{HNF}(M)$. It is easy to see that if two sets of integer vectors generate the same lattices, the Hermite normal forms of the corresponding matrices are equal except for the number of zero-columns.

⊡ It is worth noting that if M has the dimension $1 \times N$ with $N \geq 2$, the reduction to Hermite normal form is equivalent to the determination of the greatest common divisor of the entries of M. Thus, the reduction to Hermite normal form in higher dimensions can be seen as a generalization of the problem of finding the greatest common divisor of integers. Simple algorithms for the reduction can be seen as generalizations of the commonly known Euclid algorithm for obtaining the greatest common divisor of two integers.

The Hermite normal form is an important tool for dealing with lattices of high dimensions. It is widely applied in integer programming and in lattice-based cryptography. It can be used for solving a variety of problems. (E.g., 'given a lattice basis and a vector, check whether the vector belongs to the lattice' or 'given bases of two lattices, check if one lattice is a sublattice of the other'.) These problems can also be solved using the notion of the reciprocal lattice, but reduction to Hermite normal form allows for avoiding costly matrix inversion.

⊡ The Hermite normal form of the matrix M of (5.7) is given by

$$\mathrm{HNF}(M) = \begin{bmatrix} 0 & 0 & 3 & 0 & 1 \\ 0 & 0 & 0 & 3 & 0 \\ 0 & 0 & 0 & 0 & 1 \end{bmatrix} .$$

An example integer matrix $\bar{U}$ transforming M to $\mathrm{HNF}(M) = M\bar{U}$ and its inverse are, respectively,

[3] The matrix $\bar{U}$ is conventionally denoted by U. Here, the bar is added to distinguish it from the U matrix of Busing and Levy [31].

$$\bar{U} = \begin{bmatrix} 1 & 1\,1\,0\,0 \\ 0 & 1\,0\,1\,0 \\ 1 & 1\,0\,0\,0 \\ -3 & -2\,0\,0\,1 \\ 0 & -1\,0\,0\,0 \end{bmatrix} \quad \text{and} \quad \bar{U}^{-1} = \begin{bmatrix} 0\,0 & 1\,0 & 1 \\ 0\,0 & 0\,0 & -1 \\ 1\,0 & -1\,0 & 0 \\ 0\,1 & 0\,0 & 1 \\ 0\,0 & 3\,1 & 1 \end{bmatrix} . \tag{5.8}$$

The determinant of $\bar{U}$ is 1.

Computing the basis

Armed with the notion of the Hermite normal form of an integer matrix, one can return to the initial problem of computing a basis of a lattice from a set of generating lattice vectors.

The Cartesian coordinates of basis vectors of the lattice generated by the vectors $\boldsymbol{h}^n$ constituting columns of $H = [\boldsymbol{h}^1\, \boldsymbol{h}^2\, \ldots\, \boldsymbol{h}^N]$ are obtained by transforming the columns of $\mathrm{HNF}(M) = \mathrm{HNF}(m_L \mathcal{T}_0 H) = m_L \mathcal{T}_0 H \bar{U}$ back to the Cartesian system, i.e., by taking the product of $(1/m_L)\mathcal{T}_0^{-1}$ and $\mathrm{HNF}(M)$; one has

$$\frac{1}{m_L}\mathcal{T}_0^{-1}\,\mathrm{HNF}(M) = \frac{1}{m_L}\mathcal{T}_0^{-1}\,(m_L \mathcal{T}_0 H \bar{U}) = H\bar{U} = [\mathbf{0}\,\mathbf{0}\,\ldots\mathbf{0}\,\boldsymbol{a}^1\,\boldsymbol{a}^2\,\boldsymbol{a}^3] \tag{5.9}$$

with the last three columns containing the sought basis vectors $\boldsymbol{a}^i$ in the 'laboratory' Cartesian reference frame. The above relationship determines also the matrix A solving the indexing problem. It is given by

$$[\mathbf{0}\,\mathbf{0}\,\ldots\mathbf{0}\;A] = \frac{1}{m_L}\mathcal{T}_0^{-1}\,\mathrm{HNF}(m_L \mathcal{T}_0 H) = H\bar{U} .$$

Since $H = [\boldsymbol{h}^1\, \boldsymbol{h}^2\, \ldots\, \boldsymbol{h}^N] = [\mathbf{0}\,\mathbf{0}\,\ldots\boldsymbol{a}^1\,\boldsymbol{a}^2\,\boldsymbol{a}^3]\,\bar{U}^{-1}$, the last three rows of $\bar{U}^{-1}$ provide the indices h_i^n, i.e., the coefficients of the vectors $\boldsymbol{h}^n$ in the basis $\boldsymbol{a}^i$. Equivalently, these indices are elements of the $3 \times N$ matrix $A^{-1}H$. As the basis vectors $\boldsymbol{a}^i$ resulting from (5.9) may be long, the basis needs to be reduced using Buerger, Niggli or Delaunay reduction methods described in Sects. 1.2.2 and 1.2.3.

⊡ Going back to the example data of (5.6), the product $[\mathbf{0}\,\mathbf{0}\;A]$ of $(1/m_L)\mathcal{T}_0^{-1}$ and $\mathrm{HNF}(M)$ is

$$[\mathbf{0}\,\mathbf{0}\;A] = [\mathbf{0}\,\mathbf{0}\,\boldsymbol{a}^1\,\boldsymbol{a}^2\,\boldsymbol{a}^3] = \frac{1}{m_L}\mathcal{T}_0^{-1}\,\mathrm{HNF}(M) = H\bar{U} = \begin{bmatrix} 0\,0 & 3/2 & 3/2 & 1 \\ 0\,0 & \sqrt{2}/2 & \sqrt{2}/2 & \sqrt{2}/2 \\ 0\,0 & 3\sqrt{2} & 2\sqrt{2} & 3\sqrt{2}/2 \end{bmatrix}$$

The coefficients of the vectors $\boldsymbol{h}^n$ in the basis $\boldsymbol{a}^i$ are in the last three rows of the matrix $\bar{U}^{-1}$ of (5.8), i.e., one has

$$\boldsymbol{h}^1 = \boldsymbol{a}^1 , \quad \boldsymbol{h}^2 = \boldsymbol{a}^2 , \quad \boldsymbol{h}^3 = -\boldsymbol{a}^1 + 3\boldsymbol{a}^3 , \quad \boldsymbol{h}^4 = \boldsymbol{a}^3 , \quad \boldsymbol{h}^5 = \boldsymbol{a}^2 + \boldsymbol{a}^3 .$$

The basis of the direct lattice is given by

$$[\boldsymbol{a}_1\, \boldsymbol{a}_2\, \boldsymbol{a}_3] = \begin{bmatrix} -1 & 3 & -2 \\ -\sqrt{2}/2 & -3\sqrt{2}/2 & 3\sqrt{2} \\ \sqrt{2}/2 & -\sqrt{2}/2 & 0 \end{bmatrix} .$$

The Niggli reduced basis of the direct lattice and the corresponding metric are

$$[\boldsymbol{a}'_1\, \boldsymbol{a}'_2\, \boldsymbol{a}'_3] = \begin{bmatrix} 0 & -1 & -1 \\ -\sqrt{2} & -\sqrt{2}/2 & -\sqrt{2}/2 \\ 0 & \sqrt{2}/2 & -\sqrt{2}/2 \end{bmatrix} \quad \text{and} \quad g' = [\boldsymbol{a}'_i \cdot \boldsymbol{a}'_j] = \begin{bmatrix} 2 & 1 & 1 \\ 1 & 2 & 1 \\ 1 & 1 & 2 \end{bmatrix} .$$

Based on (1.31) and Tables 1.3 and 1.9, the type of the Niggli-reduced cell of the direct lattice is I, the Niggli character is 1, and the Bravais type of the lattice is cF (cubic, face-centered).

Methods of indexing exact data have little practical applications, but knowing them is useful for better understanding indexing methods applicable to error-affected data.

5.4.3 Impact of Errors

In practice, the positions $\boldsymbol{h}^n$ are known with a limited accuracy, and some of the vectors are obtained form spurious (satellite or alien) reflections that do not correspond to the investigated structure. In other words, the equalities $\boldsymbol{h}^n \approx \check{\boldsymbol{h}}^n$ and (5.1) are only approximate[4]

$$v_i^n \approx A_i{}^j h_j^n , \tag{5.10}$$

and for some n, they are not valid at all.

In principle, the process of solving (5.10) can be reduced to trial-and-error in the assignments of indices to considered reflections. With an assumed assignment, one needs to determine the matrix A such that it fits best the experimental data. This can be seen as the least-squares problem minimizing the sum of squared magnitudes of deviation vectors $d_i^n = A_i{}^k h_k^n - v_i^n$: determine A by locating the minimum of

$$\sum_n |\boldsymbol{d}^n|^2 . \tag{5.11}$$

With known v_i^n and assumed h_k^n, this linear problem can be easily solved; see, Sect. 5.14.2.[5] In practice, testing (5.11) for a wide range of indices h_i^n would be too costly, and other procedures are used.

---

[4] Errors are specific to the method used. For instance, in oscillation photographs, they are linked to the crystal oscillations (i.e., varying crystal orientation) and to partial reflections (corresponding to spots for which Bragg conditions are only approximately fulfilled).

[5] Instead of (5.11), one might consider fitting X by minimizing the sum of squared magnitudes of vectors $D_i^n = X_i{}^k v_k^n - h_i^n$; such approach differs from minimizing (5.11) because the vectors $\boldsymbol{d}^n$ are given in the Cartesian frame whereas $\boldsymbol{D}^n$ are in a space with unknown possibly very unnatural metric.

⊡ Formally, the task of indexing a single-crystal diffraction pattern can be reformulated as follows: given (experimentally obtained) components v_i^n of vectors $\boldsymbol{h}^n$, find the matrix A such that $v_i^n \approx A_i^{\ j} h_j^n$ and h_j^n are integers. As the magnitudes of the true basis vectors are limited, the reciprocal lattice metric tensor $\delta^{kl} A_k^{\ i} A_l^{\ j} = g^{ij}$ of (5.2) is bounded by $g_L^{ij} \leq \delta^{kl} A_k^{\ i} A_l^{\ j} \leq g_U^{ij}$, where g_L and g_U are some assumed symmetric matrices. The complexity of indexing can be illustrated using a formulation based on (5.11): determine the matrix A and integer h_j^n minimizing the sum of squared deviations $\boldsymbol{d}^n$, subject to $g_L^{ij} \leq \delta^{kl} A_k^{\ i} A_l^{\ j} \leq g_U^{ij}$ and $|h_j^n| \leq h_{max}$. This particular problem belongs to mixed integer nonlinear programming (MINLP). MINLP comprises nonlinear objective function of discrete and continuous variables subject to nonlinear constraints, i.e., it combines the combinatorial mixed integer programming (MIP) with nonlinear programming (NLP), and both are difficult to solve. The problem is additionally complicated by the potential presence of spurious reflections; therefore subsets of the complete set of reflections need to be considered. There exist general methods of solving constrained optimization problems, but they are rarely used for indexing. For an application of the *branch-and-bound* algorithm to indexing, see [38]. (The branch-and-bound algorithm is a repeated partitioning of the set of feasible solutions into disjoint subsets and calculation of a lower bound on the objective function for each subset; in each partitioning step, subsets with bounds larger than the value of the function for a known feasible solution are excluded from further analysis, which continues until a solution minimizing the function is found.)

The indexing of a single-crystal diffraction pattern given in the form of experimentally measured (i.e. approximate) positions of reciprocal lattice vectors can be seen as the problem of deducing a (reduced) basis of the lattice. Alternatively, it can be seen as the Pohst problem (see Sect. 1.2.1) of lattice reduction in the presence of errors, both, random and gross. With a known basis, the subsequent assignment of indices to reflections is straightforward.

Generally, the solution to the indexing problem may be not unique; cf. Fig. 5.1. There may be ambiguous solutions with primitive cells of similar volumes. In rough terms, a necessary condition for correct indexing is that the average error in positions of measured reciprocal lattice vectors $\boldsymbol{h}^n$ is smaller than half of the shortest vector of the reciprocal lattice [26].

In view of the above, obtaining strict solutions is impractical, and heuristic approaches are used with guaranteed optimality traded for speed and an easy implementation.

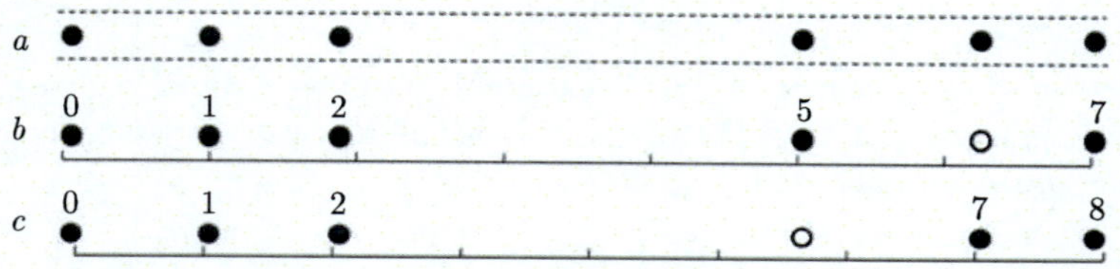

Fig. 5.1 Schematic one-dimensional illustration of the ambiguity in indexing a pattern of inaccurately positioned lattice nodes. With points given between the dashed lines (*a*), what is the right indexing (*b*) or (*c*)? In rough terms, a necessary condition for a correct assignment of indices is that the average error in node positions is smaller than half of the 'lattice' constant.

5.4.4 Some Objective Functions

At the outset, it is worth considering a simple objective function which helps to develop the right intuition. Since the component $X_i{}^j v_j^n$ ($\approx h_i^n$) of the vector $\boldsymbol{h}^n$ is expected to be close to an integer, the value of $\cos\left(2\pi X_i{}^j v_j^n\right)$ is expected to be close to 1 – the maximum of cosine. Thus, the sum over all components of all vectors is expected to reach maximum, and

$$X = \arg_Y \max \sum_{n,i} \cos\left(2\pi Y_i{}^j v_j^n\right) .$$

is expected to be a solution of the indexing problem; cf. [39].

If all recorded scattering vectors are assumed to be correct, instead of minimizing the summed deviations of $X_i{}^j v_j^n$ from integers, one may simply minimize the largest deviation, i.e.,

$$X = \arg_Y \min\left(\max_{n,i}\left\{\left|Y_i{}^j v_j^n - \lfloor Y_i{}^j v_j^n \rceil\right|\right\}\right) ;$$

$\lfloor\cdot\rceil$ denotes rounding to the nearest integer. Similarly, one may look for the matrix minimizing the largest magnitude $|X\boldsymbol{h}^n - \lfloor X\boldsymbol{h}^n \rceil|$ calculated using the Cartesian metric. These approaches, however, are not directly applicable to data sets which contain outliers or illegitimate vectors.

Unlike the above formulas, the next one is employed (in one way or another) in most indexing algorithms. In the presence of spurious vectors, one needs to count and maximize the number of vectors $X\boldsymbol{h}^n$ with each component differing from an integer by less than a certain threshold ϵ

$$X = \arg_Y \max\left(\#\{\, n\, ;\ \max_i \left|Y_i{}^j v_j^n - \lfloor Y_i{}^j v_j^n \rceil\right| \le \epsilon,\ 1 \le n \le N\,\}\right) ;$$

$\max_i$ is over the components of a single (the n-th) vector, and $\#S$ denotes the number of elements (cardinality) of the set S.

In the program *Ind_X* described below, the quality of the solution X equals

$$w_\epsilon N_\epsilon + w_{\epsilon/2} N_{\epsilon/2} \,, \tag{5.12}$$

where w_ϵ and $w_{\epsilon/2}$ are user-provided weights and N_ϵ is a number of vectors $\boldsymbol{h}^n$ with approximations of indices differing from integers by less than ϵ

$$N_\epsilon = \#\{\, n\, ;\ \max_i \left|X_i{}^j v_j^n - \lfloor X_i{}^j v_j^n \rceil\right| \le \epsilon,\ 1 \le n \le N\}$$

The formula (5.12) for the quality of a solution can be easily generalized to other weighting schemes.

Although there are numerous texts on indexing, the number of distinct ab initio approaches is not that large. One of them is known as *real-space indexing*.

5.5 Real-Space Indexing

Considerations of the previous section concerned mainly the reciprocal lattice. Some indexing methods refer explicitly to the direct lattice. They are based on the elementary fact (following from the definition of the reciprocal lattice) that the scalar product of a direct lattice vector and a reciprocal lattice vector is an integer.

Assuming one has a direct lattice test vector $\boldsymbol{t}$, the product $\boldsymbol{t} \cdot \boldsymbol{h}^n$ should be an integer for all legitimate scattering vectors $\boldsymbol{h}^n$. Thus, the legitimate vectors support the validity of $\boldsymbol{t}$, whereas vectors obtained from alien reflections are likely to fail the test of integrity of $\boldsymbol{t} \cdot \boldsymbol{h}^n$. Tests of this kind are a basis of an algorithm proposed by Duisenberg [27]. In this case, the test vectors are unit vectors $\widehat{\boldsymbol{t}}$ along zone axes (i.e., $\widehat{\boldsymbol{t}} = \boldsymbol{t}/|\boldsymbol{t}|$). The projections of vectors $\boldsymbol{h}^n$ on $\widehat{\boldsymbol{t}}$ give some values $x_n = \widehat{\boldsymbol{t}} \cdot \boldsymbol{h}^n$. If the vectors $\boldsymbol{h}^n$ correspond to legitimate reflections, the values x_n are grouped with a distribution close to a one-dimensional lattice (Fig. 5.2). The next step is to determine the period $p = 1/d$ of this lattice by minimizing the deviations of the values x_n from the integer multiples of the period. (Sect. 5.6 below is devoted to this subject.) Only d larger than the smallest expected inter-planar distance need to be considered. The period $p = 1/d$ is then used to calculate the direct lattice vector $\boldsymbol{t} = d\widehat{\boldsymbol{t}}$.[6] A given reflection $\boldsymbol{h}^n$ votes for the one-dimensional lattice and the corresponding vector $\boldsymbol{t}$ if its projection x_n is closer to the lattice node than a certain adjustable limit. The final cell of the direct lattice of the crystal is built based on $\boldsymbol{t}$ vectors with the numbers of votes not smaller than a certain 'acceptance level'.

5.5.1 *Obtaining Test Vectors*

The starting point of real space indexing based on error affected reciprocal lattice vectors $\boldsymbol{h}^n$ ($n = 1, \ldots N$) is to determine a test vector $\boldsymbol{t}$ (or $\widehat{\boldsymbol{t}}$) with a direction along a direct lattice vector. With a sufficiently large N, the products $\boldsymbol{t} \cdot \boldsymbol{h}^n$ are clustered at regular $\boldsymbol{t}$–dependent intervals. This allows for identification and elimination of those among the $\boldsymbol{h}^n$ vectors which do not contribute to the dominant periods and hence do not belong to the sought reciprocal lattice. The remaining vectors are then used to obtain a basis of the lattice.

There are numerous ways of selecting the test vectors. Some methods are suitable for certain experimental conditions. For instance, one can use experimentally acces-

---

[6] If $\boldsymbol{t}$ is a vector of the direct lattice, the coordinates of $\boldsymbol{t}$ can be easily refined. With the integer closest to $\boldsymbol{t} \cdot \boldsymbol{h}^n$ denoted by $m_n = \lfloor \boldsymbol{t} \cdot \boldsymbol{h}^n \rceil$, one looks for $\boldsymbol{t}$ minimizing the function $\sum_n (\boldsymbol{t} \cdot \boldsymbol{h}^n - m_n)^2$. The procedure for the refinement of $\boldsymbol{t}$ based on known $\boldsymbol{h}^n$ and m_n is described in the Appendix.

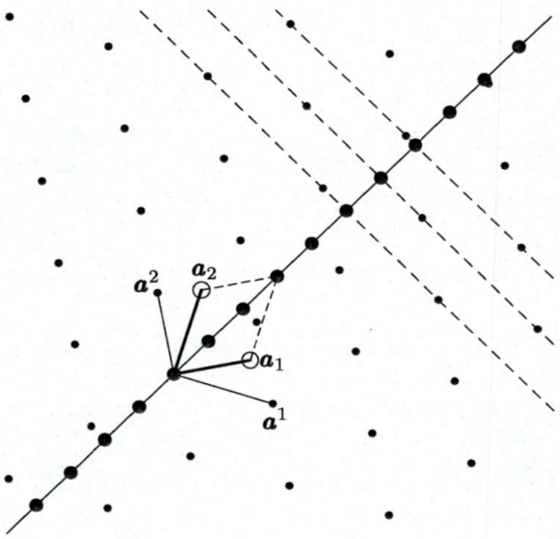

Fig. 5.2 Schematic two-dimensional illustration of projection of reciprocal lattice vectors on a direct lattice direction. With the direct lattice based on $\boldsymbol{a}_i$, the reciprocal lattice (small disks) is based on $\boldsymbol{a}^i$. The projections of the reciprocal lattice nodes on the direct lattice direction ($\boldsymbol{a}_1 + \boldsymbol{a}_2$ in this case, marked by solid line) constitute a one-dimensional lattice (large disks).

sible zones (e.g., lunes in oscillation photographs [40]) and then zone axis directions. (See Sect. 5.10.) Other methods of selecting the initial test vectors either do not use $\boldsymbol{h}^n$ vectors, or are based on individual or multiple $\boldsymbol{h}^n$ vectors. The simplest way of obtaining $\widehat{\boldsymbol{t}}$ vectors is by random [41] or systematic [24, 42] generation of points on the unit hemisphere, and a subsequent refinement of such obtained directions based on the guideline that vectors $\boldsymbol{t}$ closer to direct lattice directions result in narrower clusters of products $\boldsymbol{t} \cdot \boldsymbol{h}^n$.[7]

The number of test vectors can be limited by using properties of direct lattice vectors. Since the (ideal) direct lattice vectors are perpendicular to reciprocal lattice vectors, one can probe unit vectors $\widehat{\boldsymbol{t}}$ perpendicular to a given $\boldsymbol{h}^n$ of small magnitude. The range of test vectors can be further limited by using multiple reflections. In the simplest case, with two vectors $\boldsymbol{v}^i = \boldsymbol{h}^{n_i}$ $(i = 1, 2)$ selected from the set of vectors $\boldsymbol{h}^n$, one has the test vector $\boldsymbol{t}_{(2)} = \boldsymbol{v}^1 \times \boldsymbol{v}^2$. One can also construct a test vector using three vectors $\boldsymbol{v}^i$ $(i = 1, 2, 3)$ of small magnitude selected from among $\boldsymbol{h}^n$. This approach is well exemplified by the method of Duisenberg [27]: the test vector

$$\boldsymbol{t}_D = \boldsymbol{v}^1 \times \boldsymbol{v}^2 + \boldsymbol{v}^2 \times \boldsymbol{v}^3 + \boldsymbol{v}^3 \times \boldsymbol{v}^1 = \sum_i \varepsilon_{ijk} \boldsymbol{v}^j \times \boldsymbol{v}^k / 2 \tag{5.13}$$

is normal to the plane (in the reciprocal space) through the terminal points of the vectors $\boldsymbol{v}^i$, i.e., it has a direction of a direct lattice vector. (Duisenberg actually used normalized vector $\widehat{\boldsymbol{t}}_D$.)

---

[7] It is worth noting that the initial generation can be more sophisticated than that of [42]. The set of generated vectors should have central symmetry with vectors distributed as uniformly as possible, and there are dedicated methods for nearly uniform placing of points on a sphere with antipodal symmetry (which is a variant of the Thomson problem: uniform placing of points on a sphere); see, e.g., [43, 44] and references therein.

More general methods of generating direct space directions

A more general method for obtaining $\widehat{\boldsymbol{t}}$ from a triplet of linearly independent vectors $\boldsymbol{v}^i = \boldsymbol{h}^{n_i}$ $(i = 1, 2, 3)$ is by considering a $\boldsymbol{v}^i$–based sublattice of the reciprocal lattice. There are a number of ways of explaining the construction:
1. Let the (unknown) index of the sublattice be denoted by K_s and the (known) metric by $g_s^{ij} = \boldsymbol{v}^i \cdot \boldsymbol{v}^j$. With the matrix $[g_{ij}^s]$ inverse to $[g_s^{ij}]$, the vectors $\boldsymbol{v}_i = g_{ij}^s \boldsymbol{v}^j$ constitute a basis of a superlattice of the direct crystal lattice, and the integer combinations

$$\boldsymbol{t}_s = k^i\, \boldsymbol{v}_i = k^i\, g_{ij}^s \boldsymbol{v}^j \tag{5.14}$$

are vectors of this superlattice. Thus, $K_s \boldsymbol{t}_s \cdot \boldsymbol{h}^n$ are integers, and $\widehat{\boldsymbol{t}}_s \cdot \boldsymbol{h}^n$ are expected to be distributed at regular intervals.
2. The above scheme can also be described in a more elementary way, as was done in [45]. After selecting three linearly independent vectors $\boldsymbol{v}^i$, one calculates the basis $\boldsymbol{v}_i = \varepsilon_{ijk} \boldsymbol{v}^j \times \boldsymbol{v}^k / (2V_t^*)$ reciprocal to that formed by $\boldsymbol{v}^i$; V_t^* is the signed volume of the parallelepiped spanned by the vectors $\boldsymbol{v}^i$. Then, as in (5.14), one takes $\boldsymbol{t}_s$ as an integer combination of $\boldsymbol{v}_i$.
3. Finally, the method is also equivalent to solving the system of linear equations $\boldsymbol{v}^i \cdot \boldsymbol{x} = k^i$; one can easily verify that the system is solved by $\boldsymbol{x} = \boldsymbol{t}_s$. This approach was proposed in [41]. It follows from explanations 1 and 2 that the method based on (5.14) comprises the approach of Duisenberg; for $k^1 = k^2 = k^3 = 1$, one has

$$\boldsymbol{t}_s = \sum_i \boldsymbol{v}_i = \sum_i g_{ij}^s \boldsymbol{v}^j = \boldsymbol{t}_D / V_t^* \,,$$

i.e., this particular vector $\boldsymbol{t}_s$ is collinear with the vector $\boldsymbol{t}_D$ of (5.13).

Since an integer combination of lattice vectors is a lattice vector, one can use more than three vectors $\boldsymbol{v}^i = \boldsymbol{h}^{n_i}$ $(i = 1, 2, \ldots, J)$ to built the test vectors

$$\boldsymbol{t}_{(J)} = \sum_{j,k=1}^{J} S_{jk}\, \boldsymbol{v}^j \times \boldsymbol{v}^k / 2 \,,$$

where the integer matrix S is skew-symmetric; the symbol $\sum^J$ is explicitely written to indicate that the upper summation limit is $J \geq 2$. The vector $V_t^* \boldsymbol{t}_s$ of (5.14) is linked to $\boldsymbol{t}_{(3)}$ via $S_{jk} = k^i \varepsilon_{ijk}$, and the particular test vector $\boldsymbol{t}_{(3)} = \boldsymbol{t}_D$ is obtained if $S_{jk} = \sum_{i=1}^{3} \varepsilon_{ijk}$. There are two reasons for keeping J small. One is the combinatorial explosion of the number of possible vector combinations. The other problem is that with more complicated combinations, the danger of using spurious reflections increases.

5.5.2 Interpretations of $\boldsymbol{t} \cdot \boldsymbol{h}^n$

Given an ideal direct lattice vector, its scalar products with reciprocal lattice vectors constitute a one-dimensional lattice. Similarly, with diffraction data, the scalar products of the test vector $\boldsymbol{t}$ (nearly) parallel to a direct lattice vector, and experimental scattering vectors $\boldsymbol{h}^n$ are clustered on an axis, let it be named x, near nodes of a one-dimensional lattice. Based on these periodically clustered data, one needs to determine the interval, say, p, separating centers of the clusters. Its inverse—the frequency—will be denoted by ν $(= 1/p)$.

The parameter $x_n = \boldsymbol{t} \cdot \boldsymbol{h}^n$ may have various interpretations depending on the choice of the test vector. For instance, with $\boldsymbol{t} = \boldsymbol{t}_s$ being a direct space vector, x_n, and consequently, p and ν are dimensionless numbers. With $\widehat{\boldsymbol{t}}$ being a dimensionless unit vector, $x_n = \widehat{\boldsymbol{t}} \cdot \boldsymbol{h}^n$ and p represent distances in the reciprocal space, and $\nu = 1/p$ corresponds to the distance between planes in the direct space. Finally, if the test vector is an integer combination of vector products $\boldsymbol{v}^i \times \boldsymbol{v}^j$, i.e., $\boldsymbol{t} = \boldsymbol{t}_{(J)}$, then x_n and p have the interpretation of volume in the reciprocal space, and ν represents a volume in the direct space; the sought ν corresponds to the actual volume V of primitive cells of the direct lattice. What matters is that with integer S_{jk}, the products $\boldsymbol{t}_{(J)} \cdot \boldsymbol{h}^n$ are multiples of the volume V^*, i.e., they all can be projected on a single axis, the same for all test vectors $\boldsymbol{t}_{(J)}$ and all scattering vectors $\boldsymbol{h}^n$. This increases the reliability of the resulting period V^*. Consequently, the real space indexing is linked to indexing based on the determination of the volume of the primitive cell of the reciprocal lattice – the method proposed by Klein [26].

In summary, the above implies that three-dimensional diffraction data provide coordinates x_n of events occurring on a single axis, the events are expected to show some periodicity, and there is a need for reliable determination of the period of the data series x_n.

5.6 Period Detection

The experimentally obtained x_n series is expected to show some periodic 'pulsations'. The main goal at this point is an accurate estimation of the dominant periods. Similar problems of period determination arise in other research fields. Most notably, it is common to search for periodic components in time-series data sets. Period detection techniques for signals with arbitrary shapes include Fourier transform, Rayleigh test, various periodograms, autocorrelation, or more specific approaches like, e.g., 'phase dispersion minimization' [46, 47] or 'epoch folding' [48, 49]. Most of these methods are applicable to noise affected periodic signals.

5.6.1 Domains

The data x_n $(n = 1, \ldots, N)$ can be analyzed in the domain of the argument x, in the domain of periods ($p = 1/\nu$) or in the frequency (ν) domain. The function given in the x-domain does not directly provide the needed period; although some conclusions can be drawn based on its visual inspection, to obtain numerical results, the function must be additionally processed. The search in the p-domain directly shows the most prominent periods. The frequency domain is useful for determining components of the spectrum, and it is convenient for signal filtering or amplifying. By its nature, the frequency domain is most suitable for the analysis of long-lasting signals, particularly for the detection of weak periodic components of signals, but – as the case of indexing shows – it is also very useful for analyzing transient signals. Key for the frequency domain analysis is the transformation linking an x-domain function to a frequency domain function. The best known example is the Fourier transformation, but there exist some other approaches.

The principal tool for the identification of periodicities is a periodogram. The classic (Schuster) periodogram is defined as a squared magnitude of the Fourier transform of f. Despite its name, a periodogram is usually regarded as a function of frequency rather than of period [50].

5.6.2 Test Periods

In practice, periodograms are calculated in discrete forms. A periodogram could be evaluated at equally spaced test periods, but the relative resolution for short periods would be low compared to long periods. Moreover, the number of test periods needs to be limited for computational efficiency. On the other hand, their density must be high enough so that the actual period is not overlooked. Below is a sketchy justification for selecting the test periods at values corresponding to equally spaced frequencies.

⊡ Let the data be between $x_{\min}$ and $x_{\max}$. The true period is expected to be in the range between some $p_{\min}$ and $p_{\max}$. The test periods $p_0 < p_1 < p_2 < \ldots < p_M$ should cover this range, i.e., $p_0 \leq p_{\min}$ and $p_M \geq p_{\max}$. Let $K_0 = \lceil x_{\max}/p_{\min} \rceil$, and let $p_0 = x_{\max}/K_0$. The neighboring periods p_m and p_{m+1} must be close enough: with the same starting point, they need to be distinguishable at the end of the data series $x_{\max}$. This means that $p_m \lceil x_{\max}/p_m \rceil$ – the multiple of p_m which is near the end of the data, must differ from the multiple of p_{m+1} by a fraction of p_m not exceeding $1/2$; one can write this condition as

$$\frac{x_{\max}}{p_m} p_{m+1} - \frac{x_{\max}}{p_m} p_m = q\, p_m \,,$$

where the fraction coefficient q satisfies the inequalities $0 < q < 1/2$. The above formula can be easily dealt with if the right hand side $q\, p_m$ is approximated by $q\, p_{m+1}$. With this approximation, the test periods are

$$p_m = \frac{p_0}{1 - m\, q\, p_0/x_{\max}} \,. \tag{5.15}$$

The two largest test periods p_{M-1} and p_M satisfy $p_{M-1} < p_{\max} \leq p_M$, and one has $p_M = p_0/(1 - Mqp_0/x_{\max})$. Hence, the largest index of test periods M is determined by

$$M = (1/p_0 - 1/p_M)\ x_{\max}/q = (K_0 - K_M)/q\ , \tag{5.16}$$

where $K_M = x_{\max}/p_M = \lceil x_{\max}/p_{\max} \rceil$.

Thus, assuming one has the limits on the data $x_{\min}$ and $x_{\max}$, the limits $p_{\min}$ and $p_{\max}$ on expected period, and the coefficient q, then the test periods p_m are given by (5.15), and the maximal value M of the index m is given by (5.16). It is easy to see that the above distribution of test periods is directly linked to equispaced test frequencies $\nu_m = 1/p_m$, (as $\nu_m - \nu_{m+1} = q/x_{\max}$) with the lowest frequency $\nu_M = K_M/x_{\max}$ and the highest frequency $\nu_0 = K_0/x_{\max}$.

5.6.3 Period Determination Without Binning the Data

The distribution of events x_n $(n = 1, \ldots, N)$ is usually represented by binning them: the range covered by the products can be divided into N_b equally spaced intervals Δx or bins enumerated by $b = 0, 1..., N_b - 1$, and the number of products x_n falling into the b-th bin will be denoted by $f[x_b]$. Thus, $f[x_b]$ represents the occurrence of the observations x_n in the b-th bin. Formally,

$$f[x_b] = \sum_{n=1}^{N} \Delta_{bn}\ , \tag{5.17}$$

where Δ_{bn} equals 1 if x_n is in the b-th bin and 0 otherwise. One may think of $f[x_b]$ as an approximation of a distribution with the density $f = f(x)$. The density of events x_n can be represented in the form

$$f(x) = \sum_{n=1}^{N} \delta(x - x_n) \tag{5.18}$$

corresponding to infinitely narrow bins. This form is convenient for explaining binning-free methods, which are more suitable for the type of signal arising in indexing.

The indexing data series are of relatively short length. Such 'transient signals' from diffraction experiments are not really periodic as clusters of observations x_n may have very different cardinalities; the distributions of events x_n can be rather described as partially periodic in the sense that it has clusters of observations at nearly periodic locations but cluster 'intensities' vary. The indexing data are affected by noise of unknown distribution, some systematic errors (e.g., shift of real-space origin) and have outliers (due to spurious reflections).

It is worth noting, that some approaches to period determination require the binned 'signal' $f[x_b]$ of (5.17). For instance, this characterizes methods using circular autocorrelation $\int f(x)\,f(x-p)\,dx \approx \sum_b f[x_b]f[x_b - p]$. In indexing, however, with relatively narrow support of f, binning the data should be avoided if possible, because it blurs the data by figures of the order of the bin size. Therefore, only the binning-free approaches to period determination are considered below. The basic idea is explained on the technique known as folding.

5.6.4 *Folding*

The simplest way to check whether the data points x_n ($n = 1, ..., N$) are clustered at equal intervals p ($= 1/\nu$) is to fold them into the period. With x_n being at a node of a one-dimensional lattice of period p, the quantity $x_n/p = x_n\nu$ would be an integer. To determine the right period in the presence of errors, one needs to quantify the total deviation of all reals x_n/p from nearest integers for p in the range from some $p_{\min}$ to some $p_{\max}$. This can be formally described using a triangle wave.

Let tri be the triangle function with the support $(-1/2, 1/2)$: $\mathrm{tri}(x) = 1 - |2x|$ if $|x| < 1/2$ and $\mathrm{tri}(x) = 0$ otherwise. The doubled deviation of a real number x from the nearest integer is $1 - \psi(x)$, where

$$\psi(x) = 1 - \mathrm{tri}\,(x \bmod 1 - 1/2) = |1 - 2(x \bmod 1)|$$

is the triangle wave (Fig. 5.3). The doubled deviation of a test number $x_n\nu$ from the nearest integer is $1 - \psi(x_n\nu)$. The magnitude of the overall deviation of points x_n from lattice nodes can be quantified by the sum of the terms $\psi(x_n\nu)$. To find the right period, the experimental data need to be folded into multiple trial periods covering

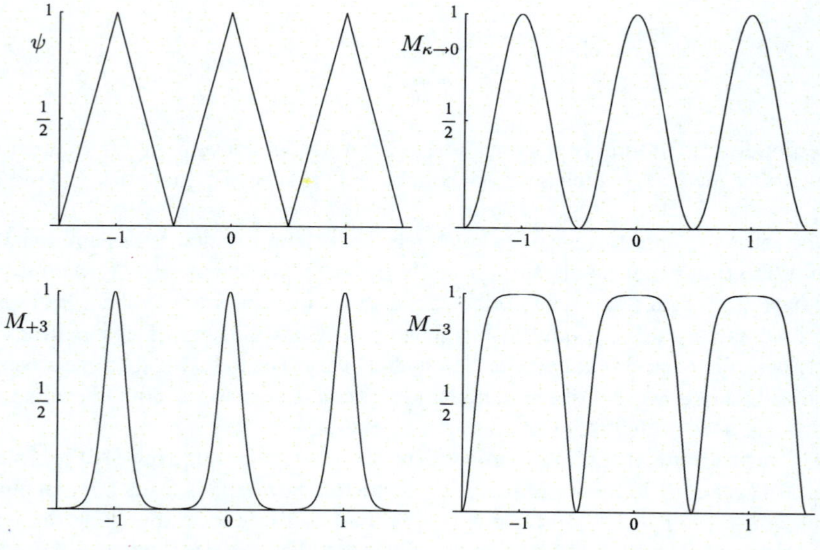

Fig. 5.3 The triangle wave function ψ and example M_κ functions.

the whole range of possible periods. Hence, one has two functions: the ν–dependent

$$S_{fd}(\nu) = \sum_{n=1}^{N} \psi(x_n \nu) , \tag{5.19}$$

and the p–dependent $P_{fd}(p) = S_{fd}(1/p)$. See, Fig. 5.4. Clearly, these functions are computed without binning the data. The maxima of S_{fd} and P_{fd} are at arguments corresponding to the period of the data series, i.e., one obtains the sought-after period by locating the maximum of S_{fd} or P_{fd}.

With f given by the distribution $f(x) = \sum_n \delta(x - x_n)$ (cf. (5.18)), one has

$$\int_{-\infty}^{\infty} f(x)\, \psi(x\nu)\, dx = S_{fd}(\nu) . \tag{5.20}$$

The folding expressed in the integral form can be seen as correlating the signal f with the triangle wave ψ.

5.6.5 Correlations with Other Functions

Other periodic functions can be correlated with the signal f using the same mechanism as in (5.20). These functions should be possibly close models of the investigated signal. In the considered case, instead of ψ, one can use the more flexible von Mises model $m_\kappa(x) = \exp(\kappa \cos(2\pi x))$ with an additional parameter κ. In *Ind_X* described below, the more convenient function is used; it is defined as

$$M_\kappa(x) = \frac{m_\kappa(x) - m_\kappa(1/2)}{m_\kappa(0) - m_\kappa(1/2)}$$

and satisfies $M_\kappa(x) = 1$ and $M_\kappa(x + 1/2) = 0$ for integer x (Fig. 5.3). As in the case of (5.20), the periodogram is given by

$$S_\kappa(\nu) = \int_{-\infty}^{\infty} f(x)\, M_\kappa(x\nu)\, dx = \sum_{n=1}^{N} M_\kappa(x_n \nu) ; \tag{5.21}$$

and the corresponding period-dependent function is $P_\kappa(p) = S_\kappa(1/p) = \sum_{n=1}^{N} M_\kappa(x_n/p)$.

By a suitable choice of the value of κ, one can adjust M_κ to the data. With $\kappa > 0$ ($\kappa < 0$), the peaks of M_κ are broad at the bottom (the top) and more suitable for detecting period in good (poor) quality data. The functions S_κ encompass the cosine transform which is defined below via (5.24); for small $|\kappa|$, the function $M_\kappa(x)$ tends to $\cos^2(\pi x) = (\cos(2\pi x) + 1)/2$, and comparison of (5.21) and (5.24) gives $S_J(\nu) + N = 2\, S_{\kappa \to 0}(\nu)$.

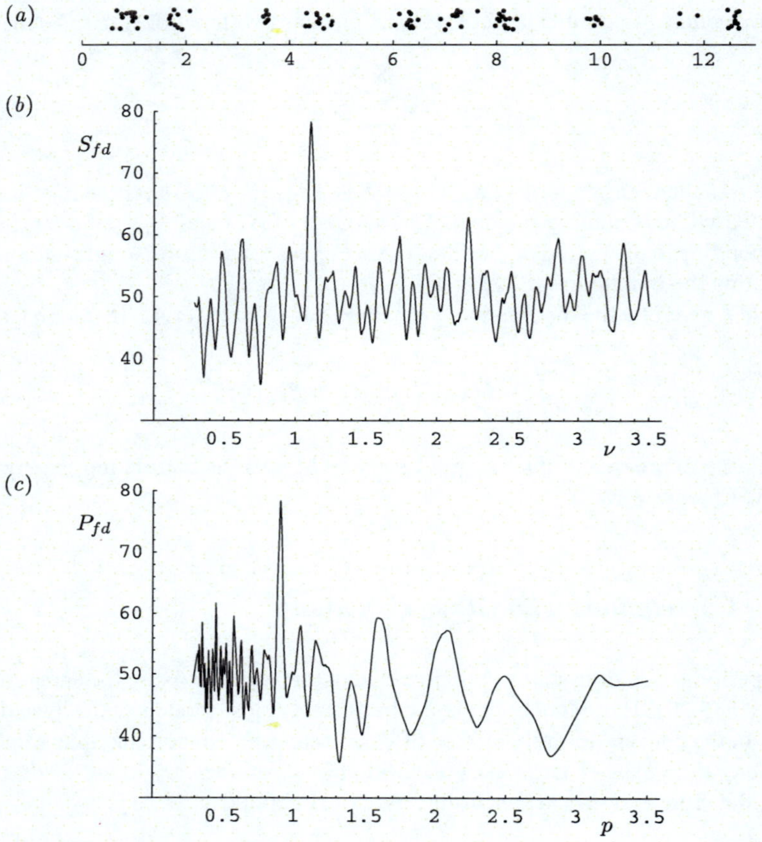

Fig. 5.4 Example data series and a periodogram. The data are 100 x-coordinates of points shown in (*a*). They are normally distributed (with $\sigma = p_0/5$) around $\{1, 2, 4, 5, 7, 8, 9, 11, 13, 14\}\, p_0$, where $p_0 = 9/10$. The uniform distribution in vertical direction is used only to reduce overlapping. (*b*) The folding-based periodogram $S_{fd}(\nu)$. (*c*) The function $P_{fd}(p)$.

5.6.6 One-Dimensional Fourier Transformation

There are a number of implementations of the Fourier transformation to ab initio indexing [19, 20, 26, 29, 41, 42]. The Fourier transformation $F(\nu) = \int_{-\infty}^{\infty} f(x) \exp(-2\pi i x \nu)\, \mathrm{d}x$ of the distribution $f(x) = \sum_n \delta(x - x_n)$ (cf. (5.18)) gives

$$F(\nu) = \sum_{n=1}^{N} \exp(-2\pi i x_n \nu) = N(\overline{c} - i\overline{s}) \,, \tag{5.22}$$

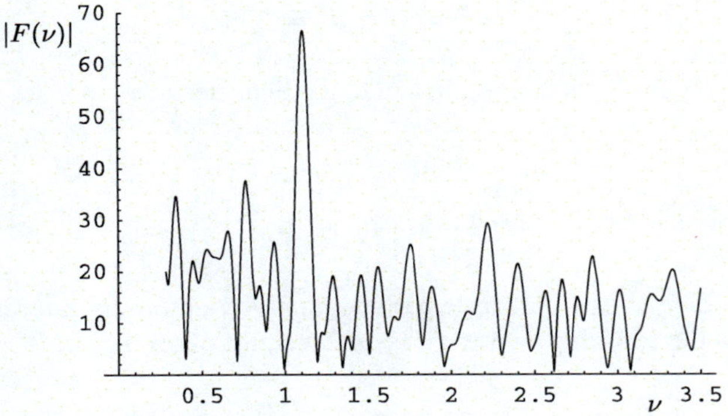

Fig. 5.5 The function $|F(\nu)| = \sqrt{S_S(\nu)}$ for the example data series of Fig. 5.4*a*.

where $\overline{c} = \sum_{n=1}^{N} \cos(2\pi x_n \nu)/N$ and $\overline{s} = \sum_{n=1}^{N} \sin(2\pi x_n \nu)/N$. Moreover, one has the frequency-dependent

$$S_S(\nu) = |F(\nu)|^2 = N^2 \left(\overline{c}^2 + \overline{s}^2\right) , \tag{5.23}$$

and the corresponding classic Schuster periodogram $S_S(1/p)$ given in the domain of periods (Fig. 5.5).

The angle (or mean direction) $\tan^{-1}(\overline{s}/\overline{c})$ is a measure of the shift of the origin of the data (zeroth node of the one-dimensional lattice) from 0. If there is no systematic shift of indexing data, for the right period $p = 1/\nu$, the products $x_n\nu$ are clustered near integers, and the term $\overline{s}$ in (5.23) should be zero. With the 'signal' function expected to be even, the Fourier transform (5.22) can be simplified to the cosine transform

$$S_J(\nu) = \sum_n \cos(2\pi\, x_n \nu) \tag{5.24}$$

proposed as a means of indexing by Jacobson [41].

5.6.7 *Rayleigh Test*

The clustering at equal intervals p can be exposed by wounding the data on a circle of radius $p/(2\pi)$. The Rayleigh test is a statistical procedure for determining whether a circular distribution is random or not [51]. Using the means $(\overline{s}, \overline{c}) = \sum_n (\sin(2\pi x_n \nu), \cos(2\pi x_n \nu))/N$, one needs to calculate the (Rayleigh) critical value $Z = N(\overline{s}^2 + \overline{c}^2)$. For a uniform distribution and large N, the probability that Z exceeds the value of β equals $e^{-\beta}$. With the above $Z = Z(\nu)$ and (5.23), one has

$$N\, Z(\nu) = |F(\nu)|^2 = S_S(\nu)\ ,$$

i.e., the critical value Z is directly linked to the Schuster periodogram.

5.6.8 Lomb-Scargle Periodogram

Worth mentioning is Lomb-Scargle periodogram accounting for unevenly sampled data series. In the considered case with equal weights of each x_n, it is given by

$$S_{LS}(\nu) = \frac{1}{2}\left(\frac{\left(\sum_n \cos(2\pi\nu(x_n - \tau))\right)^2}{\sum_n \cos^2(2\pi\nu(x_n - \tau))} + \frac{\left(\sum_n \sin(2\pi\nu(x_n - \tau))\right)^2}{\sum_n \sin^2(2\pi\nu(x_n - \tau))}\ ,\right)$$

where $\tau = \tau(\nu)$ is defined by

$$\tan(4\pi\nu\tau) = \frac{\sum_n \sin(4\pi\nu x_n)}{\sum_n \cos(4\pi\nu x_n)}\ .$$

Lomb-Scargle periodogram for the artificial data of Fig. 5.4*a* is shown in Fig. 5.6. Example Lomb-Scargle periodograms for $\boldsymbol{t}_{(2)}$-based reciprocal-space volumes x_n (cf. Sect. 5.5.2) obtained from some diffraction data are shown in Fig. 5.7; in this case, ν represents the volume in direct space.

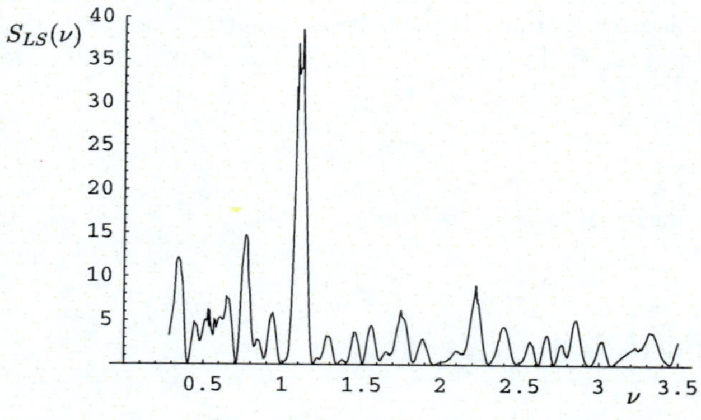

Fig. 5.6 The Lomb-Scargle periodogram $S_{LS}(\nu)$ for the example data series of Fig. 5.4*a*.

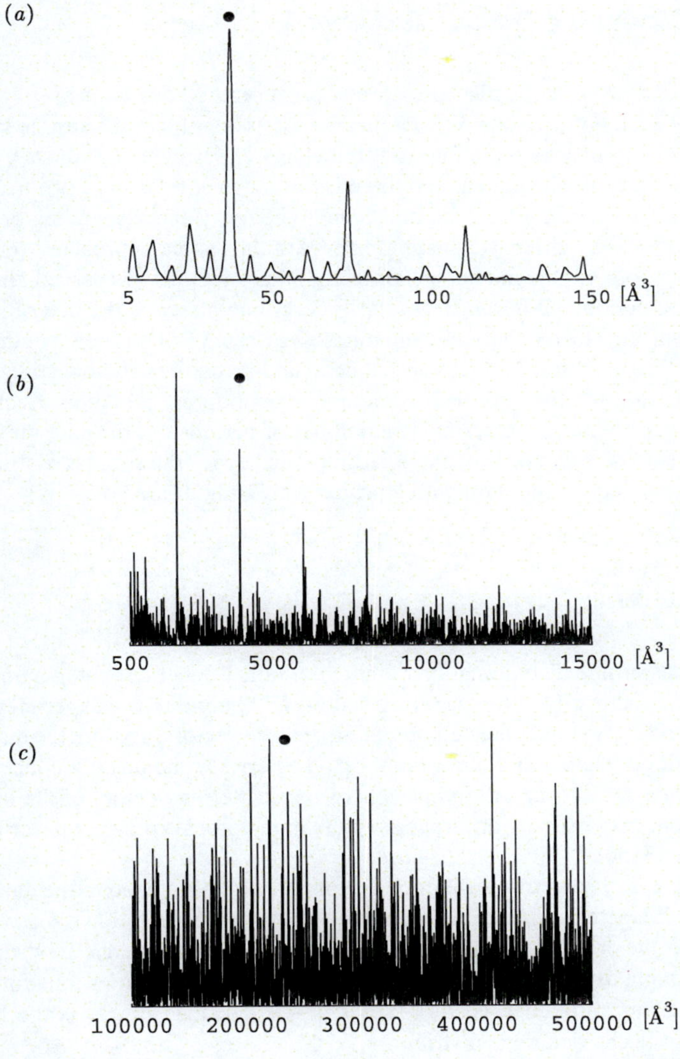

Fig. 5.7 Example Lomb-Scargle periodograms in arbitrary units versus volume in direct space. Plots are shown in ranges which one may select based on the type of the crystal or density of peaks. Disks mark volumes of parallelepipeds spanned on primitive bases of best lattices. (*a*) Periodogram obtained from sixteen $\boldsymbol{h}^n$ vectors corresponding to conics in a Kossel diffraction pattern of martensite phase of Cu-Al-Be alloy [52]. The smallest primitive cell volume allowing for indexing the pattern was about 3.7×10^1 Å^3. (*b*) Periodogram for 30 reflections from a simulated diffraction data set of a crystal with the cell of medium size 4.0×10^3 Å^3. (*c*) Periodogram based on 26 reflections from 'Example 2' of [27] representing a small protein with cell volume of about 2.3×10^5 Å^3.

5.6.9 Combining Various Techniques

The performance of particular period determination methods depends partly on the type of the data. For accurate diffraction data, periodograms have simple structures, and it is easy to determine the period. For many experimental data sets, however, the 'signals' are not sinusoidal, and there is not much similarity of profile shape; in effect, the corresponding periodograms tend to be noisy and, depending on the data, they may have very different constitutions. With this, some approaches may be less efficient than other. One may consider using other methods of spectral analysis for determining periods. Different from those described above is the determination of periodicity using cluster analysis techniques; see Sect. 5.9. The choice of the method of period determination depends on the data, and it requires some testing. Another approach is to combine various techniques of period determination. Each applied technique may 'vote' for a period or a number of periods. At the end, the solutions with the most votes will be processed further. The key for the success of this strategy is a suitable objective function with a proper weighting of the votes.

5.7 Difference Vectors

Since an integer linear combination of lattice vectors is a lattice vector, a finite initial list of lattice vectors (i.e., the scattering vectors $\boldsymbol{h}^n$ obtained from experimental data) can be extended by some linear combination of these vectors. To limit the magnitudes of the resulting vectors and the accumulation of errors, one usually considers only the *difference vectors*, i.e., differences between scattering vectors of the initial list. Implementations involving difference vectors in reciprocal space are described, e.g., in [26, 30, 53–56].

It can be easily seen that difference vectors are directly linked to the autocorrelation of the characteristic distribution of the vectors on the initial list. Consequently, vectors near the origin (i.e., short scattering vectors) are expected to be strongly represented among difference vectors. Moreover, the result of Fourier transform (5.25) calculated using difference vectors (i.e., $\boldsymbol{h}^{n_1} - \boldsymbol{h}^{n_2}$ instead of $\boldsymbol{h}^n$) is the Patterson function of the characteristic distribution of the vectors $\boldsymbol{h}^n$; cf. Sect. 2.6.

As was noted above, combining vectors leads to the accumulation of random errors. E.g., if $\boldsymbol{h}^{n_i}$ deviates from a lattice node by ϵ^{n_i}, then $\boldsymbol{h}^{n_1} + \boldsymbol{h}^{n_2}$ will deviate from a lattice node by $\epsilon^{n_1} + \epsilon^{n_2}$. Moreover, in the presence of spurious reflections, including differences is a risky process as it increases the fraction of incorrect data.

5.8 Indexing via Three-Dimensional Fourier Transformation

Fourier transformation can be applied to lattices: the Fourier transform of the Dirac comb—function characterizing an arbitrary lattice—is equal to the Dirac comb characterizing its reciprocal; cf. (1.87). The experimental data ($\boldsymbol{h}^n$ vectors) carry only fragmentary and approximate information about the reciprocal lattice, but—by the nature of Fourier transformation—the transform of the function characterizing the data is expected to expose its periodic components. In Sect. 5.6.6, Fourier transformation was used for indexing fragmentary and approximate data from a one-dimensional lattice. A similar approach can be applied to three-dimensional data.

The inverse Fourier transform of $Ш(\boldsymbol{h}) = \sum_n \delta(\boldsymbol{h} - \boldsymbol{h}^n)$ characterizing the set of experimental vectors $\boldsymbol{h}^n$ equals

$$(\mathcal{F}^{-1}Ш)(\boldsymbol{x}) = \sum_n \exp(2\pi i\, \boldsymbol{x} \cdot \boldsymbol{h}^n) = \overline{C} + i\overline{S}\,, \tag{5.25}$$

where $\overline{C} = \overline{C}(\boldsymbol{x}) = \sum_n \cos(2\pi \boldsymbol{x} \cdot \boldsymbol{h}^n)$ and $\overline{S} = \overline{S}(\boldsymbol{x}) = \sum_n \sin(2\pi \boldsymbol{x} \cdot \boldsymbol{h}^n)$. If the data were complete with $Ш(\boldsymbol{h}) = Ш_{\Lambda^\star}(\boldsymbol{h})$ the result of (5.25) would be $Ш_\Lambda$ characterizing the direct lattice Λ. With incomplete data, $\mathcal{F}^{-1}Ш$ is a smooth function of $\boldsymbol{x}$ in the direct space, and the spectral density $\left|\mathcal{F}^{-1}Ш\right|^2 = \overline{C}^2 + \overline{S}^2$ has peaks at direct-lattice nodes, among others.

The above considerations are schematically illustrated in Fig. 5.8. The reciprocal lattice shown in Fig. 5.8*b* is based on vectors with Cartesian components $(\pi, 0)$ and $(e, 1/3)$. Fifteen randomly selected vectors of the reciprocal lattice have coordinates in the aforementioned basis in the range from -2 to 4. The spectral density functions were calculated at 150^2 points. Figs. 5.8*e* and *f* need to be compared to Fig. 5.8*a*.

The highest peaks of $\left|\mathcal{F}^{-1}Ш\right|^2$ are expected to be at nodes of the direct lattice. One needs to identify the peaks and determine basis vectors of the direct lattice from peak positions. As Fig. 5.9 shows, this task may be difficult for error-affected data. The Fourier transform can be used to obtain the direct lattice vectors $\boldsymbol{t}$ for the real-space indexing [20].

Similarly to period determination without binning (Sect. 5.6.3), the expression (5.25) allows for calculating the Fourier transform without referring to any grid in the reciprocal space. If, however, the transform is calculated based on a three-dimensional grid, the latter must be sufficiently fine to contain rounding errors in assigning the $\boldsymbol{h}^n$ vectors to the nearest grid points. Indexing based on three-dimensional fast Fourier transform in relatively fine (512^3 and 256^3-point) grids was considered in [57]. The conclusions were that the calculation was affected by sparsity and asymmetry of the sampling of the reciprocal lattice, and the computational cost (governed by the fineness of the grid used by the digital Fourier transform) was relatively high [57]. Other implementations based on three-dimensional fast Fourier transform were described in [39, 58].

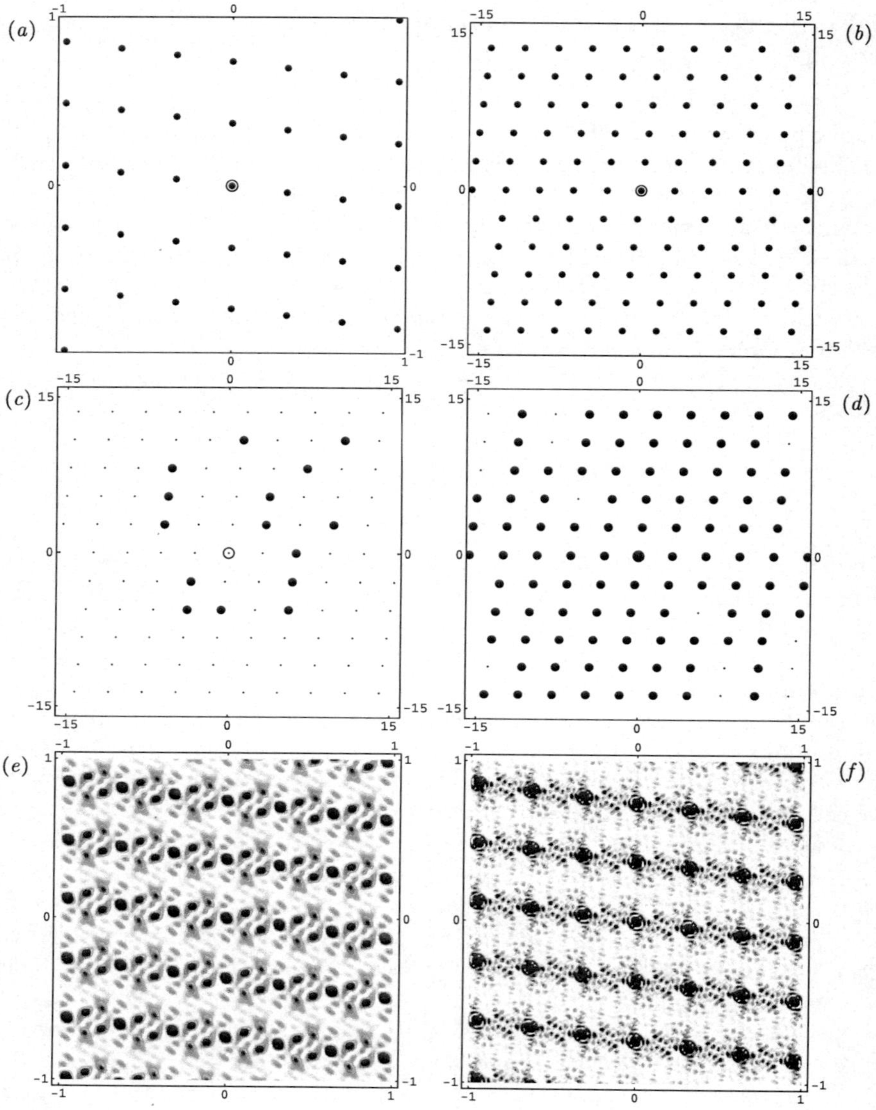

Fig. 5.8 Schematic two-dimensional illustration of Fourier-based indexing. Direct lattice and its reciprocal are shown in (*a*) and (*b*), respectively. Figure (*c*) shows selected vectors of the reciprocal lattice. The difference vectors for the set of (*c*) falling in the same plot-range are shown in (*d*). The spectral density functions obtained from the vectors of (*c*) and difference vectors of (*d*) are in (*e*) and (*f*), respectively.

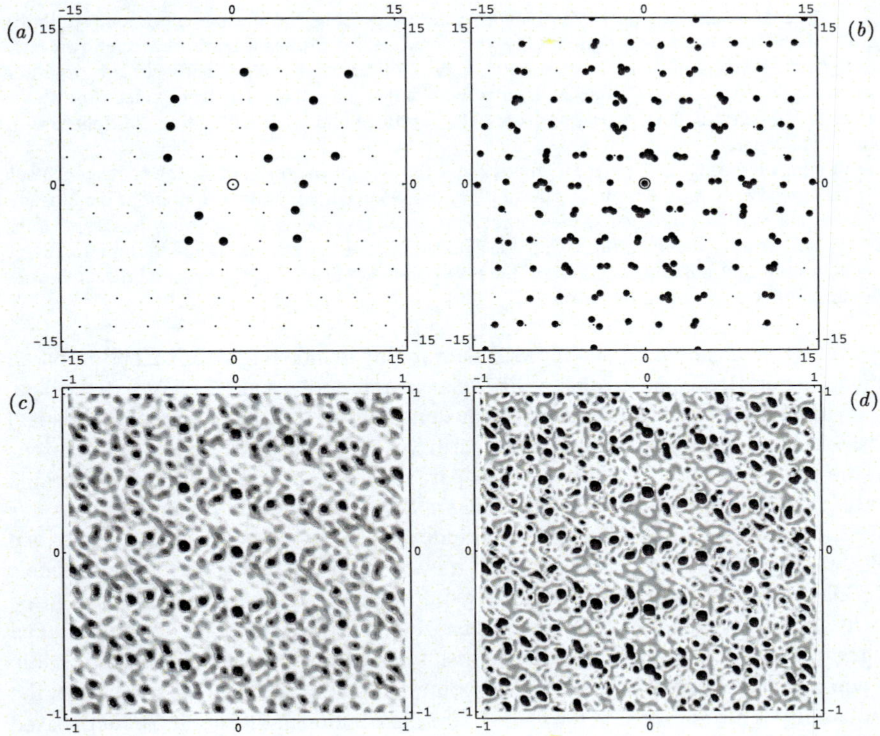

Fig. 5.9 Schematic two-dimensional illustration of the effect of errors on Fourier-based indexing. The reciprocal lattice vectors shown in (*c*) of Fig. 5.8 were modified by adding uniformly distributed errors in the range $(-1/3, 1/3)$ to their Cartesian components. The error affected vectors are shown in (*a*) and non-overlapping difference vectors falling in the same plot-range are in (*b*). The spectral density functions obtained from the error affected vectors of (*a*) and the error affected difference vectors of (*b*) are in (*c*) and (*d*), respectively.

5.9 Clustering in Reciprocal Space

If sufficiently many difference vectors are computed, they are expected to be clustered around reciprocal lattice nodes. Spatial clustering techniques can be used to detect and rank the clusters, and to calculate individual vectors (e.g. mean vectors) representing particular clusters. High rank mean vectors of small magnitude can be used to determine the actual basis of the reciprocal lattice.

Clustering is a general process of grouping data set based on a distance function. In the case of data used for indexing, this is the Euclidean distance between terminal points of vectors. There are various types of clustering algorithms. The best known are hierarchical methods and partitioning methods.

⊡ A hierarchical method provides a dendrogram representing the clusters at various similarity levels. At the root of the dendrogram is the whole data set constituting a single cluster, and the leafs are individual data points—each in separate cluster. Clusters are build gradually either by dividing or by merging. All points of a cluster are members of the 'parent' cluster closer to the root. To get the resulting clusters, the dendrogram is cut at a certain similarity level. Setting the right level is often a challenge.
Partitioning methods (like k-means method) divide the data into a pre-set number k of clusters. Each cluster is represented by a central point (e.g., by the centroid, i.e., the mean point of the cluster), and the data points are assigned to the central point based on a function measuring the 'error' of the data point; the most common function is the squared Euclidean distance. The algorithm iteratively calculates the central points and assigns the data points to clusters. In the process, it searches for the minimum of the sum of the 'errors'.

However, better suited for dealing with data arising in indexing are distribution-based clustering or density-based clustering. They are more robust against noise in the data.

With distribution-based clustering, clusters are modeled using assumed statistical distributions. All clusters of the data set are expected to be described by distributions of the same type differing only by their parameters. Thus, the method is applicable only if the type of the distribution is known. Most often the Gaussian distribution is used. The data point clusters arising in indexing are expected to be well described by the spherically symmetric Gaussian model. With the Gaussian mixture models (or GMM), a fixed number of randomly initialized Gaussian distributions is fitted to the data. The computation scheme can be similar to k-means algorithm. The algorithm iteratively repeats two steps. First, for each data point and each distribution, it computes the probability that a data point belongs to a cluster (distribution); the probability is proportional to the value of the distribution at that point. Second, based on the probabilities it fits the parameters of the distributions. The method is fuzzy in the sense that it provides the probability of belonging to a given cluster.

Density-based clustering, as the name suggests, it is based on a search for areas with a high density of data points. A flag example of a density-based algorithm is DBSCAN (density based spatial clustering of applications with noise) [59]. Given a density threshold *MinPts* and a distance threshold ε, a data point is considered a *core* point if the number of data points within the radius ε around it is larger than *MinPts*. Two data points are connected if they are separated by a distance smaller than ε. The algorithm assigns connected core points to a single cluster. A non-core data point is assigned to a cluster (as a border point) if it is connected to a core point; otherwise, it is assigned to noise. A border point may belong to more than one cluster. The performance of DBSCAN depends on the parameters ε and *MinPts*, and choosing the right values may be a challenge. The algorithm discovers clusters of arbitrary (e.g., very elongated) shapes, which in the case of data used in indexing is a disadvantage because distributions of points in these clusters are expected to be nearly spherically symmetric. Finally, it poorly tolerates large differences in densities of data points (and, in indexing, local densities of clusters can be very different). The last issue is addressed by the modification of DBSCAN called OPTICS (ordering points to identify the clustering structure) [60]. See [61] for a review of these and related methods and [56] for an application of DBSCAN to the indexing of electron diffraction patterns.

An indexing method may rely solely on clustering of reciprocal lattice vectors, thus being a pure 'reciprocal space' indexing method [56]. Some aspects of clustering have also been used in methods referring to the direct lattice. In the algorithm proposed by Kabsch [62], clusters of difference vectors are calculated, and then triplets of reciprocal lattice vectors are checked for how well they explain the difference vectors. Such a triplet is tentatively assumed to be a basis of the reciprocal lattice. It is used to calculate the basis of the direct lattice. The triplet explains a difference vector if the scalar products of the latter with vectors of the direct lattice basis are nearly integers. The algorithm searches for a triplet which explains the largest number of difference vectors, and this triplet is taken as the true basis of the reciprocal lattice.

5.10 Directions of Zone Axes from Difference Vectors

Difference vectors are convenient for identifying zones of reflections and for determining directions of zone axes. Gnomonic projections of scattering vectors and difference vectors corresponding to a set of tautozonal planes are located along a straight line. With a sufficiently large number of such vectors, the projection clearly reveals the geometry of the lattice. This is illustrated using an artificial model of 150 randomly generated scattering vectors. Gnomonic projections of the 'scattering' vectors are shown in Fig. 5.10*a*. This figure is not very revealing, but the one with projections of difference vectors (Fig. 5.10*b*) allows for an easy identification of the main zones.

Such a clear picture is obtained only if the number of the scattering vectors is sufficiently large. The impact of that number is illustrated in Fig. 5.11*a* showing projections of the difference vectors obtained from a small (30-element) subset of the original set of the scattering vectors. This figure should be compared with Fig. 5.10*b* as both show first generation difference vectors. With a small number of scattering vectors, zones are difficult to find.

For accurate data, difference vectors of second generation, i.e., differences between difference vectors, are also lattice vectors. Gnomonic projections of the second generation difference vectors based on the 30-element set scattering vectors are shown in Fig. 5.11*b*. Due to the large increase in the number of points, the zones became easily identifiable.

It must be stressed again that the use of difference vectors makes sense only with sufficiently accurate input data. Differences of low accuracy scattering vectors accumulate errors and the projections become blurry. Visually, the projections can be quite robust against spurious scattering vectors.

To illustrate the real cases, projections of the first and second generation difference vectors for Duisenberg's data referred to as 'Example 2' in [27] are shown in Fig. 5.12. The 'visibility' of zones on gnomonic projections of difference vectors may depend

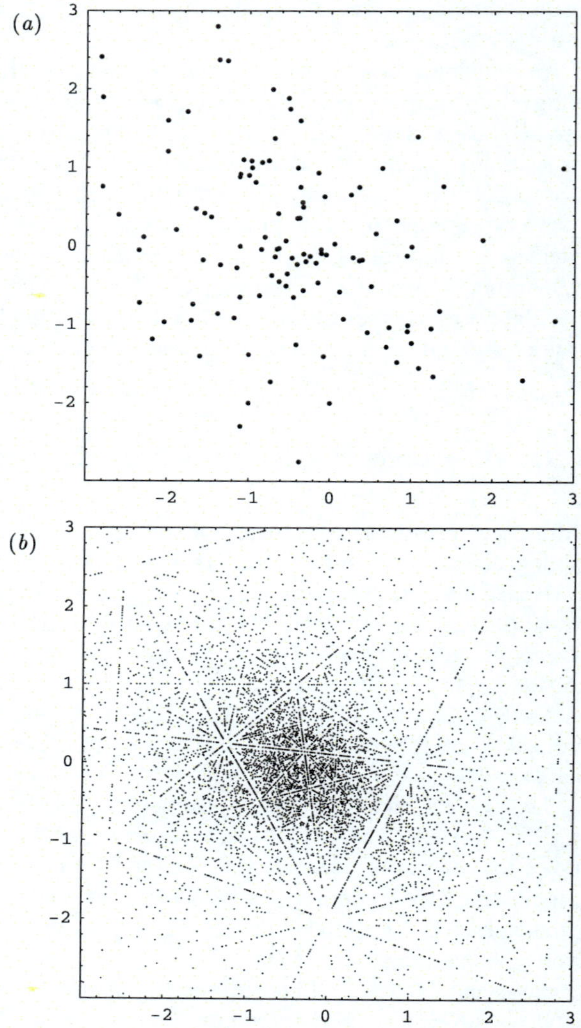

Fig. 5.10 (*a*) Gnomonic projections of scattering vectors. (*b*) Gnomonic projections of (first generation) difference vectors. To make these figures, 150 scattering vectors were generated as random vectors of a triclinic reciprocal lattice with the primitive cell of volume 1. The basis of the lattice was Niggli-reduced and the absolute values of vector coordinates did not exceed 12. The distance between the projection center and the projection plane was 1. Here and in Figs. 5.11, 5.12 and 5.13, the vectors falling out of the shown domains are discarded.

on the projection direction. The projections shown in Fig. 5.13 differ from those shown in Fig. 5.12 by the projection direction.

The determination of a zone axis direction based on the gnomonic projection of difference vectors comes down to determining parameters of a straight line on the projection. The zone axis is perpendicular to the plane through the projection center and containing the line, i.e., with $\boldsymbol{y}_1$ and $\boldsymbol{y}_2$ denoting vectors to distinct points on the line, the zone axis direction is along $\boldsymbol{y}_1 \times \boldsymbol{y}_2$. For more on determination of directions of zone axes from gnomonic projections of reciprocal lattices, see Sects. 6.4.3 and 6.4.4.

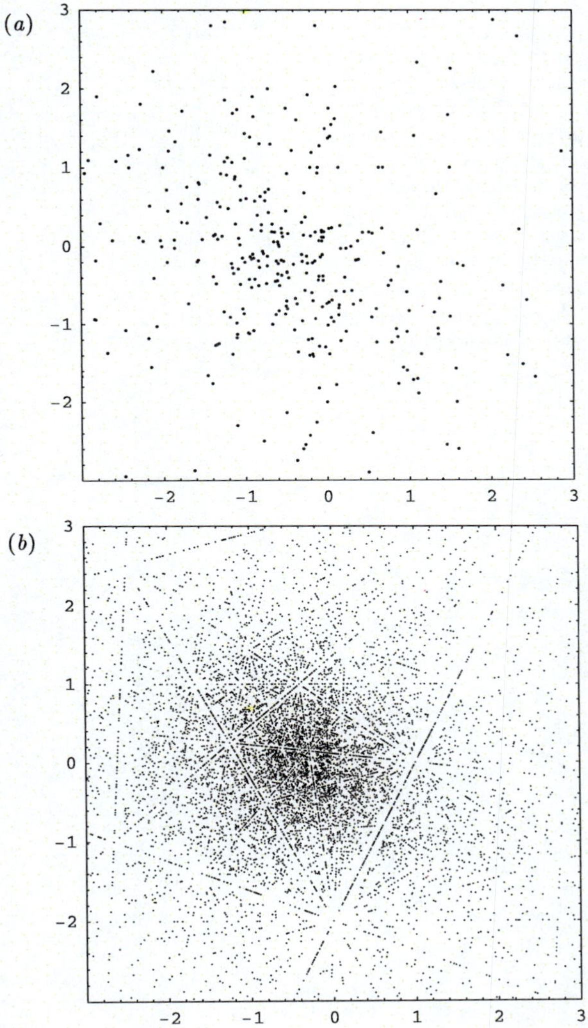

Fig. 5.11 Difference vectors for a small model of 30 scattering vectors. (*a*) Gnomonic projections of first generation difference vectors (to be compared with Fig. 5.10*b*). (*b*) Gnomonic projections of second generation difference vectors.

5.11 Constructing a Three-Dimensional Lattice

A number of important aspects of indexing are worth stressing.

1. The most essential element of indexing procedures is the deduction of the basis of the actual lattice from some lattice vectors. A triplet of independent lattice vectors, constitutes a basis of a sublattice. Hence, given a triplet of independent reciprocal lattice vectors $\boldsymbol{h}^n$, one can determine the sought lattice basis by looking at bases of the superlattices of the lattices spanned by the triplet. Since the number of the

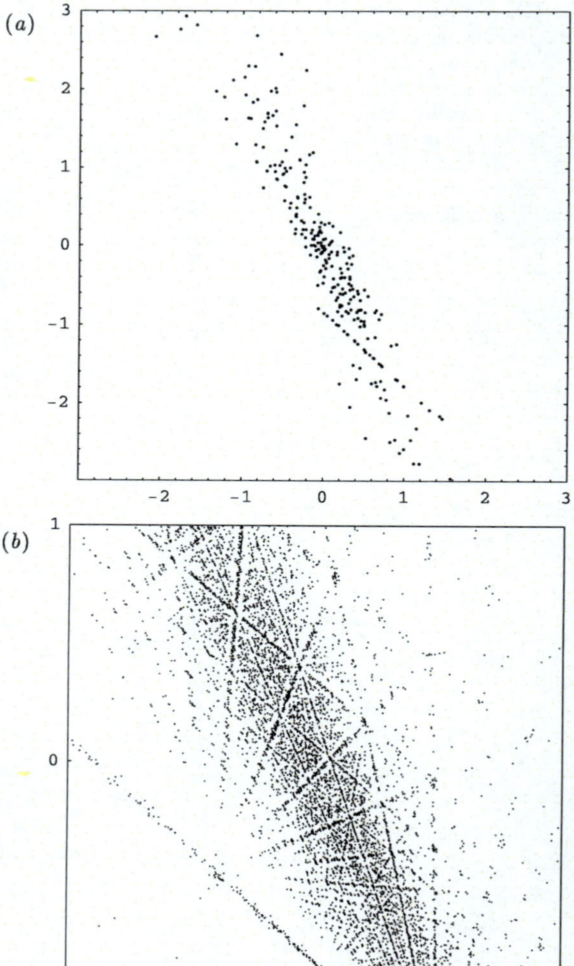

Fig. 5.12 'Example 2' of [27], i.e., X-ray data from a small protein with unit cell volume of about 2.3×10^5 Å^3; cf. Fig. 5.7. (*a*) Gnomonic projections of difference vectors. (*b*) Gnomonic projections of second generation difference vectors. The distance between the projection center and the projection plane was 1 Å^{-1}. Projection plane was perpendicular to $\boldsymbol{e}_L^3$.

superlattices grows fast with their index, in practice, only low-index superlattices can be taken into account. The scheme is illustrated two-dimensionally in Fig. 5.14. For the method of generating bases of superlattices, see Sect. 1.2.4.

2. Since some reflections may be illegitimate, one needs to work with subsets of the original set of experimentally determined vectors $\boldsymbol{h}^n$. At a given stage, the computation of tentative bases relies only on some supporting or cooperating vectors. How are these vectors selected? Generally, a quantitative measure of support is obtained from periodicity tests. Specific criteria depend on the used method of indexing. (For instance, with the method using test volumes of the primitive cell (cf. Sect. 5.12.1),

Fig. 5.13 Gnomonic projections of (first generation) difference vectors (*a*) and second generation difference vectors (*b*) for the same data as in Fig. 5.12, but here the projection plane was perpendicular to $\boldsymbol{e}_L^1$. As in the case of Fig. 5.12, the distance between the projection center and the projection plane is 1 Å^{-1}.

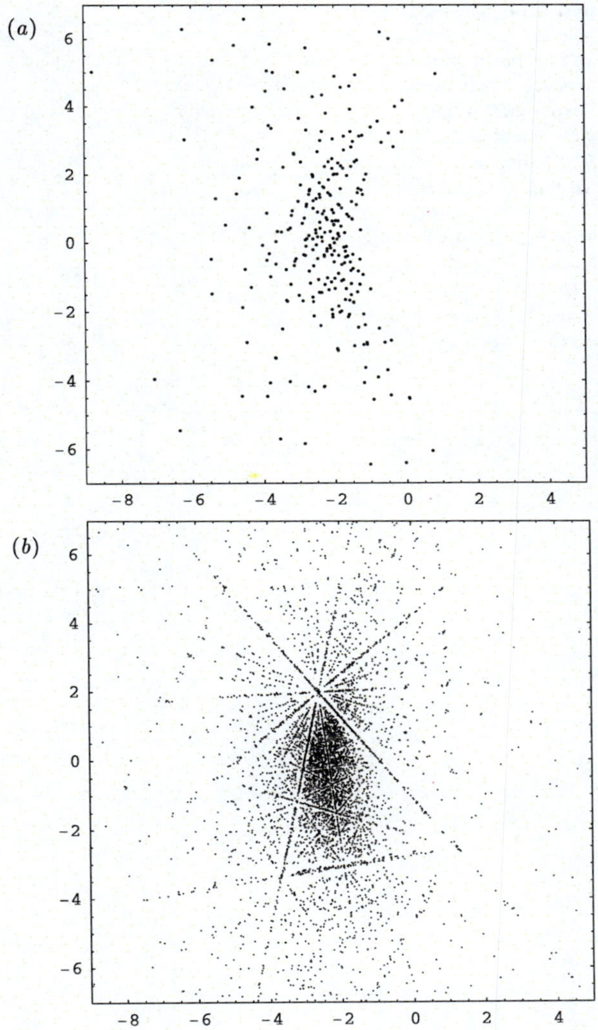

some vectors $\boldsymbol{h}^n$ will contribute to the quality of a given volume in the sense that volumes of parallelepipeds spanned by these vectors are close to integer multiples of the test volume. The amount of the contribution is a foundation for ranking the vectors, and for establishing the set of vectors $\boldsymbol{h}^n$ supporting the test volume. The same method can be applied for obtaining vectors $\boldsymbol{h}^n$ supporting a given test direct lattice vector.) The supporting vectors are processed further, e.g., they are used to construct tentative bases of the reciprocal lattice.

3. The construction of the indexing solutions includes reduction of the tentative bases. The reduction can be performed numerous times and at various stages of the

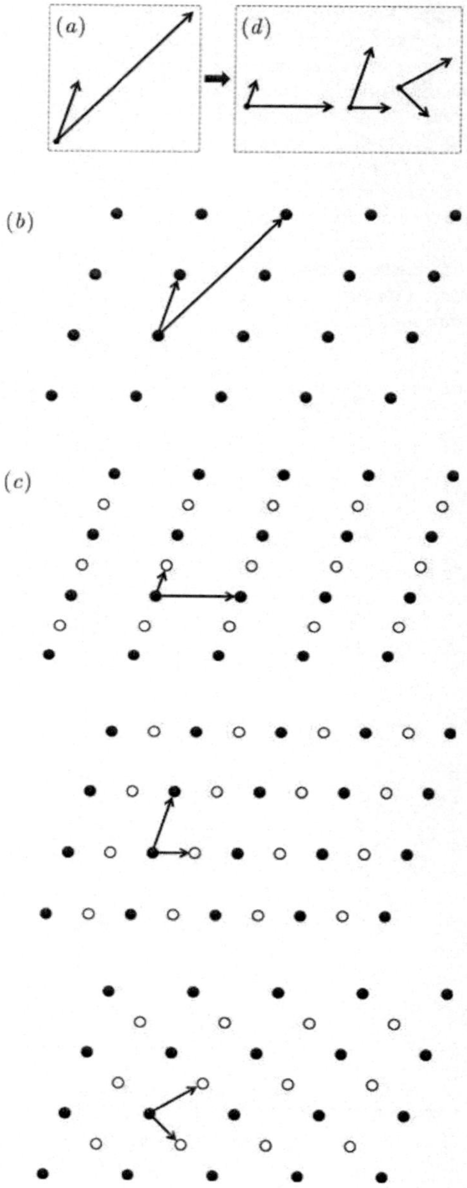

Fig. 5.14 Schematic two-dimensional illustration of the construction of tentative bases from two 'experimental' reciprocal lattice vectors drawn in (*a*). The lattice based on the two vectors is represented by disks in (*b*). Its three superlattices (disks + circles) of index 2 are shown in (*c*). The three bases of the superlattices are used as tentative bases (*d*).

indexing process. It must be also checked whether given solutions are really different. These two steps are described in Sects. 1.2.1 and 1.7, respectively.

4. Having a tentative basis in the form of $\boldsymbol{a}_i$ or equivalently X, indexing of all experimental reflections is attempted. This step needs to be combined with fitting integer combinations of basis vectors to the vectors $\boldsymbol{h}^n$ [40]. The quality of indexing can be improved by iterative indexing: the number of experimental vectors taken into account is increased, with the refinement of basis vectors after each increase. There can be different methods of selecting a vector in iteration steps. With large unit cells, indexing by assignment of $\lfloor \boldsymbol{h}^n \cdot \boldsymbol{a}_i \rceil = \lfloor X\boldsymbol{h}^n \rceil$ as reflection indices may be less reliable for long scattering vectors $\boldsymbol{h}^n$, and therefore, in each step, only the shortest of yet unindexed vectors is indexed, and the basis is refined. Alternatively, in each step, only the vector with the smallest deviation of $X\boldsymbol{h}^n$ from integers is indexed. One can also attempt local indexing, i.e., stepwise assignment of small index increments to scattering vectors close to already indexed vectors.

5. The fit of the experimental points $\boldsymbol{h}^n$ and the recalculated lattice can be improved by refinement of basic experimental parameters such as specimen-to-detector distance, pattern center (direct beam position, detector tilt) and crystal orientation. Again, this can be performed at various stages of indexing.

6. At the end, the indexing procedure provides the best solution or a list of high-quality solutions. Some additional information may also be provided. For instance, one may add *all* consistent ways of indexing within given limits and/or relationships between particular solutions. Moreover, the program may have procedures accounting for multiple crystals contributing to the pattern (in particular, twin crystals); see Sect. 10.2.

7. The final steps of the indexing process are the determination of the symmetry and Bravais lattice (as described in Sect. 1.7), and refinement of symmetry-constrained lattice parameters. One must be aware that the resulting lattice may exhibit pseudosymmetry; see Sect. 10.1.

5.12 An Example Indexing Program *Ind_X*

A number of aspects of the indexing procedure were considered above. They can be seen as building blocks of an indexing program. An (incomplete) list of these blocks includes one-dimensional period determination, reduction of lattice bases, linear optimization, clustering techniques applied to difference vectors, construction of sub- and superlattices, verification of similarity of lattices (bases), confinement to supporting (cooperating) reflections. They can be assembled into an indexing program. *Ind_X* is a program of this kind, i.e., an assembly of some of the aforementioned building blocks.

The program was written to extend capabilities of a software package for analysis of Kossel diffraction patterns [63]. The intended application was lattice parameter determination, and it concerned relatively simple crystal structures. However, additional tests on simulated diffraction data showed that the program is also applicable

Table 5.1 List of main steps of the indexing procedure implemented in *Ind_X*. Each new indent marks the beginning of a loop.

• *Ind_X* program
• Choice of test volumes
• Determination of vectors supporting a test volume
• Generation of triplets of vectors supporting the test volume
• Determination of tentative bases from a triplet
• Saving a tentative basis if its quality is high

to more complicated indexing problems. *Ind_X* can be compared to *DirAx* [27] which is considered to be most suitable for solving difficult cases. Tests indicate that *Ind_X* has a similar effectiveness in this regard. *Ind_X* is general in the sense that it is not intended for a particular diffraction experiment. It is relatively compact and easily operated, and can be conveniently used for solving small-cell problems in unconventional experiments. Finally, an important feature of *Ind_X* is the relative simplicity of the applied method.

5.12.1 Method

The procedure implemented in *Ind_X* relies on the volume of the primitive cell of the reciprocal lattice. In the ideal case, this elementary volume is a submultiple of volumes of parallelepipeds spanned by reciprocal lattice vectors, i.e., the latter volumes are distributed periodically; see Sect. 5.5.2. The crucial point is to find the period of the distribution obtained from experimental scattering vectors, i.e., to find the actual volume of the primitive cell. Test volumes can be determined using periodograms and techniques described in Sect. 5.6. The periodogram can be seen as the frequency of occurrence of particular periods among volumes of parallelepipeds spanned by triplets of the experimental scattering vectors. Alternatively, the test volumes can also be indicated by a user; they can be regularly distributed for a systematic search, or particular values can be pointed out based on an inspection of the periodogram or on other premises. The program actually checks a small number of volumes in the vicinity of each test volume.

For each considered volume, there is a subset of scattering vectors supporting this volume in the sense that volumes of parallelepipeds spanned by these vectors are close to integer multiples of the test volume. The program estimates the quality of support by an individual vector by checking triplets in which it is comprised, the vectors are ranked, and the best vectors constitute the subset. (Cf. Sect. 5.11.) The subsequent confinement to this subset eliminates potential spurious reflections and speeds up the program.

The supporting vectors are used to construct tentative solutions. The tentative bases of the reciprocal lattice are bases of low-index superlattices of lattices based on triplets of the supporting vectors (as described in Sect. 5.11). The construction of the solutions also includes two standard steps: fitting integer combinations of basis vectors to the supporting vectors [40] and Buerger reduction of the basis [64, 65]. Having a tentative basis, indexing of all experimental reflections is attempted. The result of the attempt is used to quantify the quality of the basis. The basis is saved if it is sufficiently good, i.e., better than other bases. These steps are repeated for all tentative bases constructed from a given triplet of supporting vectors, for all triplets of vectors supporting a given volume, and for all test volumes. The main steps are listed in Table 5.1. A more detailed scheme of the algorithm is given below.

Input: List of N reflections $\boldsymbol{h}^n$
Input: Optional parameters; e.g., test volumes of primitive cell, fraction of reflections supporting a test volume f, extent of search E
`/* Initialization */`
Set default values of parameters not specified in input
Other preliminary steps, e.g., get low-index superlattices, set $N_s = fN$
`/* Test volumes */`
if *test volumes not specified* **then**
    estimate test volumes of primitive cell of reciprocal lattice using periodogram
    save the volumes in *TableOfTestVolumes*
`/* Proper indexing */`
for *each volume V in TableOfTestVolumes* **do**
    for *each* V_i *from an E-dependent set* $\{V_i\}$ *of volumes around V* **do**
        get the set S of N_s trial reflections $\boldsymbol{h}^n$ supporting V_i
        for *each triplet* $(\boldsymbol{h}^{n_1}, \boldsymbol{h}^{n_2}, \boldsymbol{h}^{n_3})$ *of vectors from S* **do**
            for *tentative bases* $(\boldsymbol{a}^1, \boldsymbol{a}^2, \boldsymbol{a}^3)$ *of each low-index superlattice of the lattice based on the triplet* $(\boldsymbol{h}^{n_1}, \boldsymbol{h}^{n_2}, \boldsymbol{h}^{n_3})$ **do**
                `/* Sequential (iterative) fitting */`
                $L \leftarrow \{\boldsymbol{h}^{n_1}, \boldsymbol{h}^{n_2}, \boldsymbol{h}^{n_3}\}$
                while $S \setminus L \neq \emptyset$ **do**
                    get from $S \setminus L$ the vector $\boldsymbol{h}^{best}$ closest to a node of the lattice based on $(\boldsymbol{a}^1, \boldsymbol{a}^2, \boldsymbol{a}^3)$
                    append $\boldsymbol{h}^{best}$ to L
                    $(\boldsymbol{a}^1, \boldsymbol{a}^2, \boldsymbol{a}^3) \leftarrow$ basis $(\boldsymbol{a}^1, \boldsymbol{a}^2, \boldsymbol{a}^3)$ fitted to L
                $(\boldsymbol{a}_1, \boldsymbol{a}_2, \boldsymbol{a}_3) \leftarrow$ basis (of direct lattice) reciprocal to $(\boldsymbol{a}^1, \boldsymbol{a}^2, \boldsymbol{a}^3)$
                Buerger-reduce $(\boldsymbol{a}_1, \boldsymbol{a}_2, \boldsymbol{a}_3)$
                `/* By an attempt to index all N reflections */`
                determine *quality* of $(\boldsymbol{a}_1, \boldsymbol{a}_2, \boldsymbol{a}_3)$
                if *the quality is sufficiently good* **then**
                    store the basis $(\boldsymbol{a}_1, \boldsymbol{a}_2, \boldsymbol{a}_3)$ and its *quality*
Result: Stored (highest-quality) bases and indices of indexable reflections

Algorithm of *Ind_X*: This very simplified sketch contains all essential elements of the program.

As for the important parameter representing the quality of a basis, a number of different quantities have been tested to find a balance between effectiveness, simplicity and flexibility. The implemented approach is based on (5.12), a user-controlled limit on the allowed deviation ϵ of indices from integers, and user-controlled weights w_ϵ and $w_{\epsilon/2}$.

Ind_X can only process data sets containing a relatively small number of reflections N because the number of reflection triplets grows with N as N^3, and this has an impact on the execution time. Typically, N is expected to be about 20 to 50. With a small number of reflections and suitably chosen *Ind_X* parameters (see below), the execution time can be a fraction of a second. However, one must be aware that speed comes at a price of quality. For higher reliability of results, one can perform an extended search, but this can be a lengthy process.

⊡ *Software environment*
Ind_X is written in Fortran 90. The program uses a single input file with keywords followed by appropriate data. The file must contain a set of reflections specified by Cartesian coordinates of scattering vectors. There are also a number of optional input parameters, such as the above-mentioned limit on the allowed deviation of an index from an integer or the largest allowed absolute value of reflection indices. Besides the number of reflections, three optional parameters have an impact on the execution time. These are: (1) limits on the allowed volume of the primitive cell, (2) an upper limit on the number of supporting reflections, and (3) a single entity controlling (through a number of secondary parameters) the extent of the search for the solution. In particular, the latter entity determines the density of arguments of the periodogram, the number of automatically detected test volumes and the upper limit on the index of inspected superlattices.

A typical output file contains a list of proposed solutions. The choice of the ultimate solution from the list is left to a user. For each solution, the program lists the matrix X of basis vectors of the direct lattice. Then, there is a table of indices corresponding to particular reflections, the parameters a, b, c, α, β and γ, and the volume of the primitive cell of the direct lattice.[8] The program delivers a Buerger-reduced cell, and additional processing is needed to determine the lattice symmetry. The task can be performed using *LEPAGE* [66] or other procedures of this kind; see, e.g., [67–71]. The program can be downloaded from [72].

5.13 A Bird's Eye View on Ab Initio Indexing

The aim of this final section is to put the subject matter of this and the subsequent two chapters into a broader perspective. Ab initio indexing of diffraction patterns means assigning indexes to diffraction peaks based on known positions of the peaks, and it comes down to determination of unknown lattice parameters (and lattice orientation, in single-crystal indexing). However, 'peaks' have different counterparts in the reciprocal space for different types of diffraction patterns.

In the conceptually simplest case considered in this chapter, a peak corresponds to a node of the reciprocal lattice of the crystal. Based on the peak positions, one has all coordinates of some lattice nodes (Fig. 5.15*b*). As will be explained in the

[8] *Ind_X* also allows for saving results of a basic analysis of relationships between the proposed solutions.

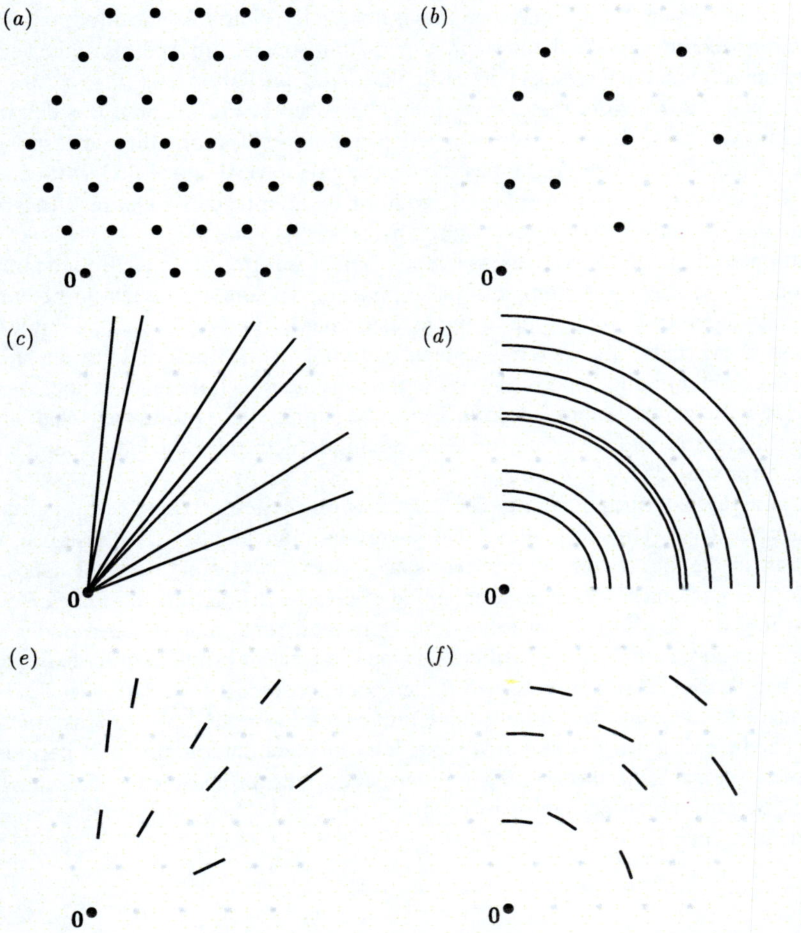

Fig. 5.15 Schematic two-dimensional illustration of smearing the reciprocal lattice nodes by experimental conditions. (*a*) A part of the reciprocal lattice. (*b*) In the simplest case, all coordinates of some scattering vectors are determined. The vectors correspond to selected nodes of the reciprocal lattice. (*c*) Only directions of scattering vectors are obtained from white-beam Laue diffraction experiments. (*d*) Only magnitudes of scattering vectors are obtained from powder diffraction. Each data point corresponds to a complete sphere in reciprocal space. (*e*) With pink-beam, the nodes extend along vector directions. The range of the extension is related to the radiation bandwidth. (*f*) The streaks corresponding to measurements by the oscillation method are one-dimensional and perpendicular to the oscillation axis.

next chapter, data from white-beam Laue diffraction are scale-independent, i.e., they correspond to reciprocal lattice vectors smeared along their directions (Fig. 5.15*c*). Analogous smearing but on short distances will occur for data with large uncertainties along directions of the scattering vectors like pink-beam diffraction or EBSD

(Fig. 5.15*e*). When the crystal orientation is uncertain or variable, the directions of the scattering vectors vary. With the oscillation method, the terminal points of the vectors cover one-dimensional streaks perpendicular to the oscillation axis (Fig. 5.15*f*), and with powder diffraction, they cover complete spheres (Fig. 5.15*d*). Besides these well-known cases, there are other experimental configurations that correspond to other (usually more complex) types of smearing reciprocal lattice nodes. (See, e.g., Fig. 2.11 or Sect. 10.5.1.) Moreover, in addition to the spread of nodes resulting from the method, there is also a spread caused by experimental errors.

Summarizing, a diffraction experiment provides information on a limited number of nodes or scattering vectors, and these nodes or vectors are known to be spread in a way dependent on the type of the measurement. In other words, the input data consist of the scattering vectors and their uncertainties comprising both the smearing of reciprocal lattice nodes due to the measurement technique and experimental errors. So the general indexing problem can be formulated as follows: based on the scattering vectors and the broad-sense uncertainties, determine the lattice parameters and ascribe indices to the vectors.

In practice, algorithms and codes dealing with particular experimental circumstances are used. However, they have some common features. For instance, one of them is the search for zones. Real-space indexing of single-crystal data (cf. Sects. 5.5, 5.10 and 6.4.2) or zone-indexing of powder diffraction patterns (cf. Sect. 7.4.3) rely on identification of reflections from tautozonal planes. Once nodes of a two-dimensional lattice are identified, one faces the easier problem of indexing two-dimensional data. After solving two-dimensional problems for a number of zones, one needs to integrate two-dimensional lattices into a single three-dimensional lattice. The indexing process may involve using tools like determination of periods in one-dimensional data, the Fourier transformation, cluster analysis, difference vectors, testing for sublattices, fitting by optimization, various 'accumulation' or 'voting' schemes, et cetera.

5.14 Appendix: Auxiliary Tools

5.14.1 Obtaining the Scattering Vector from a Kossel Line

Processing of Kossel patterns begins with determining the locations of points on Kossel conics. Based on the locations, one computes a set of unit vectors $\widehat{\boldsymbol{k}}_i$ from the source toward points on a Kossel conic. (See Sects. 2.3.2 and 2.3.3.) The wave vector corresponding to $\widehat{\boldsymbol{k}}_i$ is given by $\boldsymbol{k}_i = \widehat{\boldsymbol{k}}_i/\lambda$. The sought-after scattering vector $\boldsymbol{h}$ of the considered conic satisfies $\boldsymbol{h} \cdot (\boldsymbol{h} - 2\boldsymbol{k}_i) = 0$ or $\boldsymbol{d} \cdot \boldsymbol{k}_i = 1/2$, where $\boldsymbol{d} = \boldsymbol{h}/(\boldsymbol{h} \cdot \boldsymbol{h})$. If the vectors $\boldsymbol{k}_i$ are not coplanar, the system of linear equations $\boldsymbol{d} \cdot \boldsymbol{k}_i = 1/2$ with respect to $\boldsymbol{d}$ has a unique solution. To account for errors in $\boldsymbol{k}_i$, the system $\boldsymbol{d} \cdot \boldsymbol{k}_i = 1/2$ is solved using least squares method (problem 1 of Sect. 5.14.2). This results in

$$d = \arg_x \min \sum_i (x \cdot k_i - 1/2)^2 = D^{-1} \sum_i k_i/2 ,$$

where $D = \sum_i k_i \otimes k_i$. Hence, one can calculate the scattering vector $h = d/(d \cdot d)$ corresponding to the considered conic. The procedure works fine if the conics are sufficiently curved.

5.14.2 Linear Optimization Problem

The versions of the linear optimization problem considered below match cases described in this chapter.

General formulation

Given are two sets of corresponding vectors x^n and y^n ($n = 1, \ldots, N$). The vectors x^n may have a different number of components than the vectors y^n.
Problem 1: determine the matrix A minimizing $\sum_{n=1}^{N}(Ax^n - y^n)^2$. The solution has the form

$$A = CD^{-1} ,$$

where $C = \sum_n y^n \otimes x^n$, and $D = \sum_n x^n \otimes x^n$ is assumed to be invertible.
Problem 2: determine the matrix A and vector b minimizing $\sum_{n=1}^{N}(Ax^n + b - y^n)^2$ [40]. The solution has the form

$$A = CD^{-1} , \quad b = Y - AX ,$$

where $C = \sum_n y^n \otimes x^n - NY \otimes X$, $D = \sum_n x^n \otimes x^n - NX \otimes X$, $X = \sum_n x^n/N$, $Y = \sum_n y^n/N$, and it is assumed that the matrix D is invertible.

A vector having nearly integer scalar products with vectors of a set

Having a certain (error affected) direct lattice vector u and a set of experimental scattering vectors h^n ($n = 1, \ldots, N$), the products $\delta_n = h^n \cdot u$ are expected to be nearly integer. Vectors h^n giving products differing from an integer by more than a threshold can be ignored. Those supporting the initial u can be used for its refinement [45]. Let $m_n = \lfloor \delta_n \rceil$ be the integer nearest to δ_n. The fitted new vector u_{opt} of the direct lattice equals $\arg_x \min \sum_n (x \cdot h^n - m_n)^2$. Based on the solution of the linear optimization problem (1), it is given by

$$u_{opt} = D^{-1}C ,$$

where $\boldsymbol{C} = \sum_n m_n \boldsymbol{h}^n$ and $D = \sum_n \boldsymbol{h}^n \otimes \boldsymbol{h}^n$. The above approach works for short vectors. The products $\boldsymbol{u} \cdot \boldsymbol{h}^n$ between long vectors may jump to locations near wrong integers. To avoid such cases, the method should be applied iteratively starting from short $\boldsymbol{h}^n$ giving small $|\delta_n|$.

5.14.3 Generation of Integer Triplets

In some indexing procedures, the generation of triplets of integers is needed. Generation of such triplets is a part of simulations of diffraction patterns for test (or known) structures. Below, only the generation of plane indices is considered, but a similar approach is applicable to direction indices. In a typical case, the triplets $(h, k, l) = (h_1, h_2, h_3)$ are used to test whether a basis, say, $\boldsymbol{a}^i$, is consistent with observed reflections, which are expected to correspond to some vectors $h_i \boldsymbol{a}^i$.

The question is what are the reasonable bounds on the integers h_i. The simplest approach is to assume that the absolute values of h_i $(i = 1, 2, 3)$ are not larger than a certain limit

$$|h_i| \leq \text{ a limit },$$

e.g. [73]. Application of these straightforward conditions makes sense when it is known a priori that the indices should be in a given range. The restriction

$$|h_1| + |h_2| + |h_3| \leq \text{ a limit}$$

(see, e.g., [45]) has similar character. However, if the basis spans an elongated cell or a highly non-orthogonal cell, more sophisticated constraints involving the lattice metric are needed. The limitation can be based on the magnitude of the vector (and thus the scattering angle)

$$|\check{\boldsymbol{h}}| = \sqrt{g^{ij} h_i h_j} \leq l^*_{max} ,$$

where l^*_{max} is the maximum allowed length of the reciprocal lattice vector related to the resolution limit $d_{ResLim} = 1/l^*_{max}$ (see, Sect. 1.5.1) or the largest Bragg angle θ_{max} (via $l^*_{max} = 2 \sin \theta_{max}/\lambda$). A similar condition involving the *direct* lattice metric has the form

$$|h_i| \leq l^*_{max} |\boldsymbol{a}_i| = l^*_{max} \sqrt{g_{(i)(i)}} ,$$

but with this restriction some vectors $h_i \boldsymbol{a}^i$ longer than l^*_{max} will also be generated; cf. [45]. The limits can be initially set at low numbers and then gradually increased up to the desired final cut-off level. It must be added that the ranges of the generated indices can additionally be limited by symmetries. For instance, with the equivalence of $h_i \boldsymbol{a}^i$ and $-h_i \boldsymbol{a}^i$, it may be sufficient to generate only the vectors with terminal points in, say, upper half-space.

References

1. W. Clegg, *Crystal Structure Determination* (Oxford University Press, Oxford, UK, 1998)
2. A. Morawiec, Indexing of diffraction patterns for determination of crystal orientations. Acta Cryst. A **76**, 719–734 (2020)
3. A.W. Hull, W.P. Davey, Graphical determination of hexagonal and tetragonal crystal structures from X-ray data. Phys. Rev. **17**, 549–570 (1921)
4. T. Bjurström, Graphische Methoden zum Aufsuchen der quadratischen Form aus röntgenographischen Pulverphotogrammen. Z. Phys. **69**, 346–355 (1931)
5. R.A. Harrington, A simplified construction of Hull-Davey charts. Rev. Sci. Instr. **9**, 429–430 (1938)
6. C.W. Bunn, *Chemical Crystallography* (Clarendon Press, Oxford, UK, 1945)
7. L.K. Frevel, T.P. Blumer, Indexing rhombohedral powder diffraction patterns. J. Appl. Cryst. **7**, 621–624 (1974)
8. J.D. Bernal, On the interpretation of X-ray single crystal rotation photographs. Proc. Roy. Soc. A **113**, 117–160 (1926)
9. J. Leonhardt. VII. Die Deutung der Lauediagramme deformierter Kristalle. Z. Kristallogr. **61**, 100–112 (1924)
10. A.B. Greninger, A back-reflection Laue method for determining crystal orientation. Z. Kristallogr. **91**, 424–432 (1935)
11. J.W. Edington, *Practical Electron Microscopy in Materials Science. Monograph Two: Electron Diffraction in the Electron Microscope* (MacMillan, London, UK, 1975)
12. K.W. Andrews, D.J. Dyson, S.R. Keown, *Interpretation of Electron Diffraction Patterns* (Springer Science+Business Media, LLC, 1967). Reprint of the 1st edition by Plenum, US
13. E. Preuss, B. Krahl-Urban, R. Butz, *Laue Atlas* (Bertelsmann Universitätsverlag, Düsseldorf, 1974)
14. D.J. Dingley, K.Z. Baba-Kishi, V. Randle, *Atlas of Backscattering Kikuchi Patterns* (Institute of Physics Pub, Bristol, UK, 1995)
15. W.F. de Jong, J. Boumannn, Das Photographieren von reziproken Kristallnetzen mittels Röntgenstrahlen. Z. Kristallogr. **98**, 456–459 (1939)
16. J.M. Torroja, E. Pajares, J.L. Amorós, A single crystal X-ray camera for direct recording of the reciprocal lattice. J. Sci. Instrum. **28**, 44–46 (1951)
17. K. Weissenberg, Ein neues Röntgengoniometer. Z. Phys. **23**, 229–238 (1924)
18. M.J. Buerger, *The Photography of the Reciprocal Lattice, ASXRED Monograph No 1* (The Murray Printing Company, Cambridge, MA, 1944)
19. H.R. Powell, The Rossmann Fourier autoindexing algorithm in *MOSFLM*. Acta Cryst. D **55**, 1690–1695 (1999)
20. Z. Otwinowski, W. Minor, Processing of X-ray diffraction data collected in oscillation mode. Methods Enzymol. **276**, 307–326 (1997)
21. D. Gewirth, The *HKL* manual. A description of the programs *Denzo, XDisplayF, Scalepack.* An oscillation data processing suite for macromolecular crystallography (2003)
22. A.G.W. Leslie, H.R. Powell, G. Winter, O. Svensson, D. Spruce, S. McSweeney, D. Love, S. Kinder, E. Duke, C. Nave, Automation of the collection and processing of X-ray diffraction data - a generic approach. Acta Cryst. D **58**, 1924–1928 (2002)
23. J.W. Pflugrath, The finer things in X-ray diffraction data collection. Acta Cryst. D **55**, 1718–1725 (1999)
24. M.G. Rossmann, C.G. van Beek, Data processing. Acta Cryst. D **55**, 1631–1640 (1999)
25. W. Kabsch, *XDS*. Acta Cryst. D **66**, 125–132 (2010)
26. V.S. Klein, *USCI* - Ein Programm zur Indizierung von Einkristallreflexen. J. Appl. Cryst. **8**, 649–652 (1975)
27. A.J.M. Duisenberg, Indexing in single-crystal diffractometry with an obstinate list of reflections. J. Appl. Cryst. **25**, 92–96 (1992)
28. K. Pilz, M. Estermann, S. van Smaalen, Automatic indexing of area-detector data of periodic and aperiodic crystals. J. Appl. Cryst. **35**, 253–260 (2002)

29. N.K. Sauter, R.W. Grosse-Kunstleve, P.D. Adams, Robust indexing for automatic data collection. J. Appl. Cryst. **37**, 399–409 (2004)
30. T. Higashi, Auto-indexing of oscillation images. J. Appl. Cryst. **23**, 253–257 (1990)
31. W.R. Busing, H.A. Levy, Angle calculations for 3- and 4-circle X-ray and neutron diffractometers. Acta Cryst. **22**, 457–464 (1967)
32. W. Kabsch, A pattern-recognition procedure for scanning oscillation films. J. Appl. Cryst. **10**, 426–429 (1977)
33. C.A. Cornelius, A simple computer method for the orientation of single crystals of any structure using Laue back-reflection X-ray photographs. Acta Cryst. A **37**, 430–436 (1981)
34. G.P. Burenkov, A.N. Popov, A method of automatically indexing Laue patterns. Crystallogr. Rep. **39**, 556–561 (1994)
35. H. Cohen, *A Course in Computational Algebraic Number Theory* (Springer, Berlin, 1993)
36. M. Pohst, A modification of the LLL reduction algorithm. J. Symb. Comput. **4**, 123–127 (1987)
37. J. Buchmann, M. Pohst, Computing a lattice basis from a system of generating vectors, in *Proceedings of EUROCAL 87, Lecture Notes in Computer Science 378* (Springer, Berlin, 1989), pp. 54–63
38. W. Sun, *Computational Method for the Indexing of Unknown Powder Patterns and Unknown Rotating-Crystal Patterns*. PhD thesis, The University of Oklahoma, Norman, Oklahoma, USA, 1972
39. Z. Otwinowski, W. Minor, *DENZO* and *SCALEPACK*, in *International Tables for Crystallography, Vol. F, Section 11.4*, ed. by M.G. Rossmann, E. Arnold (Kluver Academic Publishers, Dodrecht, 2001), pp. 226–235
40. S. Kim, Auto-indexing oscillation photographs. J. Appl. Cryst. **22**, 53–60 (1989)
41. R.A. Jacobson, A cosine transform approach to indexing. Z. Kristallogr. **212**, 99–102 (1997)
42. I. Steller, R. Bolotovsky, M.G. Rossmann, An algorithm for automatic indexing of oscillation images using Fourier analysis. J. Appl. Cryst. **30**, 1036–1040 (1997)
43. E.B. Saff, A.B.J. Kuijlaars, Distributing many points on a sphere. Math. Intell. **19**, 5–11 (1997)
44. C.G. Koay, A simple scheme for generating nearly uniform distribution of antipodally symmetric points on the unit sphere. J. Comput. Sci. **2**, 377–381 (2011)
45. W. Clegg, Enhancement of the 'auto-indexing' method for cell determination in four-circle diffractometry. J. Appl. Cryst. **17**, 334–336 (1984)
46. R.F. Stellingwerf, Period determination using phase dispersion minimization. Astrophys. J. **224**, 953–960 (1978)
47. S.R. Davies, An improved test for periodicity. Mon. Not. R. Astr. Soc. **244**, 93–95 (1990)
48. D.A. Leahy, R.L. Elsner, M.C. Weisskopf, On searches for periodic pulsed emission. The Rayleigh test compared to epoch folding. Astrophys. J. **272**, 256–258 (1983)
49. S. Larsson, Parameter estimation in epoch folding analysis. Astron. Astrophys. Suppl. **117**, 197–201 (1996)
50. T.W. Anderson, *The Statistical Analysis of Time Series* (Wiley, New York, 1971)
51. K.V. Mardia, *Statistics of Directional Data* (Academic Press, New York, 1972)
52. D. Bouscaud, S. Berveiller, R. Pesci, E. Patoor, A. Morawiec, Local stress analysis in an SMA during stress-induced martensitic transformation by Kossel microdiffraction. Adv. Mat. Res. **996**, 45–51 (2014)
53. W. Kabsch, Automatic indexing of rotation diffraction patterns. J. Appl. Cryst. **21**, 67–71 (1988)
54. R. Müller, G. Roth, Matrix-free integration of image-plate diffraction data. J. Appl. Cryst. **38**, 280–290 (2005)
55. U. Kolb, T. Gorelik, M.T. Otten, Towards automated diffraction tomography. Part II – Cell parameter determination. Ultramicroscopy **108**, 763–772 (2008)
56. S. Schlitt, T.E. Gorelik, A.A. Stewart, E. Schömer, T. Raasch, U. Kolb, Application of clustering techniques to electron-diffraction data: determination of unit-cell parameters. Acta Cryst. A **68**, 536–546 (2012)
57. J.W. Campbell, The practicality of using a three-dimensional fast Fourier transform in auto-indexing protein single-crystal oscillation images. J. Appl. Cryst. **31**, 407–413 (1998)

58. F.R.N.C. Maia, C. Yang, S. Marchesini, Compressive auto-indexing in femtosecond nanocrystallography. Ultramicroscopy **111**, 807–811 (2011)
59. M. Ester, H.P. Kriegel, J. Sander, X. Xu, A density based algorithm for discovering clusters in large spatial databases with noise, in *Proceedings of the 2nd ACM International Conference on Knowledge Discovery and Data Mining (KDD), Portland, OR* (AAAI Press, 1996), pp. 226–231
60. M. Ankerst, M.M. Breunig, H.P. Kriegel, J. Sander, OPTICS: ordering points to identify the clustering structure, in *Proceedings of the ACM International Conference on Management of Data (SIGMOD), Philadelphia, PA* (ACM Press, 1999), pp. 49–60
61. H.P. Kriegel, P. Kröger, J. Sander, A. Zimek, Density-based clustering. WIREs Data Mining Knowl. Discov. **1**, 231–240 (2011)
62. W. Kabsch, Automatic processing of rotation diffraction data from crystals of initially unknown symmetry and cell constants. J. Appl. Cryst. **26**, 795–800 (1993)
63. A. Morawiec, A program for refinement of lattice parameters and strain determination using Kossel diffraction patterns. J. Appl. Cryst. **49**, 322–329 (2016)
64. M.I. Buerger, Reduced cells. Z. Kristallogr. **109**, 42–60 (1957)
65. M.I. Buerger, Note on reduced cells. Z. Kristallogr. **113**, 52–56 (1960)
66. A.L. Spek, *LEPAGE* - an MS-DOS program for the determination of the metrical symmetry of a translation lattice. J. Appl. Cryst. **21**, 578–579 (1988)
67. W. Clegg, Cell reduction and lattice symmetry determination. Acta Cryst. A **37**, 913–915 (1981)
68. Y. Le Page, The derivation of the axes of the conventional unit cell from the dimensions of the Buerger-reduced cell. J. Appl. Cryst. **15**, 255–259 (1982)
69. H. Zimmermann, H. Burzlaff, *DELOS* - a computer program for the determination of a unique conventional cell. Z. Kristallogr. **170**, 241–246 (1985)
70. T. Higgins, R. Dark, P. McArdle, J. Simmie, *BRVCEL* - a computer program for cell reduction and Bravais lattice determination. Computers Chem. **14**, 33–36 (1990)
71. J. Macíček, A. Yordanov, *BLAF* – a robust program for tracking out admittable Bravais lattice(s) from the experimental unit-cell data. J. Appl. Cryst. **25**, 73–80 (1992)
72. http://imim.pl/personal/adam.morawiec/. Accessed Aug 2022
73. R.A. Jacobson, A single-crystal automatic indexing procedure. J. Appl. Cryst. **9**, 115–118 (1976)

Chapter 6
Ab-Inito Indexing of Laue Patterns

When polychromatic X-ray radiation is used, pattern acquisition times are much shorter than with monochromatic techniques [1]. The high speed of data collection is an important advantage in some applications [2–4]. The Laue method, however, is not commonly used because of difficulties with data processing [4, 5].

6.1 Geometry of Laue Patterns

In the Laue method, with a range of wave spectrum contributing to a diffraction pattern, only directions of scattering vectors can be computed from positions of diffraction spots; their magnitudes are indeterminable. As was explained in Sect. 2.3.5, a Laue spot corresponding to a reciprocal lattice vector $\check{\boldsymbol{h}}$ is at the location indicated by the normalized wave vector

$$\widehat{\boldsymbol{k}}_{\check{h}} = \widehat{\boldsymbol{k}}_0 + \lambda_{\check{h}} \check{\boldsymbol{h}} \, , \tag{6.1}$$

where the wavelength $\lambda_{\check{h}} = -2\widehat{\boldsymbol{k}}_0 \cdot \check{\boldsymbol{h}}/\check{\boldsymbol{h}}^2$ is in a range $[\lambda_{min}, \lambda_{max}]$. Substitution of $\lambda_{\check{h}}$ in (6.1) leads to

$$\widehat{\boldsymbol{k}}_{\check{h}} = \widehat{\boldsymbol{k}}_0 - 2\,(\widehat{\boldsymbol{k}}_0 \cdot \widehat{\boldsymbol{h}})\,\widehat{\boldsymbol{h}} \, ,$$

and hence one has

$$\widehat{\boldsymbol{k}}_{S\check{h}} = \widehat{\boldsymbol{k}}_{\check{h}}$$

A. Morawiec, *Indexing of Crystal Diffraction Patterns*, Springer Series in Materials Science 326, https://doi.org/10.1007/978-3-031-11077-1_6

for a positive scaling factor S. The above relationship implies that, first, reflections from planes $(h\,k\,l)$ and $(nh\,nk\,nl)$ contribute to the same spot[1] (the circumstance referred to as *harmonic* spot overlap), and second, two crystal lattices differing by a scale lead to the same pattern. In consequence of the second fact, ab initio indexing of Laue patterns can provide a lattice up to a scale, i.e., the absolute volume of the primitive cell is indeterminable.

⊡ The absolute volume of the primitive cell can be estimated by identifying wavelengths of some reflections. This could be done based on λ_{min} [7], but λ_{min} is usually known only approximately as the intensity of the incident beam drops smoothly with decreasing wavelength. Another method is to use a metal foil with an absorption edge at known wavelength [8].

6.1.1 Experimentally Accessible Part of the Reciprocal Space

The experimentally accessible part of the reciprocal lattice is bounded by $\lambda_{min} < \lambda_{\check{h}} < \lambda_{max}$ and by position and physical dimensions of the detector. In the case of transmission and complex crystal structures, there is the additional restriction on the maximal acceptable $|\check{\boldsymbol{h}}|$ caused by the resolution limit: $|\check{\boldsymbol{h}}| \leq 1/d_{ResLim}$.

It is easy to see the impact of the size of a round detector on the accessible region of the reciprocal space when the detector is perpendicular to the incident beam, and the pattern center is at the center of the detector. In transmission ($\widehat{\boldsymbol{L}} = +\widehat{\boldsymbol{k}}_0$), if the edge of the detector is at the angle $2\theta_x$ from the vector $\boldsymbol{L}$, the wave vectors $\boldsymbol{k}_{\check{h}}$ pointing toward the detector satisfy the condition $\widehat{\boldsymbol{k}}_{\check{h}} \cdot \widehat{\boldsymbol{L}} > \cos(2\theta_x)$. (See Fig. 6.1.) Based on (6.1), this leads to $|\widehat{\boldsymbol{h}} \cdot \widehat{\boldsymbol{k}}_0| < \sin\theta_x = \cos(\pi/2 - \theta_x)$. Thus, in transmission, for a spot corresponding to the reciprocal lattice vector $\check{\boldsymbol{h}}$ to be in the field of view, the angle between $\check{\boldsymbol{h}}$ and $-\widehat{\boldsymbol{k}}_0$ must be larger than $\pi/2 - \theta_x$.

In back-reflection ($\widehat{\boldsymbol{L}} = -\widehat{\boldsymbol{k}}_0$), with the edge of the detector at the angle $\pi - 2\theta_x$ from the vector $\boldsymbol{L}$, the wave vectors $\boldsymbol{k}_{\check{h}}$ pointing toward the detector satisfy the condition $\widehat{\boldsymbol{k}}_{\check{h}} \cdot \widehat{\boldsymbol{L}} > \cos(\pi - 2\theta_x)$. Hence, one has $|\widehat{\boldsymbol{k}}_0 \cdot \widehat{\boldsymbol{h}}| > \cos(\pi/2 - \theta_x)$. Thus, in back-reflection, for a spot corresponding to the reciprocal lattice vector $\check{\boldsymbol{h}}$ to be in the field of view, the angle between $\check{\boldsymbol{h}}$ and $-\widehat{\boldsymbol{k}}_0$ must be smaller than $\pi/2 - \theta_x$.

6.2 Gnomonic Projection of Reciprocal Lattice Nodes

In a Laue pattern, spots corresponding to tautozonal planes are located on a conic; see, Sect. 2.3.5. The pattern can be transformed so these spots are on a straight line. This section explains technical details of such transformation.

---

[1] For this reason, some authors make a distinction between 'Bragg reflection' and 'Laue reflection' [6]. The latter corresponds to a spot on a Laue pattern. It may arise from superposition of several Bragg reflections differing by n. Laue reflections (spots) are customarily ascribed relatively prime indices.

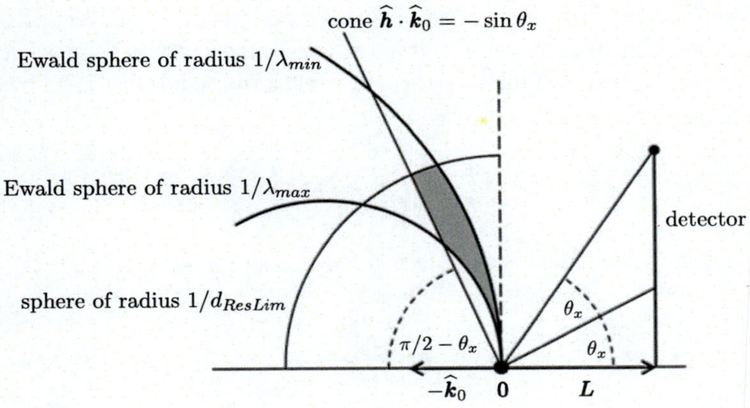

Fig. 6.1 Two-dimensional schematic of the accessible part of the reciprocal space (marked in gray) for Laue diffraction in transmission with round detector perpendicular to the incident beam ($\widehat{L} = +\widehat{k}_0$) and pattern center at the center of the detector. To get the three-dimensional region, the figure needs to be rotated about the drawn horizontal axis. Only reciprocal lattice points between Ewald spheres of radii $1/\lambda_{min}$ and $1/\lambda_{max}$ can contribute to the diffraction pattern. The additional bounds are the **0**-centered sphere of radius $1/d_{ResLim}$ and the cone resulting from the finite radius of the detector.

The transformation is related to gnomonic projection of (normalized vectors of) crystal reciprocal lattice. With the pattern center at $\boldsymbol{L}$, a Laue spot corresponding to a reciprocal lattice vector $\check{\boldsymbol{h}}$ is at the location given by the gnomonic projection (2.11)

$$\boldsymbol{y}_{Laue}(\check{\boldsymbol{h}}) = \frac{L^2}{\widehat{\boldsymbol{k}}_{\check{h}} \cdot \boldsymbol{L}} \widehat{\boldsymbol{k}}_{\check{h}} - \boldsymbol{L} \, . \tag{6.2}$$

Based on (6.1), one has $\check{\boldsymbol{h}} = (\widehat{\boldsymbol{k}}_{\check{h}} - \widehat{\boldsymbol{k}}_0)/\lambda_{\check{h}}$ and the unit vector along $\check{\boldsymbol{h}}$ is given by $\widehat{\boldsymbol{h}} = (\widehat{\boldsymbol{k}}_{\check{h}} - \widehat{\boldsymbol{k}}_0)/|\widehat{\boldsymbol{k}}_{\check{h}} - \widehat{\boldsymbol{k}}_0|$. If the vector $\widehat{\boldsymbol{k}}_{\check{h}}$ in the expression (6.2) for the position of Laue spot on the pattern is replaced by $\widehat{\boldsymbol{h}}$, i.e.,

$$\widehat{\boldsymbol{k}}_{\check{h}} \leftarrow \widehat{\boldsymbol{h}} = \frac{\widehat{\boldsymbol{k}}_{\check{h}} - \widehat{\boldsymbol{k}}_0}{|\widehat{\boldsymbol{k}}_{\check{h}} - \widehat{\boldsymbol{k}}_0|} \, , \tag{6.3}$$

one obtains the formula for gnomonic projection (2.11) of the point $\widehat{\boldsymbol{h}}$

$$\frac{L^2}{\widehat{\boldsymbol{h}} \cdot \boldsymbol{L}} \widehat{\boldsymbol{h}} - \boldsymbol{L} \, .$$

Since, in this case, the projection plane is just a geometric (i.e., not a real) object, the vector $\boldsymbol{L}$ can be replaced by any non-zero vector. The most natural approach is to use a vector collinear with $\boldsymbol{L}$. With $\boldsymbol{L} \to s\boldsymbol{L}$, the point corresponding to $\check{\boldsymbol{h}}$ is projected at

$$\boldsymbol{y}_{proj}(\check{\boldsymbol{h}}) = \frac{(s\boldsymbol{L})^2}{\widehat{\boldsymbol{h}} \cdot (s\boldsymbol{L})}\widehat{\boldsymbol{h}} - s\boldsymbol{L} = s\left(\frac{\boldsymbol{L}^2}{\widehat{\boldsymbol{h}} \cdot \boldsymbol{L}}\widehat{\boldsymbol{h}} - \boldsymbol{L}\right) = s\left(\frac{\boldsymbol{L}^2}{(\widehat{\boldsymbol{k}}_{\check{h}} - \widehat{\boldsymbol{k}}_0) \cdot \boldsymbol{L}}(\widehat{\boldsymbol{k}}_{\check{h}} - \widehat{\boldsymbol{k}}_0) - \boldsymbol{L}\right),$$

i.e., s is simply a non-zero scaling factor. The cases of transmission ($\iota = +1$) and back-reflection ($\iota = -1$) correspond to

$$\widehat{\boldsymbol{k}}_0 = \iota\widehat{\boldsymbol{L}},$$

and by eliminating $\widehat{\boldsymbol{k}}_{\check{h}} \cdot \boldsymbol{L}$ using (6.2), $\boldsymbol{y}_{proj}(\check{\boldsymbol{h}})$ can be expressed via $\boldsymbol{y}_{Laue}(\check{\boldsymbol{h}})$; with the argument $\check{\boldsymbol{h}}$ omitted, one has

$$\boldsymbol{y}_{proj} = \frac{s\,\boldsymbol{y}_{Laue}}{1 - \iota\left(1 + \boldsymbol{y}_{Laue}^2/\boldsymbol{L}^2\right)^{1/2}}.$$

It is convenient to use the convention $s = -\iota S$, where $S > 0$; one has $(s\widehat{\boldsymbol{L}}) \cdot \widehat{\boldsymbol{k}}_0 = -S < 0$, and the formula for the transformation of the Laue pattern takes the form

$$\boldsymbol{y}_{proj} = \frac{S\,\boldsymbol{y}_{Laue}}{\left(1 + \boldsymbol{y}_{Laue}^2/\boldsymbol{L}^2\right)^{1/2} - \iota}. \tag{6.4}$$

With the above convention, one has $\boldsymbol{y}_{Laue} \cdot \boldsymbol{y}_{proj} \geq 0$, i.e., a spot and its projection are on the same side of the pattern center in both transmission and back-reflection.

In the case of transmission, a point $\boldsymbol{y}_{Laue}$ close to the pattern center is projected on $\boldsymbol{y}_{proj}$ distant from the center (Fig. 6.2). In order to see such points on the projection, S must satisfy the inequality $\boldsymbol{y}_{proj}^2 > (S\boldsymbol{L})^2$. Similarly, in the case of back-reflection (Fig. 6.3), to see the projections of points located at the distance of the order of $|\boldsymbol{L}|$, there must occur $\boldsymbol{y}_{proj}^2 < (S\boldsymbol{L})^2$. With these conditions satisfied, one has

$$\boldsymbol{y}_{Laue} = \frac{2\,\iota\,\boldsymbol{y}_{proj}/S}{\boldsymbol{y}_{proj}^2/(S\boldsymbol{L})^2 - 1}. \tag{6.5}$$

This is the transformation inverse to (6.4).

⊡ The transformations (6.4) and (6.5) can also be easily expressed using the Bragg angle. The cosine of 2θ is given by $\widehat{\boldsymbol{k}}_0 \cdot \widehat{\boldsymbol{k}}_{\check{h}} = \cos(2\theta)$; cf. (2.10). On the other hand, the corresponding expression after transformation, i.e., after substitution of $\widehat{\boldsymbol{k}}_{\check{h}}$ by $\widehat{\boldsymbol{h}}$ is

$$(-\widehat{\boldsymbol{k}}_0) \cdot \widehat{\boldsymbol{h}} = -\frac{\widehat{\boldsymbol{k}}_0 \cdot (\widehat{\boldsymbol{k}}_{\check{h}} - \widehat{\boldsymbol{k}}_0)}{|\widehat{\boldsymbol{k}}_{\check{h}} - \widehat{\boldsymbol{k}}_0|} = \frac{1 - \cos(2\theta)}{2\sin\theta} = \sin\theta = \cos(\pi/2 - \theta).$$

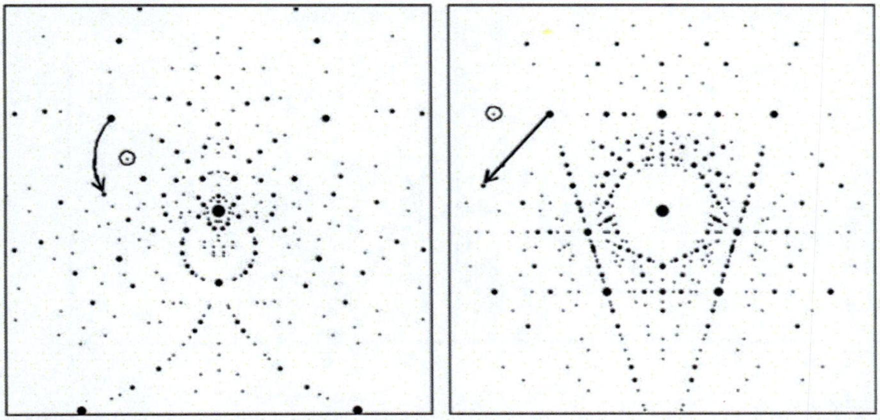

Fig. 6.2 Simulated transmission Laue pattern of Si ([113] zone axis, incident beam perpendicular to the screen) and its projection. The corresponding spots are marked by a circle and an arrow. The central part of the pattern is not represented on the projection and vice versa.

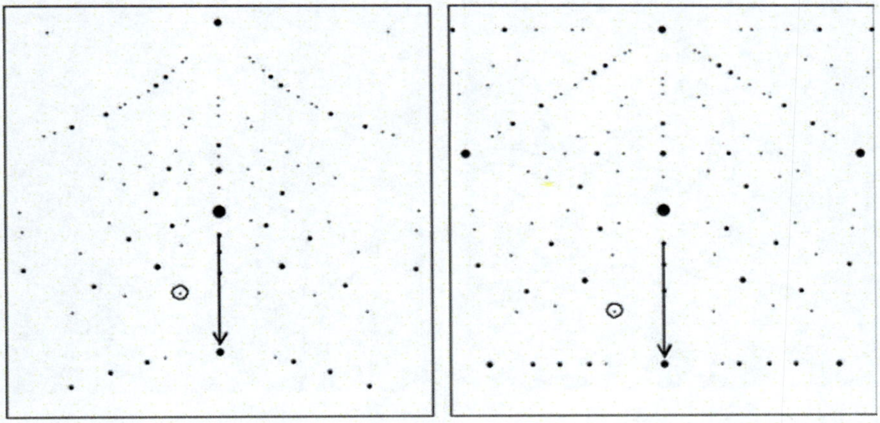

Fig. 6.3 Simulated back-reflection Laue pattern of Si ([113] zone axis, incident beam perpendicular to the screen) and its projection. The corresponding spots are marked by a circle and an arrow.

Thus, in the transformed pattern, the spot corresponding to $\check{\boldsymbol{h}}$ is at the angle $\pi/2 - \theta$ to $-\widehat{\boldsymbol{k}}_0$, and the azimuth angle is the same as in the original pattern. See Fig. 6.4. Alternatively, this conclusion can be reached by using (6.4) and (6.5) and the observations that in transmission $|\boldsymbol{y}_{Laue}|/|\boldsymbol{L}| = \tan(2\theta)$ with $0 < \theta < \pi/4$, in back-reflection $|\boldsymbol{y}_{Laue}|/|\boldsymbol{L}| = \tan(\pi - 2\theta)$ with $\pi/4 < \theta < \pi/2$, and in both cases $|\boldsymbol{y}_{proj}|/|\boldsymbol{SL}| = \tan(\pi/2 - \theta)$.

As was explained in Sect. 2.3.5, the diffracted beams corresponding to reciprocal lattice vectors of one zone lie on a cone, and—with a planar detector—the observed diffraction spots are on a conic section. On the other hand, the gnomonic projections of reciprocal lattice vectors of one zone lie on a straight line. It is easy to see this.

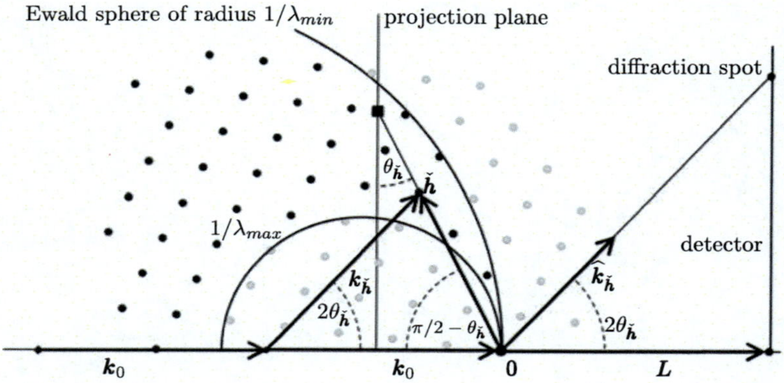

Fig. 6.4 Schematic illustration of the geometry of Laue diffraction in transmission. Reciprocal lattice points between Ewald spheres of radii $1/\lambda_{min}$ and $1/\lambda_{max}$ are marked in black. The point representing gnomonic projection of the reciprocal lattice vector $\check{\boldsymbol{h}}$ is marked by a square.

(a) (b)

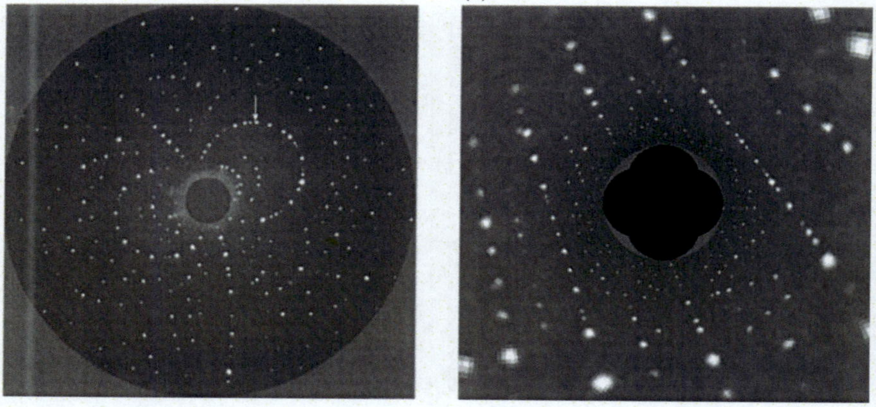

Fig. 6.5 (*a*) Experimental (transmission) Laue pattern of an organic compound published in [10]. Reproduced with permission of the International Union of Crystallography. (*b*) Part of the pattern transformed with respect to fitted pattern center. Arrows in (*a*) and (*b*) indicate a conic and the corresponding line.

Vectors $\widehat{\boldsymbol{h}}$ of the zone $\check{\boldsymbol{t}}$ satisfy the condition $\check{\boldsymbol{t}} \cdot \widehat{\boldsymbol{h}} = 0$. Let $\check{\boldsymbol{t}}_{proj}$ be $\check{\boldsymbol{t}}$ projected along $\widehat{\boldsymbol{L}}$ on the projection plane, i.e., $\check{\boldsymbol{t}}_{proj} = \check{\boldsymbol{t}} - (\check{\boldsymbol{t}} \cdot \widehat{\boldsymbol{L}})\widehat{\boldsymbol{L}}$. For all vectors $\widehat{\boldsymbol{h}}$ satisfying the zone law $\check{\boldsymbol{t}} \cdot \widehat{\boldsymbol{h}} = 0$, one has $\boldsymbol{y}_{proj}(\widehat{\boldsymbol{h}}) \cdot \check{\boldsymbol{t}}_{proj} = -s\,\check{\boldsymbol{t}} \cdot \boldsymbol{L} =$ const, i.e., the projections $\boldsymbol{y}_{proj}(\widehat{\boldsymbol{h}})$ of the vectors $\widehat{\boldsymbol{h}}$ of the zone lie on a straight line. This means that a Laue pattern can be transformed via (6.4) to a pattern in which spots of zones are located along straight lines[2] (Fig. 6.5).

[2] This was first noticed by Friedrich W.B. Rinne [9].

The above property is useful for devising indexing procedures. The geometry of the reciprocal lattice is in a sense visible on the transformed Laue pattern which is a gnomonic projection of a part of the reciprocal lattice. Formally, the gnomonic projection of a lattice is dense with points, but clearly, a transformation of a real Laue pattern is not dense because of the limited size of the experimentally accessible part of the reciprocal space.

6.3 Gnomonic Projection of a Cell

Before proceeding to indexing procedures, one needs to explain the gnomonic projection of a cell from its vertex at $\mathbf{0}$. The projections p_i, $i = 1, \ldots, 7$ of the remaining vertices on the projection plane are arranged in a certain pattern. With the numbering of the points p_i as in Fig. 6.6, the triplets (p_1, p_2, p_6), (p_2, p_3, p_4), (p_3, p_1, p_5), (p_4, p_5, p_7), (p_5, p_6, p_7) and (p_6, p_4, p_7) are collinear. Let $\boldsymbol{p}^i$ ($i = 1, \ldots, 7$) denote a non-zero vector (of arbitrary magnitude) directed toward p_i, and let $d_{ijk} = \det[\boldsymbol{p}^i\ \boldsymbol{p}^j\ \boldsymbol{p}^k]$. Formally, the collinearity implies that

$$d_{126} = d_{234} = d_{315} = d_{457} = d_{567} = d_{647} = 0 \,. \tag{6.6}$$

With proper ordering of the indices 1, 2 and 3, one has $d_{123} > 0$.

Now, assuming the cell has a certain volume v and the projection points are known, what are the cell parameters? The cell is based on the vectors $x_1\boldsymbol{p}^1$, $x_2\boldsymbol{p}^2$ and $x_3\boldsymbol{p}^3$,

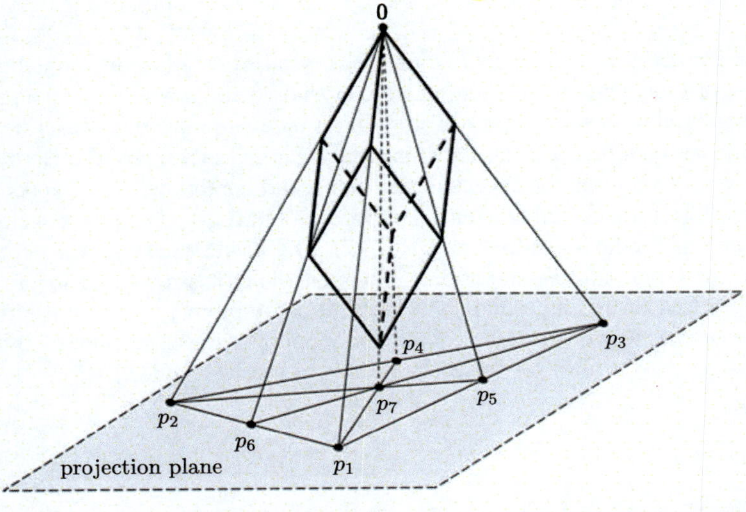

Fig. 6.6 Gnomonic projection from cell vertex $\mathbf{0}$. Projections of the other vertices are at points p_i, $i = 1, \ldots, 7$.

and the cell vertex opposite to $\mathbf{0}$ is given by

$$x_7 \boldsymbol{p}^7 = x_1 \boldsymbol{p}^1 + x_2 \boldsymbol{p}^2 + x_3 \boldsymbol{p}^3 \; .$$

This equation and the condition that the volume of the cell is

$$\det[x_1 \boldsymbol{p}^1 \; x_2 \boldsymbol{p}^2 \; x_3 \boldsymbol{p}^3] = x_1 x_3 x_2 \, d_{123} = v$$

can be solved with respect to x_1, x_3, x_2 and x_7. Assuming $d_{123} > 0$, one obtains

$$(x_1, \; x_2, \; x_3, \; x_7) = \alpha^{1/3} \, (d_{723}, \; d_{173}, \; d_{127}, \; d_{123}) \; ,$$

where $\alpha = v/(d_{723}\, d_{173}\, d_{127}\, d_{123})$. Summarizing, with an arrangement of projection points satisfying (6.6), one can get parameters of the cell up to its volume.[3] The above construction applies to vectors in direct space or in reciprocal space. In the latter case, the cell parameters are

$$\boldsymbol{a}^i = x_{(i)} \boldsymbol{p}^{(i)} \; , \quad i = 1, 2, 3 \; . \tag{6.7}$$

For convenience, one can take vectors $\boldsymbol{p}^i$ of unit magnitude and assume that $v = 1$.

6.4 Laue Indexing

6.4.1 Indexing Software

There are numerous computer programs for crystal orientation determination and indexing of Laue patterns originating from crystals with known lattice parameters. In contrast, only a few procedures for ab initio indexing have been described, and generally, the literature on the subject is scanty. Carr, Cruickshank and Harding [7] described indexing via visual recognition of special configurations of spots in the gnomonic projection. Building on that approach, Ravelli *et al.* [10] developed a semi-automatic indexing software (*LaueCell*). The *XMAS* package (for multigrain indexing of Laue patterns and strain refinement) [11] has a routine for ab initio indexing based on the algorithm of Ravelli et al. A prototype indexing program *IndX_Laue* was written to test additional steps improving efficiency of indexing (Sect. 6.4.5). A Windows version of *IndX_Laue* is available at [12].

[3] For alternative expressions, see [1] or [10].

6.4.2 An Approach Referring to Direct Space

The methods described in [7] and [10] were intended for transmission Laue patterns, but a similar approach can be used for back-reflection patterns. Reflections from tautozonal planes correspond to conics in the Laue pattern. (See Sect. 2.3.5.) A unit vector, say, $\widehat{\boldsymbol{t}}$, along the zone axis has the direction of a certain direct lattice vector $\check{\boldsymbol{t}}$. A conic in with Laue spots, or equivalently, a line with spots in the projected pattern, allows for determining $\widehat{\boldsymbol{t}}$. The procedures described in [7, 10] are based on "main" conics (lines) in the pattern (the projected pattern); these are the conics (lines) with small average spacing between spots and with wide clear regions bordering them. They are expected to correspond to low index zone axes, i.e., to low index direct lattice vectors [6].

Thus, the first step is to look for the "main" conics corresponding to "main" reciprocal lattice planes and low-index direct lattice vectors [7, 10]. An important role is played by a conic-search algorithm. It needs to account for the number of spots contributing to the conic and the quality of fit. Detected conics provide a set of unit vectors $\widehat{\boldsymbol{t}}_m$. All possible triplets of linearly independent vectors $\widehat{\boldsymbol{t}}_m$ are checked whether the vectors have directions of basis vectors of the direct lattice. With a given triplet $\widehat{\boldsymbol{t}}_{m_1}, \widehat{\boldsymbol{t}}_{m_2}, \widehat{\boldsymbol{t}}_{m_3}$, the list of remaining vectors is searched for $\widehat{\boldsymbol{t}}_{m_4}, \widehat{\boldsymbol{t}}_{m_5}, \widehat{\boldsymbol{t}}_{m_6}, \widehat{\boldsymbol{t}}_{m_7}$ which could be arranged in pattern representing gnomonic projection of a cell considered in Sect. 6.3. If such vectors are on the list, one has a septuplet of vectors determining (via a formula analogous to (6.7)) a scaled cell of the direct lattice. The resulting cells are reduced and sorted by frequency of occurrence. The final solution is chosen based on the root-mean-square distance between the predicted and observed spot positions.

6.4.3 Getting Zone Axes via Integral Transforms

A vector $\widehat{\boldsymbol{t}}$ along a zone axis can be determined and refined using some integral transforms considered in the field of integral geometry (which investigates measures and transformations of functions on geometrical spaces). These transforms are considered in this chapter because, they fit here naturally: with Laue diffraction, the magnitudes of the scattering vectors are unknown, and data are given on a unit sphere or its projections. Similar methods are applicable in the case of zone axes used in indexing of single-crystal data from monochromatic experiments; see Sects. 5.5 and 5.10. The transforms are also used in crystal orientation determination.

Integral Transforms

The Radon transform $\mathcal{R}f$ is a function on the set of hyperplanes in R^n: for a suitable real-valued function f on R^n, the $\mathcal{R}f$ function at the hyperplane L^{n-1} is given by

$(\mathcal{R}f)(L^{n-1}) = \int_{L^{n-1}} f$. In the simplest version, the Radon transform is a function on the set of lines in R^2. It turns out that a smooth function on R^n is actually determined by its integrals over the hyperplanes, and hence, there exists a Radon inversion formula, i.e., one can get f from Rf. Related to the Radon transform is a the X-ray transform $\mathcal{X}$. In R^n, it is defined by integrals over lines L^1: $(\mathcal{X}f)(L^1) = \int_{L^1} f$. There also exists an X-ray inversion formula. In two-dimensions, the X-ray transform coincides with the Radon transform.

Similarly defined are Funk transformations: in the simplest form, R^2 and a line of the Radon transform are replaced by the sphere S^2 and a great circle, respectively. The Funk transform is an even function on the sphere: for a suitable real-valued function f on S^2, the value of the Funk transform at $\widehat{\boldsymbol{n}}$ is $\int_{C(\widehat{\boldsymbol{n}})} f$, where $C(\widehat{\boldsymbol{n}})$ is the great circle on the plane normal to $\widehat{\boldsymbol{n}}$. The kernel of the Funk transform consists of odd functions. Funk transform of even continuous functions is invertible. When generalized to S^n, the integration is over S^{n-1} or over geodesics (great circles S^1), i.e., they are analogous to Radon transform and to X-ray transform, respectively.

Relationship Between Zone Axes and Peaks on Integral Transforms of Patterns

It is worth to recapitulate briefly Sect. 6.2. The pattern recorded on a planar detector represents the distribution of intensities as a function of $\boldsymbol{y}_{Laue}$ (as in Fig. 6.5*a*). The recorded pattern projected on the unit sphere corresponds to the distribution of intensities versus normalized wave vectors $\widehat{\boldsymbol{k}}_{\check{h}}$, and the recorded pattern is its gnomonic projection given by (6.2). Reflections of a zone are located on a conic on the recorded pattern.

The intensities can also be ascribed to normalized vectors $\widehat{\boldsymbol{h}} = (\widehat{\boldsymbol{k}}_{\check{h}} - \widehat{\boldsymbol{k}}_0)/|\widehat{\boldsymbol{k}}_{\check{h}} - \widehat{\boldsymbol{k}}_0|$. This leads again to a function on the unit sphere. In this case, reflections of zones are located on great circles. The gnomonic projection of the function gives the distribution of intensities versus $\boldsymbol{y}_{proj}$. In this case, reflections of zones are projected on straight lines.

The standard technique for line detection is the Radon transformation. In digital image analysis, the discrete form of the Radon transform in R^2 is known as the Hough transform. The integration over lines is seen as an accumulation (or voting) procedure. Each line on the image is represented by two line parameters which represent a pixel in the 'accumulator space'. The Hough transform is a function over the accumulator space. To get the Hough transform of an image, intensity at a pixel of the image is added to that function at all accumulator space pixels representing lines through the image pixel, and that step is performed for all image pixels. If there is a high intensity line in the image, there will be a corresponding high intensity peak in the Hough transform of the image. The detection of line in the image is carried out by detection of local maxima of its Hough transform.

With high intensity spots (of reflections from tautozonal planes) located on straight lines in gnomonic projections of Laue patterns, Hough transforms of the projections can be used for analysis of these patterns. Having parameters of a line on a projection of Laue pattern, one can calculate the corresponding $\hat{t}$ vector. The Hough transform of the projected pattern of Fig. 6.5*b* is shown in Fig. 6.7*a*.

Instead of using the gnomonic projection (with spots of reflections from tautozonal planes on straight lines), one can use the original function on the unit sphere (with those spots on great circles). Analogously to the former case, high intensity great circles lead to maxima on the Funk transform of this function. The Funk transform

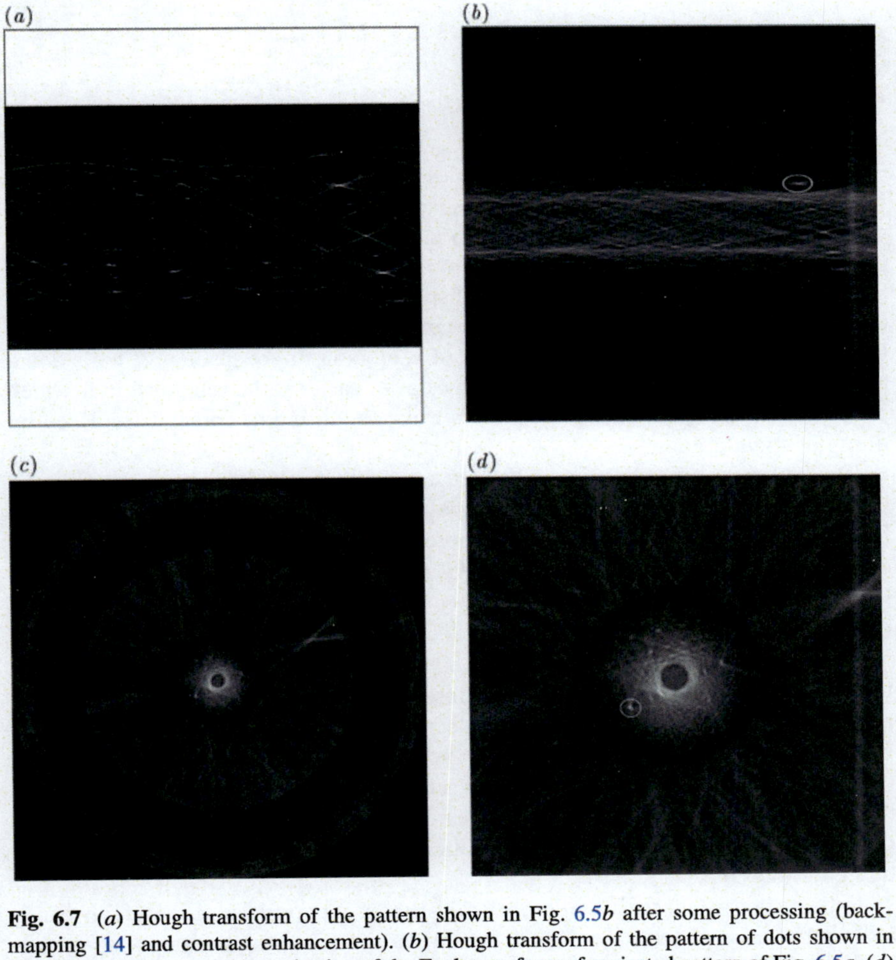

Fig. 6.7 (*a*) Hough transform of the pattern shown in Fig. 6.5*b* after some processing (back-mapping [14] and contrast enhancement). (*b*) Hough transform of the pattern of dots shown in Fig. 6.8*a*. (*c*) Stereographic projection of the Funk transform of projected pattern of Fig. 6.5*a*. (*d*) Enlarged central part of (*c*). In (*a*) and (*b*), the usual (Duda-Hart) coordinates are used: the horizontal axis corresponds to the line inclination angle in the range from 0 to π and the vertical axis represents the signed distance from the pattern center. The spots marked in (*b*) and (*d*) represent the conic indicated by an arrow in Fig. 6.5*a*.

of the spherical function correspnding to the pattern of Fig. 6.5 is shown in Figs. 6.7*c* and *d*.

The relationship between zone axes and maxima of Radon or Funk transforms can be used in various ways. The simplest application is for detection of spots; see, e.g., [13]. The other are determination of zone axes, ascribing Laue spots to zones and refinement of spot position. Instead of applying the transform to a pattern of intensities, one can detect spots on the recorded pattern, and apply the transform to the binary image with detected spots as in Fig. 6.7*b*. The transforms can also be used for refinement of the position of the pattern center.

6.4.4 Fitting a Consistent Mesh

The arrangement of reciprocal lattice nodes can be to some extent determined at the stage of spot detection. Standardly, peak-search routines are used to get spot positions on the pattern, and the distance between measured and predicted spot positions is minimized at the late stage of refinement when approximate lattice parameters are already known. However, the positions of spots on the pattern can be corrected at the outset of processing by fitting conics. The problem is linear (see Sect. 5.14.1) so the fitting can be easily performed. The use of vectors in direct space [10] boils down to fitting conics to spots. The task becomes even easier if the relationship between Laue pattern and gnomonic projection is used: the fitting of conics is replaced by fitting of straight lines.

The key point is to fit conics or lines in a consistent way. For simplicity, only the lines are considered. Generally, n lines may intersect at $n(n-1)/2$ distinct points. However, a diffraction spot ascribed to more than two lines provides additional constraints on the fitted lines as they should intersect at a *single* point near the spot (i.e., they are concurrent). Therefore, rather than fitting individual lines sequentially, they can be fitted simultaneously, using the constraints imposed by the identity of such intersection points. Ultimately, this approach gives spot positions more consistent and better matching a lattice than those resulting from the application of only a peak-search algorithm. In particular, positions of points at intersections of lines match a certain reciprocal lattice (and its superlattices). In other words, by fitting the consistent mesh, i.e., a mesh comprising sets of concurrent lines, one partly refines the projection of the lattice without knowing lattice parameters, and the refined part of projection practically determines the lattice. Figure 6.8 illustrates such processing of a Laue pattern.

Fitting a consistent mesh and proper adjusting of spot positions is helpful because indexing of Laue patterns involves searching for special configurations of spots (see Sect. 6.4.1), and this is done by testing for coplanarity of scattering vectors. Clearly, vectors corresponding to spots ascribed to one zone are coplanar, i.e., adjusting spot positions to a consistent mesh greatly simplifies the search for special spot configurations. Moreover, the mesh can be used for dealing with *spatial* spot overlap

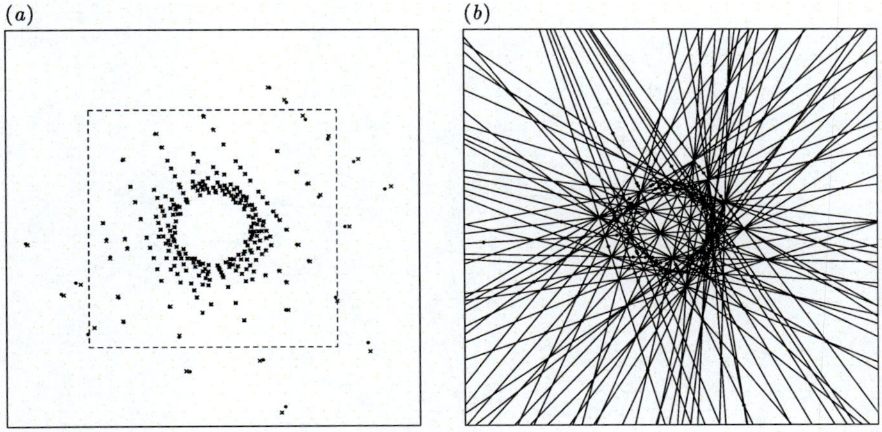

Fig. 6.8 (*a*) Gnomonic projection of Laue spots detected on the pattern shown in Fig. 6.5*a* (crosses) and and their positions (disks) matching straight lines. Dashed square marks approximate location of the image shown in Fig. 6.5*b*. (*b*) The mesh of lines fitted to seven or more points. The mesh was created in a simple iterative way: lines passing near sufficiently large number of points were determined, they were fitted to points (with weighting accounting for projection), and then positions of the points were slightly moved toward the fitted lines. The root-mean-square deviation of spot positions observed on the 512×512 pixel diffraction pattern from the positions after fitting was 0.42 of a pixel. Of 246 detected spots, 181 were ascribed to intersections of lines. (Adapted from [15].)

arising for structures with large unit cells and for automatic identification of nodal spots, i.e., low index spots surrounded by clear spaces.

6.4.5 Indexing Limited to Reciprocal Space

The use of vectors in direct space is convenient for fitting conics to spots, but indexing of reflections can be performed without reference to the direct space, i.e., can be based on points of the reciprocal space. The simplest approach is to identify a cell of the reciprocal lattice, (assuming the cell is primitive) get a reciprocal lattice basis, and check how well the reflections can be indexed with the basis. Clearly, one needs to test numerous cells. There are two simple methods to enlarge the set of cells tested as potential primitive lattice cells. The first one is to use configurations of spots smaller than septuplets, e.g., sextuplets. Moreover, the likelihood of detecting a primitive cell can be increased by testing subcells of the cells based on these special spot configurations. (These two methods are applicable in both direct and reciprocal space indexing.)

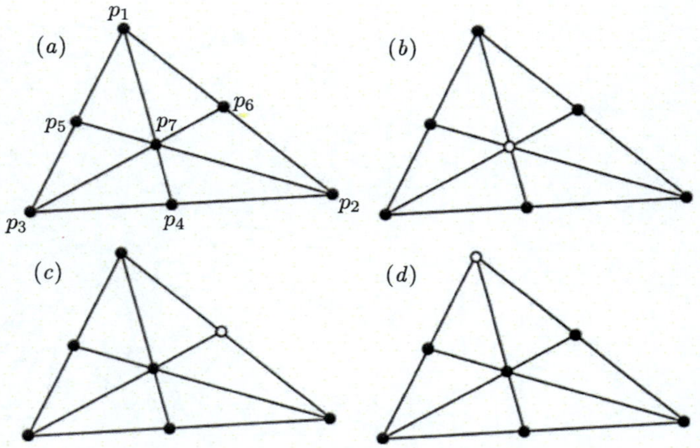

Fig. 6.9 Illustration of the projection of a parallelepipedal cell from one of its vertices. In general, the other vertices are projected on seven specially arranged points $p_1, \; p_2, \ldots, \; p_7$ (*a*). The cell is determined by a smaller number of points, in particular, by sextuplets of the types shown in (*b*)–(*d*). Points marked by circles are missing.

6.4.6 Using Sextuplets of Points

The key task is to get the parameters of a primitive cell of the reciprocal lattice. The cell can be derived from a septuplet of specially arranged spots. However, some spots may be missed in automatic detection and some are absent, and the septuplets are scarce. The problem can be addressed by noting that the cell is actually determined by a smaller number of spots. Instead of looking for septuplets, one can get a cell from a smaller number of spots and test whether it indexes the pattern. These can be specially arranged sextuplets (Fig. 6.9). (One may also consider getting cells from quintuplets or quadruplets of points.) Using such smaller configurations significantly increases the number of tested cells and the likelihood of encountering the correct one.[4]

6.4.7 Testing Superlattices

A cell resulting from the previous steps is likely to be non-primitive. To increase the probability of finding a true primitive cell, besides testing the reciprocal lattice

[4] An incomplete configuration can also be used for additional peak detection: having, say, a sextuplet of spots, the software can compute the coordinates of the seventh point, and then go back to the diffraction pattern to check for the missing spot. Similarly, with the mesh described in Sect. 6.4.4, having a line intersection, one can check for a missing Laue spot.

Table 6.1 Example indexing rates for different numbers of points used to get a cell and different indices of tested superlattices. N is the number of reflections used in indexing. $M_{k,i}$ means that indexing was performed using k-tuplets and all superlattices up to index i. The symbols 6b and 6c denote sextuplets of the types shown in Figs. 6.9b and c, respectively. The data were obtained using *IndX_Laue* from a set of 100 patterns simulated based on data given in frame number 3 in [10] (and corresponding to the pattern shown in Fig. 6.5a). The crystal orientations were random, uniformly distributed random errors were added to spot positions, and N randomly selected spots were used for indexing. Only the crystal lattice was accounted for, i.e., intensities (in particular reflection absences) were ignored. Results depend on simulation parameters (magnitude of errors, the largest allowed reflection index) and on thresholds used in indexing; for all cases listed in Table 6.1, the same simulation parameters and thresholds were used. In particular, the largest allowed reflection index in both simulation and indexing was 12.

N	$M_{7,1}$	$M_{7,3}$	$M_{7,8}$	$M_{7,16}$	$M_{6b,1}$	$M_{6c,1}$	$M_{6c,3}$	$M_{6c,8}$
90	0.02	0.04	0.14	0.18	0.31	0.13	0.34	0.68
120	0.02	0.14	0.43	0.60	0.57	0.23	0.67	0.96
150	0.12	0.45	0.75	0.91	0.75	0.43	0.89	0.99
180	0.20	0.66	0.93	0.97	0.88	0.66	0.94	1.00

basis spanning a given cell, one can also test bases of low-index superlattices of that lattice. This step is exactly the same as that described in Sect. 5.11.

The impact of using superlattices and sextuplets is illustrated in Table 6.1. The table contains example indexing rates (i.e., fractions of instances when the correct solution is listed) for septuplets and sextuplets of two types and a number of superlattices.

6.4.8 Indices of an Individual Reflection

In practice, the key step in indexing is to ascribe indices to an individual Laue spot. The data are a trial basis $\boldsymbol{a}^i$ of reciprocal lattice (obtained from a septuplet or a sextuplet of points via (6.7)) and a normalized scattering vector $\widehat{\boldsymbol{h}}$ (obtained from the spot position); both the trial basis $\boldsymbol{a}^i$ and the vector $\widehat{\boldsymbol{h}}$ are given in a laboratory Cartesian reference frame based on some vectors $\boldsymbol{e}_i^L$, i.e., one has components $A_j^{\ i}$ and v_i such that $\boldsymbol{a}^i = A_j^{\ i}\,\boldsymbol{e}_L^j$ and $\widehat{\boldsymbol{h}} = v_i\,\boldsymbol{e}_L^i$. With $X = A^{-1}$, the triplet $X\widehat{\boldsymbol{h}} = \boldsymbol{r}$ contains components of $\widehat{\boldsymbol{h}}$ in the tested basis $\boldsymbol{a}^i$. If the basis is close to the true basis of the crystal, then the triplet $\boldsymbol{r} = [r_1\, r_2\, r_3]^T$ is nearly proportional to integer indices of the reflection. The dimensionless triplet $\bar{\boldsymbol{r}} = \boldsymbol{r}/\sqrt{r_1^2 + r_2^2 + r_3^2}$ is not affected by the arbitrarily chosen scale factor of the trial basis $\boldsymbol{a}^i$. Let

$$\mathsf{h}(x) = \lfloor x\,\bar{\boldsymbol{r}} \rceil\,, \quad \chi(x) = A\,\mathsf{h}(x)$$

and

$$x_{min} = \text{argmin}_x \arccos(\widehat{\chi}(x) \cdot \widehat{\boldsymbol{h}}) ,$$

where x is larger than 0, and it does not exceed $\sqrt{3}\times$ the maximal allowed value of a reflection index. At the right x, $\mathsf{h}(x)$ is a triplet of integers close to $x\,\bar{\boldsymbol{r}}$, i.e., these are (integer) components of a node of reciprocal lattice in the basis $\boldsymbol{a}^i$. This node is represented by $\chi(x)$ with components in the Cartesian basis $\boldsymbol{e}_i^L$, $\arccos\left(\widehat{\chi}(x) \cdot \widehat{\boldsymbol{h}}\right)$ is the angular deviation between the experimental and computed normalized scattering vectors ($\widehat{\boldsymbol{h}}$ and $\widehat{\chi}(x)$, respectively), and x_{min} is the argument at which the deviation takes minimal value. If the value of $\arccos\left(\widehat{\chi}(x_{min}) \cdot \widehat{\boldsymbol{h}}\right)$ is smaller than a threshold, then the three integers

$$[h_1\, h_2\, h_3]^T = \mathsf{h}(x_{min})$$

(after division by their largest common divisor) are assumed to be the indices of the reflection corresponding to $\widehat{\boldsymbol{h}}$.

Besides the angular deviation $\arccos\left(\widehat{\chi}(x_{min}) \cdot \widehat{\boldsymbol{h}}\right)$, the quality of the individual reflection $\widehat{\boldsymbol{h}}$ is measured by the magnitude of the difference between the real experiment-based approximation of the indices and the actual integer values of the indices. With $\bar{r}_k$ representing the k-th component of $\bar{\boldsymbol{r}}$, a real $\widehat{\boldsymbol{h}}$-based approximation of the indices is $h_i^r = \left(\sum_k h_k \bar{r}_k\right) \bar{r}_i$. As above (see Sect. 5.4.4), the quantities $\sum_i \cos\left(2\pi h_i^r\right)$, $\sqrt{\sum_i \left(h_i^r - h_i\right)^2}$ or

$$q(\widehat{\boldsymbol{h}}) = \max_i \left\{\left|h_i^r - h_i\right|\right\} \tag{6.8}$$

are simple examples of measures of quality of individual reflections.

6.4.9 *Quality of Solution—Figure of Merit*

Having a tentative solution, i.e., the basis of the reciprocal lattice and indices of reflections, one can refine the lattice parameters. Critical for the performance of indexing programs is a figure of merit, i.e., quantitative measure of quality of solutions. As in indexing of single-crystal patterns collected using monochromatic techniques, it needs to incorporate the number of indexed reflections, the largest absolute value of the reflection index and the quality of the fit between (directions of) simulated scattering vectors and corresponding detected vectors.

In *IndX_Laue*, the refinement is performed only at the very end and it applies only to stored solutions. It is done by fitting integer combinations of basis vectors to the experimental scattering vectors with magnitudes of scattering vectors calculated based on the indices obtained for these vectors. The measure of quality of a given solution is analogous to that applied in *Ind_X* (cf. Sect. 5.4.4)

$$w_\epsilon N_\epsilon + w_{\epsilon/2} N_{\epsilon/2} , \tag{6.9}$$

where w_ϵ and $w_{\epsilon/2}$ are user-provided weights, and N_ϵ is the number of scattering vectors $\widehat{\boldsymbol{h}}^n$ $(n = 1, 2, \ldots N)$ with approximations of indices differing from integers by less than ϵ, i.e., $N_\epsilon = \#\{\, n\, ;\; q\left(\widehat{\boldsymbol{h}}^n\right) \leq \epsilon,\; 1 \leq n \leq N\}$ with $q\left(\widehat{\boldsymbol{h}}^n\right)$ given by (6.8).

Clearly, if a diffraction pattern can be indexed using a given reciprocal lattice, it can also be indexed using its superlattices. In conventional indexing of single-crystal patterns, when the magnitudes of scattering vectors are known, if the actual reciprocal lattice is a proper sublattice of the determined lattice, at least one index of all reflections is a multiple of an integer larger than 1. (E.g., if, say, the true basis vector $\boldsymbol{a}^3 = \mathbf{c}^\star$ equals $2\times$ the determined vector, then all indices $h_3 = l$ ascribed to legitimate reflections will be even.) This allows for an easy automatic identification and elimination of such lattices. In the case of Laue patterns, the magnitudes of scattering vectors are not known, and indices hkl ascribed to spots are relatively prime even if the true indices of the reflection are not, i.e., the actual indices of a reflection are divided by their largest common divisor. This eliminates the aforementioned property; the actual reciprocal lattice may be a proper sublattice of the determined lattice with none of h, k or l being a multiple of a fixed integer (> 1) for all reflections. Therefore, the criterion based on the presence of a common divisor of an index for all reflections needs to be replaced by a mechanism modifying the figure of merit based on the frequency of occurrence of such a divisor. In practice, the key point is to check what are the fractions of odd indices h, k and l for a given solution, and to reduce its quality if there is a large disparity between the fractions.

In the program *IndX_Laue*, a simple quality-modifying factor was used to account for superlattices of the true lattice: with n_h, n_k and n_l denoting the number of odd h, k and l indices, respectively, if $x = \min\{n_h, n_k, n_l\}/\max\{n_h, n_k, n_l, 1\} < a = 1/2$, the quality of a solution was simply reduced by the factor $(2a - x)x/a^2$. The algorithm of *IndX_Laue* is sketched below.

6.5 Indexing of Pink-Beam Diffraction Patterns

In principle, data obtained by energy-dispersive Laue diffraction (EDLD) methods employing polychromatic beam and energy-sensitive detectors which give access to magnitudes of scattering vectors (e.g., [16], [17] or [18]) can be indexed using procedures of Chap. 5 [19]. In practice, these indexing methods are applicable if the energy resolution is sufficiently high. Similar situation occurs in the cases of pink-beam data[5] and monochromatic data with inaccurate magnitudes of scattering vectors. A convenient example of a technique providing scattering vectors with low-accuracy magnitudes is EBSD. The magnitudes of the scattering vectors obtained from EBSD patterns correspond directly to the widths of EBSD bands; because of vague band intensity profiles, it is difficult to get accurate band widths, and the magnitudes are poorly defined. For pink-beam data, the ranges of the uncertainties

[5] Pink-beam serial crystallography emerges as a promising method for determination of structures of macromolecules [4].

Input: List of N reflections $\widehat{\boldsymbol{h}}^{n}$
Input: Optional parameters; e.g., assumed volume of primitive cell, thresholds, extent of search E
`/* Initialization */`
Set default values of parameters not specified in input
`/* Proper indexing */`
for *each triplet* $(\widehat{\boldsymbol{h}}^{n_1}, \widehat{\boldsymbol{h}}^{n_2}, \widehat{\boldsymbol{h}}^{n_3})$ **do**
    `/*` $\widehat{\boldsymbol{h}}^{n_i}$ $(i = 1, 2, 3)$ `corresponds to` $\boldsymbol{p}^i$ `of Sect. 6.3 */`
    `/* and to points` p_i `of Fig. 6.9 */`
    look for remaining $\boldsymbol{p}^i$ of a septuplet (sextuplet) which would constitute projections of vertices of a cell
    for *each septuplet (sextuplet)* **do**
        get the triplet $(\boldsymbol{a}^1, \boldsymbol{a}^2, \boldsymbol{a}^3)$ by using (6.7)
        for *tentative bases* $(\boldsymbol{b}^1, \boldsymbol{b}^2, \boldsymbol{b}^3)$ *of each low-index superlattice of the lattice based on the triplet* $(\boldsymbol{a}^1, \boldsymbol{a}^2, \boldsymbol{a}^3)$ **do**
            Buerger-reduce $(\boldsymbol{b}^1, \boldsymbol{b}^2, \boldsymbol{b}^3)$
            `/* By an attempt to index all reflections (Sect. 6.4.8) */`
            get basis quality by eq.(6.9)
            if *the quality is sufficiently good* **then**
                $(\boldsymbol{b}_1, \boldsymbol{b}_2, \boldsymbol{b}_3) \leftarrow$ basis (of direct lattice) reciprocal to $(\boldsymbol{b}^1, \boldsymbol{b}^2, \boldsymbol{b}^3)$
                Buerger-reduce $(\boldsymbol{b}_1, \boldsymbol{b}_2, \boldsymbol{b}_3)$
                store the basis $(\boldsymbol{b}_1, \boldsymbol{b}_2, \boldsymbol{b}_3)$ and its *quality*

for *all stored highest-quality bases* **do**
    refine the basis
    modify basis quality (end of Sect. 6.4.9)
sort the bases
Result: Highest-quality bases and indices of indexable reflections

Algorithm of *IndX_Laue*: This is a sketch of the prototype program showing its key steps.

are related to the spectral width of the radiation. Some pink-beam X-ray data are indexed using software designed for monochromatic data; see, e.g, [20, 21]. Also reasonably accurate EBSD-based data can be indexed by programs designed for monochromatic data [22].

If the uncertainties of magnitudes of the scattering vectors are large, the magnitudes can be discarded by normalizing the vectors, and the problem becomes analogous to ab initio Laue indexing. The input consists of normalized scattering vectors. As was noted above, ab initio indexing provides a lattice up to a scale, i.e., the absolute volume of the primitive cell is indeterminable. Moreover, there is no ground for specifying the order of a reflection, and only relatively prime indices are ascribed to reflections. However, having also crude estimates of the magnitudes of the scattering vectors, one can fit the scaling factor and the orders of the indexed reflections. An algorithm for that is described below.

6.5.1 *Algorithm for Fitting the Scaling Factor and Orders of Reflections*

Let X be the true matrix with direct basis vectors (in rows) in the Cartesian laboratory reference frame. The basis resulting from indexing normalized scattering vectors is known up to a scaling factor; with the factor denoted by S (> 0), one knows the matrix $X_S = X/S$ such that $|\det X_S| = 1$. The true volume of the direct lattice cell is $V = |\det X| = S^3$. Let $\mathtt{h}_i$ in typewriter font denote the column matrix with relatively prime indices of the i-th reflection obtained from the Laue indexing. The relationship (5.3) between the scattering vector $\boldsymbol{g}_i$ and the indices $\mathtt{h}_i$ is $X\,\boldsymbol{g}_i = S\,X_S\,\boldsymbol{g}_i \approx m_i \mathtt{h}_i$, where $m_i \geq 1$ is a small integer. (Summation convention is suspended in this subsection.) With $\check{\boldsymbol{h}}_i = X_S^{-1}\,\mathtt{h}_i$, one has

$$S\,\boldsymbol{g}_i \approx m_i\,\check{\boldsymbol{h}}_i\,. \tag{6.10}$$

Given N Laue-indexed experimental scattering vectors $\boldsymbol{g}_i$ and corresponding vectors $\check{\boldsymbol{h}}_i$, one can determine the common factor S and the orders m_i of individual reflections by the following procedure based on the relationship (6.10).

For a selected reflection j (i.e., a selected $\boldsymbol{g}_j$ and $\check{\boldsymbol{h}}_j$), assuming that $m_j = 1$:

1. Get the scaling factor σ_j of the j-th reflection by minimizing $|\sigma \boldsymbol{g}_j - \check{\boldsymbol{h}}_j|^2$ with respect to σ; it is given by $\sigma_j = \boldsymbol{g}_j \cdot \check{\boldsymbol{h}}_j / |\boldsymbol{g}_j|^2$.

2. With the obtained σ_j, for each $i = 1, 2, \ldots, N$, get the integer factor $\mu_i(j) \geq 1$ by minimizing $|\sigma_j\,\boldsymbol{g}_i - \mu\,\check{\boldsymbol{h}}_i|^2$ with respect to μ; it is given by $\mu_i(j) = \max\left(1, \left\lfloor \sigma_j \boldsymbol{g}_i \cdot \check{\boldsymbol{h}}_i / |\check{\boldsymbol{h}}_i|^2 \right\rceil\right)$, where $\lfloor x \rceil$ denotes the integer nearest to x.

3. Knowing $\mu_i(j)$ for $i = 1, 2, \ldots, N$, compute the scaling factor $\Sigma(j)$ applicable to all indexed reflections by minimizing $\sum_{i=1}^{N} \left|\sigma\,\boldsymbol{g}_i - \mu_i(j)\,\check{\boldsymbol{h}}_i\right|^2$ with respect to σ; it is given by

$$\Sigma(j) = \frac{\sum_{i=1}^{N} \mu_i(j)\,\boldsymbol{g}_i \cdot \check{\boldsymbol{h}}_i}{\sum_{i=1}^{N} |\boldsymbol{g}_i|^2}\,.$$

4. Compute the residue $R(j) = \sum_{i=1}^{N} \left|\Sigma(j)\,\boldsymbol{g}_i - \mu_i(j)\,\check{\boldsymbol{h}}_i\right|^2$.

Since the initial assumption $m_j = 1$ may be false, the steps 1–4 must be repeated for all $j = 1, \ldots, N$ (or at least for a sufficient number of them). The actual scaling factor S corresponds to $j = J$ leading to minimal R, i.e., with $J = \arg\min_j R(j)$, one has $S = \Sigma(J)$ and $m_i = \mu_i(J)$. The actual indices of the i-th reflection are $m_i\,\mathtt{h}_i$.

References

1. J.L. Amorós, M.J. Buerger, M.C. de Amorós, *The Laue Method* (Academic Press, New York, 1975)
2. Z. Ren, D. Bourgeois, J.R. Helliwell, K. Moffat, V. Šrajer, B.L. Stoddard, Laue crystallography: coming of age. J. Synchrotron Rad. **6**, 891–917 (1999)
3. D. Popov, N. Velisavljevic, M. Somayazulu, Mechanisms of pressure-induced phase transitions by real-time Laue diffraction. Crystals **9**, 672 (2019)
4. A. Förster, C. Schulze-Briese, A shared vision for macromolecular crystallography over the next five years. Struct. Dyn. **6**, 064302 (2019)
5. S. Cornaby, D.M.E. Szebenyi, D.M. Smilgies, D.J. Schuller, R. Gillilan, Q. Hao, D.H. Bilderback, Feasibility of one-shot-per-crystal structure determination using Laue diffraction. Acta Cryst. D **66**, 2–11 (2010)
6. D.W.J. Cruickshank, J.R. Helliwell, K. Moffat, Angular distribution of reflections in Laue diffraction. Acta. Cryst. A **47**, 352–373 (1991)
7. P.D. Carr, D.W.J. Cruickshank, M.M. Harding, The determination of the unit-cell parameters from the Laue diffraction patterns using their gnomonic projection. J. Appl. Cryst. **25**, 294–308 (1992)
8. P.D. Carr, I.M. Dodd, M.M. Harding, The determination of unit-cell parameters from a Laue diffraction pattern. J. Appl. Cryst. **26**, 384–387 (1993)
9. R.W.G. Wyckoff, The crystal structures of some carbonates of the calcite group. Am. J. Sci. **50**, 317–360 (1920)
10. R.B.G. Ravelli, A.M.F. Hezemans, H. Krabbendam, J. Kroon, Towards automatic indexing of the Laue diffraction pattern. J. Appl. Cryst. **29**, 270–278 (1996)
11. N. Tamura, *XMAS*: a versatile tool for analyzing synchrotron X-ray microdiffraction data, in *Strain and Dislocation Gradients from Diffraction*. ed. by R. Barabash, G. Ice (Imperial College Press, London, 2014), pp.125–155
12. http://imim.pl/personal/adam.morawiec/. Accessed Aug 2022
13. H.R. Wenk, F. Heidelbach, D. Chateigner, F. Zontone, Laue orientation imaging. J. Synchrotron Rad. **4**, 95–101 (1997)
14. G. Gerig, F. Klein, Fast contour identification through efficient Hough transform and simplified interpretation strategy, in *Proceedings of the 8th International Joint Conference on Pattern Recognition* (Paris, 1986), pp. 498–500
15. A. Morawiec, On ab initio indexing of Laue diffraction patterns. J. Appl. Cryst. **54**, 333–337 (2021)
16. K.F. Fischer, H.G. Krane, W.H.W. Morgenroth, A single crystal X-ray diffractometer for white synchrotron radiation with solid state detectors: energy dispersive Laue (EDL) instrument at HASYLAB, Hamburg/Germany. Nucl. Instrum. Methods Phys. Res. A **369**, 306–311 (1996)
17. S. Send, M. von Kozierowski, T. Panzner, S. Gorfman, K. Nurdan, A.H. Walenta, U. Pietsch, W. Leitenberger, R. Hartmann, L. Strüder, Energy-dispersive Laue diffraction by means of a frame-store pnCCD. J. Appl. Cryst. **42**, 1139–1146 (2009)
18. F. Kurdzesau, Energy-dispersive Laue experiments with X-ray tube and PILATUS detector: determination of lattice constants. J. Appl. Cryst. **52**, 72–93 (2019)
19. R.A. Jacobson, An orientation-matrix approach to Laue indexing. J. Appl. Cryst. **19**, 283–286 (1986)
20. J.M. Martin-Garcia, L. Zhu, D. Mendez, M.Y. Lee, E. Chun, C. Li, H. Hu, G. Subramanian, D. Kissick, C. Ogata, R. Henning, A. Ishchenko, Z. Dobson, S. Zhang, U. Weierstall, J.C.H. Spence, P. Fromme, N.A. Zatsepin, R.F. Fischetti, V. Cherezov, W. Liu, High-viscosity injector-based pink-beam serial crystallography of microcrystals at a synchrotron radiation source. IUCrJ **6**, 412–425 (2019)
21. A. Tolstikova, *Development of Diffraction Analysis Methods for Serial Crystallography*. PhD thesis, Universität Hamburg, Hamburg, Germany, 2020
22. A. Morawiec, A remark on ab initio indexing of EBSD patterns. J. Appl. Cryst. **54**, 1844–1846 (2021)

Chapter 7
Indexing of Powder Diffraction Patterns

Not all crystalline materials can be obtained as single crystals of sizes suitable for diffraction experiments. Their structures are investigated using powder diffraction methods [1–3]. As was already mentioned in Sect. 2.3, typical powder diffraction patterns are recorded using standard laboratory X-ray setups and the Bragg-Brentano technique, but they also arise in numerous other circumstances. When a monochromatic beam is diffracted by a polycrystalline material, diffracted rays constitute coaxial cones. The intersections of these cones with an area detector perpendicular to the incident beam (and to the axis of the cones) are concentric circles known as Debye-Scherrer rings. In the field of electron microscopy, they are called electron ring patterns.

Solving a crystal structure from powder diffraction data begins with the determination of peak (line) positions and then indexing, i.e., reconstruction of the unknown lattice of the crystal. Further steps are space group determination, calculation of atom positions, refinement of parameters and a validation confirming that the obtained structural model is right. It is clear that a reliable indexing procedure is crucial for the correctness of the resulting structure. The reliability strongly depends on the quality of experimental data and on the complexity of the investigated structures. (There are difficulties in indexing patterns from triclinic and monoclinic crystals).

7.1 Link Between Peaks Positions and Reflection Indices

As was described in Sect. 2.3, with crystals of various orientations contributing to a powder diffraction pattern, the pattern is one-dimensional: it shows the intensity of the scattered radiation versus the scattering angle. In the simplest approach, indexing of

A. Morawiec, *Indexing of Crystal Diffraction Patterns*, Springer Series in Materials Science 326, https://doi.org/10.1007/978-3-031-11077-1_7

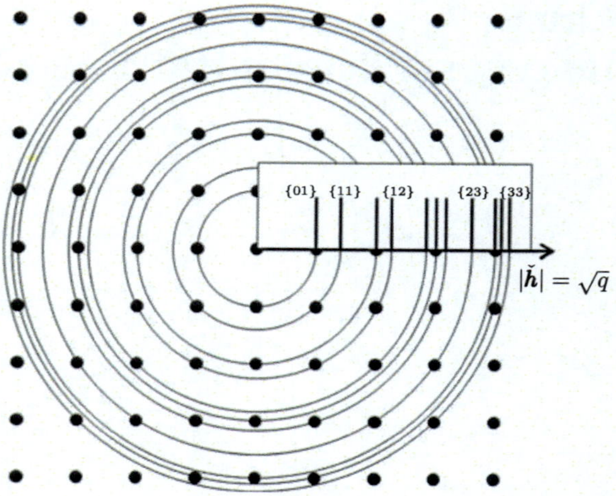

Fig. 7.1 Schematic of a two-dimensional square reciprocal lattice and corresponding stick pattern. Cf. Fig. 2.6.

powder diffraction patterns involves only peak positions.[1] In this case, the input data set is not an experimental pattern but a line (or 'stick') pattern, i.e., the line positions plus the tolerances for the positions[2]; see Fig. 7.1. Instead of using the Bragg's angle, it is convenient to characterize the line positions by squared reciprocals of inter-planar distances

$$1/d^2_{(h\,k\,l)} = (2\sin\theta/\lambda)^2 = q\ ; \tag{7.1}$$

cf. (2.2). The powder indexing is based on a list of experimentally obtained squared reciprocals of inter-planar distances $q_n, n = 1, ..., N$, sorted according to $q_n < q_{n+1}$. Using (1.48), one has

$$1/d^2_{(h_1\,h_2\,h_3)} = \check{\boldsymbol{h}}^2 = \check{\boldsymbol{h}} \cdot \check{\boldsymbol{h}} = g^{ij} h_i h_j\ . \tag{7.2}$$

The quadratic form on the right-hand side of this expression will be denoted by $Q(\check{\boldsymbol{h}}) = \check{\boldsymbol{h}} \cdot \check{\boldsymbol{h}} = g^{ij} h_i h_j$ or by $Q(hkl)$ (where $h = h_1, k = h_2$ and $l = h_3$). Formulas (7.1) and (7.2) are fundamental for indexing of powder diffraction patterns. The point is to determine a lattice metric g and a set of reciprocal lattice vectors $\check{\boldsymbol{h}}_m$ ($m = 1 \ldots M$) such that for each experimental coordinate q_n there is an m such that

$$q_n = Q(\check{\boldsymbol{h}}_m)\ . \tag{7.3}$$

For q_n known with an experimental uncertainty δ_n, one actually expects that

$$|Q(\check{\boldsymbol{h}}_m) - q_n| < \delta_n\ . \tag{7.4}$$

---

[1] Peak positions are obtained from peak profiling procedures, and peak widths are used to estimate the precision of the positions.

[2] More sophisticated approaches are described below at the end of 7.4.

If arbitrarily large integers were allowed, the problem of assigning $\check{\boldsymbol{h}}_m$ to q_n with inequalities (7.4) satisfied would have an infinite number of solutions. However, there are physical limitations on the unit cell dimensions and the largest reflection index. Moreover, ideally, there should be no $Q(\check{\boldsymbol{h}}_m)$ in the range $[q_1, q_N]$ without a corresponding diffraction peak, and no peak without a corresponding $\check{\boldsymbol{h}}_m$. Thus, one looks for a lattice metric with an integer solution of (7.3) (within a known limited domain).

7.2 Ambiguities

A question arises whether the indexing problem as described above has a unique solution. The answer is no; different lattices (with different reduced cells) may give the same stick patterns [4, 5]. If two metrics are related by a uni-modular integer matrix (cf. (1.8)), then they are equivalent in the sense that they correspond to the same lattice. The ambiguity arises when two non-equivalent metrics lead to the same set of $Q(\check{\boldsymbol{h}}_m)$ numbers.

A simple example of such ambiguity involves primitive cubic and primitive tetragonal lattices. With the reciprocal metric of the cubic lattice given by $\mathrm{diag}(1, 1, 1)/a^2$, the quantity $a^2 Q(hkl) = h^2 + k^2 + l^2$ takes all values which can be constructed as a sum of three squares of integers, i.e, all positive integers except the numbers of the form $4^{n_1}(8n_2 + 7)$ with integer $n_1, n_2 \geq 0$, (i.e., except numbers like 7, 15, 23, 28, 31 et cetera, Sloane's A004215, [6]); see Fig. 7.2. For the primitive tetragonal lattice with the reciprocal lattice metric of the form $\mathrm{diag}(2, 2, 1)/a^2$, one has $a^2 Q(hkl) = 2(h^2 + k^2) + l^2 = h'^2 + k'^2 + l^2$, where $h' = h + k$ and $k' = h - k$, i.e., the list of values of $2(h^2 + k^2) + l^2$ is the same as that for $h^2 + k^2 + l^2$. This shows that peak positions for the considered tetragonal lattice are exactly the same as those for the cubic lattice. There are more such ambiguous cases. Some patterns correspond to more than two different lattices [7]. An algorithm for listing all lattices with the same sequence of $Q(\check{\boldsymbol{h}})$ values as a given lattice is reported in [8].

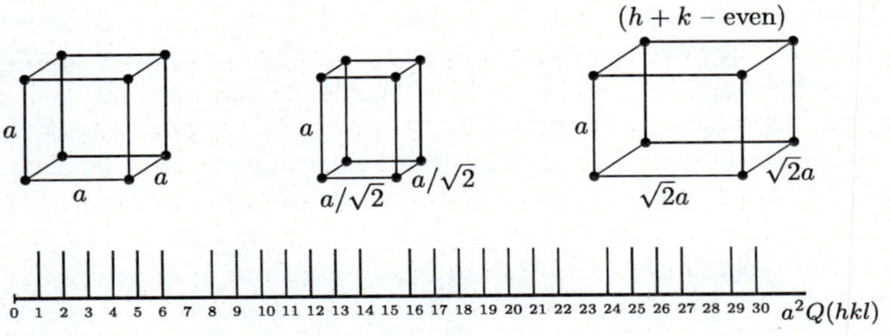

Fig. 7.2 Ambiguous pattern corresponding to primitive cubic and primitive tetragonal lattices, and unit cells of direct lattices leading to this pattern.

The situation is additionally complicated by absences of reflections. For instance, for the primitive tetragonal lattice with the reciprocal lattice metric $\mathrm{diag}(1/2, 1/2, 1)/a^2$, one has $a^2 Q(hkl) = (h^2 + k^2)/2 + l^2$. With $h' = (h + k)/2$ and $k' = (h - k)/2$, there occurs $(h^2 + k^2)/2 + l^2 = h'^2 + k'^2 + l^2$. Thus, if the conditions for reflections by the planes (hkl) are that $h + k$ are even (e.g., the special conditions for the space group $P4 = C_4^1$, No. 75 in *ITC-A*), the list of values $a^2 Q(hkl)$ for this tetragonal lattice is again the same as that considered above for the primitive cubic lattice.

Extra information is needed to resolve ambiguous cases. Prior knowledge may help in selecting the right solution. If there is no supplementary information, wrong solutions must be identified and eliminated in further steps of structure determination process. Modern programs link the indexing stage with subsequent stages to resolve the indexing problem in a unique way or to reduce the number of proposed solutions.

7.3 Figures of Merit

The experimental data are affected by errors. The coordinates q_n are given with limited accuracy and precision. Common is the systematic 'zero-point' error; it is a shift of the q coordinates from $(2 \sin\theta/\lambda)^2$ to $(2 \sin(\theta + \delta_0)/\lambda)^2$. Peak broadening increases uncertainty of peak positions. The indexing is additionally complicated by spurious (impurity) or overlapping peaks and reflection absences. In effect, the assignment between the magnitudes $Q(\check{\boldsymbol{h}}_m)$ and the lines in the range $[q_1, q_N]$ may not necessary be one-to-one.

In the context of the above, the expectation is to formulate the powder indexing as an optimization problem. First, one wants to minimize the number of experimental lines q_n without assignment and the number of low index vectors $\check{\boldsymbol{h}}_m$ without assignment. Moreover, a correct solution should give small deviations between the experimental q_n and the corresponding calculated $Q(\check{\boldsymbol{h}}_m)$. Available software calculates numerous indicators of the quality of the proposed solutions.

It is worth noting here, that in powder indexing, it is not necessary to exploit the entire available pattern. At lower Bragg angles, the diffraction lines are better separated (so it is easier to get peak positions), line positions are less sensitive to deviations from true lattice parameters, and indices of the peaks have low absolute values. Therefore, in most cases, 20 to 30 low-angle lines are used [9].

Some criteria for accepting a resulting lattice metric as a solution have been formulated based on experimental practice and their use became a standard. They are known as *figures of merit*. Two classical figures of merit were proposed by de Wolff [10, 11] and Smith and Snyder [12].

De Wolff's figure of merit [10] is defined as

$$M_{20} = \frac{Q_{20}}{\Delta_Q} \frac{1}{2\mathrm{N}_{20}} ,$$

where Q_{20} is the q value for the 20-th *indexed* experimental peak, N_{20} is the number of different calculated $Q(\check{\boldsymbol{h}})$ values not larger than Q_{20}, and Δ_Q is the average deviation between calculated $Q(\check{\boldsymbol{h}})$ and the corresponding experimental q_n values for the first 20 indexed peaks. Results of indexing with $M_{20} < 10$ are doubtful and those with $M_{20} > 20$ are likely to be correct. The de Wolff indicator is sometimes generalized in an obvious way to M_N with the limit N other than 20, but M_{20} is a widely used standard.

Similarly, the Smith–Snyder's figure of merit [12] is given by

$$F_N = \frac{1}{\Delta_{2\theta}} \frac{N}{\mathsf{N}_N} ,$$

where N_N is the number of different calculated $Q(\check{\boldsymbol{h}})$ values not larger than q for the N-th indexed experimental peak, and $\Delta_{2\theta}$ is the average deviation between calculated and the corresponding experimental 2θ angles (in degrees) for the first N indexed peaks. The units of F_N are reciprocal degrees. Since N/N_N never exceeds unity, there occurs $\Delta_{2\theta} \leq 1/F_N$, and with correct indexing, F_N allows for an immediate evaluation of the accuracy of experimental data. However, practice shows that, as a tool for assessing the indexing results, de Wolff's figure of merit is superior [9].

These and other figures of merit (e.g., [13–15]) involve only positions of lines, and they may fail for complicated structures or error-affected data. More sophisticated criteria are based on the complete pattern: with assumed lattice metrics, the intensities are estimated, and the new figures of merit are designed to weigh the calculated profiles against the experimental one [14, 16].

7.4 Indexing Procedures

The indexing procedures usually start with the assumption of cubic symmetry and than go to lower symmetries (possibly down to the triclinic system). The number of independent parameters and the number of feasible solutions grow with the decrease of symmetry. The problem of indexing is solved by scanning the integer triplets $hkl \sim \check{\boldsymbol{h}}$ or by scanning the continuous 'space' of lattice parameters, or a combination of both.

⊡ It is worth noting that very rough estimates of lattice parameters can be made based on the density of diffraction lines. As an example, consider the method proposed in [17]. Let the pattern contain N lines with the coordinates q_n not exceeding q_{max}. A ball of radius $\sqrt{q_{max}}$ in the reciprocal space has the volume of $V_b = 4\pi q_{max}^{3/2}/3$. The volume of the unit cell of the reciprocal lattice is $V_r = 1/\sqrt{\det(g)}$. The number of reciprocal lattice points N in the ball is approximately equal to $V_b/(m V_r) = 4\pi q_{max}^{3/2} \sqrt{\det(g)}/(3m)$, where m is introduced to take into account reciprocal lattice points giving overlapping lines and those corresponding to absent or very weak reflections. The factor m is structure-dependent and difficult to estimate quantitatively but some ball-park figures can be assigned for particular symmetries. In [17], m is assumed to be the symmetry-induced multiplicity factor for a general reflection multiplied by 2 to account for absent reflections. Hence, the volume of the unit cell of the direct lattice is roughly $V = 1/V_r = \sqrt{\det(g)} \approx 3mN/(4\pi q_{max}^{3/2})$. Knowing the volume, one can estimate lengths of edges of the primitive cell (under the assumption that its shape does not deviate too much from a cube).

7.4.1 Search in the Continuous Parameter Space

The "dichotomy method" of Louërs [18] is the best known example of the search in the parameter space. The point is to eliminate ranges of the entries of g, for which there is no correspondence between the simulated and experimental patterns. The software based on this principle (*DICVOL*), initially applicable to high symmetries [18], was later extended to lower symmetries [19–21]. The procedure starts with the assumption of cubic system and than goes to lower symmetries. The full domain of independent parameters is the Cartesian product of intervals of particular parameters applicable to a given symmetry. The intervals are then probed with certain steps (e.g., 0.4Å for lengths and 5° for angles). This divides the full domain into small domains. Such a small domain is retained if the observed lines q_n lie within the bounds imposed on $Q(\check{\boldsymbol{h}})$ by the choice of the domain and experimental accuracies; otherwise, the domain is discarded. Subsequently, the retained domain is investigated in more detail. It is subdivided by halving the intervals of parameters involved, and each sub-domain is tested whether it imposes bounds on $Q(\check{\boldsymbol{h}})$ such that they bracket the observed lines. This dichotomous division is repeated a number of times. In some respects, the method of Louërs is similar to the branch-and-bound paradigm; cf. [22].

Parameter space can be scanned in other ways. E.g., in *McMaille* [23] the lattice parameters are selected randomly (or—straightforwardly—by a grid search). The recalculated diagram with some spread added to lines is compared to the experimental pattern. If they are close enough, the parameters are refined by random testing with small parameter changes. Random search in the parameter space combined with trial assignments of indices and least-squares solutions of (7.3) with respect to g^{ij} is described in [24]. Searches in the parameter space were reported to be implemented in other programs; see, e.g., [25, 26].

7.4.2 Search in the Discrete Index Space

This approach is based on a tentative assignment of reflection indices to some lines (referred to as base-lines). With the assumed assignments of $\check{\boldsymbol{h}}_m$ to q_n, (7.2) and (7.3) are turned into a system of linear equations with respect to entries of the metric tensor. In the most general (triclinic) case, one needs a system of six independent equations with respect to six independent entries g^{ij}. If symmetry is higher than triclinic, solving the problem requires smaller number of equations, and consequently, the number of required base-lines is smaller. Each system of independent equations leads to a lattice metric. Such obtained metric can be tested against the rest of the experimental pattern. It indexes the base-lines but not necessary the remaining lines. If it indexes all lines, the metric and the assignment constitute one of the solutions of the indexing problem.

Search through *all* possible combinations of vectors with integer entries requires prohibitive times, and therefore, an exhaustive trial-and-error with respect to the indices is not a viable solution. The crucial point is to limit the number of tried combinations of $\check{\boldsymbol{h}}_m$ vectors. The number is smaller for higher symmetries, and they are tried first. Some assignments are unfeasible (e.g., very high indices assigned to low q_n). Further restrictions may be imposed based on constraints on the values of lattice parameters and the volume of the primitive cell. Identification of such cases, allows for reduction of execution times.

There are a number of programs based on the search in the discrete index 'space'. To speed up the computations, pre-calculated tables of possible and non-redundant combinations of vectors with integer entries for particular symmetries are used [27–29]. In [30, 31], to solve the triclinic case, the computation times are reduced on account of the assumption that the first two lines can be indexed as (100) and (010) or as (200) and (010) or as (100) and (110) or as (020) and $(\bar{1}10)$; with two lattice parameters known, the system $q_n = g^{ij} h_i h_j$ is reduced to four equations. Finally, the search through feasible assignments is used in the indexing program *TREOR* [16, 32, 33].

7.4.3 Relationships Between Line Positions

In the cubic case, with $g^{ij} = a^{*2}\delta^{ij}$ and a^* denoting the unknown reciprocal lattice parameter, the experimental peak positions q_n are expected to be approximately equal to $Q(\check{\boldsymbol{h}}) = a^{*2}(\delta^{ij} h_i h_j)$, i.e., to a^{*2} times integers being sums of squares of the indices h_i; see Sect. 7.2. Thus, for low value q_n coordinates, their ratios should be fractions of small integers. In this simple indexing case, the main problem is to identify the integers based on the fractions. For tetragonal and hexagonal lattices with conventional metrics (Sect. 1.4) and integer indices (hkl), one has $Q(\check{\boldsymbol{h}}) = ma^{*2} + l^2c^{*2}$, where m equals $h^2 + k^2$ and $h^2 + k^2 + hk$, in the tetragonal and hexagonal cases, respectively. The differences between $Q(\check{\boldsymbol{h}})$ (and between experimental q_n) which happen to correspond to the same l (or the same (h, k)) are integers multiplied by the common a^{*2} (or c^{*2}). This may suffice to index some simple patterns, but in general, more sophisticated relationships between q_n coordinates are needed.

Relationships between the quantities $Q(\check{\boldsymbol{h}})$ (which are reflected in relationships between the experimental quantities q_n if the lattice metric and the assignment are correct) are a basis of an approach sometimes referred to as Runge–Ito–de Wolff method or as zone-indexing; see, e.g., [34, 35]. The simplest example of such a relationship is

$$Q(\eta\check{\boldsymbol{h}}) = \eta^2 Q(\check{\boldsymbol{h}}) .$$

Diffraction lines at positions q_{n_1} and q_{n_2} such that $q_{n_1} = \eta^2 q_{n_2}$ for some integer $\eta > 1$ are likely to correspond to different orders of the same reflection. More useful relations involve two reciprocal lattice vectors. For instance, with $\check{\boldsymbol{h}}_1$ and $\check{\boldsymbol{h}}_2$ denoting such vectors, since $(\check{\boldsymbol{h}}_1 + \check{\boldsymbol{h}}_2)^2 + (\check{\boldsymbol{h}}_1 - \check{\boldsymbol{h}}_2)^2 = 2\check{\boldsymbol{h}}_1^2 + 2\check{\boldsymbol{h}}_2^2$, one has

$$Q(\check{\boldsymbol{h}}_1 + \check{\boldsymbol{h}}_2) + Q(\check{\boldsymbol{h}}_1 - \check{\boldsymbol{h}}_2) = 2Q(\check{\boldsymbol{h}}_1) + 2Q(\check{\boldsymbol{h}}_2) . \tag{7.5}$$

The vectors $\check{\boldsymbol{h}}_1, \check{\boldsymbol{h}}_2, \check{\boldsymbol{h}}_1 + \check{\boldsymbol{h}}_2$ and $\check{\boldsymbol{h}}_1 - \check{\boldsymbol{h}}_2$ are coplanar, i.e., the corresponding planes are tautozonal, and the approach can be seen as a search for a zone based on $\check{\boldsymbol{h}}_1$ and $\check{\boldsymbol{h}}_2$. A quadruplet of diffraction lines at positions q_{n_i} $(i = 1, \ldots, 4)$ such that $q_{n_1} + q_{n_2} = 2q_{n_3} + 2q_{n_4}$ is likely to correspond to such coplanar vectors. A case more general than (7.5) is considered in [36].

⊡ With K vectors $\check{\boldsymbol{h}}_i$ and integers η_i, $(i = 1, \ldots K)$, one has $Q\left(\sum_i \eta_i \check{\boldsymbol{h}}_i\right) = \sum_i \eta_i^2 Q(\check{\boldsymbol{h}}_i) + 2\sum_{i<j} \eta_i \eta_j \check{\boldsymbol{h}}_i \cdot \check{\boldsymbol{h}}_j$; it is more convenient to write these expressions using the $\sum$ sign, and the summation convention is suspended to the end of this subsection. One may consider multiple, say, L $(1 < L \leq K)$, relations of this type for the same set of vectors $\check{\boldsymbol{h}}_i$

$$Q\left(\sum_i \eta_i^l \check{\boldsymbol{h}}_i\right) = \sum_i \left(\eta_i^l\right)^2 Q(\check{\boldsymbol{h}}_i) + 2\sum_{i<j} \eta_i^l \eta_j^l \check{\boldsymbol{h}}_i \cdot \check{\boldsymbol{h}}_j \,, \quad l = 1, \ldots, L \,, \quad \text{no summation over } l \,.$$

A relationship between the $Q(\check{\boldsymbol{h}})$ values is obtained if the last terms $(2\sum \eta_i^l \eta_j^l \check{\boldsymbol{h}}_i \cdot \check{\boldsymbol{h}}_j)$ are eliminated. The point is to find a set of non-zero coefficients S_l such that

$$\sum_l S_l \, \eta_i^l \eta_j^l = 0 \tag{7.6}$$

for all i and $j > i$. With such coefficients, one has

$$\sum_l S_l \, Q\left(\sum_i \eta_i^l \check{\boldsymbol{h}}_i\right) = \sum_{l,i} S_l \left(\eta_i^l\right)^2 Q(\check{\boldsymbol{h}}_i) \,.$$

For the important case of $K = 2 = L$, the coefficients $S_1 = \eta_1^2 \eta_2^2$ and $S_2 = -\eta_1^1 \eta_2^1$ satisfy the condition (7.6), and thus

$$\eta_1^2 \eta_2^2 \, Q(\eta_1^1 \check{\boldsymbol{h}}_1 + \eta_2^1 \check{\boldsymbol{h}}_2) - \eta_1^1 \eta_2^1 \, Q(\eta_1^2 \check{\boldsymbol{h}}_1 + \eta_2^2 \check{\boldsymbol{h}}_2) = \left(\eta_1^1 \eta_2^2 - \eta_2^1 \eta_1^2\right)\left(\eta_1^1 \eta_1^2 \, Q(\check{\boldsymbol{h}}_1) - \eta_2^2 \eta_2^1 \, Q(\check{\boldsymbol{h}}_2)\right) .$$

In particular, with $\eta_1^1 = \eta_2^1 = \eta_1^2 = -\eta_2^2 = 1$, one obtains (7.5).

Relationships between line positions are exploited in the classic program *ITO* [37], and in a number of other packages; see, e.g., [38, 39]. The software determines zones—sets of coplanar reciprocal lattice vectors; see Sect. 1.5.1. With multiple zones, the two-dimensional solutions are combined into a three-dimensional lattice, i.e., a full solution of the indexing problem.

7.4.4 *Metric in Conventional Crystallographic Setting*

In general, the basis resulting from indexing is different form the conventional one for a given Bravais class, and the transformation between these two bases needs to be determined. The first step in this direction is to check whether the obtained basis is primitive. Then, one of the methods described in Sect. 1.7 can be used to get the lattice symmetries and the lattice parameters in the conventional setting. Subsequently, systematic absences are identified, and the remaining steps of structure determination are carried out.

7.4.5 *Indexing Based on Complete Pattern*

Indexing programs propose multiple solutions, and a user must decide which of them are to be considered further. As was mentioned before, such decisions can be based on further steps of structure determination. Thus, indexing and structure solution are linked and mutually dependent. This fact is taken into account in more recent indexing programs. Fitting the whole profile and optimization with respect to lattice parameters is reported in [40]. Whole profile is also used in the recent version of *TREOR* (*N-TREOR09*) [16] and in *TOPAS* (of Bruker AXS) [41]. The new software uses figures of merit relying on the complete pattern and factors determined by models feasible for the considered structure [14, 16, 41]. Additionally, in experimental practice, prior knowledge of structure fragments, isostructural characteristics, expected bond lengths *et cetera* is used to increase the chances of selecting the right solution.

7.5 Integrated Software Packages

As was mentioned above, the reliability of indexing is affected by impurity lines,[3] inaccuracies (mainly zero-point error) and imprecision in peak positions, and it strongly depends on the complexity of the structure (size of the asymmetric unit, crystal symmetry).[4] In the presence of errors, the indexing problem frequently cannot be resolved using just one approach. Since the methods and algorithms used by some of the indexing programs are different, it is logical to apply multiple programs to increase the chances of getting a correct solution. There exist integrated software packages containing several indexing programs (e.g., [16, 33, 43, 44]). For compre-

[3] In most programs, provisions are made for a small number of impurity lines. Programs designed to process data from multiphase samples allow for numerous spurious lines. For instance, *McMaille* [23] and *AUTOX* [42] are reported to have this capability.

[4] Moreover, indexing programs are vulnerable in the presence of dominant zone axes. Some of them are equipped in dominant-zone-axis tests, e.g. *TREOR* [16, 32].

hensive lists of powder indexing programs the reader is referred to [45–48] and to more recent papers listed in the bibliography.

The efficiency of powder diffraction software was checked in round robin tests [49, 50]. The conclusions concerning indexing are that the process is far from being fully automatic, and that judging the quality of solutions proposed by software "remains a task for a well-trained crystallographer" [47] but one must take into account high complexity of investigated structures and high expectations (raised by comparisons with single-crystal methods). In such difficult cases, indexing is not fully automatic but a computer-assisted interactive process. This means that there is still space for improvement of the indexing software at least to the point at which "a well-trained crystallographer" will be unlikely to do better than a computer program.

References

1. J.I. Langford, D. Louër, Powder diffraction. Rep. Prog. Phys. **59**, 131–234 (1996)
2. Structure determination from powder diffraction data, in *International Union of Crystallography Monographs on Crystallography 13*. ed. by W.I.F. David, K. Shankland, L.B. McCusker, Ch. Baerlocher (Oxford University Press, Oxford, 2002)
3. V.K. Pecharsky, P.Y. Zavalij, *Fundamentals of Powder Diffraction and Structural Characterization of Materials* (Springer, New York, 2005)
4. A.D. Mighell, A. Santoro, Geometrical ambiguities in the indexing of powder patterns. J. Appl. Cryst. **8**, 372–374 (1975)
5. A.D. Mighell, J.K. Stalick, The reliability of powder indexing procedures. NBS Special Publication **567**, 393–403 (1980)
6. N.J.A. Sloane, The on-line encyclopedia of integer sequences. http://oeis.org/. Accessed Aug 2022
7. A.D. Mighell, Ambiguities in powder indexing: conjunction of ternary and binary lattice metric singularity in the cubic system. J. Res. Natl. Inst. Sand. Technol. **109**, 569–579 (2004)
8. R. Oishi-Tomiyasu, Method to generate all the geometrical ambiguities of powder indexing solutions. J. Appl. Cryst. **47**, 2055–2059 (2014)
9. R. Shirley, Data accuracy for powder indexing. NBS Special Publication **567**, 361–382 (1980)
10. P.M. de Wolff, A simplified criterion for the reliability of a powder pattern indexing. J. Appl. Cryst. **1**, 108–113 (1968)
11. P.M. de Wolff, The definition of the indexing figure of merit M_{20}. J. Appl. Cryst. **5**, 243 (1972)
12. G.S. Smith, R.L. Snyder, F_N: A criterion for rating powder diffraction patterns and evaluating the reliability of powder-pattern indexing. J. Appl. Cryst. **12**, 60–65 (1979)
13. T. Ishida, Y. Watanabe, A criterion method for indexing unknown powder diffraction patterns. Z. Kristallogr. **160**, 19–32 (1982)
14. J. Bergmann, *EFLECH/INDEX*–another try of whole pattern indexing. Z. Kristallogr. Suppl. **26**, 197–202 (2007)
15. R. Oishi-Tomiyasu, Reversed de Wolff figure of merit and its application to powder indexing solutions. J. Appl. Cryst. **46**, 1277–1282 (2013)
16. A. Altomare, G. Campi, C. Cuocci, L. Eriksson, C. Giacovazzo, A. Moliterni, R. Rizzi, P.E. Werner, Advances in powder diffraction pattern indexing: *N-TREOR09*. J. Appl. Cryst. **42**, 768–775 (2009)
17. H. Lipson, Indexing powder photographs of orthorhombic crystals. Acta Cryst. **2**, 43–45 (1949)
18. D. Louër, M. Louër, Méthode d'essais et erreurs pour l'indexation automatique des diagrammes de poudre. J. Appl. Cryst. **5**, 271–275 (1972)

19. D. Louër, R. Vargas, Indexation automatique des diagrammes de poudre par dichotomies successives. J. Appl. Cryst. **15**, 542–545 (1982)
20. A. Boultif, D. Louër, Indexing of powder diffraction patterns for low-symmetry lattices by the successive dichotomy method. J. Appl. Cryst. **24**, 987–993 (1991)
21. D. Louër, A. Boultif, Indexing with the successive dichotomy method, DICVOL04. Z. Kristallogr. Suppl. **23**, 225–230 (2006)
22. W. Sun, *Computational Method for the Indexing of Unknown Powder Patterns and Unknown Rotating-Crystal Patterns*. PhD thesis, The University of Oklahoma, Norman, Oklahoma, USA, 1972
23. A. Le Bail, Monte Carlo indexing with *McMaille*. Powder Diffr. **19**, 249–254 (2004)
24. A.A. Coelho, Indexing of powder diffraction patterns by iterative use of singular value decomposition. J. Appl. Cryst. **36**, 86–95 (2003)
25. J. Bergmann and R. Kleeberg. *EFLECH/INDEX*—a program for peak search/fit and indexing. *IUCr CPD Newsletter*, 21:Art. 5, 1999
26. M.A. Neumann, *X-Cell*: a novel indexing algorithm for routine tasks and difficult cases. J. Appl. Cryst. **36**, 356–365 (2003)
27. D. Taupin, Une méthode générale pour l'indexation des diagrammes de poudres. J. Appl. Cryst. **1**, 178–181 (1968)
28. D. Taupin, A powder-diagram automatic-indexing routine. J. Appl. Cryst. **6**, 380–385 (1973)
29. D. Taupin, Enhancements in powder-indexing. J. Appl. Cryst. **22**, 455–459 (1989)
30. F. Kohlbeck, E.M. Hörl, Indexing program for powder patterns especially suitable for triclinic, monoclinic and orthorhombic lattices. J. Appl. Cryst. **9**, 28–33 (1976)
31. F. Kohlbeck, E.M. Hörl, Trial and error indexing program for powder patterns of monoclinic substances. J. Appl. Cryst. **11**, 60–61 (1978)
32. P.-E. Werner, L. Eriksson, M. Westdahl, *TREOR*, a semi-exhaustive trial-and-error powder indexing program for all symmetries. J. Appl. Cryst. **18**, 367–370 (1985)
33. A. Altomare, C. Giacovazzo, A. Guagliardi, A.G.G. Moliterni, R. Rizzi, P.E. Werner, New techniques for indexing: *N-TREOR* in *EXPO*. J. Appl. Cryst. **33**, 1180–1186 (2000)
34. P.M. de Wolff, On the determination of unit-cell dimensions from powder diffraction patterns. Acta. Cryst. **10**, 590–595 (1957)
35. P.M. de Wolff, Detection of simultaneous zone relations among powder diffraction lines. Acta Cryst. **11**, 664–665 (1958)
36. D. Stöckelmann, H. Kroll, W. Hoffmann, R. Heinemann, A system of metrically invariant relations between the moduli squares of reciprocal-lattice vectors in one-, two- and three-dimensional space. J. Appl. Cryst. **43**, 269–275 (2010)
37. J.W. Visser, A fully automatic program for finding the unit cell from powder data. J. Appl. Cryst. **2**, 89–95 (1969)
38. Jr. R.B. Roof, *INDX*: A computer program to aid in the indexing of X-ray powder patterns of crystal structures of unknown symmetry. *Los Alamos Report*, LA-3920, 1968
39. G.S. Smith, E. Kahara, Automated computer indexing of powder patterns: the monoclinic case. J. Appl. Cryst. **8**, 681–683 (1975)
40. B.M. Kariuki, S.A. Beimonte, M.I. McMahon, R.L. Johnston, K.D.M. Harris, R.J. Nelmes, A new approach for indexing powder diffraction data based on whole-profile fitting and global optimization using a genetic algorithm. J. Synchrotron Rad. **6**, 87–92 (1999)
41. A.A. Coelho, An indexing algorithm independent of peak position extraction for X-ray powder diffraction patterns. J. Appl. Cryst. **50**, 1323–1330 (2017)
42. V.B. Zlokazov, *MRIAAU* - a program for auto-indeing multiphase polycrystals. J. Appl. Cryst. **25**, 69–72 (1992)
43. R. Shirley, *The Crysfire 2002 System for Automatic Powder Indexing: Users Manual* (The Lattice Press, Guilford, UK, 2002)
44. M. Bortolotti, I. Lonardelli, *ReX.Cell:* a user-friendly program for powder diffraction indexing. J. Appl. Cryst. **46**, 259–261 (2013)
45. P.E. Werner, Autoindexing. pp. 118–135. in [2]

46. R. Shirley, Overview of powder-indexing program algorithms (history and strengths and weaknesses). IUCr Computing Commission Newsletter **2**, 48–54 (2003)
47. J. Bergmann, A. Le Bail, R. Shirley, V. Zlokazov, Renewed interest in powder diffraction data indexing. Z. Kristallgr. **219**, 783–790 (2004)
48. Collaborative Computational Project No. 14 (CCP14) in Powder and Small Molecule Single Crystal Diffraction. http://ccp14.cryst.bbk.ac.uk/solution/indexing/index.html. Accessed Aug 2022
49. A. Le Bail, L.M.D. Cranswick, Structure determination by powder diffractometry round robin. 2002 – http://www.cristal.org/sdpdrr2/ and 2008 – http://www.cristal.org/SDPDRR3/. Accessed Aug 2022
50. A. Le Bail, L.M.D. Cranswick et al., Third structure determination by powder diffractometry round robin (SDPDRR-3). Powder Diffr. **24**, 254–262 (2009)

Chapter 8
Indexing for Crystal Orientation Determination

The indexing described in previous chapters is a process of ascribing indices to diffraction reflections without knowing the crystal structure. Here, we consider the end-indexing, i.e., indexing of patterns originating from a priori known crystal structures. In particular, it is assumed that unit cell parameters are known, and one knows which reflections are forbidden and which can show up in the pattern. Indexing of patterns from crystals of known structures is used mainly for determination of crystal orientations. Knowing reflection indices, one can easily get the orientation. The issue of crystal orientation determination arises in many fields, e.g., in conventional electron microscopic analysis of individual crystallites or interfaces and in diffractometric measurements [1–3]. Orientation determination is a key element of *orientation mapping*.

8.1 Orientation Mapping

The technique of orientation mapping is widely applied in studies on microstructures and crystallographic textures of polycrystalline materials. Automatic systems collect diffraction patterns originating from small areas of a specimen, determine local orientations, and provide digital maps with orientation parameters ascribed to individual pixels or grains. The automatically generated orientation maps complement conventional contrast images with quantitative data: besides showing topography of crystallites, the map contains information on grain orientations.

Contemporary orientation mapping methods stem from techniques relying on Kossel patterns [4–6]. Nowadays, most orientation maps are created based on a step-by-step scans on computer controlled instruments equipped with digital cameras. At each step, a diffraction pattern is acquired and an orientation is determined. Particularly successful are orientation mapping systems using scanning electron microscopy and electron backscattering diffraction patterns. EBSD-based mapping is a well

A. Morawiec, *Indexing of Crystal Diffraction Patterns*, Springer Series in Materials Science 326, https://doi.org/10.1007/978-3-031-11077-1_8

established technique of characterization of polycrystalline materials [7]. Although less often, orientation maps are also created using other types of diffraction patterns. Similar to EBSD is the technique relying on transmission Kikuchi diffraction in SEM [8, 9]. There are maps based on TEM Kikuchi patterns [10], and spot patterns [11, 12]. TEM orientation maps were also obtained by the so-called 'conical dark-field scanning' with orientation at a given point determined from images corresponding to various incident beam directions [13]; with this approach, the step-by-step beam scanning is avoided.

⊡ In conical dark-field scanning, the incident beam direction is altered (as in centered dark-field imaging [14]). The beam moves in steps on Debye–Scherrer rings, and at each step dark-field microstructure images are recorded. A grain is bright in an image if diffraction occurred, and the incident beam direction determines a scattering vector for that grain. Based on multiple images and multiple scattering vectors, the crystal orientation is determined [13, 15]. With this technique, the conical beam scan replaces the conventional beam scan over pixels of a map.

Orientation maps are also acquired using diffraction of X-rays from both synchrotron [16, 17] and laboratory sources [18–21]. In most cases, polychromatic radiation is used [17–21]. For completeness, one may add orientation mappings in SEM using electron channeling patterns [22, 23].

Besides the 2D maps, diffraction-based determination of local orientations is also used for reconstruction of 3D microstructures, either by combining 2D sections (e.g., [24]) or by resolving diffraction patterns generated by synchrotron beam transmitted through a multi-grain specimen (e.g., [16]). This chapter is focused on indexing of patterns from individual crystallites; multigrain indexing is briefly described in Chap. 11.

The orientation mapping methods differ considerably in spatial resolution, orientation resolution, level of automation and accessibility. Any deeper comparisons are beyond the scope of the book. Let us only note that spatial resolution of a given method is frequently indicated by a proper prefix to the word 'diffraction'; hence, one has microdiffraction and nanodiffraction (or micro-beam diffraction and nano-beam diffraction).

Clearly, computer programs for orientation determination are expected to be general, so they can be applied to arbitrary crystal structures. The other important aspect are orientation resolution and accuracy. However, the two critical features which determine the quality of programs for automatic orientation determination are robustness and speed. Crystallites of polycrystalline materials are relatively challenging, especially in the case of small grain size or large density of defects, and the quality of diffraction patterns may be poor. Moreover, with thousands or millions of diffraction patterns used for creating a pixelized orientation map, what matters is the time-efficiency of pattern processing. For that, the patterns are greatly reduced in size. In other words, for time-efficiency, the pattern-solving software is expected to be capable of dealing with coarse patterns. Generally, the software used for computing orientations must be able to handle poorly localized and spurious reflections. Orientation determination algorithms are as good as their ability to cope with imperfections of input data.

The text below is focused on the computational aspect of the determination of crystal orientations from EBSD and TEM Kikuchi patterns, but since K-line diffraction patterns of different types have the same diffraction geometry, similar principles are applicable to indexing in all these cases.[1] Differences lie in the type of input data (bands, line pairs, single lines) and in the pattern acquisition angle, i.e., the solid angle covered by the detector. This angle is related to the sample-to-detector distance (effective camera length); it is usually relatively small for TEM-based patterns, and it is large for SEM patterns. Roughly, the smaller the acquisition angle, the higher index reflections must be taken into account, and this has an impact on the reliability of the indexing results.

Crystal orientation can be determined without identification of individual reflections, by matching an experimental pattern as a whole to simulated patterns. This method (referred to as direct pattern matching) differs considerably from those involving indexing. The direct matching is presented only briefly in Sect. 8.9. The part on indexing (Sects. 8.2–8.8.1) is more detailed. It describes ideas and building blocks for constructing algorithms, and some example indexing algorithms.

8.2 Orientation via Pattern Indexing

8.2.1 Scattering Vectors and Reciprocal Lattice Vectors

Orientation determination systems based on explicit pattern indexing perform detection and identification of individual reflections. Methods for detecting reflection positions depend on the pattern type. Clearly, locations of diffraction spots are determined using peak detection techniques. To get parameters of bands or lines in EBSD or Kikuchi patterns, various variants of Hough transform are applied. Positions of detected diffraction peaks are used to get coordinates of corresponding scattering vectors. To get indices of reflections, these scattering vectors are matched with reciprocal lattice vectors calculated from crystallographic data. Ultimately, indexing algorithms rely on two lists: a list of scattering vectors and a list of reciprocal lattice vectors.

Before going to a more detailed description of the lists, one needs to note the issue of vector discrimination. If two vectors on one of the lists are closer than a permissible uncertainty in vector determination, they could be matched with the same vector on the other list. This complication can be avoided by eliminating such close vectors at the outset when constructing the lists.

List of Scattering Vectors

The scattering vector $\boldsymbol{g}$ corresponding to a diffraction spot are calculated by simple application of the Laue equation $\boldsymbol{g} = \boldsymbol{k} - \boldsymbol{k}_0$, where $\boldsymbol{k}_0$ and $\boldsymbol{k}$ are wave vectors of the incident beam and the reflected beam, respectively; cf. (2.3). If magnitudes of wave

[1] For older methods of determining crystal orientation (or misorientation, or primary beam direction) from Kikuchi patterns based on 'manual' indexing of patterns see [25–40].

vectors are not accessible (as in the case of spots in Laue patterns), the direction $\widehat{\boldsymbol{g}}$ of the scattering vector can be obtained from a location of a spot by normalizing the difference $\widehat{\boldsymbol{k}} - \widehat{\boldsymbol{k}}_0$ ($\propto \widehat{\boldsymbol{g}}$). As before, symbols with the hat $\widehat{}$ denote normalized vectors.

For K-line patterns, the primary issue is getting scattering vectors from individual K-lines. The relevant procedure described in Sect. 5.14.1 works only for sufficiently curved conics, e.g., for Kossel lines. With electron diffraction, due to the shortness of wavelength, K-lines are usually seen as straight lines, and the expression given in Sect. 5.14.1 is not applicable, but getting the scattering vector from line positions is simple. Take a Kikuchi pattern with pairs of deficit and excess (dark and bright) lines (Fig. 8.1). If $\boldsymbol{L}$ denotes the vector from the radiation source to the pattern center, and $\boldsymbol{l}$ is the vector perpendicular to the Kikuchi line, from the pattern center to the line, then $\boldsymbol{k} = (\boldsymbol{L} + \boldsymbol{l})/(\lambda|\boldsymbol{L} + \boldsymbol{l}|)$ is the wave vector to the point of the line nearest to the pattern center; see, Fig. 8.1. Having two such wave vectors $\boldsymbol{k}_1$ and $\boldsymbol{k}_2$ for the bright and the dark lines of one pair, one can calculate the scattering vector $\boldsymbol{g} = \boldsymbol{k}_1 - \boldsymbol{k}_2$. In EBSD patterns, locations of lines approximating the whole bands are usually determined. In such cases, only directions $\widehat{\boldsymbol{g}}$ of the scattering vectors can be obtained. For a given band, $\widehat{\boldsymbol{g}}$ is the normal to the plane through the line approximating the band and the source of radiation.

⊡ Derivation of the formula for determination of the scattering vector $\boldsymbol{g}$ from positions of Kikuchi lines is simple. The lines are generated by divergent beams, i.e., directions of incident beams are arbitrary. The wave vectors of diffracted beams are governed by (2.5), i.e., the wave vectors $\boldsymbol{k}$ corresponding to the scattering vector $\boldsymbol{g}$ are on the cone described by $\boldsymbol{g} \cdot (\boldsymbol{g} - 2\boldsymbol{k}) = 0$. Similarly, the wave vectors corresponding to $-\boldsymbol{g}$ are on the cone $(-\boldsymbol{g}) \cdot ((-\boldsymbol{g}) - 2\boldsymbol{k}) = \boldsymbol{g} \cdot (\boldsymbol{g} + 2\boldsymbol{k}) = 0$. Let the wave vector $\boldsymbol{k}_1$ be on the first cone and $\boldsymbol{k}_2$ on the second cone, i.e.,

$$\boldsymbol{g} \cdot (\boldsymbol{g} - 2\boldsymbol{k}_1) = 0 \,, \quad \boldsymbol{g} \cdot (\boldsymbol{g} + 2\boldsymbol{k}_2) = 0 \,.$$

By adding and subtracting the sides of these equations, one gets

$$\boldsymbol{g} \cdot (\boldsymbol{g} - (\boldsymbol{k}_1 - \boldsymbol{k}_2)) = 0 \,, \quad \boldsymbol{g} \cdot (\boldsymbol{k}_1 + \boldsymbol{k}_2) = 0 \,. \tag{8.1}$$

If $\boldsymbol{g}$, $\boldsymbol{k}_1$ and $\boldsymbol{k}_2$ are coplanar, one can express the scattering vector as a linear combination of $\boldsymbol{k}_1$ and $\boldsymbol{k}_2$. With $\boldsymbol{g} = \alpha_1 \boldsymbol{k}_1 + \alpha_2 \boldsymbol{k}_2$, (8.1) can be written in the form

$$\boldsymbol{g} \cdot ((\alpha_1 \boldsymbol{k}_1 + \alpha_2 \boldsymbol{k}_2) - (\boldsymbol{k}_1 - \boldsymbol{k}_2)) = 0 \,, \quad (\alpha_1 \boldsymbol{k}_1 + \alpha_2 \boldsymbol{k}_2) \cdot (\boldsymbol{k}_1 + \boldsymbol{k}_2) = 0 \,.$$

Using $\boldsymbol{g} \cdot \boldsymbol{k}_1 = \boldsymbol{g}^2/2$, $\boldsymbol{g} \cdot \boldsymbol{k}_2 = -\boldsymbol{g}^2/2$ and $\boldsymbol{k}_1^2 = 1/\lambda^2 = \boldsymbol{k}_2^2$, one obtains

$$(\alpha_1 - \alpha_2 - 2)\boldsymbol{g}^2 = 0 \,, \quad (\alpha_1 + \alpha_2)(1/\lambda^2 + \boldsymbol{k}_1 \cdot \boldsymbol{k}_2) = 0 \,. \tag{8.2}$$

Both $\boldsymbol{g}^2$ and $1/\lambda^2 + \boldsymbol{k}_1 \cdot \boldsymbol{k}_2$ are positive for physical reasons, and this means that (8.2) are solved by $\alpha_1 = 1$ and $\alpha_2 = -1$. Thus, the scattering vector can be expressed as

$$\boldsymbol{g} = \boldsymbol{k}_1 - \boldsymbol{k}_2 = (\widehat{\boldsymbol{k}}_1 - \widehat{\boldsymbol{k}}_2)/\lambda \,, \tag{8.3}$$

where $\widehat{\boldsymbol{k}}_j = \boldsymbol{k}_j/|\boldsymbol{k}_j| = \lambda \boldsymbol{k}_j$ ($j = 1, 2$), and the vectors $\widehat{\boldsymbol{k}}_1$ and $\widehat{\boldsymbol{k}}_2$ are coplanar with $\boldsymbol{g}$. When $\widehat{\boldsymbol{k}}_j$ is directed toward the foot of the perpendicular from the pattern center of the jth edge, the condition of coplanarity is satisfied. Thus, knowing the vectors $\widehat{\boldsymbol{k}}_j$ for the Kikuchi lines and the wavelength λ, one can determine the scattering vector corresponding to the lines by using (8.3).

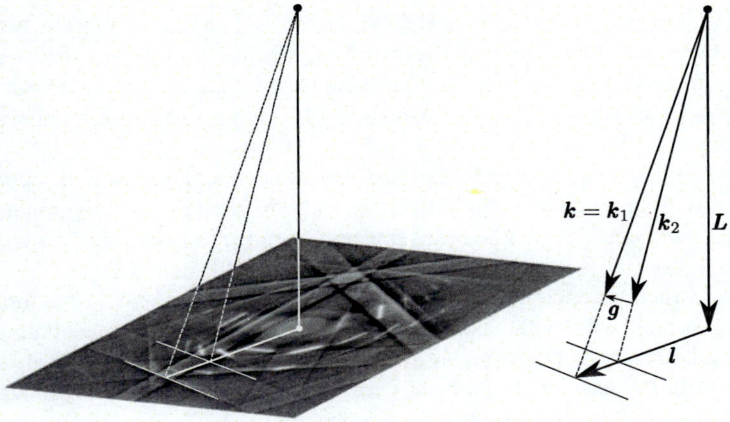

Fig. 8.1 Construction used for calculating of the scattering vector $\boldsymbol{g}$ corresponding to a pair of Kikuchi lines enumerated by $i = 1, 2$. $\boldsymbol{k}_i$ is the wave vector to the point nearest to the pattern center. The scattering vector is given by $\boldsymbol{g} = \boldsymbol{k}_1 - \boldsymbol{k}_2$ (Adapted from [41].)

List of Reciprocal Lattice Vectors

Knowing the crystal structure, one can easily get the reciprocal lattice vectors. Because of crystal symmetry there are families of symmetrically equivalent reflecting planes. To get all symmetrically equivalent members of the family, one needs to apply the symmetry operations of the crystal's Laue group to a given reciprocal lattice vector. The primary criterion for including a family of reciprocal lattice vectors are the expected reflection intensities obtained from the structure factors. In practice, it makes sense to support this criterion by visual inspection of indexing results, to find out which reflections are actually detectable and which are not. Clearly, if only directions of vectors are to be matched, there is no distinction between the families corresponding to parallel lattice planes, i.e., between families $\{nh\, nk\, nl\}$ differing by n.

In order to match the scattering vectors to reciprocal lattice vectors, it is convenient to express the latter in a Cartesian coordinate frame $\boldsymbol{e}_i$ ascribed to the crystal lattice. With a reciprocal lattice vector $\check{\boldsymbol{h}}$ specified by indices $(h\,k\,l) = (h_1\,h_2\,h_3)$, the ith Cartesian coordinate of $\check{\boldsymbol{h}}$ is $\sum_j h_j \boldsymbol{a}^j \cdot \boldsymbol{e}_i$, where $\boldsymbol{a}^j$ is the jth basis vector of the reciprocal lattice.

8.2.2 *Vector Magnitudes and Reflection Intensities*

For complex crystal structures, to match scattering vectors to reciprocal lattice vectors, one needs to take into account vector magnitudes. With some patterns, e.g. Laue patterns or most EBSD patterns, the magnitudes of scattering vectors are not accessible. Depending on accessibility of vector magnitudes, there are two basic types of input data. One type with measured coordinates of complete scattering vectors, the other with *directions* of the scattering vectors. Clearly, the case with known vector magnitudes can be reduced to the one with known directions by normalizing all vectors.

Moreover, in some cases, reflection intensities need to be taken into account (cf. [42]), but it is common to index diffraction patterns based on their geometry with intensities used only for dichotomous division of reflections into the 'on' and 'off' types.

Crystal structures encountered in metallurgy and materials science are frequently relatively simple, and both vector magnitudes and reflection intensities can be ignored. Initial considerations below allow for vector magnitudes, but the algorithms are limited to the case of normalized vectors. This simplifies the procedures and their description. It is also clear that such procedures still allow for indexing a broad range of diffraction patterns. In particular, they index bands in patterns obtained via EBSD—the technique that is nowadays the main tool for orientation mapping.

8.3 Formal Aspects of End-Indexing

8.3.1 Basic Relationships

As was mentioned before, the basic idea for most indexing procedures is to match the scattering vectors obtained from a diffraction pattern and the vectors of the crystallographic reciprocal lattice. The lattice vectors corresponding to detectable reflections will be denoted by $\check{\boldsymbol{h}}_m$, with m ranging from 1 to a certain M. The coordinates of these vectors are given in the Cartesian frame attached to the crystal. On the other hand, by analyzing diffraction reflections one obtains, say, N, scattering vectors $\boldsymbol{g}_n$, with Cartesian coordinates in the sample reference system.

Each reflection is a consequence of the diffraction from a crystal plane. Thus, each legitimate vector $\boldsymbol{g}_n$ corresponds to a certain $\check{\boldsymbol{h}}_m$. Now, the task is to match the scattering vectors $\boldsymbol{g}_n$ with some properly oriented reciprocal lattice vectors $\check{\boldsymbol{h}}_m$ related to potentially detectable reflections. In principle, each $\boldsymbol{g}_n$, $(n = 1, \ldots, N)$, is to be matched to a unique $\check{\boldsymbol{h}}_m$, $(m = 1, \ldots, M)$, and indexing can be seen as a procedure of determining an injective mapping $\varpi : \{1, \ldots, N\} \to \{1, \ldots, M\}$; to each of subscripts enumerating experimental vectors $\boldsymbol{g}_n$, the function ϖ ascribes one of the subscripts enumerating vectors $\check{\boldsymbol{h}}_m$, and it does not map distinct elements of the set $\{1, \ldots, N\}$ to the same element of $\{1, \ldots, M\}$. Point symmetry operations map the set of $\check{\boldsymbol{h}}_m$ vectors onto itself and induce permutations of the indices $\{1, \ldots, M\}$. If σ is such a permutation, the assignment ϖ is equivalent to $\sigma\varpi$. Knowing one representative of the class of equivalent assignments, it is straightforward to get all other elements of that class. With known ϖ, indices $(h\,k\,l) = (h_1\,h_2\,h_3)$ of the nth reflection (or the scattering vector $\boldsymbol{g}_n$) are $h_i = \boldsymbol{a}_i \cdot \check{\boldsymbol{h}}_{\varpi(n)}$, where $\boldsymbol{a}_i$ is the ith basis vector of the direct lattice.

Apart from experimental errors, there exists a rotation transforming all vectors $\boldsymbol{g}_n$ on vectors $\check{\boldsymbol{h}}_m$, i.e., $\check{\boldsymbol{h}}_{\varpi(n)} = O\boldsymbol{g}_n$.[2] Let the vectors $\boldsymbol{g}_n$ and $\check{\boldsymbol{h}}_m$ constitute columns of matrices G and H, respectively. The relationship between $\boldsymbol{g}_n$ and $\check{\boldsymbol{h}}_m$ vectors can be briefly expressed as

$$OG = HP \ ,$$

where P is an unknown $M \times N$ matrix representing ϖ. P has zero entries everywhere except the value of 1 at each entry (mn) such that the vector $\check{\boldsymbol{h}}_m$ corresponds to the vector $\boldsymbol{g}_n$, i.e., there is exactly one '1' in each column of P. The search for O corresponds to the orientation determination problem, and the search for P (representing ϖ) solves the indexing problem.

It is obvious that random errors in positions of detected reflections are inevitable. These errors affect the scattering vectors and the entries of G, and the matrix OG is only approximately equal to HP. Therefore, one can portray the considered problem (with error affected but legitimate scattering vectors) in the following form: find P and O minimizing

$$\sum_{n=1}^{N} \left| \check{\boldsymbol{h}}_{\varpi(n)} - O\boldsymbol{g}_n \right|^2 = \|HP - OG\|^2 \ , \tag{8.4}$$

where $\|\cdot\|$ denotes the Frobenius norm. In this formulation, the problem lies in the domain of combinatorial (P or ϖ) and continuous (O) optimization.

8.3.2 Related Solvable Problems

It is instructive to break down the above problem into simpler problems concerning the case of $N = M$, i.e., with the sets of matched vectors having the same cardinality, and the matrices G on H having the same dimensions. Two simplified partial problems are how to get the permutation matrix P minimizing $\|HP - G\|^2$, and how to get the orthogonal matrix O minimizing $\|H - OG\|^2$.

Linear Assignment Problem

With a real $N \times N$ cost matrix C, the (linear assignment) problem is to find

$$P = \arg\min_X \text{tr}(XC^T)$$

[2] With the matrix O in front of $\boldsymbol{g}$ the convention of [43] is followed. It is usually used in texture analysis and in orientation mapping systems.

subject to the condition that X is an $N \times N$ permutation matrix, i.e., the entries of X are 0 or 1, and there is exactly one '1' in each row and exactly one '1' in each column. The linear assignment problem is solved by classic Kuhn's Hungarian algorithm.

⊡ It is executed by subsequent transformations of the C matrix:
1. in each row, subtract the smallest entry from all entries of the row,
2. in each column, subtract the smallest entry from all entries of the column,
3. mark complete rows and complete columns so that all zeros are marked; the number N_x of marked rows and columns must be the smallest possible,
4. if $N_x < N$, get the smallest unmarked entry, subtract it from each unmarked row, and subsequently add it to each marked column; then go to step 3;
otherwise, i.e., if $N_x = N$, select N zeros, one in each column and in each row; their positions are the same as positions of the entries '1' in the permutation matrix P minimizing $\text{tr}(XC^T)$.

Since permutation matrices satisfy the relationship $XX^T = I$, one has $\arg\min_X \|X - A\|^2 = \arg\max_X \text{tr}\left(XA^T\right)$, i.e., the permutation matrix nearest to the square matrix A can be obtained by solving the linear assignment problem with the cost matrix $C = -A$. Similarly, for two $3 \times N$ matrices G and H, one has the equality

$$\arg\min_X \|HX - G\|^2 = \arg\max_X \text{tr}\left(XG^T H\right) ,$$

i.e., the minimization of $\|HX - G\|^2$ is reduced to the linear assignment problem with the cost matrix $C = -H^T G$.[3]

Orthogonal Matrix Nearest to a Square Matrix

The problem is to find an orthogonal matrix nearest to a square matrix. More precisely, given a square matrix A determine

$$\mathcal{O}(A) = \arg\min_X \|A - X\|^2 \quad \text{subject to} \quad XX^T = I \tag{8.5}$$

with entries of X being real numbers. A general way of getting $\mathcal{O}(A)$ is by singular value decomposition (SVD)[4] of A. With $A = U\Sigma V^T$, one has

$$\mathcal{O}(A) = UV^T .$$

---

[3] It is worth noting that based on $OG = HP$ with orthogonal O, one has the relationship $P^T H^T HP = G^T G$ involving only the assignment P. With $N = M$, i.e., with quadratic matrices P, the minimization of $\|P^T H^T HP - G^T G\|^2$ with respect to P is reduced to one of the fundamental issues of combinatorial optimization known as the quadratic assignment problem or QAP [44].

[4] SVD of a rectangular $m \times n$ matrix A is the factorization

$$A = U\Sigma V^T ,$$

where U is an $m \times m$ orthogonal matrix, Σ is an $m \times n$ diagonal matrix with non-negative entries in descending order, and V is $n \times n$ orthogonal matrix. The columns of U (V) are known as the left-singular vectors (right-singular vectors) of A. The left-singular vectors (right-singular vectors) of A are orthonormal eigenvectors of AA^T ($A^T A$). The diagonal entries of Σ are called singular values of A. The singular values of A are the square roots of eigenvalues of AA^T and $A^T A$.

An alternative expression, applicable to invertible A, has the form

$$\mathcal{O}(A) = AA_s^{-1} \,,$$

where $A_s^2 = A^T A$. The matrix A_s is calculated by solving the eigenproblem for $A^T A$: one has $A^T A V = V\Sigma^2$, where Σ^2 is the diagonal matrix with eigenvalues of $A^T A$, and V is the orthogonal matrix built of corresponding eigenvectors. With Σ being the diagonal matrix with square roots of eigenvalues of $A^T A$, one has $A_s = V\Sigma V^T$.

Related to the above subject is the problem of determining the nearest orthogonal matrix with the additional condition of special orthogonality: given an invertible matrix A find a *special* orthogonal matrix $\mathcal{SO}(A)$ minimizing $\|A - X\|^2$. With $A = U\Sigma V^T$, one has

$$\mathcal{SO}(A) = USV^T \,,$$

where $S = \text{diag}[+1, +1, \text{sign}(\det(A))]$. As above, one can alternatively use AA_s^{-1}, where the determinant of A_s must have the same sign as the determinant of A. Thus, if $\det(A)$ is positive, the procedure is as above, but if $\det(A) < 0$, one needs to change the sign of the smallest diagonal element of Σ to minus. With eigenvalues sorted in order of decreasing value, one has $A_s = V\Sigma S V^T$.

Orthogonal Matrix Matching Corresponding Vectors

Determination of an orthogonal matrix matching corresponding vectors is known under various names, e.g., as the orthogonal Procrustes problem. The problem is linear and its solution has been described numerous times [45–49]. Given two sets of vectors, $\boldsymbol{g}_n$ and corresponding $\boldsymbol{h}_n$ ($n = 1, \ldots, N$), find an orthogonal matrix $X = O$ minimizing

$$\sum_{n=1}^{N} |\check{\boldsymbol{h}}_n - X\boldsymbol{g}_n|^2 \,. \tag{8.6}$$

It is easy to see that this is equivalent to matrix-based formulation: given two matrices G $(= [\boldsymbol{g}_1\ \boldsymbol{g}_2\ \ldots\ \boldsymbol{g}_N])$ and H $(= [\boldsymbol{h}_1\ \boldsymbol{h}_2\ \ldots\ \boldsymbol{h}_N])$ find an orthogonal matrix minimizing

$$\|H - XG\|^2 \,. \tag{8.7}$$

Solutions to these problems follow directly from expressions listed in the previous section. By transforming (8.5), one has $\arg\min_X \|A - X\|^2 = \arg\max_X \text{tr}\left(A^T X\right)$. Similarly, by transforming (8.7), one obtains $\arg\min_X \|H - XG\|^2 = \arg\max_X \text{tr}\left((HG^T)^T X\right)$. This implies that the orthogonal matrix matching corresponding vectors is given by

$$O = \mathcal{O}(HG^T) \,,$$

i.e., the problem is reduced to finding the orthogonal matrix nearest to HG^T.

The analogous problem with the condition of *special* orthogonality is frequently referred to as determination of absolute orientation. Given two sets of corresponding vectors $\boldsymbol{g}_n$ and $\boldsymbol{h}_n$ or the matrices G and H, find a *special* orthogonal matrix O minimizing (8.6) and (8.7). In this case, if HG^T is invertible,

$$O = \mathcal{SO}(HG^T) ,$$

i.e., the problem is reduced to finding the special orthogonal matrix nearest to HG^T.

SVD-Based Matching Without Correspondence

Let H and G be $3 \times N$ matrices linked (like in Sect. 8.3) by the exact relationship

$$HP = OG , \tag{8.8}$$

where P is an unknown $N \times N$ permutation matrix, and O is an unknown 3×3 orthogonal matrix. In error-free cases, the matrices O and P can be calculated directly from H and G using a procedure described in [50].

⊡ Let SVD of the matrix H be

$$H = U_H \Sigma_H V_H^T . \tag{8.9}$$

Similarly, let $G = U_G \Sigma_G V_G^T$. Since $HH^T = HPP^T H^T = OGG^T O^T$ and the eigenvalues are invariant under similarity transformations, one has

$$\Sigma_H = \Sigma_G = \Sigma .$$

For simplicity, the eigenvalues in Σ are assumed to be distinct.

Based on (8.9), one has $H^T H = V_H \Sigma^T U_H^T U_H \Sigma V_H^T = V_H \Sigma^T \Sigma V_H^T$ or

$$(H^T H) V_H = V_H \Sigma^T \Sigma . \tag{8.10}$$

On the other hand, by using $H^T H = PG^T O^T OGP^T = PG^T GP^T$ and $G^T G = V_G \Sigma^T \Sigma V_G^T$, one obtains $H^T H = PV_G \Sigma^T \Sigma V_G^T P^T$ or

$$(H^T H)(PV_G) = (PV_G)\Sigma^T \Sigma . \tag{8.11}$$

It follows from (8.10) and (8.11) that columns of both V_H and PV_G are eigenvectors of $H^T H$. The first three eigenvectors are unique up to their sign. Therefore, one can write

$$V_H \overline{S} = PV_G S , \tag{8.12}$$

where S and $\overline{S}$ are $N \times 3$ matrices with zero entries except $S_{11} = S_{22} = S_{33} = 1$ and $\overline{S}_{11} = \pm 1$, $\overline{S}_{22} = \pm 1$, $\overline{S}_{33} = \pm 1$. Let $\text{abs}(X)$ be the matrix of absolute values of elements of the matrix X. Application of abs to both sides (8.12) eliminates $\overline{S}$. With $W_H = \text{abs}(V_H \overline{S}) = \text{abs}(V_H) S$ and $W_G = \text{abs}(V_G S) = \text{abs}(V_G) S$, one has the relationship

$$W_H = PW_G , \tag{8.13}$$

from which one can get P and than use (8.8) to get O.

The exact equalities used above can be replaced by approximate relationships, i.e., one has $HP \approx OG$ instead of (8.8) and $W_H \approx PW_G$ instead of (8.13). The correspondence matrix P can be calculated by solving the linear assignment problem with the cost matrix $C = -W_H W_G^T$, and subsequently O is obtained from $O = \mathcal{O}(HPG^T)$. [An alternative way to get O is by checking all eight combinations of signs in the matrix $\overline{S}$ and all six orders of *rows* in U_G, i.e., altogether 48 cases. This can be seen as using all 48 matrices R^{α}_{cub} ($\alpha = 1, \ldots, 48$) representing octahedral group O_h given by (1.43) to get 48 orthogonal matrices $O^{\alpha} = U_H R^{\alpha}_{cub} U_G^T$. Then, for each O^{α} one needs to get the corresponding P^{α} based on $HP^{\alpha} \approx O^{\alpha}G$, and check which pair (O^{α}, P^{α}) corresponds to the smallest $\|HP^{\alpha} - O^{\alpha}G\|$.] However, the results are optimal only if the deviations from congruency between the figures built of vectors contained in G and H are small.

8.3.3 Rotations Versus Proper Rotations

There is an issue of the domain of O in (8.4). One may ask whether this domain comprises all rotations or just proper rotations. I.e., one needs to choose between using $\mathcal{O}$ or $\mathcal{SO}$ of Sect. 8.3.2. A couple of facts need to be recalled. First, orientation of an object (a crystallite) corresponds to a proper rotation and is determined by a special orthogonal matrix. Second, almost all materials studied by orientation mapping have centrosymmetric crystal structures. Third, the figure built of vectors $\check{\boldsymbol{h}}_m$ is centrosymmetric due to Friedel's law. In the case of centrosymmetric crystals, with the matrix O relating G to HP assumed to be special orthogonal (i.e., the matrix may be a result of using $\mathcal{SO}$), one gets a single solution modulo *proper* operations of the Laue group. If the matrix relating G to HP was allowed to be orthogonal, one would get a single solution modulo *all* operations of the Laue group. If the crystal structure is non-centrosymmetric, because of Friedel's law, the unequivocal crystallite orientation is undeterminable by conventional orientation mapping methods; cf. Sect. 8.9.

8.3.4 Computational Context

The crystallographic problem of orientation determination is formally similar to the problem of registering two figures of spatially distributed points, which arises in other fields. In computer science, with points as graph vertices and distances between points as attributes (weights) of edges, the problem can be seen as edge-attributed *subgraph matching*; see, e.g., Chap. 2 of [51]. If reflection intensities are known, it turns into edge-attributed (by distance) vertex-attributed (by intensity) subgraph matching. In computational geometry, one considers geometric *point-pattern matching*. It is usually formulated in terms of point 'sets' and in a more general framework of rigid motion: Given two sets $\mathcal{G}$ and $\mathcal{H}$ of spatially distributed points, determine the smallest ε_H such that there is a rigid motion of the 'query' figure $\mathcal{G}$ that brings each of its points within distance ε_H of a point in $\mathcal{H}$ (e.g., [52]). In other words, the task is to match the complete figure $\mathcal{G}$ to a subfigure of $\mathcal{H}$.[5]

[5] Formally, ε_H is the one-sided Hausdorff distance from transformed $\mathcal{G}$ to $\mathcal{H}$. With $\boldsymbol{g}_i$ and $\boldsymbol{h}_j$ representing points in $\mathcal{G}$ and $\mathcal{H}$, respectively, and T denoting the rigid motion, one can write $\varepsilon_H = \max_i \min_j |T\boldsymbol{g}_i - \boldsymbol{h}_j|$.

8.4 Spurious Scattering Vectors

With $\mathcal{G}$ being a figure composed of terminal points of the scattering vectors $\boldsymbol{g}_n$ and $\mathcal{H}$ denoting a figure composed of terminal points of reciprocal lattice vectors $\check{\boldsymbol{h}}_m$, the 'indexing problem' of Sect. 8.3 would be to determine an optimal subfigure of $\mathcal{H}$ nearly congruent to the complete figure $\mathcal{G}$. However, the account given in Sect. 8.3 ignores an important aspect of automatic detection of reflections: the input data are affected by *gross errors* in reflection positions, and these errors lead to illegitimate scattering vectors. (For a two-dimensional illustration of the indexing problem, see Figs. 8.2, 8.3 and 8.4.) With spurious reflections present, instead of (8.4), it is more suitable to consider the function

$$N_c^{-1} \sum_{n=1}^{N} w_n | \check{\boldsymbol{h}}_{\varpi(n)} - O\boldsymbol{g}_n |^2 , \tag{8.14}$$

where the binary weight $w_n = 1$ if $\boldsymbol{g}_n$ is legitimate, $w_n = 0$ otherwise, and $N_c = \sum_{n=1}^{N} w_n$ is the number of non-zero terms. The simplest approach to this complication is to eliminate spurious scattering vectors by considering subsets of the complete set of these vectors, and to use some additional criteria of accepting a solution for a given subset.

Allowing for spurious vectors is an integral part of an important issue of computational geometry known as the *largest common point set* (LCPS) problem: Given two point sets $\mathcal{G}$ and $\mathcal{H}$ and a positive number ϵ, determine the largest subfigure $\mathcal{G}'$ of $\mathcal{G}$ such that there is an isometry carrying $\mathcal{G}$ to a position in which each point of $\mathcal{G}'$ is not further than ϵ from a point of $\mathcal{H}$ (e.g., [53]). The subfigure of $\mathcal{H}$ congruent to $\mathcal{G}'$ will be denoted by $\mathcal{H}'$. With the isometry being a proper rotation and points of $\mathcal{G}$ and $\mathcal{H}$ identified with their position vectors bound to the fixed center of rotation, the indexing problem can be seen as a simplified version (with isometry replaced by rotation, i.e., without translation) of the algorithmic LCPS problem.

In the simplest case, e.g., when EBSD bands or Laue-pattern spots are indexed, the magnitudes of the scattering vectors are unknown, and all involved vectors are assumed to be of unit magnitude. The corresponding restricted case of LCPS is referred to as the *constellation problem* [54]: given two sets $\mathcal{G}$ and $\mathcal{H}$ of unit vectors, determine a rotation carrying the largest subset of $\mathcal{G}$ to a position approximating a subset of $\mathcal{H}$. The constellation problem is also portrayed as star identification problem which arises in determination of satellite orientation (attitude).

In the field of molecular science, there is a considerable interest in optimal alignment of structures because of their applications to determination of similarities among chemical structures and recognition of molecules (for pharmaceutical purposes). The considered problem is also close to the problem of registering two clouds of points, which arises in computer vision.

⊡ The matching problems in particular fields have specific features. There are similarities and there are some essential differences. Most notably, unlike other matching problems, the indexing problem is very much affected by the crystal symmetry. But also other aspects are worth considering.

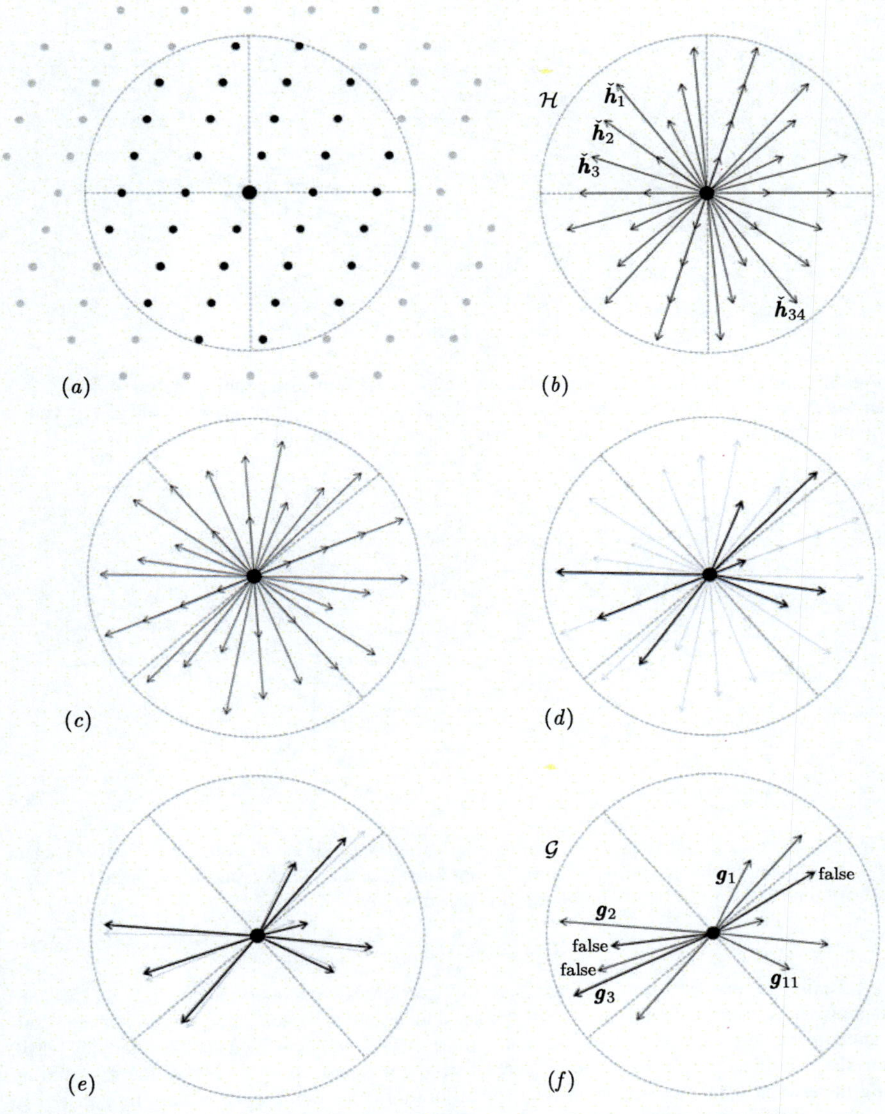

Fig. 8.2 Two-dimensional illustration of the relationship between the reciprocal lattice and scattering vectors corresponding to detected reflections. (*a*) Nodes of reciprocal lattice. (*b*) Vectors $\check{\boldsymbol{h}}_m$ with terminal points at nodes corresponding to detectable reflections, i.e., figure $\mathcal{H}$. (*c*) Rotated vectors $\check{\boldsymbol{h}}_m$. (*d*) Rotated vectors $\check{\boldsymbol{h}}_m$ corresponding to *detected* reflections, i.e., error-free scattering vectors. (*e*) The scattering vectors affected by experimental errors. (*f*) Error-affected and false scattering vectors, i.e., figure $\mathcal{G}$ (Adapted from [41].)

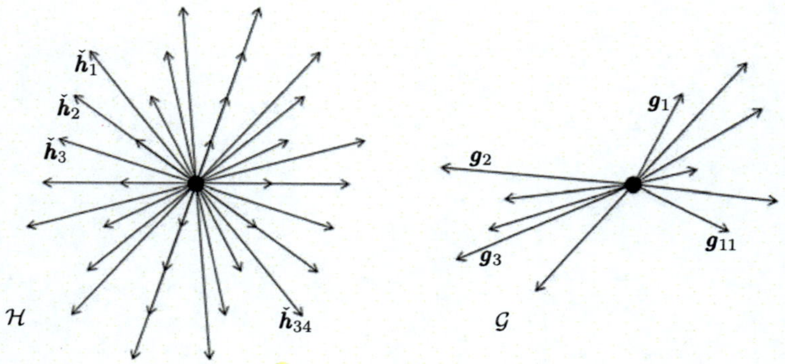

Fig. 8.3 Schematic 2D illustration of the orientation determination problem based on Fig. 8.2. It shows input data available when the magnitudes of scattering vectors g_n are known. The task is find the rotation of $\mathcal{G}$ leading to the 'best match' between 'large' subfigures of $\mathcal{G}$ and $\mathcal{H}$.

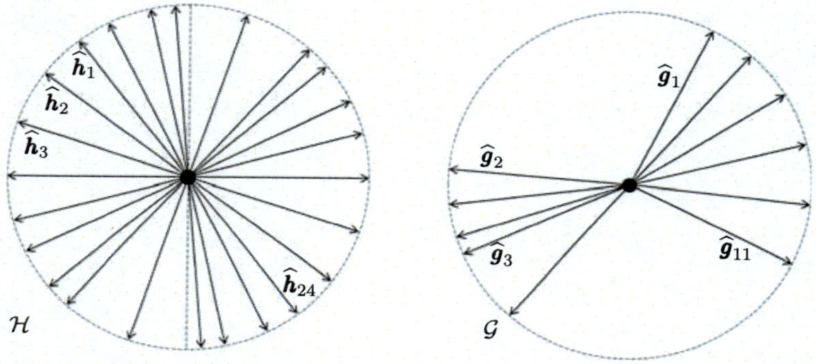

Fig. 8.4 Schematic 2D illustration of the orientation determination problem in the case when the magnitudes of the scattering vectors are not known and all vectors are normalized to 1 (i.e., illustration of the constellation problem). It is based on Fig. 8.2.

As an example, one may compare indexing of diffraction patterns to star identification. The solid angle (field of view) covered by a star tracker is small, and in consequence, angles between star-indicating vectors are small. Moreover, the number of false or undetected stars is relatively small, and the identification can be based on nearest-neighbor separation. This is not true in indexing; angles between scattering vectors in $\mathcal{G}$ can be arbitrarily large, the fraction of false reflections can be large, and many reflections can be undetected. In star identification, frequently the scale of star patterns is to be determined [55]; in indexing, there is no need to scale the patterns. If known, the reflection intensity would correspond to star brightness (magnitude); the latter is taken into account in some star identification algorithms.

8.4.1 Accumulation

As was noted above, to deal with spurious scattering vectors, one needs to match test subfigures of $\mathcal{G}$ and $\mathcal{H}$ and specify criteria for accepting a match as the solution. In principle, one may use k-tuples from $\mathcal{G}$ and $\mathcal{H}$ as matching elements. Since the fraction of spurious scattering vectors in $\mathcal{G}$ can be large, the number of vectors in test subfigures must be small. In practice, the algorithms are based on pairs, triplets, or quadruplets of vectors as matching elements. The most reliable matching algorithms rely on accumulation. Matched elements, e.g., triplets, cast votes [55, 56], i.e., they contribute to an accumulation space, and the final solution is based on results of the voting. The advantage of the accumulation-based methods lies in their robustness and conceptual simplicity. The algorithms are relatively easy to implement. The task is particularly simple when pairs of vectors are matched. Accumulation can be made in the discrete space of the number of matched vectors (i.e., the cardinality of the sets $\mathcal{G}'$ and $\mathcal{H}'$) or in the continuous space of crystal orientations. For simplicity, the strategies considered below are limited to the case of normalized $\widehat{\boldsymbol{h}}_m$ and $\widehat{\boldsymbol{g}}_n$ vectors, i.e., they allow for indexing bands in EBSD patterns.

Rotation Relating Two k-Tuplets of Vectors

With $k \geq 3$, the rotation transforming k-tuplet $(\widehat{\boldsymbol{g}}_{n_1}, \ldots, \widehat{\boldsymbol{g}}_{n_k})$ to position closest to that of $(\widehat{\boldsymbol{h}}_{m_1}, \ldots, \widehat{\boldsymbol{h}}_{m_k})$ is given by the formula for absolute orientation[6]

$$O^{(k)} = \mathcal{SO}\left([\widehat{\boldsymbol{h}}_{m_1}\ \widehat{\boldsymbol{h}}_{m_2}\ \ldots\ \widehat{\boldsymbol{h}}_{m_k}\,]\,[\widehat{\boldsymbol{g}}_{n_1}\ \widehat{\boldsymbol{g}}_{n_2}\ \ldots\ \widehat{\boldsymbol{g}}_{n_k}\,]^T\right) . \tag{8.15}$$

The above expression cannot be directly extended to get rotations matching 2-tuples. The easy way of circumventing this problem is to use (8.15) for $k = 3$ with the third linearly independent vectors being the cross products $\boldsymbol{g}_\times = \widehat{\boldsymbol{g}}_{n_1} \times \widehat{\boldsymbol{g}}_{n_2}$ and $\boldsymbol{h}_\times = \widehat{\boldsymbol{h}}_{m_1} \times \widehat{\boldsymbol{h}}_{m_2}$; one has

$$O^{(2)} = \mathcal{SO}\left([\widehat{\boldsymbol{h}}_{m_1}\ \widehat{\boldsymbol{h}}_{m_2}\ \boldsymbol{h}_\times\,]\,[\widehat{\boldsymbol{g}}_{n_1}\ \widehat{\boldsymbol{g}}_{n_2}\ \boldsymbol{g}_\times\,]^T\right) = [\widehat{\boldsymbol{r}}_1^h\ \widehat{\boldsymbol{r}}_2^h\ \widehat{\boldsymbol{r}}_3^h\,]\,[\widehat{\boldsymbol{r}}_1^g\ \widehat{\boldsymbol{r}}_2^g\ \widehat{\boldsymbol{r}}_3^g\,]^T , \tag{8.16}$$

where $\widehat{\boldsymbol{r}}_1^g$, $\widehat{\boldsymbol{r}}_2^g$ and $\widehat{\boldsymbol{r}}_3^g$ represent normalized versions of $\boldsymbol{g}_\times$, $\widehat{\boldsymbol{g}}_{n_1} + \widehat{\boldsymbol{g}}_{n_2}$ and $\widehat{\boldsymbol{g}}_{n_1} - \widehat{\boldsymbol{g}}_{n_2}$, respectively, and the $\widehat{\boldsymbol{r}}_i^h$ vectors are defined in analogous way based on $\widehat{\boldsymbol{h}}_{m_i}$ vectors. $O^{(2)}$ transforms the pair $\widehat{\boldsymbol{g}}_{n_1}$, $\widehat{\boldsymbol{g}}_{n_2}$ to the position closest to that of $\widehat{\boldsymbol{h}}_{m_1}$, $\widehat{\boldsymbol{h}}_{m_2}$. With the last part of (8.16), the computation of this optimal rotation matrix involves only elementary arithmetic operations, i.e., there is no need for applying the operator $\mathcal{SO}$.

The case of $k = 2$, i.e., the pairwise matching, is particularly convenient. It is in a sense elemental among those providing discrete solutions. In matchings based on k larger than two, additional vectors sharpen the criteria for the congruence of given subsets. On the other hand, using higher order k-tuples increases the computational costs. To avoid combinatorial explosion, k needs to be kept small. Moreover,

---

[6] Equation (8.15) and its interpretation remain valid if $\widehat{\boldsymbol{h}}_{m_i}$ and $\widehat{\boldsymbol{g}}_{n_i}$ are replaced by $\check{\boldsymbol{h}}_{m_i}$ and $\boldsymbol{g}_{n_i}$, respectively.

numerical experiments show that with error affected data, higher order k-tuples do not really improve the quality of indexing. (Cf. Fig. 8.8.)

Number of k-Tuples

Critical for the temporal performance of the above algorithms is the number of k-tuplets. To test all potential matches, one needs ordered k-tuplets from one set (say, from N-element $\mathcal{G}$) and k-element combinations of vectors constituting the other set (M-element $\mathcal{H}$). The number of partial permutations (k-permutations) of N-element set equals $N!/(N-k)!$, and the number of k-combinations of M-element set equals $\binom{M}{k}$. Hence, the total number of k-tuple matches would be $M!\ N!/(k!\,(M-k)!\,(N-k)!)$. In practice, the number of actual matchings is reduced by considering only k-tuples with certain characteristics. One needs to use features allowing for an early elimination of as many redundant k-tuples as possible. Most of the matchings can be omitted by observing that the vectors $\widehat{\boldsymbol{g}}_{n_i}$ can match $\widehat{\boldsymbol{h}}_{m_i}$ only if the angles between vectors of the k-tuples differ by less than a threshold; if not, the k-tuples are rejected as a possible match. This is particularly simple when $k = 2$: the vectors $\widehat{\boldsymbol{g}}_{n_1}$ and $\widehat{\boldsymbol{g}}_{n_2}$ can match $\widehat{\boldsymbol{h}}_{m_1}$ and $\widehat{\boldsymbol{h}}_{m_2}$ only if the angle between $\widehat{\boldsymbol{g}}_{n_1}$ and $\widehat{\boldsymbol{g}}_{n_2}$ is close to that between $\widehat{\boldsymbol{h}}_{m_1}$ and $\widehat{\boldsymbol{h}}_{m_2}$. Additionally, due to crystal symmetry, some k-tuples of reciprocal lattice vectors are equivalent, and only one representative of these k-tuples needs to be considered. In the simplest approach, it is enough to use only one vector of a family of symmetrically equivalent reciprocal lattice vectors as the first vector $\widehat{\boldsymbol{h}}_{m_1}$ of a k-tuple. Clearly, when vector magnitudes are available, some matchings can be excluded at the outset if magnitudes of $\boldsymbol{g}_n$ and $\check{\boldsymbol{h}}_m$ differ beyond a threshold.

8.5 Accumulation in Discrete Space

Of all possible assignments, one looks for the one with the largest support from matching of k-tuples. Briefly, pairs of k-tuples vote for assignments, and the one with the largest number of votes is assumed to solve the problem. The quality of the solutions is reflected in the number of matched vectors (i.e., the cardinality of the sets $\mathcal{G}'$ and $\mathcal{H}'$). Knowing the assignment P, the orientation is $O = \mathcal{SO}(HPG'^T)$, where G' is the matrix with vectors of $\mathcal{G}'$. This approach is similar to *geometric hashing*.

⊡ Geometric hashing is a voting-based method used for object recognition [57]. The task is to recognize an object matching one of a number of transformed models. Both the models and the object are represented by points. The algorithm consists of two phases: preprocessing and recognition. In the preprocessing phase, for each model and for each pair of its points used as a 'geometric basis', compute coordinates of all points in this basis, and in the corresponding entry of a hash table indexed by (discretized) coordinates insert information on the model and the geometric basis. In the recognition phase, for each pair of object's points used as a 'geometric basis' compute coordinates of all points in this basis, access the corresponding entry of the hash table and cast votes for models and bases found there. The votes are ultimately used to select the model and to determine the model-to-object transformation.

To give an example, three strategies based on pair-matching and accumulation in the discrete space are described in detail. As was noted above, the strategies are limited to the case of normalized $\widehat{\boldsymbol{h}}_m$ and $\widehat{\boldsymbol{g}}_n$ vectors.

Strategy A

The preliminary step is to create a list (hereafter referred to as `LIST`) of angular distances between $\widehat{\boldsymbol{h}}_m$ vectors; with the angle between $\widehat{\boldsymbol{h}}_{m_1}$ and $\widehat{\boldsymbol{h}}_{m_2}$ being $d_h = d(\widehat{\boldsymbol{h}}_{m_1} \cdot \widehat{\boldsymbol{h}}_{m_2}) = \arccos(\widehat{\boldsymbol{h}}_{m_1} \cdot \widehat{\boldsymbol{h}}_{m_2})$, the entry added to the list has the form of the quintuplet $(d_h, f_1, k_1, f_2, k_2)$, where f_i indicates the family (i.e., a set indexed as $\{h\,k\,l\}$) of symmetrically equivalent vectors to which $\check{\boldsymbol{h}}_{m_i}$ belongs, and k_i indicates the particular vector of the family. Clearly, each pair (f_i, k_i) identifies a reciprocal lattice vector. The list is shortened thanks to crystal symmetry; for symmetrically equivalent pairs separated by the same distance, only one representative quintuplet $(d_h, f_1, k_1, f_2, k_2)$ needs to be saved.[7] The list is sorted by the distances d_h. With this, determining the items matching the angular distance between a pair of $\widehat{\boldsymbol{g}}_n$ vectors is fast. `LIST` is created once for all patterns of a given phase.

For a given pattern, an accumulation (vote count) table $V(n, f)$ will be needed. It is indexed by n enumerating the scattering vectors $\widehat{\boldsymbol{g}}_n$ obtained from the pattern, and by f $(= 1, \ldots, M_f)$ enumerating the families of equivalent reciprocal lattice vectors. The table V is initialized with zero entries.

The indexing of the pattern begins with accumulation loops. For each pair $(\widehat{\boldsymbol{g}}_{n_1}, \widehat{\boldsymbol{g}}_{n_2})$ such that $n_1 < n_2$ get the angular distance $d(\widehat{\boldsymbol{g}}_{n_1}, \widehat{\boldsymbol{g}}_{n_2})$. From the `LIST` of quintuplets $(d_h, f_1, k_1, f_2, k_2)$ retrieve all items with d_h deviating from $d(\widehat{\boldsymbol{g}}_{n_1}, \widehat{\boldsymbol{g}}_{n_2})$ by less than a threshold. This threshold, say, p_1, is a parameter of the strategy A. The families f_i which appear in the retrieved items constitute a set S (without duplicates). Each element of the set casts votes for both $\widehat{\boldsymbol{g}}_{n_1}$ and $\widehat{\boldsymbol{g}}_{n_2}$, i.e., for each f_i in S, the values of $V(n_1, f_i)$ and $V(n_2, f_i)$ are increased by 1.

With the accumulation completed, the larger the value of $V(n, f)$ for a given n, the larger the probability that $\widehat{\boldsymbol{g}}_n$ matches a vector from the fth family. At this point, one has the families (or indices $\{h\,k\,l\}$) expected to correspond to each scattering vector, whereas the task is to get the concrete vector (or indices $(h\,k\,l)$) modulo Laue symmetry of the crystal. A simple approach is to assume that $\widehat{\boldsymbol{g}}_n$ matches a vector of the single family $\arg\max_f V(n, f)$, but clearly, one can ascribe two or more families with the highest number of votes. The number p_d of ascribed families

[7] The above issue is simple, but a word of caution is in order. Matrices $O^{(2)}$ given by (8.16) represent proper rotations. This affects the allowed relationships of equivalence between pairs of $\widehat{\boldsymbol{h}}_m$ vectors. Take the $m\overline{3}m$ crystal symmetry. The angular distance between the vectors $\check{\boldsymbol{h}}_1 = (1\,0\,1)$ and $\check{\boldsymbol{h}}_2 = (2\,2\,1)$ is $\pi/4$. The same distances separate the vectors $\check{\boldsymbol{h}}_1' = (0\,1\,1)$ and $\check{\boldsymbol{h}}_2' = (2\,1\,2)$, and the vectors $\check{\boldsymbol{h}}_1'' = (0\,1\,1)$ and $\check{\boldsymbol{h}}_2'' = (2\,2\,1)$, but of these two only the pair $(\check{\boldsymbol{h}}_1', \check{\boldsymbol{h}}_2')$ is related to $(\check{\boldsymbol{h}}_1, \check{\boldsymbol{h}}_2)$ by a symmetry operation being a proper rotation; the same concerns the pair $(-\check{\boldsymbol{h}}_1'', -\check{\boldsymbol{h}}_2'')$ but not the pair $(\check{\boldsymbol{h}}_1'', \check{\boldsymbol{h}}_2'')$.

is another parameter of the strategy A.[8] In the particular implementation described below, this parameter is fixed at $p_d = \max(\lfloor M_f/2 \rfloor, 1)$.

In the final stage, one needs to discriminate between members of families of equivalent vectors. This stage can be performed in various ways. One could look for a consistent solution by considering only angular distances, but more convenient is a method involving orientation. For every pair $(\widehat{\boldsymbol{g}}_{n_1}, \widehat{\boldsymbol{g}}_{n_2})$ satisfying $n_1 < n_2$ construct all pairs of vectors $(\widehat{\boldsymbol{h}}_{m_1}, \widehat{\boldsymbol{h}}_{m_2})$ such that only the vectors $\widehat{\boldsymbol{h}}_{m_i}$ from the family (families) ascribed to $\widehat{\boldsymbol{g}}_{n_i}$ in the previous step are allowed to correspond to $\widehat{\boldsymbol{g}}_{n_i}$. Based on $(\widehat{\boldsymbol{g}}_{n_1}, \widehat{\boldsymbol{g}}_{n_2})$ and $(\widehat{\boldsymbol{h}}_{m_1}, \widehat{\boldsymbol{h}}_{m_2})$, using (8.16), get a trial orientation $O^{(2)}$. Having the trial orientation, one can perform a trial indexing: for each vector $O^{(2)}\boldsymbol{g}_n$ $(n = 1, \ldots, N)$ one finds the closest of unmatched vectors $\widehat{\boldsymbol{h}}_m$. If the distance between $O^{(2)}\boldsymbol{g}_n$ and this closest vector, say, $\widehat{\boldsymbol{h}}_{m_x}$, does not exceed a threshold (parameter p_2 of the strategy A), the closest vector is ascribed to $\boldsymbol{g}_n$, i.e., $\varpi(n) = m_x$; otherwise, $\boldsymbol{g}_n$ is considered to be spurious. The problem is solved by the assignment ϖ leading to the smallest number of spurious $\boldsymbol{g}_n$ vectors. At the end, one needs to apply $\mathcal{SO}$ to legitimate $\boldsymbol{g}_n$ vectors and their partners $\widehat{\boldsymbol{h}}_{\varpi(n)}$ to get the orientation O. In the case of a draw (i.e., the same number of spurious vectors for different orientations), the result with the smallest residue is used as the final solution.[9] The residue can be defined as, e.g., $\sum_n w_n |\widehat{\boldsymbol{h}}_{\varpi(n)} - O\widehat{\boldsymbol{g}}_n|^2$ or $\sum_n w_n \arccos\left(\widehat{\boldsymbol{h}}_{\varpi(n)} \cdot O\widehat{\boldsymbol{g}}_n\right)$; like in (8.14), $w_n = 1$ if $\boldsymbol{g}_n$ is legitimate, and $w_n = 0$ if $\boldsymbol{g}_n$ is spurious.

Strategy B

At the outset, as in strategy A, the preliminary step is to create the `LIST` of quintuplets $(d_h, f_1, k_1, f_2, k_2)$.[10] The indexing strategy B is similar to the final stage of the strategy A. It relies on trial orientations from the beginning. The steps are as follows. For each pair $(\widehat{\boldsymbol{g}}_{n_1}, \widehat{\boldsymbol{g}}_{n_2})$ $(n_1 < n_2)$ get the angular distance $d(\widehat{\boldsymbol{g}}_{n_1}, \widehat{\boldsymbol{g}}_{n_2})$ and retrieve from the `LIST` all items with distances d_h within a threshold p_1 from $d(\widehat{\boldsymbol{g}}_{n_1}, \widehat{\boldsymbol{g}}_{n_2})$. For every retrieved item $(d_h, f_1, k_1, f_2, k_2)$, using (8.16), get a trial orientation $O^{(2)}$ based on the pairs $(\widehat{\boldsymbol{g}}_{n_1}, \widehat{\boldsymbol{g}}_{n_2})$ and $(\widehat{\boldsymbol{h}}_{m_1}, \widehat{\boldsymbol{h}}_{m_2})$, where m_i is determined by the pair (f_i, k_i). Having the trial orientation, perform a trial indexing: for each vector $O^{(2)}\widehat{\boldsymbol{g}}_n$ $(n = 1, \ldots, N)$ get the closest of unmatched vectors $\widehat{\boldsymbol{h}}_m$. If the distance between $O^{(2)}\widehat{\boldsymbol{g}}_n$ and the closest vector $\widehat{\boldsymbol{h}}_{m_x}$ does not exceed a threshold p_2, $\widehat{\boldsymbol{h}}_{m_x}$ is ascribed to $\boldsymbol{g}_n$, i.e., $\varpi(n) = m_x$; otherwise, $\widehat{\boldsymbol{g}}_n$ is considered to be spurious. The problem is solved by the assignment ϖ leading to the smallest number of spurious $\boldsymbol{g}_n$ vectors. At the end, one needs to apply $\mathcal{SO}$ to legitimate $\boldsymbol{g}_n$ vectors and their partners $\widehat{\boldsymbol{h}}_{\varpi(n)}$ to get the orientation O. In the case of a draw (the same number of spurious vectors

---

[8] In this step, $M_f - p_d$ families are rejected as potential ascriptions to $\widehat{\boldsymbol{g}}_n$. One may built a more sophisticated rejection algorithm, in which the accumulation stage is repeated, but this time only the pairs consisting of vectors from earlier-accepted families are allowed to cast votes. The results of the second voting round are saved in V. There is an analogy between the second voting and an extension to Hough transform known as back-mapping [58]. See also [59].

[9] If one wants to save a number of best solutions, an additional parameter is needed to discriminate between close solutions.

[10] In this case, a simpler list built of triplets (d_h, m_1, m_2) can be used; the indices m_1 and m_2 identify items on the complete list of vectors $\widehat{\boldsymbol{h}}_m$, $(m = 1, \ldots, M)$.

Input: lattice parameters, M_f sets of indices of reflectors, symmetry operations, parameters p_i, $(i = 1, 2, d)$
/* Initialization */
get M reciprocal lattice vectors $\check{\boldsymbol{h}}_m$ grouped in M_f families
normalization $\widehat{\boldsymbol{h}}_m = \check{\boldsymbol{h}}_m / |\check{\boldsymbol{h}}_m|$
create LIST

for *each pattern* **do**
  Input: N normalized scattering vectors $\widehat{\boldsymbol{g}}_n$
  $V = 0$
  /* Accumulation */
  for *each pair* $(\widehat{\boldsymbol{g}}_{n_1}, \widehat{\boldsymbol{g}}_{n_2})$ *such that* $n_1 < n_2$ **do**
    $d(\widehat{\boldsymbol{g}}_{n_1}, \widehat{\boldsymbol{g}}_{n_2}) = \arccos(\widehat{\boldsymbol{g}}_{n_1} \cdot \widehat{\boldsymbol{g}}_{n_2})$
    retrieve from LIST quintuplets $(d_h, f_1, k_1, f_2, k_2)$ such that $|d(\widehat{\boldsymbol{g}}_{n_1}, \widehat{\boldsymbol{g}}_{n_2}) - d_h| \leq p_1$
    $S = \bigcup$ (indices f of families present in the retrieved quintuplets)
    for *each* f *in* S *and* $i = 1, 2$ **do**
      $V(n_i, f) \longleftarrow V(n_i, f) + 1$
  /* Only p_d (≥ 1) high-vote families are ascribed to $\widehat{\boldsymbol{g}}_n$ */
  for *each* n **do**
    Set the $M_f - p_d$ lowest values of $V(n, \cdot)$ to 0
  /* Final stage */
  HighestNumberOfAssignedVects $\longleftarrow 0$
  for *each pair* $(\widehat{\boldsymbol{g}}_{n_1}, \widehat{\boldsymbol{g}}_{n_2})$ *such that* $n_1 < n_2$ **do**
    Retrieve vectors $\widehat{\boldsymbol{h}}_{m_i}$ $(i = 1, 2)$ of the families f for which $V(n_i, f) \neq 0$
    /* Getting trial orientation */
    for *each pair* $(\widehat{\boldsymbol{h}}_{m_1}, \widehat{\boldsymbol{h}}_{m_2})$ *of the retrieved vectors* **do**
      $O^{(2)} \longleftarrow$ result of (8.16) applied to $(\widehat{\boldsymbol{g}}_{n_1}, \widehat{\boldsymbol{g}}_{n_2})$ and $(\widehat{\boldsymbol{h}}_{m_1}, \widehat{\boldsymbol{h}}_{m_2})$
    /* Trial assignment */
    NumberOfAssignedVects $\longleftarrow 0$; H' and $G' \longleftarrow$ empty matrices
    for *each* n **do**
      $\widehat{\boldsymbol{h}}_{m_x} \longleftarrow$ vector $\widehat{\boldsymbol{h}}_m$ $(m = 1, \ldots, M)$ nearest to $O^{(2)} \widehat{\boldsymbol{g}}_n$
      if $d(\widehat{\boldsymbol{h}}_{m_x}, O^{(2)} \widehat{\boldsymbol{g}}_n) \leq p_2$ **then**
        append $\widehat{\boldsymbol{g}}_n$ and $\widehat{\boldsymbol{h}}_{m_x}$ to matrices G' and H', respectively
        NumberOfAssignedVects $\longleftarrow$ *NumberOfAssignedVects* $+ 1$
    /* Saving the result */
    if *NumberOfAssignedVects* $>$ *HighestNumberOfAssignedVects* **then**
      $O_{res} \longleftarrow \mathcal{SO}(H' G'^T)$; $quality(O_{res}) \longleftarrow \mathrm{tr}(H'^T O_{res} G')$
      HighestNumberOfAssignedVects $\longleftarrow$ *NumberOfAssignedVects*
    else if *NumberOfAssignedVects* $=$ *HighestNumberOfAssignedVects* **then**
      $O \longleftarrow \mathcal{SO}(H' G'^T)$; $quality(O) \longleftarrow \mathrm{tr}(H'^T O G')$
      if $quality(O) > quality(O_{res})$ **then**
        $O_{res} \longleftarrow O$; $quality(O_{res}) \longleftarrow quality(O)$
  Result: Orientation O_{res} and $quality(O_{res})$

Algorithm A: Simplified description of strategy A for computing crystal orientation from normalized scattering vectors $\widehat{\boldsymbol{g}}_n$.

for different orientations), the result with the smallest residue is used as the final solution. This strategy is suitable for early exit. If a solution reached at a certain stage of matching has a satisfactory figure of merit, further search for even better solutions can be abandoned. If the parameter p_d of strategy A is set to M_f, strategy A becomes equivalent to strategy B.

Strategy C

In this last example, the first steps are like in strategy B. One constructs the `LIST`, and one gets the admissible trial orientations $O^{(2)}$ based on pairs of vectors. As above, this involves a threshold p_1. Moreover, like in strategy A, one needs an accumulation table $V(n, f)$.

Having the trial orientation, a trial indexing is performed in the following way: for each vector $O^{(2)}\boldsymbol{g}_n$ $(n = 1, \ldots, N)$ get the closest of vectors $\widehat{\boldsymbol{h}}_m$. If the distance between $O^{(2)}\boldsymbol{g}_n$ and this closest vector, say, $\widehat{\boldsymbol{h}}_{m_x}$, does not exceed a threshold p_2, $\boldsymbol{g}_n$ is considered legitimate and $\varpi(n) = m_x$. Let N_L denote the number of legitimate vectors obtained for this trial orientation. If $N_L \geq 3$, there is the accumulation step

$$V(n, f) \leftarrow V(n, f) + N_L^{p_3} ,$$

where f represents the family of the vector $\widehat{\boldsymbol{h}}_{\varpi(n)}$, and p_3 is a parameter. At the end of this stage, having the accumulation table V, p_d families with the highest number of votes in $V(n, \cdot)$ are ascribed to $\widehat{\boldsymbol{g}}_n$. (In the implementation *KiKoCh*2 described in Sect. 8.5.2, $p_d = 1$.) The final stage of getting the orientation and assignment is the same as in the case of strategy A; this involves another parameter (which in *KiKoCh*2 is set equal to p_2).

8.5.1 *Triplet Voting*

A natural extension of pair-based voting is voting by triplets. Such method was applied in a program for indexing of Laue patterns [56]. There is a family of algorithms used for autonomous star recognition based on triplets of stars (e.g., [55]); they are also known as the "triangle algorithms" [60]. Triplet voting is used for indexing EBSD patterns in the proprietary OIM Analysis™ system [7, 61].

Triplet based accumulation can be implemented by simply modifying a pair based method. To cast votes, instead of pairs, vectors $\widehat{\boldsymbol{g}}_n$ must constitute triplets nearly congruent to triplets of $\widehat{\boldsymbol{h}}_m$ vectors. Other aspects of indexing are mutatis mutandis the same as in pairwise voting.

Strategy D

This is an example strategy based on triplets of vectors. It is a modified version of strategy A, and the preliminary steps are the same in both cases. The accumulation stage is different. For each triplet $(\widehat{\boldsymbol{g}}_{n_1}, \widehat{\boldsymbol{g}}_{n_2}, \widehat{\boldsymbol{g}}_{n_3})$ get the angular distances $d(\widehat{\boldsymbol{g}}_{n_1}, \widehat{\boldsymbol{g}}_{n_2})$, $d(\widehat{\boldsymbol{g}}_{n_2}, \widehat{\boldsymbol{g}}_{n_3})$ and $d(\widehat{\boldsymbol{g}}_{n_3}, \widehat{\boldsymbol{g}}_{n_1})$. For each of these three distances, retrieve from the `LIST` the items such that the deviation between the distance and d_h is smaller than a threshold. The threshold, say, p_1, is a parameter of the strategy D. The families which appear in the items retrieved based on $d(\widehat{\boldsymbol{g}}_{n_i}, \widehat{\boldsymbol{g}}_{n_j})$ are collected without duplicates in a set S_{ij}. One obtains three such sets S_{12}, S_{23} and S_{31}. Now, the vector $\widehat{\boldsymbol{g}}_{n_1}$ is supported by families common to S_{12} and S_{31}, i.e., by those in $S_1 = S_{31} \cap S_{12}$. Similarly, $S_2 = S_{12} \cap S_{23}$ and $S_3 = S_{23} \cap S_{31}$ support $\widehat{\boldsymbol{g}}_{n_2}$ and $\widehat{\boldsymbol{g}}_{n_3}$, respectively. Each element of the set S_i casts a vote for $\widehat{\boldsymbol{g}}_{n_i}$, i.e., for each f in S_i, the value of $V(n_i, f)$ is increased by 1. As in A, one ultimately gets the resulting accumulator table $V(n, f)$. For a given n, the larger the number $V(n, f)$, the larger the probability that $\widehat{\boldsymbol{g}}_n$ matches a vector from the fth family. The final stage is the same as in the case of strategy A. The strategy D is controlled by the same parameters as strategy A.

8.5.2 *Example Implementation*

The algorithms A–D have been implemented in a program called *KiKoCh2*. Applicability of the procedures and the program for indexing of various patterns is illustrated in figures below. *KiKoCh2* can be used to index EBSD and Kikuchi patterns (Fig. 8.5). It is also applicable to indexing of Kossel patterns (Fig. 8.7) and Laue patterns from simple structures (Fig. 8.6).

8.6 Accumulation in Rotation Space

Contributions can also be accumulated in the rotation space (or more precisely, in the symmetry induced fundamental region in the rotation space [63]). The contributions can be made by individual vectors (one vector from each $\mathcal{G}$ and $\mathcal{H}$), and in this case accumulation takes place along continuous curves in the parameter space. With two or more vectors from each set, accumulation takes place at discrete points in the parameter space [64]. These two cases are described in more details in separate sections below. At the end, the rotation with the largest total accumulation is considered to solve the problem; the issue of getting this rotation is addressed in Sect. 8.6.3.

8.6.1 *Accumulation at Points of the Rotation Space*

For a k-tuple ($k > 1$) of vectors $\widehat{\boldsymbol{g}}_i$ from $\mathcal{G}$ ($i = 1, \ldots, k$) and the k-tuple $\widehat{\boldsymbol{h}}_i$ from $\mathcal{H}$, the special orthogonal matrix representing the best rotation of $\widehat{\boldsymbol{g}}_i$ to $\widehat{\boldsymbol{h}}_i$ is given by (8.15) and (8.16). To determine the rotation carrying the largest subset of $\mathcal{G}$ to a subset of $\mathcal{H}$, all k-tuples from $\mathcal{G}$ and from $\mathcal{H}$ are considered, and each pair of k-tuples contributes to a point in the rotation space. Assuming exact congruence between the largest matching subsets of $\mathcal{G}$ and $\mathcal{H}$, the rotation relating them gets the

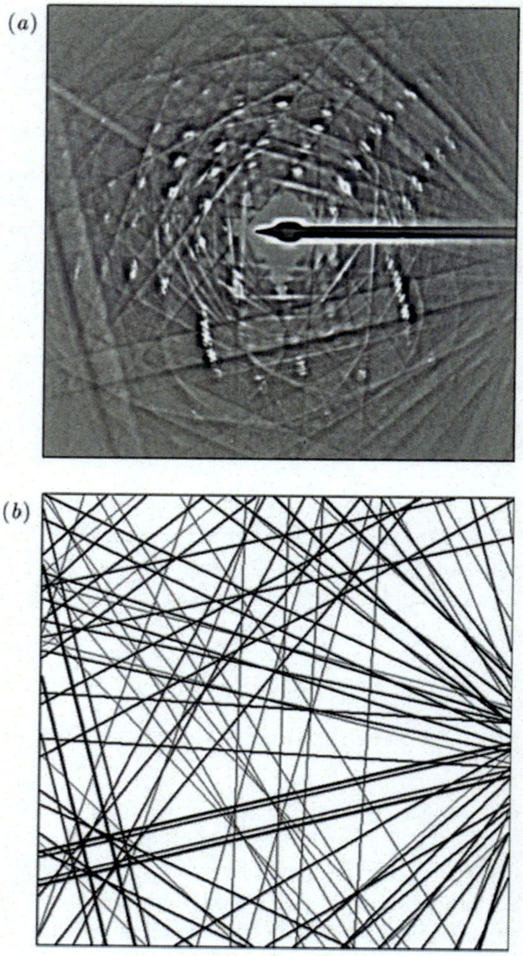

Fig. 8.5 (*a*) Experimental Kikuchi pattern of chromium carbide Cr_3C_2 alloy. Courtesy of E. Bouzy. (*b*) Simulated Kikuchi lines in orientation determined by *KiKoCh*2 based on marking lines manually positioned on the experimental pattern.

largest number of contributions, and the problem is reduced to locating a maximum of the contributions. The same approach is applicable to experimental patterns, but a tolerance needs to be allowed. The amount of contribution made to the parameter space by a given pair of k-tuples can be weighted by the quality of the match between the k-tuples. In the simplest case, one can use the rectangular function, i.e., the contribution will be non-zero only if the match is better than a threshold. This method was implemented a number of times; see, e.g., strategy 4 for indexing Kikuchi patterns in [44] or a strategy for indexing Laue patterns described in [65].

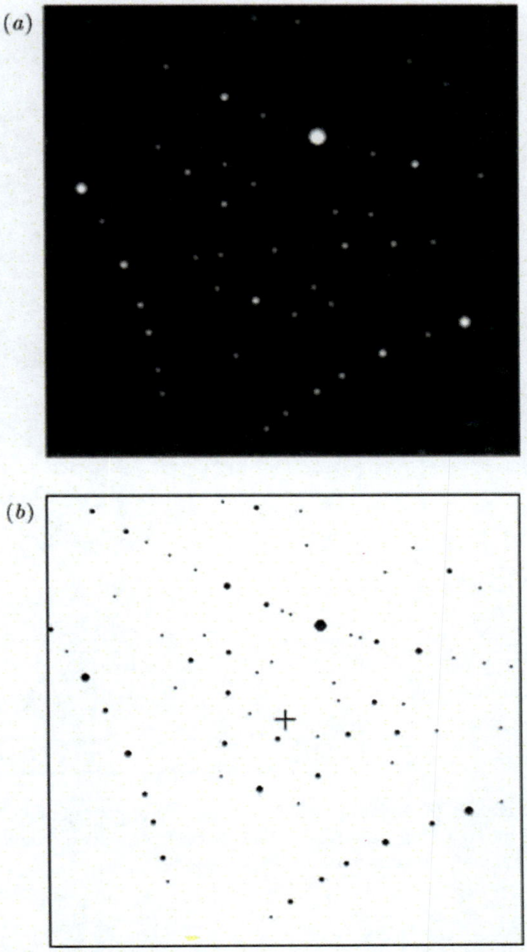

Fig. 8.6 (*a*) Simulated reflection Laue pattern of Si. (*b*) Laue spots in orientation determined by *KiKoCh*2 based on spots manually marked on the pattern shown in (*a*).

8.6.2 Accumulation Along Curves in the Space of Rotations

There is a procedure for accumulating contributions based on individual vectors, i.e., with $k = 1$ [64]. Let $R(\boldsymbol{v}, \omega)$ be the special orthogonal matrix of the rotation by the angle ω about the axis determined by $\boldsymbol{v}$ ($\neq \mathbf{0}$). Given two vectors $\widehat{\boldsymbol{g}}$ and $\widehat{\boldsymbol{h}}$ in $\mathcal{G}$ and $\mathcal{H}$, respectively, one looks for all rotations transforming $\widehat{\boldsymbol{g}}$ on $\widehat{\boldsymbol{h}}$. Assuming that $\widehat{\boldsymbol{g}} \neq -\widehat{\boldsymbol{h}}$, these rotations are represented by the matrices

$$O^{(1)}(\omega) = R(\widehat{\boldsymbol{g}} + \widehat{\boldsymbol{h}}, \pi)\, R(\widehat{\boldsymbol{g}}, \omega)\ . \tag{8.17}$$

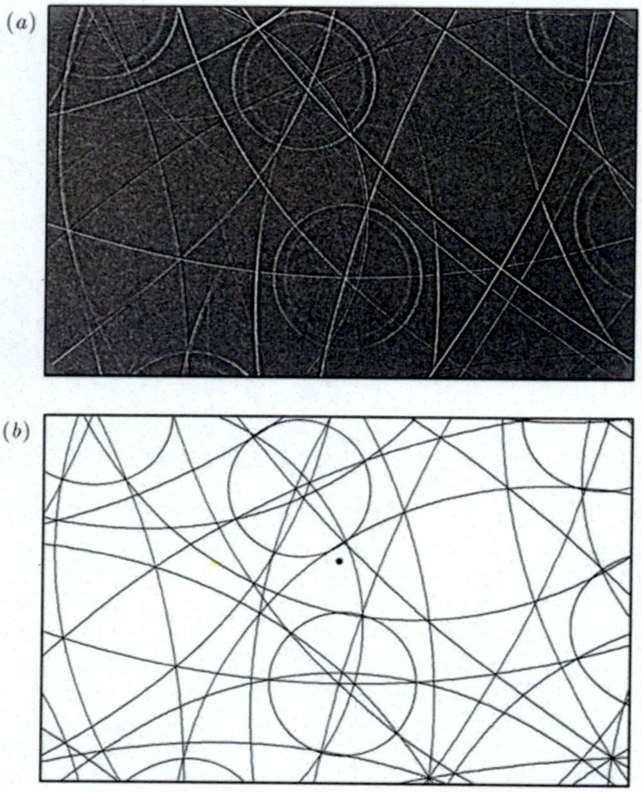

Fig. 8.7 (*a*) Experimental Kossel pattern of pattern of a CuBeAl alloy [62]. Courtesy of D. Bouscaud. (*b*) Simulated Kossel lines in orientation determined by *KiKoCh2* based on marking points manually positioned on the experimental pattern.

They constitute an ω-parameterized geodesic in the rotation space. Each pair from $\mathcal{G} \times \mathcal{H}$ contributes to rotations located on one of such curves.[11] In practice, the contributions are made along the curves at some small intervals of ω.

As in the case of $k \geq 2$, the rotation relating the largest matching subsets of $\mathcal{G}$ and $\mathcal{H}$ gets the largest number of contributions, and again, the problem is reduced to locating the global maximum of the contributions in the accumulator space. The maximum corresponds to the parameters of the sought orientation. This approach to orientation determination and indexing is a form of the generalized Hough transform. Implementations of variants of this method for multigrain indexing are described in [66, 67]; see also [68, 69].

---

[11] Since the group of unit quaternions (forming the S^3 sphere) is homeomorphic to the group of rotations, the accumulation along geodesics in the space of rotations given by (8.17) corresponds to the Funk transform on S^3. See Sect. 6.4.3.

8.6.3 Maxima in Rotation Space

As in all accumulation-based methods, at the end, one needs to determine the locations of maxima in the accumulator space, i.e., in the symmetry-induced fundamental region in the space of proper rotations. Since the match between vectors is only approximate, some tolerance must be allowed. A straightforward way to evaluate the contributions is by partitioning the space into equivolume bins of size linked to the accuracy of experimental data and the resolution of resulting orientations. The center of the bin with the largest accumulation is the sought rotation matching the largest subsets of $\mathcal{G}$ and $\mathcal{H}$. With this approach, extra measures are needed to take account of contributions distributed in neighboring bins or near borders of the fundamental region.

Instead of binning, methods of cluster analysis can be implemented. One approach (particularly suitable for resolving the discrete cases with $k \geq 2$) is to use a list of potential solutions; every new orientation obtained from (8.15) or (8.16) is compared to already saved potential solutions. If the orientation deviates from an earlier-saved solution by an angle smaller than a threshold, a ranking number of that solution is increased, and the solution is corrected by taking a weighted average [70] of the solution and the new orientation. Otherwise, the orientation is appended to the list as a new potential solution.

Contributions can also be made by adding continuous orientation components, e.g., represented by coefficients of an expansion of a peak into (symmetrized) generalized spherical harmonics. In this case (and also in cases described in Sect. 8.6.4), one needs a method of searching for global extrema of functions on $SO(3)$ given by analytical expressions. One way is to search for local extrema using, say, one of gradient methods, starting from points of a nearly-uniform grid on $SO(3)$. The highest of these local maxima is considered to be the global maximum. Clearly, the grid density must correspond to the width of peaks of the function.

8.6.4 Other Orientation-Based Algorithms

As was already mentioned, the problem of orientation determination is close to the problem of registering two clouds of points studied in the field of computer vision. Algorithms used in computer vision for object recognition and for aligning three-dimensional shapes are not necessarily directly applicable to indexing, but numerous approaches are considered in that field, and ideas used there can be inspiring. Therefore, the subject is worth a look. Again, in general, the methods for object recognition and alignment involve translation, whereas in orientation determination the translation term is fixed at zero. In standard formulations, these methods do not account for symmetries.

⊡ *Iterative closest point (ICP)*
The most popular algorithm for 3D registration is the iterative closest point (ICP) algorithm [71]. It

is designed to perform registration of 3D shapes, but it reduces the problem to registration of "point sets"; in the parlance of [71], our problem would be to match "data" point set $\mathcal{G}$ to a "model" point set $\mathcal{H}$. The algorithm is iterative: first, it finds points of the model $\mathcal{H}$ closest to points of $\mathcal{G}$, i.e., one has a certain trial assignment ϖ, and second, with this assignment, the standard registration is applied (minimization of squared distances between corresponding points, and in our case, that would be the $\mathcal{SO}$ operation). It is easy to see that the above procedure leads to a local minimum. To find global minimum, the algorithm is run multiple times for a number of appropriate initial states, but still it is susceptible to local minima. As noted in [71], the ICP algorithm "will have difficulty correctly registering 'see urchins' and 'planets'", and these are the shapes of $\mathcal{G}$ and $\mathcal{H}$ figures in the indexing problem.[12] There are, however, extensions of ICP such that each point of the "model" $\mathcal{H}$ interacts with all "data" points of $\mathcal{G}$ [72]. E.g., one has the 'softassign matching' algorithm of [73]: The mixed (combinatorial + continuous) problem is replaced by a purely continuous nonlinear problem. The matrix representing combinatorial part of the problem is allowed to have continuous entries, but its evolution in the optimization process is guided by additional constraints and barrier functions.

⊡ *Kernel correlation*

Kernel correlation is a simple way of turning the registration of point-sets into minimization of a continuous function. In its general framework, the method is clearly presented in [72]. Briefly, in the case restricted to rotations, the problem of orientation determination is reduced to finding the proper rotation O minimizing

$$f(O) = -\sum_{m=1}^{M}\sum_{n=1}^{N} \exp\left(-|\check{\boldsymbol{h}}_m - O\boldsymbol{g}_n|^2/(2\sigma^2)\right) . \tag{8.18}$$

With unit vectors $\widehat{\boldsymbol{h}}_m$ and $\widehat{\boldsymbol{g}}_n$, this function takes the form $-\sum_m \sum_n \exp\left(\widehat{\boldsymbol{h}}_m \cdot O\widehat{\boldsymbol{g}}_n/\sigma^2\right)$, i.e., it is a sum of O-dependent von Misses–Fisher functions specified by the pairs $(\widehat{\boldsymbol{h}}_m, \widehat{\boldsymbol{g}}_n)$ [63]. The above expressions are based on the Gaussian kernel, but clearly other kernels can be used.

⊡ *Trial-and-error approach*

It is worth noting that similarly to (8.18), one could get the orientation by locating the global minimum of the function

$$f(O) \propto N_c^{-1} \sum_{n=1}^{N} w_n(O) \min_m |\check{\boldsymbol{h}}_m - O\boldsymbol{g}_n|^2 ,$$

where $w_n(O) = 1$ if $\min_m |\check{\boldsymbol{h}}_m - O\boldsymbol{g}_n| < \epsilon$, and $w_n(O) = 0$ otherwise, and $N_c = \sum_{n=1}^{N} w_n(O)$ is the number of non-zero terms. With unit vectors $\widehat{\boldsymbol{h}}_m$ and $\widehat{\boldsymbol{g}}_n$, the function can be expressed as $f(O) \propto N_c^{-1} \sum_{n=1}^{N} w_n(O) \max_m \widehat{\boldsymbol{h}}_m \cdot O\widehat{\boldsymbol{g}}_n$, where $w_n(O) = 1$ if $\max_m \widehat{\boldsymbol{h}}_m \cdot O\widehat{\boldsymbol{g}}_n > 1 - \epsilon$, and $w_n(O) = 0$ otherwise. This is in a sense a trial-and-error approach accounting for all orientations (f as a function of O) and all assignments (search for $\min_m |\check{\boldsymbol{h}}_m - O\boldsymbol{g}_n|^2$ or $\max_m \widehat{\boldsymbol{h}}_m \cdot O\widehat{\boldsymbol{g}}_n$ for each $O\widehat{\boldsymbol{g}}_n$). The strategy B and final stage of strategy A described in Sect. 8.4.1 are based on the same principle, but instead of considering all orientations, only those matching pairs of vectors are tested. Clearly, with this and similar approaches, an additional computational device promoting solutions with large N_c is needed.

[12] Additionally, in its original formulation [71], the algorithm uses complete "data" point sets i.e., it does not account for spurious data points.

8.7 Testing of Indexing Algorithms

The standard method to check the quality of indexing algorithms is to generate artificial data based on known initial orientations, add errors, run the program, and check the deviations between the resulting orientations and the true initial orientations.

Both the indexing success rate and speed depend on the types of patterns and lengths of the lists matched vectors. Example results (success rate versus random errors) of tests on simulated EBSD bands for α-Ti are illustrated in Fig. 8.8. The figure shows the performance of algorithms A, B C, D and a triplet-based accumulation in orientation space [44]. For these particular data, the algorithms A–D outperform the accumulation in the orientation space in the robustness to large random errors.[13]

The indexing algorithms usually involve some parameters. Particular values of these parameters are to be fixed for given diffraction data. The numbers will depend on the type of patterns (e.g. Kossel vs. Laue) settings of pattern acquisition system (e.g., sample-detector distance), crystal structure, level of errors affecting the scattering vectors et cetera. Optimal values of parameters can be determined by checking the quality of indexing on simulated data having characteristics similar to those of experimental data. Clearly, with a properly defined goal function, this can be done in an automatic way by using a minimization software. In the case of data shown in Fig. 8.8, the parameters of strategies A–D were the same; the success rate could be improved by tuning the parameters to the strategy and error level.

In orientation mapping, the quality of indexing can be evaluated based on real experimental maps. The better the routine, the smaller number of unsolved cases and misindexing 'spikes' on maps, and the smaller the average misorientation between neighboring pixels within individual grains. To estimate reliability of indexing a single pattern, one needs a figure of merit.

8.8 Figures of Merit and Other Issues

Ultimately, after indexing a diffraction pattern, one has the crystal orientation and the assignment between some vectors of $\mathcal{G}$ and $\mathcal{H}$. With this, one can specify nearly congruent subfigures $\mathcal{G}'$ and $\mathcal{H}'$ of $\mathcal{G}$ and $\mathcal{H}$, respectively. There is a question about the quality of this congruency and the reliability of the solution of a given diffraction pattern. A number of factors can be used and combined to get a final figure of merit. The simplest measures of quality are the number of detected reflections which were actually indexed, (i.e., the cardinality of the sets $\mathcal{G}'$ and $\mathcal{H}'$), the ratio of the number of indexed reflections to the number of detected reflections, (i.e., the ratio of the

---

[13] The speeds of indexing with strategies A–D were similar with the fastest (A) processing more than 15 thousand patterns per second and the slowest (C) nearly 10 thousand patterns per second. (CPU times on a 2.6 GHz personal computer. Computation without any form of parallelism.) The indexing by accumulation in the orientation space was much slower; the speed was about 200 patterns per second.

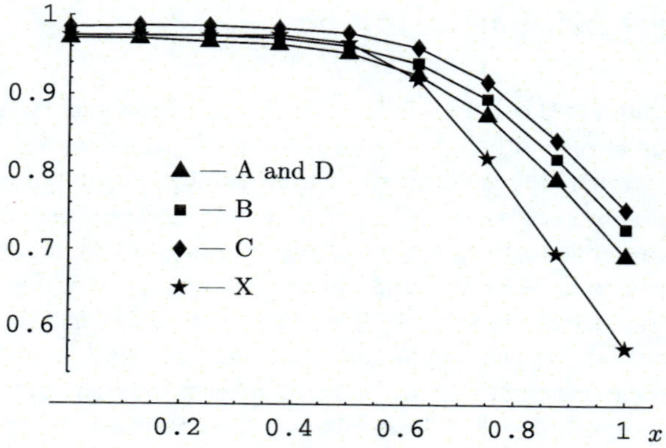

Fig. 8.8 Example results of tests on simulated EBSD bands for α-Ti. Success rate versus random errors. The total number of bands was seven of which three were spurious. The error range for the distance of a band from the pattern center (band shift) was $x \times 8\%$ of the pattern diameter. The error range for band inclination (band slope) was $x \times 4°$. The errors were uniformly distributed in their ranges. Results for strategies A and D are not the same but visually indistinguishable. The data marked by X (star) were obtained using an algorithm based on accumulation in orientation space [44].

cardinalities of the sets $\mathcal{G}'$ and $\mathcal{G}$), and the quality of the match between $\mathcal{G}'$ and $\mathcal{H}'$ (e.g., the actual value of the expression (8.14)). One may also use indicators specific to the used method, e.g., for methods based on voting, this can be the maximal number of votes.

The other issue is how the best solution compares to other potential solutions. In particular, how the best solution compares to the second best solution. This can be quantified by parameters constructed from the above factors. A good example of such parameter is the *confidence index* [74] used in combination with the voting algorithm of [61, 75]. It is defined as

$$C = \frac{v_1 - v_2}{v_1 + v_2} ,$$

where v_i is the number of votes for the ith best solution. Clearly, the confidence index is equally suitable in all accumulation-based algorithms.

Frequently, figures of merit are correlated to some physical aspects of diffraction from differently oriented crystallites, and this is visible on corresponding maps. In particular, the figures of merit are indispensable for phase discrimination; see, e.g., [76].

8.8.1 *Three Remarks*

First, for brevity, the text above was focused on the constellation problem, with all vectors having the same magnitude, but it is worth noting that the described procedures can be relatively easily generalized to matching vectors with arbitrary magnitudes. One way to account for vector magnitudes is by using weighted contributions: the larger the difference between magnitudes of potentially matching vectors the smaller the contribution. It is also sensible to ban contributions when the difference is larger than a threshold.

Second, the accumulation-based indexing methods differ by the nature of contributions, the parameter space, and the ways of collecting and counting the contributions. Depending on demands, one may consider using various combinations of these methods. In particular, to improve reliability of indexing, one may apply multiple contributions of different types and multiple ways of counting the contributions. Moreover, whenever the accumulation can be seen as the generalized Hough transform, its reliability can be improved by adapting known enhancements of that transform [77].

Third, in the above considerations, it was assumed that $\mathcal{G}$ and $\mathcal{H}$ are built of vectors bound to a fixed center of rotation which corresponds to the position of the source of diffraction. This position is known only approximately, and additionally, it changes during mapping as the incident beam scans the specimen. With known indexing, the experimental scattering vectors can be used for refining the position of the source. In this case, the optimized function depends on $\boldsymbol{L}$, i.e., pattern center and sample-detector distance (camera length) and on orientation parameters.

8.9 Orientation Determination via Direct Pattern Matching

Crystal orientation can be determined without indexing. One can get the orientation by matching an experimental pattern to simulated patterns. With a simple crystal structure and only a couple of experimental patterns, this can be done by human inspection of an atlas of simulated patterns; see, e.g., [78]. In large-scale mappings, experimental patterns can be matched to pre-computed simulated patterns automatically [11, 79, 80]. The simulations must be carried out for orientations constituting a sufficiently dense grid in the orientation space.[14] For each experimental pattern, the best match is obtained based on a suitably chosen similarity measure. An effective and standard method to match intensity images is correlation. The method is applicable to diffraction patterns of various types; see, e.g., [12, 80].

[14] It is not necessary to have a grid covering all possible orientations. Because of crystal symmetry, it is sufficient to use an asymmetric domain (fundamental region) of the orientation space plus a margin of a width resulting from the pattern acquisition angle. For methods of generating a grid in orientation space, see [81–83].

The natural domain for the complete simulated diffraction pattern is the sphere. That is why it is natural to match patterns projected on the sphere with spherical cross-correlation as the pattern similarity measure [84–86]. The correlation of real functions f_s (simulated intensity function) and f_p (experimental pattern) on the unit sphere S^2 is given by

$$(f_s \star f_p)(O) = \int_{S^2} f_s(\widehat{\boldsymbol{n}})\, f_p(O\widehat{\boldsymbol{n}})\, \mathrm{d}\widehat{\boldsymbol{n}}\,.$$

The correlation reaches maximum at orientation corresponding to the best match of the patterns, and the problem is reduced to finding the maximum of $f_s \star f_p$. The above objective function is defined via an integral over the entire sphere whereas the domain of the experimental pattern f_p is limited to a window of the detector. The problem is addressed first by modifying f_p so there is a smooth decay from the values inside the window to zero outside it. Moreover, there is a correction to the spherical cross-correlation accounting for the shape of the window

$$f(O) = (f_s \star f_p)(O) - (N_{f_p}/N_\chi)\,(f_s \star \chi)(O)\,.$$

where χ is the characteristic (cut-off) function of the window, $N_\chi = \int_{S^2} \chi(\widehat{\boldsymbol{n}})\mathrm{d}\widehat{\boldsymbol{n}}$, and the normalization coefficient N_{f_p} of f_p is defined in analogous way [85]. The integrals can be estimated by direct numerical computation, but this approach is slow. It is more time-efficient to use expansion of the functions f_s and f_p into spherical harmonics, and to express $f_s \star f_p$ as a series with coefficients given by expansion coefficients of f_s and f_p. See [84, 85] for details.

Clearly, after direct matching, knowing the crystal structure and its orientation, one can make additional effort and ascribe indices to reflections, but this step is unnecessary when the goal is to ascribe the orientation to a pixel of an orientation map.

Direct matching of patterns has some advantages over the orientation determination via reflection indexing. It may be preferable in the case of diffraction patterns with diffuse reflections, e.g., patterns from highly deformed samples. As the quality of reflection detection deteriorates in the presence of noise in diffraction patterns, the direct matching may work where the indexing-based orientation determination may fail. Moreover, direct matching accounts for intensities and this makes a difference in phase discrimination. In particular, it can discriminate between phases with the same types of Bravais lattices (e.g., [87]). In principle, it allows for discriminating chiralities [88] or polarities of crystallites [89].

The main disadvantage of the direct matching is high computational cost of the exhaustive comparison of patterns. In the case of EBSD patterns, this issue has been addressed by an approach based on machine learning and artificial neural networks [90, 91].[15] A neural network is trained on simulated diffraction patterns, and when

[15] In a related development, neural networks have been used for identification of Bravais type of crystal lattices from EBSD patterns [92].

applied to experimental patterns, it quickly finds the matching images. Generally, the neural network needs to be trained for different crystal structures and different measurement geometries, and the training times may be long. The issue of computational cost is also alleviated by parallelization and use of graphics processing units (GPUs) [90].

A number of names have been used for this brute-force trial-and-error method of orientation determination; the list includes "template matching" [11], "forward projection" [93] and "dictionary indexing" [80].

8.9.1 Direct Matching Limited by a Detected Reflection

Worth noting is a modified version of the direct pattern matching which has been used for computer-aided orientation determination [94]. Given an experimental diffraction pattern, one needs to chose or detect the strongest or most pronounced reflection and get the corresponding scattering vector $\boldsymbol{g}$. Assuming its indices are $\check{\boldsymbol{h}}$, as in Sect. 8.6.2, one gets possible crystal orientations along a line in the orientation space. They are given by (8.17). The orientations are tested by moving along the line in steps. At each step, i.e., at each of the selected orientations, the simulated diffraction pattern is compared to the experimental pattern. (In other versions, positions of reflections are calculated and compared with positions of observed diffraction peaks.) Since the actual indices of the chosen reflection are not known, one needs to test a representative of each potential family of equivalent reciprocal lattice vectors, i.e., one needs to test all lines determined by $\boldsymbol{g}$ and these representatives. The orientation corresponding to the best match is regarded as the result. This is again a trial-and-error method, but test orientations do not cover the whole asymmetric domain but are limited to a number of lines in the orientation space.

References

1. J.R. Milch, T.C. Minor, The indexing of single-crystal X-ray rotation photographs. J. Appl. Crystallogr. **7**, 502–505 (1974)
2. G. Vriend, M.G. Rossmann, Determination of the orientation of a randomly placed crystal from a single oscillation photograph. J. Appl. Cryst. **20**, 338–343 (1987)
3. W. Kabsch, Automatic indexing of rotation diffraction patterns. J. Appl. Cryst. **21**, 67–71 (1988)
4. Y. Inokuti, Y. Shimizu, H. Shimanaka, Crystallographic orientation determination of small areas in grain oriented silicon steel by transmission Kossel technique. Kawasaki Steel Giho **12**, 89–98 (1980)
5. Y. Inokuti, C. Maeda, Y. Ito, Transmission Kossel study of the formation of (110)[001] grains after an intermediate annealing of grain oriented silicon steel containing a small amount of Mo. Metall. Trans. A **16**, 1613–1623 (1985)

6. Y. Inokuti, C. Maeda, Y. Ito, Computer color mapping of configuration of Goss grains after transverse cold rolling in grain oriented silicon steel. Trans. Iron Steel Inst. Jpn. **27**, 302–311 (1987)
7. B.L. Adams, S.I. Wright, K. Kunze, Orientation imaging: the emergence of a new microscopy. Metall. Trans. A **24**, 819–831 (1993)
8. R.R. Keller, R.H. Geiss, Transmission EBSD from 10 nm domains in a scanning electron microscope. J. Microsc. **245**, 245–251 (2012)
9. J.J. Fundenberger, E. Bouzy, D. Goran, J. Guyon, H. Yuan, A. Morawiec, Orientation mapping by transmission-SEM with an on-axis detector. Ultramicroscopy **161**, 17–22 (2016)
10. J.J. Fundenberger, A. Morawiec, E. Bouzy, J.S. Lecomte, Polycrystal orientation maps from TEM. Ultramicroscopy **96**, 127–137 (2003)
11. E.F. Rauch, L. Dupuy, Rapid diffraction patterns identification through template matching. Arch. Metall. Mater. **50**, 87–99 (2005)
12. E.F. Rauch, M. Véron, J. Portillo, D. Bultreys, Y. Maniette, S. Nicolopoulos, Automatic crystal orientation and phase mapping in TEM by precession diffraction. Microsc. Anal. **22**, S5–S8 (2008)
13. S.I. Wright, D.J. Dingley, Orientation imaging in the transmission electron microscope. Mater. Sci. Forum **273–275**, 209–214 (1998)
14. D.B. Williams, C.B. Carter, *Transmission Electron Microscopy. A Textbook for Materials Science* (Springer, New York (USA), 2009)
15. D.J. Dingley, Orientation imaging microscopy for the transmission electron microscope. Microchim. Acta **155**, 19–29 (2006)
16. H.F. Poulsen, *Three-Dimensional X-Ray Diffraction Microscopy* (Springer, Berlin, 2004)
17. G.E. Ice, J.W.L. Pang, Tutorial on X-ray microLaue diffraction. Mater. Charact. **60**, 1191–1201 (2009)
18. A. Miyamoto, I. Weikusat, T. Hondoh, Complete determination of ice crystal orientation using Laue X-ray diffraction method. J. Glaciol. **57**, 103–110 (2011)
19. T. Lehmann, M. Trempa, E. Meissner, M. Zschorch, C. Reimann, J. Friedrich, Laue scanner: a new method for determination of orientations and grain boundary types of multicrystalline silicon on full wafer scale. Acta Mater. **69**, 1–8 (2014)
20. W. Whitley, C. Stock, A.D. Huxley, A laboratory-based Laue X-ray diffraction system for enhanced imaging range and surface grain mapping. J. Appl. Cryst. **48**, 1342–1345 (2015)
21. P. Lynch, M. Barnett, A. Stevenson, B. Hatchinson, Characterizing grain-oriented silicon steel sheet using high-resolution Laue X-ray diffraction. Metall. Mater. Trans. A **48**, 5206–5210 (2017)
22. W. Lorenz, H. Hougardy, SAECP combined with quantitative image analysis - an improved tool for texture analysis. Textures Microstruct. **8–9**, 579–598 (1988)
23. N.H. Schmidt, N.O. Olesen, Computer-aided determination of crystal-lattice orientation from electron channeling patterns in the SEM. Can. Mineral. **27**, 15–22 (1989)
24. D.M. Saylor, A. Morawiec, G.S. Rohrer, Distribution of grain boundaries in magnesia as a function of five macroscopic parameters. Acta Mater. **51**, 3663–3674 (2003)
25. H. Wilman, The interpretation and application of electron-diffraction "Kikuchi-line" patterns - part II. The methods of indexing the patterns. Proc. Phys. Soc. **61**, 416–430 (1948)
26. H.M. Otte, J. Dash, H.F. Schaake, Electron microscopy and diffraction of thin films: interpretation and correlation of images and diffraction patterns. Phys. Status Solidi (b) **5**, 527–549 (1964)
27. M. von Heimendahl, W. Bell, G. Thomas, Applications of Kikuchi line analyses in electron microscopy. J. Appl. Phys. **35**, 3614–3616 (1964)
28. P.L. Ryder, W. Pitsch, On the accuracy of orientation determination by selected area electron diffraction. Philos. Mag. **18**, 807–816 (1968)
29. P.H. Pumphrey, K.M. Bowkett, An accurate method for determining crystallographic orientations by electron diffraction. Phys. Status Solidi (a) **2**, 339–346 (1970)
30. M. von Heimendahl, Precise orientation determination by electron diffraction single-pole Kikuchi patterns. Phys. Status Solidi (a) **5**, 137–146 (1971)

31. C.T. Young, J.H. Steele Jr., J.L. Lytton, Characterization of bicrystals using Kikuchi patterns. Metall. Trans. **4**, 2081–2089 (1973)
32. R. Bonnet, F. Durand, Precise determination of the relative orientation of two crystals from the analysis of two Kikuchi patterns. Phys. Status Solidi (a) **27**, 543–549 (1975)
33. H. Helfmeier, M. Feller-Kniepmeier, Analytical determination of the exact primary beam direction from Kikuchi patterns. J. Appl. Phys. **48**, 3997–3997 (1977)
34. J.A. Kozubowski, Determination of the crystal orientation from intersections of Kikuchi lines. Phys. Status Solidi (a) **43**, 535–539 (1977)
35. M.J. Goringe, B. Loberg, D.A. Smith, On the determination of intergranular orientation relationships by Kikuchi electron diffraction. Phys. Status Solidi (a) **55**, 569–572 (1979)
36. C.J. Ball, Accurate determination of crystallographic orientation from Kikuchi patterns. Philos. Mag. A **44**, 1307–1317 (1981)
37. P. Heilmann, W.A.T. Clark, D.A. Rigney, Computerized method to determine crystal orientations from Kikuchi patterns. Ultramicroscopy **9**, 365–372 (1982)
38. H.O. Martikainen, M.A. Korhonen, V.K. Lindroos, Determination of grain boundary rotation parameters from Kikuchi patterns. I. A simplified method for determining the accurate orientation relationship of two grains. Phys. Status Solidi (a) **75**, 559–566 (1983)
39. F.R. Chen, A.H. King, Determination of the crystallographic directions and planes of features and of the misorientations of crystals with high accuracy and internal estimation of errors. J. Electron Microsc. Tech. **6**, 55–61 (1987)
40. H. Jang, D. Farkas, J.T.M. De Hosson, Determination of grain boundary geometry using TEM. J. Mater. Res. **7**, 1707–1717 (1992)
41. A. Morawiec, Indexing of diffraction patterns for determination of crystal orientations. Acta Crystallogr. A **76**, 719–734 (2020)
42. K. Chen, C. Dejoie, H.R. Wenk, Unambiguous indexing of trigonal crystals from white-beam Laue diffraction patterns: application to Dauphiné twinning and lattice stress mapping in deformed quartz. J. Appl. Cryst. **45**, 982–989 (2012)
43. H.J. Bunge, *Texture Analysis in Materials Science* (Butterworth's, London, 1982)
44. A. Morawiec, Automatic orientation determination from Kikuchi patterns. J. Appl. Cryst. **32**, 788–798 (1999)
45. J.K. MacKenzie, The estimation of an orientation relationship. Acta Cryst. **10**, 61–62 (1957)
46. G. Wahba, J.L. Farrell, J.C. Stuelpnagel, Problem 65–1, a least squares estimate of satellite attitude. SIAM Rev. **8**, 384–386 (1966)
47. W. Kabsch, A solution for the best rotation to relate two sets of vectors. Acta Cryst. A **32**, 922–923 (1976)
48. W. Kabsch, A discussion of the solution for the best rotation to relate two sets of vectors. Acta Cryst. A **34**, 827–828 (1978)
49. M.A. Stephens, Vector correlation. Biometrika **66**, 41–48 (1979)
50. J.N. Juang, H.Y. Kim, J.L. Junkins, An efficient and robust singular value method for star pattern recognition and attitude determination. Technical report NASA/TM-2003-212142, National Aeronautics and Space Administration (2003)
51. E. Bengoetxea, Inexact graph matching using estimation of distribution algorithms. Ph.D. thesis, Ecole Nationale Supérieure des Télécommunications, Paris, France (2002)
52. M.T. Goodrich, J.S.B. Mitchell, M.W. Orletsky, Approximate geometric pattern matching under rigid motions. IEEE Trans. Pattern Anal. Mach. Intell. **21**, 371–379 (1999)
53. C. Ambühl, S. Chakraborty, B. Gärtner, Computing largest common point sets under approximate congruence, *ESA*. Lecture Notes in Computer Science, vol. 1879 (Springer, Berlin, 2000), pp. 52–64
54. D.E. Cardoze, L.J. Schulman, Pattern matching for spatial point sets, in *Proceedings of 39th Annual Symposium on Foundations of Computer Science* (IEEE, Los Alamitos, 1998), pp. 156–165
55. E. Groth, A pattern-matching algorithm for two-dimensional coordinate lists. Astron. J. **91**, 1244–1248 (1986)

56. R. Ohba, I. Uehira, T. Hondoh, Computer-aided spot indexing for X-ray Laue patterns. Jpn. J. Appl. Phys. **20**, 811–816 (1981)
57. H.J. Wolfson, I. Rigoutsos, Geometric hashing: an overview. IEEE Comput. Sci. Eng. **4**, 10–21 (1997)
58. G. Gerig, F. Klein, Fast contour identification through efficient Hough transform and simplified interpretation strategy, in *Proceedings of the 8th International Joint Conference on Pattern Recognition* (Paris, 1986), pp. 498–500
59. M. Kolomenkin, S. Polak, I. Shimshoni, M. Lindenbaum, A geometric voting algorithm for star trackers. IEEE Trans. Aerosp. Electron. Syst. **44**, 441–456 (2008)
60. G. Zhang, *Star Identification: Methods, Techniques and Algorithms* (Springer, Berlin, 2017)
61. S.I. Wright, B.L. Adams, Automatic analysis of electron backscatter diffraction patterns. Metall. Trans. A **23**, 759–767 (1992)
62. D. Bouscaud, S. Berveiller, R. Pesci, E. Patoor, A. Morawiec, Local stress analysis in an SMA during stress-induced martensitic transformation by Kossel microdiffraction. Adv. Mater. Res. **996**, 45–51 (2014)
63. A. Morawiec, *Orientations and Rotations. Computations in Crystallographic Textures* (Springer, Berlin, 2004)
64. A. Morawiec, M. Bieda, On algorithms for indexing of K-line diffraction patterns. Arch. Metall. Mater. **50**, 47–56 (2005)
65. J.A. Kalinowski, A. Makala, P. Coppens, The *LaueUtil* toolkit for Laue photocrystallography. I. Rapid orientation matrix determination for intermediate-size-unit-cell Laue data. J. Appl. Crystallogr. **44**, 1182–1189 (2011)
66. S. Schmidt, *GrainSpotter*: a fast and robust polycrystalline indexing algorithm. J. Appl. Cryst. **47**, 276–284 (2014)
67. J.V. Bernier, N.R. Barton, U. Lienert, M.P. Miller, Far-field high-energy diffraction microscopy: a tool for intergranular orientation and strain analysis. J. Strain Anal. Eng. Des. **46**, 527–547 (2011)
68. K.R. Beyerlein, T.A. White, O. Yefanov, C. Gati, I.G. Kazantsev, N.F.G. Nielsen, P.M. Larsen, H.N. Chapman, S. Schmidt, *FELIX*: an algorithm for indexing multiple crystallites in X-ray free-electron laser snapshot diffraction images. J. Appl. Cryst. **50**, 1075–1083 (2017)
69. Y. Gevorkov, A. Barty, W. Brehm, T.A. White, A. Tolstikova, M.O. Wiedorn, A. Meents, R.R. Grigat, H.N. Chapman, O. Yefanov, *pinkIndexer* - a universal indexer for pink-beam X-ray and electron diffraction snapshots. Acta Cryst. A **76**, 121–131 (2020)
70. A. Morawiec, A note on mean orientation. J. Appl. Cryst. **31**, 818–819 (1998)
71. P. Besl, N.D. McKay, A method for registration of 3-D shapes. IEEE Trans. Pattern Anal. Mach. Intell. **14**, 239–256 (1992)
72. Y. Tsin, T. Kanade, A correlation-based approach to robust point set registration, in *Proceedings of European Conference on Computer Vision*, vol. Part III, ed. by T. Pajdla, J. Matas (Springer, Berlin, 2004), pp. 558–569
73. A. Rangarajan, H. Chui, F.L. Bookstein, The softassign Procrustes matching algorithm, in *Information Processing in Medical Imaging*, ed. by J. Duncan, G. Gindi (Springer, Berlin, 1997), pp. 29–42
74. D.P. Field, Recent advances in the application of orientation imaging. Ultramicroscopy **67**, 1–9 (1997)
75. S.I. Wright, A review of automated orientation imaging microscopy. J. Comput.-Assist. Microsc. **5**, 207–221 (1993)
76. J. Pan, Phase identification of single crystals using a self-consistent indexing scheme based on the Laue method. J. Appl. Cryst. **25**, 195–198 (1992)
77. V.F. Leavers, Which Hough transform?, in *Computer Vision Graphics and Image Understanding: Image Understanding*, vol. 58 (Springer, Berlin, 1993), pp. 250–264
78. J.W. Edington, *Practical Electron Microscopy in Materials Science. Monograph Two: Electron Diffraction in the Electron Microscope* (MacMillan, London, 1975)
79. V.K. Gupta, S.R. Agnew, Indexation and misorientation analysis of low-quality Laue diffraction patterns. J. Appl. Cryst. **42**, 116–124 (2009)

80. Y. Chen, S.U. Park, D. Wei, G. Newstadt, M. Jackson, J. Simmons, M. De Graef, A. Hero, A dictionary approach to EBSD indexing. Microsc. Microanal. **21**, 739–752 (2015)
81. D. Roşca, A. Morawiec, M. De Graef, A new method of constructing a grid in the space of 3D rotations and its applications to texture analysis. Model. Simul. Mater. Sci. Eng. **22**, 075013 (2014)
82. P.M. Larsen, S. Schmidt, Improving orientation sampling for indexing diffraction patterns of polycrystalline materials. J. Appl. Cryst. **50**, 1571–1582 (2017)
83. R. Quey, A. Villani, C. Maurice, Nearly uniform sampling of crystal orientations. J. Appl. Cryst. **51**, 1162–1173 (2018)
84. A. Makadia, K. Daniilidis, Rotation recovery from spherical images without correspondences. IEEE Trans. Pattern Anal. Mach. Intell. **28**, 1170–1175 (2006)
85. R. Hielscher, F. Bartel, T.B. Britton, Gazing at crystal balls: electron backscatter diffraction indexing and cross correlation on a sphere. Ultramicroscopy **207**, 112836 (2019)
86. W.C. Lenthe, S. Singh, M. De Graef, A spherical harmonic transform approach to the indexing of electron back-scattered diffraction patterns. Ultramicroscopy **207**, 112841 (2019)
87. F. Ram, M. De Graef, Phase differentiation by electron backscatter diffraction using the dictionary indexing approach. Acta Mater. **144**, 352–364 (2018)
88. A. Winkelmann, G. Nolze, Chirality determination of quartz crystals using electron backscatter diffraction. Ultramicroscopy **149**, 58–63 (2015)
89. G. Naresh-Kumar, A. Vilalta-Clemente, H. Jussila, A. Winkelmann, G. Nolze, S. Vespucci, S. Nagarajan, A.J. Wilkinson, C. Trager-Cowan, Quantitative imaging of anti-phase domains by polarity sensitive orientation mapping using electron backscatter diffraction. Sci. Rep. **7**, 10916 (2017)
90. D. Jha, S. Singh, R. Al-Bahrani, W. Liao, A. Choudhary, M. De Graef, A. Agrawal, Extracting grain orientations from EBSD patterns of polycrystalline materials using convolutional neural networks. Microsc. Microanal. **24**, 497–502 (2018)
91. Y.F. Shen, R. Pokharel, T.J. Nizolek, A. Kumar, T. Lookman, Convolutional neural network-based method for real-time orientation indexing of measured electron backscatter diffraction patterns. Acta Mater. **170**, 118–131 (2019)
92. K. Kaufmann, C. Zhu, A.S. Rosengarten, D. Maryanovsky, T.J. Harrington, E. Marin, K.S. Vecchio, Crystal symmetry determination in electron diffraction using machine learning. Science **367**, 564–568 (2020)
93. H. Sharma, R.M. Huizenga, S.E. Offerman, A fast methodology to determine the characteristics of thousands of grains using three-dimensional X-ray diffraction. II. Volume, centre-of-mass position, crystallographic orientation and strain state of grains. J. Appl. Cryst. **45**, 705–718 (2012)
94. B. Krahl-Urban, R. Butz, E. Preuss, A simplified procedure for orientation of single crystals of any structure. Acta Cryst. **A29**, 86–90 (1973)

Chapter 9
Indexing of Electron Spot-Type Diffraction Patterns

As was already noted in Sect. 2.3.4, a spot diffraction pattern observed in high-voltage TEM is an arrangement of spots of various intensity; see, Fig. 2.10. The term 'spot pattern' is informal. In most cases, it means selected area electron diffraction (SAED) pattern. SAED patterns are formed with parallel illumination. Similar in appearance are TEM microdiffraction or nanodiffraction patterns formed with low-angle convergent beam. Moreover, slightly different SAED and nanodiffraction patterns are observed using the precession diffraction technique; see Sect. 9.2.1. Here, spot patterns encompass all these groups.

Indexing of spot-type patterns is a special case of general problem of indexing single-crystal diffraction patterns, but electron microscopists often see it as a separate field because of specific features of these patterns. Due to low curvature of the Ewald sphere, spot patterns are close to magnified planar cuts through crystal reciprocal space. Since in most cases the specimen is a thin foil, positions of diffraction peaks are affected not only by measuring errors, but also by excitation errors; see Sect. 2.3.4 and Fig. 2.11. Moreover, the patterns are frequently complicated by double diffraction. Finally, with spot pattern 'interpretation', often the goal is not the assignment of indices or determination of crystal orientation or lattice parameters but phase discrimination.

9.1 Conventional Indexing of Zone Axis Patterns

In the most basic approach, indexing concerns patterns originating from a known phase. The standard procedure is to tilt the specimen so the optical axis is parallel to a prominent zone axis [1]. Reflections of a zeroth-order Laue zone form a grid or two-dimensional lattice. The grid is a magnified near-**0** part of a reciprocal lattice plane. For two-dimensional indexing, it is enough to ascribe indices to two shortest

A. Morawiec, *Indexing of Crystal Diffraction Patterns*, Springer Series in Materials Science 326, https://doi.org/10.1007/978-3-031-11077-1_9

independent lattice vectors $\boldsymbol{a}^1$ and $\boldsymbol{a}^2$, and the indices of remaining nodes are obtained via integer combinations of these vectors.

The conventional indexing procedure relies on the approximate formula

$$\lambda L \approx r d \ ,$$

where L is the effective camera length, λ is the wavelength, r is the distance of the spot to the pattern center, and d is the interplanar distance (d-spacing) corresponding to the spot [1]. The formula follows directly from Bragg's law (2.9) for small Bragg angle θ: $rd = dL \tan(2\theta) \approx 2Ld \sin\theta = \lambda L$. The quantity λL is referred to as camera constant.

The d-spacings obtained from measured distances r are used for identifying the indices of reflections by comparing them to theoretical interplanar spacings. Clearly, an interplanar distance is not always sufficient for identification of the reflection as, generally, there are nonequivalent reflections with equal interplanar spacings. (The simple example are the planes (2 2 1) and (0 0 3) in a cubic lattice; in both cases, the spacing (1.48) is $d = a/3$, where a denotes the direct lattice parameter.) In such cases, also the angle between basis vectors is taken into account, and sometimes multiple zone axis patterns must be analyzed.

The 'interpretation' of zone axis patterns originating from a known phase may comprise determination of the direct beam direction or specimen orientation. Knowing indices of reflections of the zeroth-order Laue zone, one can easily get indices of the zone axis. They are given by (1.49). However, such data do not determine the sign of the axis. In general, they do not unequivocally determine the lattice orientation.

9.1.1 180°-*Ambiguity*

The problems with orientation determination are caused by the so–called 180°-ambiguity; see, e.g., [2]. It concerns zone patterns with low-index axes. The issue can be illustrated using the Cu [112] pattern shown in Fig. 9.1. The zeroth-order Laue zone contains reflections (hkl) satisfying $h + k + 2l = 0$. This pattern has two–fold symmetry with respect to the microscope axis: the rotation about [112] direction by 180° is represented by the matrix

$$R = R^T = \frac{1}{3}\begin{bmatrix} -2 & 1 & 2 \\ 1 & -2 & 2 \\ 2 & 2 & 1 \end{bmatrix} ,$$

and one has $(hkl)R = (\overline{h}\,\overline{k}\,\overline{l})$ for all reflections the zone. On the other hand, that rotation is not a crystal symmetry operation. Therefore, there are two non–equivalent orientations leading to the same pattern; if the first one is represented by $O_{c|m}$, the other one is $RO_{c|m}$. The first order [112] zone is missing due to systematic absences;

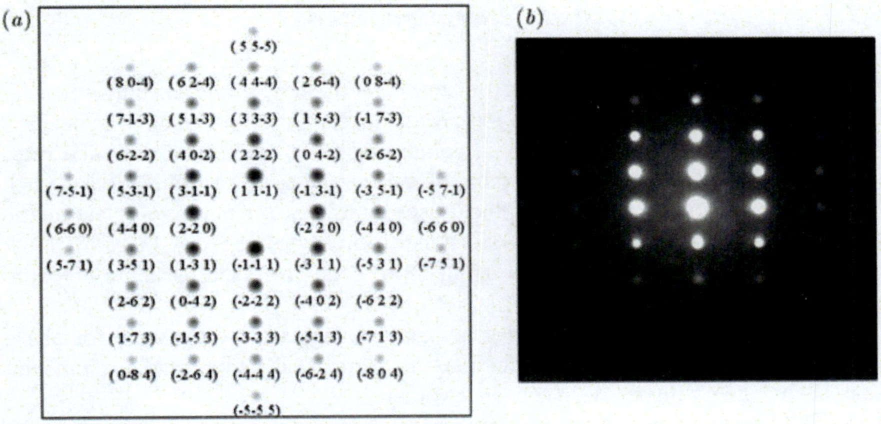

Fig. 9.1 (*a*) Cu [112] diffraction pattern simulated for the voltage of 200 kV, effective camera length of 180mm, detector size of 1 in × 1 in, and large maximal excitation error of 0.125Å^{-1}. Logarithm of intensities was taken and inverted gray scale is used to enhance visibility of weak spots. (*b*) An experimental near [112] diffraction pattern of Cu.

the zone consists of reflections (*hkl*) such that $h + k + 2l = 1$, and this condition is satisfied only by triads h, k, l containing both odd and even numbers—a recipe for reflection absence in the case of crystals with fcc lattices.

In general, the entries of the rotation matrix R' of the half-turn about the zone axis $\check{\boldsymbol{u}}$ in the crystal reference frame are

$$R'{}_i{}^j = 2n_i n^j - \delta_i{}^j \ ,$$

where n_i are components of $\widehat{\boldsymbol{n}} = \check{\boldsymbol{u}}/|\check{\boldsymbol{u}}|$; see, (1.19). With a reflection of the zone represented by $\check{\boldsymbol{h}}$, one has the zone law $\check{\boldsymbol{u}} \cdot \check{\boldsymbol{h}} = 0 = \widehat{\boldsymbol{n}} \cdot \check{\boldsymbol{h}} = n^i h_i$ and $R'{}_i{}^j\, h_j = -h_i$. There are two orientations leading to the same zone axis pattern; one is obtained from the other by composition with the above rotation. With R representing the same rotation in the Cartesian frame ascribed to the crystal, the two orientations leading to the same zone axis pattern are $O_{c|m}$ and $R\,O_{c|m}$. If the rotation happens to belong to the point group of the crystal, the two orientations are equivalent. Thus, the 180°-ambiguity does not affect zone axis patterns with the zone axis being a two-fold symmetry axis.

In conventional indexing involving human inspection it is easy to address the 180°-ambiguity problem by tilting the specimen [3]. Correctness of indexing can be relatively easily verified if a small set of patterns is considered, and an interactive approach is used. Such verification is difficult in orientation mapping as it requires thousands of patterns, and the orientation determination procedure is fully automatic.

9.1.2 Computer-Assisted Conventional Indexing

In principle, conventional indexing of zone axis patterns could be performed by hand. In some cases it is enough to look for a similar pattern in an atlas of patterns [1, 4]. However, in most cases, some computing tools are used, from slide rules [5, 6] to dedicated computer programs of various level of sophistication [7–14]. Nowadays, the programs have graphical user interfaces for marking or automatic detection of spots and measurement of distances and angles; see, e.g., [12, 14]. Some micoroscopists index spot patterns using more general crystallographic software (e.g., *CaRIne Crystallography* [15]).

As was mentioned above, the d-spacings and angles are often used for phase discrimination or phase identification. These data are usually supported by elemental analysis by energy dispersive X-ray spectrometry.

In TEM practice, most common is phase *discrimination*. Knowing the elements present, one gets magnitudes of two shortest vectors (largest interplanar spacings) and the angle between the vectors from a prominent zone axis pattern and compares these data to the data for the compounds on a (short) list of expected phases. (For a slightly more elaborate approach, see [16].)

A good example of computer-aided phase *identification* is the procedure described in [17]. Phases are identified by comparing collected data with NIST Crystal Data database [18]. Again, the input consists of information on elements present, magnitudes of two shortest vectors and the angle between them. For earlier programs of this kind, see, e.g., [19–21]. Since reciprocal lattices of multiple phases may have similar two-dimensional sections, phase identification based on a single pattern is generally not reliable. The issue can be addressed by checking several patterns recorded at different tilts.

Even if the zone axis is aligned with the optical axis, some extra spots not fitting the lattice may be visible at a distance from pattern center. They appear due to finite radius of the Ewald sphere, and they belong to Laue zones of order higher than zero. Even more complicated form have off-axis patterns, i.e., patterns that depart from any primary zone. All such patterns are indexable, but the three-dimension character of the crystal lattice must be taken into account. Solving off-axis patterns is essential for spot-pattern-based automatic orientation mappings.

9.2 Automatic Orientation Determination

With known or assumed crystal structure, nanodiffraction spot patterns can be used for constructing high (spatial) resolution orientation maps of polycrystalline materials [22]. Since the number of patterns needed for constructing a map is large, they are acquired automatically. Orientations are determined by "template matching" [22], i.e., matching experimental patterns to computer simulated patterns; see Sect. 8.9. (For an attempt to index spot patterns via detection of spots, see [23].) Template matching is suitable for dealing with electron spot patterns because the dependence of pattern geometry and spot intensities on crystal orientation is relatively weak, and therefore the number of needed templates is relatively small. On the other hand, this weak dependence on crystal orientation has an impact on the precision of orientations on the maps, which is smaller than that of maps based on Kikuchi patterns [24]. Moreover, with typical nanodiffraction patterns obtained from metallic samples, for certain crystallographic directions of the incident electron beam, an automatic procedure leads to false crystal orientations due to the 180°-ambiguity [25].

One may argue that the 180°-ambiguity concerns only prominent zone axes in a large continuous domain of parameters and the probability of encountering ambiguities in orientation mapping is low. However, the problem with indexing concerns not just a particular zone axis pattern but also patterns nearby because the match between an experimental pattern and a corresponding template is never perfect – neither in terms of peak intensities nor in their locations. Fast automatic indexing relies on simple kinematic approximation of intensities. On the other hand, measured intensities are influenced by noisy background; the influence is particularly strong for weak reflections, which include large–angle reflections, and the latter are frequently crucial for discriminating between potential solutions. Moreover, the template easily 'misses' narrow large–angle reflections because the pattern center and camera length are known with a limited accuracy. Finally, the 180°-ambiguity is often not solvable for a relatively wide range of microscope settings.

The reliability of solutions is measured by a properly defined (confidence) index; see Sect. 8.8. By computing the index as a function of direct beam direction for simulated data, one can identify crystal orientations difficult to determine in actual experiments. Example results of such a scan through all beam directions for Cu are shown in Fig. 9.2. Most of the patterns causing problems are near the $\langle 112 \rangle$ zone axis. The reliability index versus beam direction calculated for Ferrite is shown Fig. 9.2*b*. The low reliability orientations are spread around $\langle 111 \rangle$ and $\langle 113 \rangle$ directions. As in the case of the fcc $\langle 112 \rangle$ pattern, the first order zones in bcc $\langle 111 \rangle$ and $\langle 113 \rangle$ patterns are missing due to systematic absences. These figures are not universal. The diameter of the spread of low reliability areas depends on the experimental setup including the pattern acquisition system and the indexing program.

Generally, the problem can be alleviated by increasing the curvature of the Ewald sphere, i.e., by reducing the accelerating voltage. However, simulations show that relatively large reductions are needed, and such voltage changes may be limited by other aspects of the measurement. A more convenient way to address the problem is to 'enlarge' the spot patterns by using precession diffraction.

9.2.1 *Precession Electron Diffraction*

With this technique, the incident beam is deflected (twice) and rotated around the optical axis leading to the same effect as if the specimen was precessed. An extra device is used to control the electron beam (but see [26] for description of a

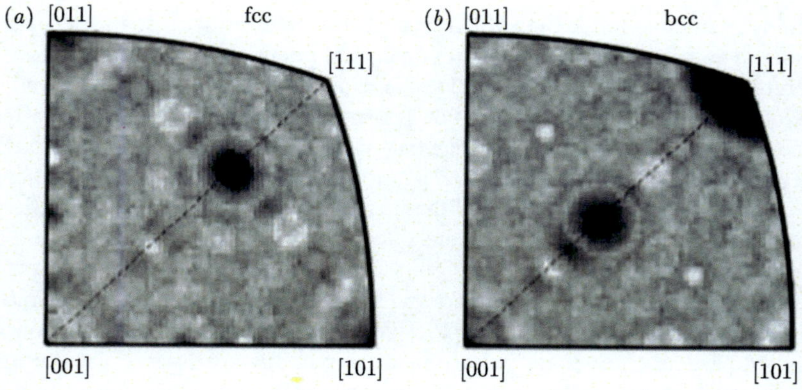

Fig. 9.2 The dependence of the reliability index on direct beam direction for simulated patterns of Cu (*a*) and Ferrite (*b*). Simulation conditions were the same as those listed in caption to Fig. 9.1. Equal area projection. The reliability index is defined as $1 - m_2/m_1$, where m_i is a measure of the match between the pattern and the i-th best template, and the orientations corresponding to m_1 and m_2 differ by at least 7°; cf. [22]. Arbitrary units are used because maximal values on the graph depend on the parameters of the indexing program, and what matters here is only the location of low values of the index (dark areas).

software solution for modern microscopes). Precession decreases dynamical effects and increases the number of diffraction spots in patterns. In effect, precession diffraction alleviates the 180°-ambiguity problem [27, 28]. With small precession angles ($\sim 0.5°$), the beam broadening and the loss in orientation accuracy are negligible.

9.3 Three-Dimensional Ab Initio Indexing

It must be stressed that with spot patterns, lattice parameters can be determined only approximately. To reconstruct three-dimensional reciprocal lattice, one needs to combine results from several spot diffraction patterns acquired from the same crystallite at different orientations. Standardly, the sample orientation is changed using double-tilt or rotation-tilt holders. (See Sect. 4.2.1.)

One needs to determine three linearly independent vectors such that their integer combinations match spots on patterns corresponding to different crystal orientations. In the simplest approach, to get vectors spanning a primitive cell, one can use two patterns corresponding to prominent zone axes related by a tilt about a low-index reciprocal lattice direction (i.e., a direction with high density of spots); spots along this direction are common for both patterns (Fig. 9.3).[1] Clearly, having more data,

[1] If the sample is thick enough to produce Kikuchi patterns, positions of adjacent intersections of Kikuchi bands can be used to select tilt axes and tilt angles.

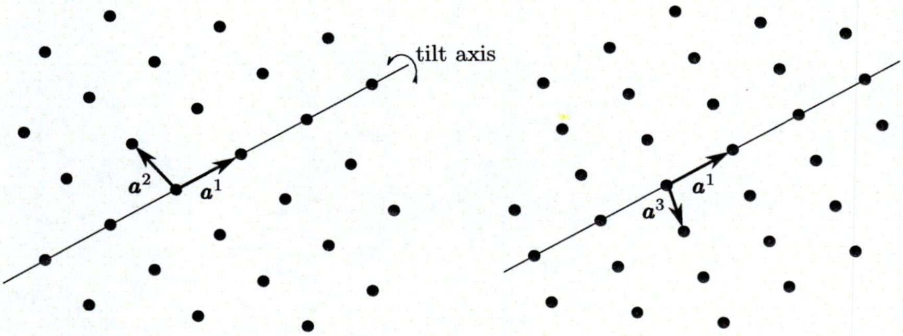

Fig. 9.3 Schematic of two zone axis patterns at different crystal orientations. The orientations are related by a tilt about a low-index reciprocal lattice direction. Three shortest independent lattice vectors constitute the lattice basis.

e.g., three patterns with common directions among each pair of the patterns, improves the reliability of the resulting lattice parameters [29].

⊡ In some cases, one may attempt to determine lattice basis from a single pattern containing both zeroth-order Laue zone and high-order Laue zone reflections [30]. Besides two shortest linearly independent vectors of the zeroth-order Laue zone, a third vector to a spot in the high-order Laue zone is needed. For that, the spacing between layers o the zones is estimated based on the curvature of the Ewald sphere, and the accuracy of the spacing is affected by the excitation error.

Once a basis is determined, one can proceed with other steps of lattice parameter determination: basis reduction, symmetry determination, selection of conventional cell.

Computer-aided determination of bases of three-dimensional crystal lattices can be used for phase discrimination [31], but currently, most of the software for three-dimensional indexing is intended for electron crystallography [32–38].

Electron crystallography is a research field concerned with using electron diffraction for crystal structure determination. Methods of electron crystallography are similar to those of X-ray analysis. Their key advantage is that electrons can be focused and therefore, first, much smaller crystals can be analyzed, and second, electrons can form images allowing for selecting objects of interest. On the other hand, electron diffraction intensities tend to depend on experimental parameters; they are not always reproducible. The intensities are affected by dynamic effects and crystal geometry (small dimensions and bending).

9.3.1 Automatic Recording of Tilt Series

One of the goals of electron crystallography is developing reasonably fast procedures for collecting data of quality sufficient for crystal structure determination. High

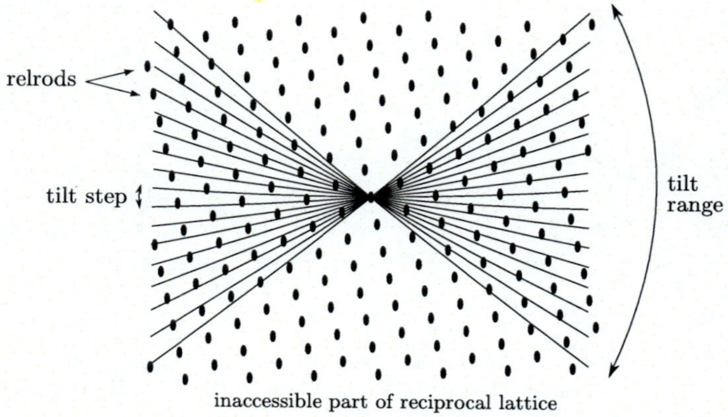

Fig. 9.4 Schematic illustration of acquisition of a tilt series. Two-dimensional section perpendicular to the tilt axis. The lines represent slices through reciprocal space. In reality, not the lines but the crystal and the lattice are tilted. The slices are recorded on diffraction patterns.

degree of automation is particularly important in TEM studies on beam-sensitive materials, when the time of expose to electron beam needs to be short. One of the steps in such procedures is automatic collection of tilt series; see [39].

In manual collection of tilt series, an operator checks whether a pattern originates from the investigated crystal. This task is complicated in automatic mode. In order to keep track of the investigated crystal while tilting the specimen, the automatic system must combine acquisition of diffraction patterns with imaging. A tilt series is collected in repeated sequence of tilting the specimen by a given angle, switching to imaging mode to track the crystal, and switching to diffraction mode to acquire a pattern [39].

The patterns of the series are usually not perfectly centered as the pattern center moves during acquisition of the series. After correcting for the pattern center, one ultimately gets a series of slices through the reciprocal space. The slices intersect each other at the common tilt axis (Fig. 9.4). Clearly, the density of slices is determined by the tilt step. The volume of the tested three-dimensional region in reciprocal space is determined by the range of the tilt and the solid angle covered by the diffraction patterns. Having a tilt series of patterns, positions of spots are found within this three-dimensional region. With a sufficiently small tilt step, most reflections within the tilt range can be detected.[2] It must be taken into account that spots corresponding to a single reflection may be visible in multiple patterns. In the implementation described in [39], the detected reflections are indexed using difference vectors between extracted peak positions and a search for clusters in a ball of limited radius centered at the center of the patterns; see Sects. 5.7 and 5.9.

[2] If the tilt axis is arbitrary, prominent zone axis patterns may be present in the series only by accident. Zone axis patterns will be present if tilting is around a low-index axis and the tilt angle is sufficiently small.

Various modifications of the system of [39] have been implemented, e.g., it was combined with precession [40, 41]. With tilting speed tuned to detector parameters, diffraction patterns can be acquired while the specimen is tilted in continuous way [41, 42].

9.4 Note on Other TEM-Based Patterns

Other types of diffraction patterns often acquired using TEM are single-crystal Kikuchi patterns and polycrystal ring patterns.

Kikuchi patterns recorded with relatively short effective camera lengths are usually used for orientation determination, and this subject is discussed in Chap. 8. However, there is an issue of end-indexing K-line patterns in which only one of the lines of the pair (the deficiency line) can be detected and the other one is out of view. The problem may occur for Kikuchi patterns collected with a large value of camera constant, and it usually arises in the case of central disks of CBED patterns (like the one shown in Fig. 3.8). A time-efficient program capable of reliable end-indexing of such lines is yet to be written.[3]

With a large number of crystallites contributing to a recorded pattern, one gets rings of a Debye-Scherrer (or ring) pattern. Integration over rings gives (electron) powder diffraction patterns. They are analyzed based on principles described in Chap. 7. There are a number of computer programs for interpretation of ring patterns [12–14, 44–47]. The main tasks of the software are to identify individual rings and to correct distortions. The resulting data can be used for phase identification.

References

1. K.W. Andrews, D.J. Dyson, S.R. Keown, *Interpretation of Electron Diffraction Patterns*, 1st edn. (Springer Science+Business Media, LLC, 1967)
2. P.L. Ryder, W. Pitsch, On the accuracy of orientation determination by selected area electron diffraction. Phil. Mag. **18**, 807–816 (1968)
3. W. Prantl, A computer program for unique indexing of electron diffraction patterns. J. Appl. Cryst. **17**, 39–42 (1984)
4. J.W. Edington, *Practical Electron Microscopy in Materials Science. Electron Diffraction in the Electron Microscope* (MacMillan, London, UK, Monograph Two, 1975)
5. W.R. Roser, G. Thomas, Slide rule method for indexing electron indexing patterns. Rev. Sci. Instr. **35**, 613–615 (1964)
6. R. Andrew, A method for the solution of electron diffraction patterns. J. Phys. E: Sci. Instrum. **10**, 216–218 (1977)
7. M. Booth, M. Gittos, P. Wilkes, A general program for interpreting electron diffraction patterns. Metall. Trans. **5**, 775–776 (1974)

[3] The procedure for indexing lines in CBED patterns of cubic crystals described in [43] requires zone axis indices as input parameters, which means that it solves only the simple problem of getting the angle of rotation about the axis.

8. P. Wilkes, Complete indexing of electron diffraction patterns by computer. J. Mater. Sci. **9**, 517–518 (1974)
9. R.P. Goehner, P. Rao, Computer-aided indexing of transmission electron diffraction patterns. Metallography **10**, 415–424 (1977)
10. C. Narayan, A sorting and searching computer program to index electron diffraction patterns from crystals of low symmetry. J. Electron Microsc. Tech. **3**, 151–158 (1986)
11. J. Brink, M.W. Tam, Processing of electron diffraction patterns acquired on a slow-scan CCD camera. J. Struct. Biol. **116**, 144–149 (1996)
12. J.L. Lábár, Consistent indexing of a (set of) single crystal SAED pattern(s) with the Process-Diffraction program. Ultramicroscopy **103**, 237–249 (2005)
13. M. Klinger, A. Jäger, Crystallographic Tool Box (*CrysTBox*): automated tools for transmission electron microscopists and crystallographers. J. Appl. Cryst. **48**, 2012–2018 (2015)
14. H.L. Shi, M.T. Luo, W.Z. Wang, *ElectronDiffraction tools*, a DigitalMicrograph package for electron diffraction analysis. Comput. Phys. Commun. **243**, 166–173 (2019)
15. C. Boudias, D. Monceau, *CaRIne* Crystallography Software (2021), http://carine.crystallography.pagespro-orange.fr/. Accessed Aug 2022
16. W. Griem, P. Schwaab, Computer assisted indexing of electron diffraction patterns. Prakt. Metallogr. **14**, 389–409 (1977)
17. H.V. Hart, *ZONES*: a search/match database for single crystal electron diffraction. J. Appl. Cryst. **35**, 552–555 (2002)
18. A.D. Mighell, V.L. Karen, NIST crystallographic databases for research and analysis. J. Res. Natl. Inst. Stand. Technol. **101**, 273–280 (1996)
19. M.J. Carr, W.F. Chambers, D. Melgaard, A search/match procedure for electron diffraction data based on pattern matching in binary bit maps. Powder Diffr. **1**, 226–234 (1986)
20. A.D. Mighell, V.L. Himes, R. Anderson, M.J. Carr, *d*-spacing/formula index for compound identification using electron diffraction data, in *Proceedings of the 46th Annual Meeting of Electron Microscopy Society of America*, ed. by G.W. Bailey (1988) (San Francisco Press, San Francisco, CA), pp. 912–913
21. V. Dimov, V. Iamakov, K. Bozhilov, Automated identification of monocrystal microphases in transmission electron microscopy (TEM). Comput. Geosci. **20**, 1267–1273 (1994)
22. E.F. Rauch, L. Dupuy, Rapid diffraction patterns identification through template matching. Arch. Metall. Mater. **50**, 87–99 (2005)
23. V. Kumar, Orientation imaging microscopy with optimized convergence angle using CBED patterns in TEMs. IEEE Trans. Image Process. **22**, 2637–2645 (2013)
24. A. Morawiec, E. Bouzy, H. Paul, J.J. Fundenberger, Orientation precision of TEM-based orientation mapping techniques. Ultramicroscopy **136**, 107–118 (2014)
25. A. Morawiec, E. Bouzy, On the reliability of fully automatic indexing of electron diffraction patterns obtained in a transmission electron microscope. J. Appl. Cryst. **39**, 101–103 (2006)
26. C.T. Koch, P. Bellina, P. van Aken, Software precession electron diffraction, in *Proceedings of the 14th European Microscopy Congress (EMC)*, vol. 2 (Springer, Berlin, 2008), pp. 201–202
27. E.F. Rauch, M. Véron, J. Portillo, D. Bultreys, Y. Maniette, S. Nicolopoulos, Automatic crystal orientation and phase mapping in TEM by precession diffraction. Microsc. Anal. **22**, S5–S8 (2008)
28. P. Moeck, S. Rouvimov, E.F. Rauch, M. Véron, H. Kirmse, I. Häusler, W. Neumann, D. Bultreys, Y. Maniette, S. Nicolopoulos, High spatial resolution semi-automatic crystallite orientation and phase mapping of nanocrystals in transmission electron microscopes. Cryst. Res. Technol. **46**, 589–606 (2011)
29. Y. Yang, C. Cai, J. Lin, L. Gong, Q. Yang, Accurate determination of lattice parameters based on Niggli reduced cell theory by using digitized electron diffraction micrograph. Micron **96**, 9–15 (2017)
30. H.L. Shi, Z.A. Li, *UnitCell Tools*, a package to determine unit-cell parameters from a single electron diffraction pattern. IUCrJ **8**, 805–813 (2021)
31. P. Fraundorf, Stereo analysis of single crystal electron diffraction data. Ultramicroscopy **6**, 227–236 (1981)

32. M. Wołcyrz, M. Andruszkiewicz, *WINREKS* - a computer program for the reciprocal lattice reconstruction from a set of electron diffractograms, in *Electron Crystallography*. ed. by D.L. Dorset, S. Hovmöller, X. Zou. Proceedings of the NATO Advanced Study Institute. (Springer Science, Dordrecht, 1997), pp. 427–430
33. D. Belletti, G. Calestani, M. Gemmi, A. Migliori, *QED* V 1.0: a software package for quantitative electron diffraction data treatment. Ultramicroscopy **81**, 57–65 (2000)
34. X.D. Zou, A. Hovmöeller, S. Hovmöeller, *TRICE* - a program for reconstructing 3D reciprocal space and determining unit-cell parameters. Ultramicroscopy **98**, 187–193 (2004)
35. X.Z. Li, Computer programs for unit-cell determination in electron diffraction experiments. Ultramicroscopy **102**, 269–277 (2005)
36. H. Zhao, D. Wu, J. Yao, A. Chang, *QtUCP* - a program for determining unit-cell parameters in electron diffraction experiments using double-tilt and rotation-tilt holders. Ultramicroscopy **108**, 1540–1545 (2008)
37. L. Jiang, D. Georgieva, J.P. Abrahams, *EDIFF*: a program for automated unit-cell determination and indexing of electron diffraction data. J. Appl. Cryst. **44**, 1132–1136 (2011)
38. X.Z. Li, *TEMUC3*, a computer program for unit-cell determination of crystalline phases in TEM experiments. Micron **117**, 1–7 (2019)
39. U. Kolb, T. Gorelik, M.T. Otten, Towards automated diffraction tomography. Part II—Cell parameter determination. *Ultramicroscopy* **108**, 763–772 (2008)
40. E. Mugnaioli, T. Gorelik, U. Kolb, Ab initio structure solution from electron diffraction data obtained by a combination of automated diffraction tomography and precession technique. Ultramicroscopy **109**, 758–765 (2009)
41. M. Gemmi, M.G.I. La Placa, A.S. Galanis, F. Rauch E, S. Nicolopoulos, Fast electron diffraction tomography. *J. Appl. Cryst.* **48**, 718–727 (2015)
42. B.L. Nannenga, D. Shi, A.G.W. Leslie, T. Gonen, High-resolution structure determination by continuous-rotation data collection in microED. Nat. Methods **11**, 927–930 (2014)
43. D. Fournier, G. L'Esperance, R.G. Saint-Jacques, Systematic procedure for indexing HOLZ lines in convergent beam electron diffraction patterns of cubic crystal. J. Electron Micr. Tech. **13**, 123–149 (1989)
44. J.L. Lábár, *ProcessDiffraction*: A computer program to process electron diffraction patterns from polycrystalline or amorphous samples, in *Proceedings of the 12th European Congress on Electron Microscopy (EUREM)*, eds. by L. Frank and F. Čiampor (2000) vol. 3 (Czechoslovak Society for Electron Microscopy, Brno, Czech Republic, 2000), pp. I379–I380
45. X.Z. Li, *JECP/PCED* - a computer program for simulation of polycrystalline electron diffraction pattern and phase identification. Ultramicroscopy **99**, 257–261 (2004)
46. X.Z. Li, Quantitative analysis of polycrystalline electron diffraction patterns. Microsc. Microanal. **13**, 966–967 (2007)
47. D.R.G. Mitchell, Circular Hough transform diffraction analysis: A software tool for automated measurement of selected area electron diffraction patterns within Digital Micrograph^{TM}. Ultramicroscopy **108**, 367–374 (2008)

Chapter 10
Example Complications in Indexing

It was already noted that, in most cases, indexing is simple when positions of diffraction peaks are sufficiently accurate. But there are exceptions from this rule. For some crystals, complications arise even if the positions are free from errors. Below, selected departures from simple crystal indexing are described. This chapter comprises short accounts on the mutually related subjects of pseudosymmetry which arises in structure determination, diffraction by twins and ambiguities in diffraction based orientation determination, a section on modulated structures, and a section on indexing data from some non-conventional diffraction methods.

10.1 Pseudosymmetry

Geometry of a diffraction pattern is not sufficient for determination of the crystal symmetry. The space group can only be postulated, and then it must be verified by comparison of integrated intensities of reflections expected to be symmetrically equivalent. When the symmetry suggested by experimental observations is higher than the actual crystal symmetry, the former is called pseudosymmetry. Pseudosymmetry may have structural character, or it may be with respect to a crystal lattice (metric pseudosymmetry).

Structural pseudosymmetry arises when the actual atomic positions are close to positions of a virtual structure of higher symmetry. Structural pseudosymmetry is rotational if the additional operation implied by the experiment is a rotation. Translational pseudosymmetry, with that extra operation being a translation, can be seen as a false lattice centering. If intensities of certain reflections are low, one may get a sublattice of the actual reciprocal lattice, i.e., the unit cell of the direct lattice that would be too small. (The subject is related to modulated structures considered in Sect. 10.4.)

A. Morawiec, *Indexing of Crystal Diffraction Patterns*, Springer Series in Materials Science 326, https://doi.org/10.1007/978-3-031-11077-1_10

Ab initio indexing seen as determination of crystal lattice ultimately provides information on lattice symmetry. It is quite common that the point-group symmetry of a crystal is different than the point-group symmetry of its lattice. Metric (i.e., lattice) symmetry can be higher but not lower than the symmetry of the crystal. A merohedral crystal, by definition, has point symmetry lower than that of its lattice. In most cases, crystal point symmetry can be determined if reflection intensities are taken into account, but there are exceptions. A simple example of the latter are chiral crystals when the diffraction data obey Friedel's law.

⊡ The actual spatial arrangement of atoms of a chiral (or more generally, non-centrosymmetric) crystal in a cell based on right-handed coordinate frame is know as absolute structure. Determination of absolute structures requires special techniques. It is routinely carried out using resonant scattering of X-rays with the wavelength near an absorption edge of a particular atom. Resonant scattering leads to different intensities of reflexions of Bijvoet pairs. Absolute structures are also studied using electron diffraction. In this case, the decisive asymmetries in reflection intensities are due to dynamic effects. See, e.g., [1, 2] and references therein.

Opposite to pseudosymmetry is the problem of missed symmetries. A structure determination software my provide symmetry lower than the actual crystal symmetry. In the determination of crystal structures, the higher symmetry gives the advantage of smaller number of parameters, but the main point is to determine the actual symmetry. Symmetry check and verification of the ascribed space group is a part of structure validation procedures [3–6].

The need for deciding about presence or absence of a symmetry operator relates pseudosymmetry to twinning. Moreover, pseudosymmetry facilitates twinning in the sense that twinning is easy for crystals showing pseudosymmetry. For more insight into pseudosymmetries, it is worth to consider twin types and their diffraction patterns.

10.2 Indexing of 'Multi-lattice' Diffraction Patterns

Twins are particular types multi-grain aggregates. In some cases, instead of a single crystal, a cluster of crystals contributes to a diffraction pattern. If the number of crystals is large, one may want or may be forced to use powder diffraction methods; see, e.g. [7]. If the sample consists of a small number of single crystals, in particular, if it is a bi-crystal, an extension of single-crystal indexing methods may be appropriate.

A distinction needs to be made between various multi-crystal samples. One may have crystallites of just one type or the more general situation with crystallites of different species. The latter situation happens, but it is usually avoided on the level of diffraction experiment; for an attempt to provide a tool for solving diffraction patterns in this general case, see [8]. In the case of a single-phase aggregate, there may be no specific orientation relationship between crystallites, or one may have a twin.

As for the diffraction patterns generated by general single-phase clusters of grains, a number of methods for dealing with such cases have been described; see, e.g., [9–11]. Their effectiveness very much depends on the complexity of the crystal structure and the expected number of contributing grains. A simple approach to the problem the iterative 'index-and-subtract' method: a single-crystal indexing software determines the lattice of dominant crystallite, the reflections associated with that lattice are subtracted, and indexing is retried based on the remaining reflections [10, 11].

10.2.1 Twins

A twin is composed of (usually two) structurally identical single crystals (individuals or twin components or twin domains) related by (proper or improper) rotations leaving some nodes of the crystal lattices invariant. In other words, intersection of lattices of twin components is a lattice. When the common sublattice of the lattices of twin components is one-, two-, three-dimensional, the twin is called mono-, di-, triperiodic, respectively. Only the latter class is considered here as actual or potential triperiodic twinning seriously affects analysis of diffraction patterns. Rotations relating orientations of the twin components are called twin operations. The twin law is the complete set of these operations.

The diffraction pattern acquired from a twin is a superposition of patterns of the twin domains. If volume fractions of the domains are nearly equal, diffraction pattern of the twin reflects the symmetry incorporating twin operations, and the density of diffraction peaks is the same or higher than that in a pattern originating from a single component. Standard indexing of such patterns may lead to an erroneous unit cell larger than that of the single crystal.

10.2.2 Types of Twins

A number of terminologies are used in classifications of twins; see, e.g., [12]. The following names of twinning types are known as Friedel's nomenclature:
– Twinning by merohedry can occur only when point symmetry of the crystal is not holohedral, i.e., when the point symmetry of the crystal lattice exceeds point symmetry of the crystal. In the case of twinning by merohedry, twin law operations are symmetry operations of the crystal lattice, i.e., the twin domains share the same direct lattice.
– With twinning by reticular merohedry, the intersection of lattices of twin components is a proper sublattice of each of these lattices. This construction is analogous to that of coincidence 'site' lattice or CSL used in materials science.
– In twinning by pseudo-merohedry, the twin law operations approximate rotational symmetry operations of the crystal lattice, i.e., the lattices of twin domains slightly deviate one from the other. When this deviation is a rotation, it is quantitatively

represented by a twin obliquity angle; in other cases, a more general lattice misfit parameter is needed [13].
– Twinning by reticular pseudo-merohedry is related to twinning by reticular merohedry in the same way as twinning by pseudo-merohedry is related to twinning by merohedry. The lattices of twin components are close to having a common sublattice.

10.2.3 Diffraction Patterns Originating From Twins

The key question is how having a twin instead of a single crystal affects a diffraction pattern. The answer depends on the type of the twin.

In the case of a twin by merohedry, twin components share the same direct lattice. Thus, their reciprocal lattices are also identical with respect to both metric and orientation. This implies that diffraction patterns originating from twin components are geometrically identical, i.e., positions of diffraction peaks from the components overlap exactly. Consequently, geometric information is not sufficient for discriminating a single crystal from a twin.

The intensities of reflections from a twin by merohedry may depend on fractional volume of twin domains. After [14], twins by merohedry are classified as twins of class I with twin operations belonging to the crystal Laue group, and twins of class II for which at least one of twin operations does not belong to the crystal Laue group. Twins of class II are possible only for trigonal, tetragonal, hexagonal and cubic crystals.

If Friedel's law is valid, reflexion intensities from an individual crystal satisfy Laue symmetry. For twins of class I, the nodes of reciprocal lattices of individuals giving rise to a reflexion are equivalent by Laue symmetry. In other words, the intensity of a reflexion is contributed by two reflexions, one from each crystal, which are equivalent by Laue symmetry. Diffraction pattern of a twin of class I is identical to the pattern of single crystals, and it does not depend on fractional volume of twin domains.

In the case of twins of class II, some nodes of reciprocal lattices of individuals giving rise to a reflexion are not equivalent by crystal Laue symmetry. In effect, diffraction pattern of a twin of class II depends on fractional volume of individuals.

In a twin by reticular merohedry, direct lattices of twin components have a common sublattice of index, say ι (> 1). The lattice reciprocal to the sublattice is a superlattice of the reciprocal lattices of the components, and its index is also ι (Fig. 10.1). In consequence, the primitive cell of the reciprocal lattice satisfying the conditions imposed by two crystals is smaller than that of single crystal. Hence, the primitive cell of the direct lattice derived from a pattern of a twin is a multiple of the actual primitive cell of the crystal lattice.

Finally, twinning by (reticular) pseudo-merohedry causes splitting of diffraction spots. Separations between individual spots of doublets is determined by the twin misfit (obliquity).

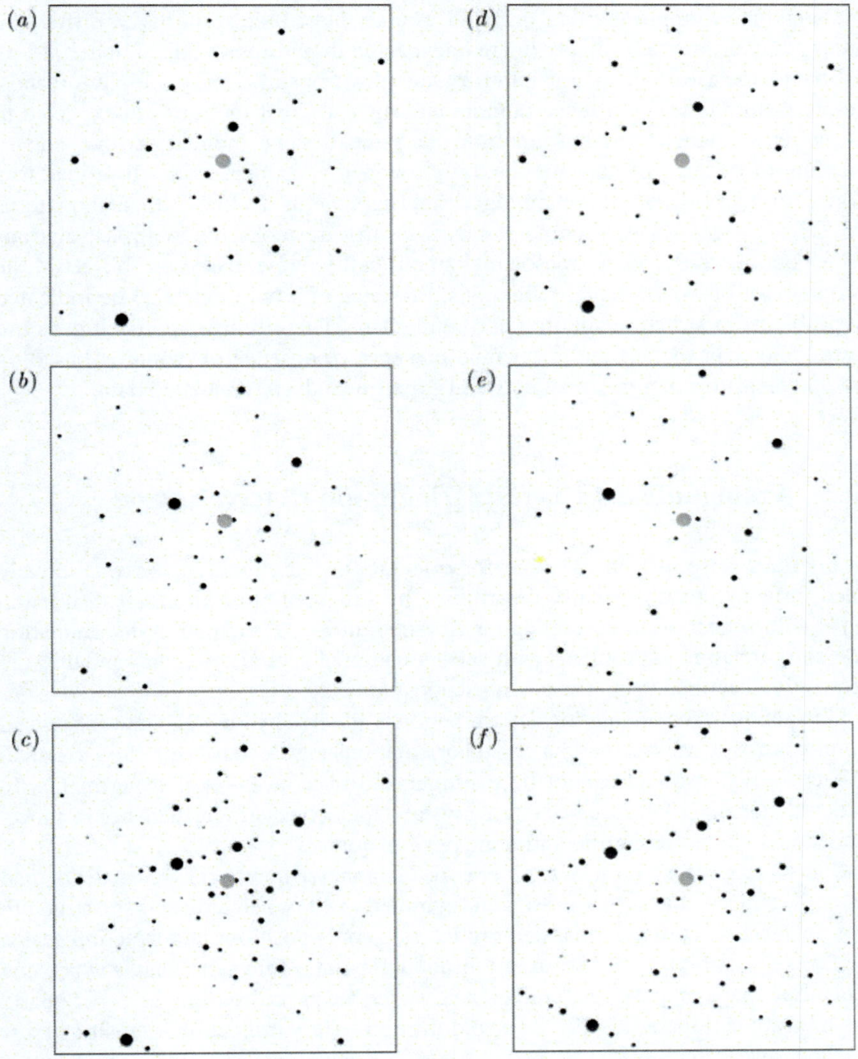

Fig. 10.1 Illustration of consequences of twinning by reticular merohedry on diffraction patterns. (*a*), (*b*) Simulated Laue patterns of Cu crystals related by the twin operation (half-turn about [1 1 1]). (*c*) Normalized sum of (*a*) and (*b*), i.e., the diffraction pattern of the twin with equal contributions of twin domains. (*d, e, f*) Gnomonic projections (cf. Sect. 6.2) corresponding to patterns shown in (*a*), (*b*), (*c*), respectively.

Detecting twinning is often difficult. The simplest indication that a diffraction pattern may originate from a twin are unexpected dimensions of the unit cell. More involved is the use of diffraction intensity statistics: For sufficiently complex crystals, atomic positions are in a sense random and intensity distributions are expected to follow certain statistics; a departure from these statistics may indicate a twin. Standard indexing of patterns of non-merohedrally twinned specimens will fail or lead to a large primitive cell, but there are algorithms capable of dealing with such patterns [15, 16]. To identify twin components, the software looks for multiple solutions corresponding to the same (reasonably small) cell volume, based on subsets of the complete set of experimental reflections. Presence of a twin may also be indicated by problems in structure solution and refinement. The ultimate approach is to use crystal and twin models and determine presence or absence of twinning based on suitable measures of agreement between experimental and simulated data.

10.3 Ambiguities in Crystal Orientation Determination

Ambiguities arise also in indexing for orientation determination. As was already noted, orientations are usually determined by fitting only positions of diffraction peaks with reflection intensities ignored. Ambiguities in orientation determination have similar nature as problems with determination of crystal symmetry and they are frequently described using the term ‘pseudosymmetry’.

Diffraction based orientation determination is affected by Friedel’s law. If a crystal is non-centrosymmetric, distinct crystal orientations may lead to metrically identical diffraction patterns and cannot be distinguished by conventional experiments. In particular, left and right forms of chiral crystals cannot be distinguished. For instance, handedness is ignored in orientation maps of α-quartz.

Quartz is a good example of another issue. Trigonal α-quartz belongs to hexagonal lattice system, i.e., the lattice has six-fold symmetry axis, whereas the crystal structure does not have it. If orientation is determined using only positions of diffraction peaks, one faces the ambiguity of 180° rotation about the three-fold axis. That is why some orientation maps of quartz are collected using the hexagonal setting, i.e., as if quartz was a hexagonal material [17]. To avoid this ambiguity, reflection intensities must be taken into account [18].

Another typical reason for ambiguities is the approximate metric pseudosymmetry. The accuracy in detection of reflection positions may be insufficient for distinguishing different orientations. E.g., for a tetragonal material with c/a close to 1, crystallites in three distinct orientations may produce nearly identical patterns; see Fig. 10.2.

Indexing ambiguities are quite frequent and troublesome in the case of orientation mappings. Misindexing in mappings is usually manifested by speckled appearance of grains or checkerboard patterns inside crystallites in special orientations. It is quite common that input data are insufficient for unambiguous indexing because of automated high speed reflection detection. In particular, this occurs when

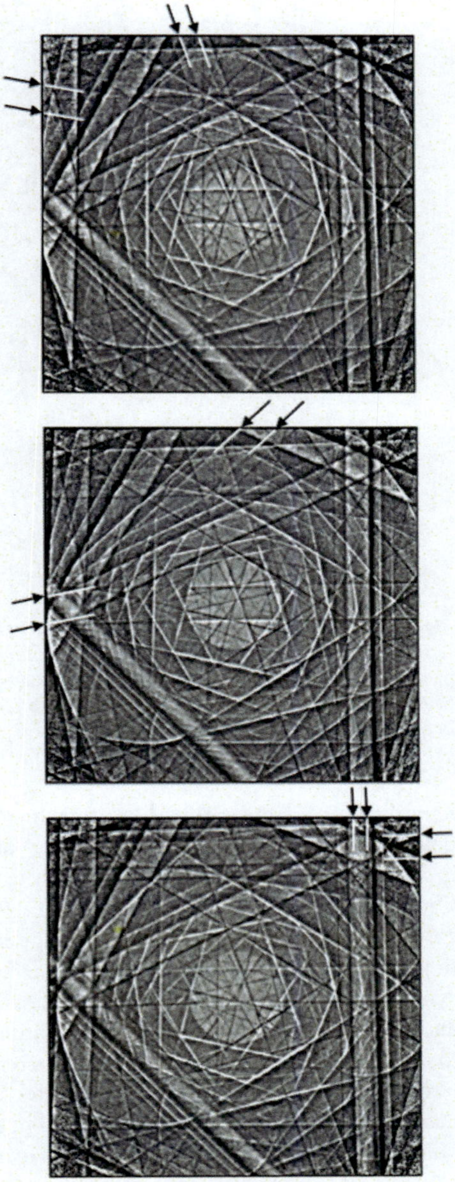

Fig. 10.2 Kikuchi patterns of variants of γ phase of lamellar TiAl. The phase is tetragonal with $c/a \approx 1.02$. The geometry of Kikuchi lines is nearly the same. There are differences between the patterns due to presence of weak lines some of which are marked by arrows and white segments [19]. For methods of resolving this pseudosymmetry using simulated EBSD patterns see [20].

the detected reciprocal lattice vectors are coplanar. The flag example is the 180°-ambiguity described in Chap. 9, but this ambiguity is not limited to spot diffraction patterns. Analogous problem may arise in orientation determination based on EBSD or Kikuchi patterns. Kikuchi lines or bands of reasonably low indices allowing for

Fig. 10.3 Simulated wide angle ⟨111⟩ Kikuchi pattern and two near–⟨111⟩ experimental patterns of ferrite for the accelerating voltage of 200 kV. (Adapted from [21].) The arc in one of the experimental patterns marks area of the pattern in which Kikuchi lines are detected. The central part of the ⟨111⟩ pattern has six-fold symmetry. To determine the orientation, some peripheral lines must be taken into account.

identification of the right orientation may have intensities too low for automatic detection. With K-line patterns, the 180°-ambiguity typically arises when a pattern is perpendicular to a three-fold symmetry axis, but the arrangement of detected reflections shows six-fold symmetry (see Fig. 10.3). Clearly, the same problem concerns cases with a two-fold axis and an arrangement with four-fold symmetry, or a direction which is not a symmetry axis but the arrangement has two-fold symmetry. The symmetry of a pattern appears to be higher than the corresponding crystal symmetry, and the indexing software finds two solutions to be equally likely. The simplest way to avoid these ambiguities, or at least increase the fraction of correctly indexed patterns, is by increasing the number of detected reflections. One might simply change the corresponding software parameter, but this may lead to an increased number of false reflections, and consequently, to an increased number of unsolved or incorrectly solved cases. From the experimental viewpoint, increasing the number of detected reflections means increasing the solid angle covered by the patterns (i.e., decreasing sample-detector distance), or improving the quality of the patterns (i.e., increasing pattern acquisition time or pattern resolution).

Some of the ambiguities in orientation determination can be avoided by abandoning indexing and using the time consuming direct pattern matching. (See Sect. 8.9).

10.4 Indexing of Satellite Reflections

If a crystal structure with a given primitive cell is modified by a periodic disturbance (modulation) extending over several primitive cells, its diffraction patterns show some extra reflections accompanying the main reflections originating from the undisturbed crystal. After [22], the extra reflections are called satellites.

The structural modulations may arise due to displacement of atoms from equilibrium positions, spatially varying composition and occupation probabilities, periodic distribution of defects et cetera. Satellites in neutron diffraction patterns reveal modulated magnetic ordering in some metals.

A different but related class of materials are intergrowth or composite crystal structures with two (or more) components having distinct lattices. Accordingly, diffraction involves multiple reciprocal lattices, and diffraction patterns show multiple sets of main reflections. Due to mutual interaction, the lattice of one component is distorted with the periodicity of the modulation related to the periodicity of the other lattice. In effect, a diffraction pattern of a composite crystal is a superposition of patterns of the components and satellites caused by the modulation.

Satellite reflections could be indexed using integers if the original primitive cell was replaced by a larger cell restoring the true periodicity. Formally, this cannot be done if the period of the modulation is incommensurate with the period of the undisturbed crystal (in the same direction), i.e., if the ratio of the two periods is irrational. Assuming the commensurate case with a rational ratio of the periods, a special 'satellite indexing' is used when the truly periodic lattice has no or little physical significance. It is frequently convenient to ignore the satellites and index the pattern using main reflections and lattice parameters of the undisturbed crystal. With such lattice parameters, the satellites are at positions represented by fractional indices.

The periodic disturbance in modulated crystal can be of a sinusoidal type, but equally well the modulating function may have a different shape. It is instructive to consider the simple models of sinusoidally modulated one-dimensional 'crystals'.

10.4.1 *Sinusoidally Commensurately Modulated One-Dimensional 'Crystals'*

Sinusoidally Modulated Scattering Power

Let M (> 1) be the number of cells over which the single modulation extends, an let the modulation function have the form $f(x) = 1 + A\sin(2\pi x/M)$, where $0 < A < 1$ (Fig. 10.4). The 'crystal' is assumed to consist of scattering centers at points $\ldots, -1, 0, 1, 2, \ldots$, i.e., the modulated density of scatterers can be expressed as

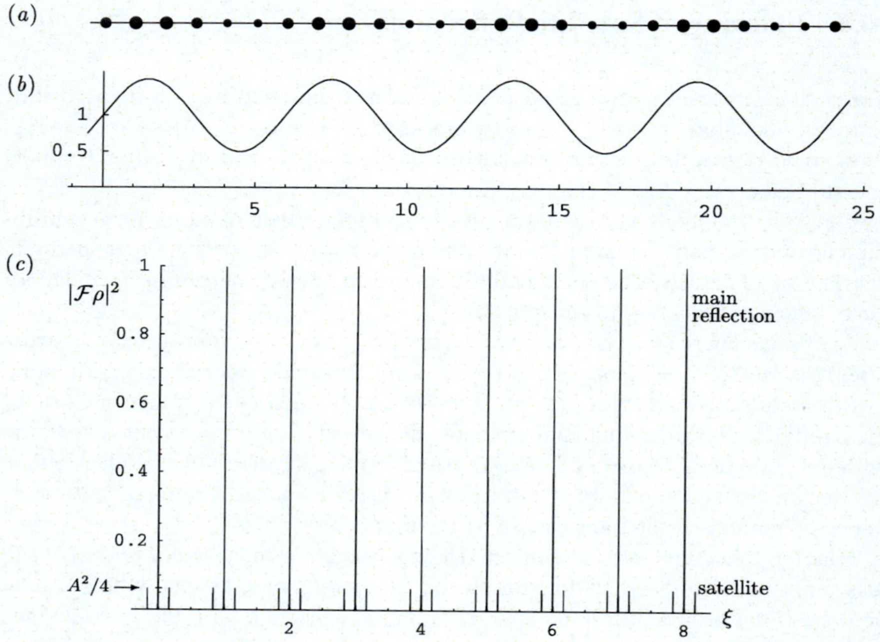

Fig. 10.4 Example of one-dimensional 'crystal' with sinusoidally modulated scattering power for $A = 1/2$ and $M = 6$. (*a*) Distribution of scatterers with areas of spots proportional to their scattering powers. (*b*) Modulation function $f(x) = 1 + (1/2)\sin(2\pi x/6)$. (*c*) The corresponding spectrum $|\mathcal{F}\rho|^2$. The figure was drawn using the approximation $\mathcal{F}\rho(\xi) \mathrel{\tilde{\propto}} \sum_{n=-n_{\max}}^{n_{\max}} f(n)\exp(-2\pi i\xi n)/(2n_{\max}+1)$ with $n_{\max} = 250$.

$$\rho(x) = \sum_{n=-\infty}^{\infty} (1 + A\sin(2\pi x/M))\,\delta(x-n)\,; \tag{10.1}$$

see Fig. 10.4. The undisturbed or average structure has the period of 1 whereas the period of the truly periodic structure is M. The Fourier transform $\mathcal{F}_x[\rho(x)](\xi) = \int \rho(x)\exp(-2\pi i\xi x)\mathrm{d}x$ of ρ given by (10.1) has the form of the series

$$\mathcal{F}\rho(\xi) = \sum_{n=-\infty}^{\infty} (1 + A\sin(2\pi n/M))\,\exp(-2\pi i\xi n)\,, \tag{10.2}$$

and with $\sin(x) = (e^{ix} - e^{-ix})/(2i)$, one obtains

$$\mathcal{F}\rho(\xi) = \sum_n \left(\exp(-2\pi i\xi n) - \frac{iA}{2}\exp(2\pi i n(\xi - 1/M)) + \frac{iA}{2}\exp(-2\pi i n(\xi + 1/M)\right).$$

Using (1.80), the sums over n can be replaced by Dirac combs so the amplitude is given by

$$\mathcal{F}\rho(\xi) = Ш(\xi) - \frac{iA}{2}Ш(\xi - 1/M) + \frac{iA}{2}Ш(\xi + 1/M) ;$$

cf. [22, 23]. It is the sum of the Dirac comb of main reflections and two Dirac combs of satellites. Each main reflection has a single pair of satellites. The satellites are at the distance of $1/M$ from the main reflections. The intensity of the satellites relative to that of the main reflection is $A^2/4$. Example spectrum for one-dimensional 'crystal' with sinusoidally modulated scattering power is shown in Fig. 10.4*c*. It was computed based on (10.2) using a finite number of terms.

Sinusoidally Modulated Longitudinal Displacement

As in the example above, let the original undisturbed configuration be described by the Dirac comb. With sinusoidally modulated longitudinal displacement, the density of scatterers is

$$\rho(x) = \sum_{n=-\infty}^{\infty} \delta(x - x_n) ,$$

where $x_n = n + B\sin(2\pi n/M)$, M is an integer reasonably larger than 1 and the modulation amplitude B satisfies the inequalities $0 < B \ll 1$. The Fourier transform $\mathcal{F}_x[\rho(x)](\xi) = \int \sum_n \delta(x - x_n)\exp(-2\pi i\xi x)\mathrm{d}x$ has the form of the series

$$\mathcal{F}\rho(\xi) = \sum_n \exp(-2\pi i\xi x_n) = \sum_n \exp(-2\pi i\xi n)\exp(-2\pi i\xi B\sin(2\pi n/M)) .$$

Using the Jacobi-Anger expansion (1.64) and the identity (1.80), one gets

$$\mathcal{F}\rho(\xi) = \sum_{m=-\infty}^{\infty} J_m(-2\pi\xi B)\sum_n \exp(-2\pi i n(\xi - m/M)) = \sum_m J_m(-2\pi\xi B)\,Ш(\xi - m/M) ,$$

where J_m is the m-th Bessel function of the first kind. The transform $\mathcal{F}\rho$ is a sum of weighted Dirac combs (Fig. 10.5). The Dirac comb for $m = 0$ corresponds to main reflections. The satellites of the $|m|$-order are at the distance $|m|/M$ from their main reflections. The intensity of the n-th main reflection is $J_0^2(-2\pi nB)$. Since $\sum_{m=-\infty}^{\infty} J_m^2(-2\pi nB) = 1$, the sum of intensities of the main reflection and all its satellites is constant and equal to the intensity of the reflections from the undisturbed crystal [22, 24].

The functions J_m with $|m| \geq 1$ are nearly zero for arguments of absolute value small compared to $|m|$; see, Fig. 12.3. Therefore, with small B and small $|n|$, the intensities of high $|m|$ satellites are low. In the regime of small modulation amplitudes, the intensities of satellites of low-index main reflections relative to the intensities of the main reflections grow with the amplitude of the modulation. When $|2\pi nB|$ takes

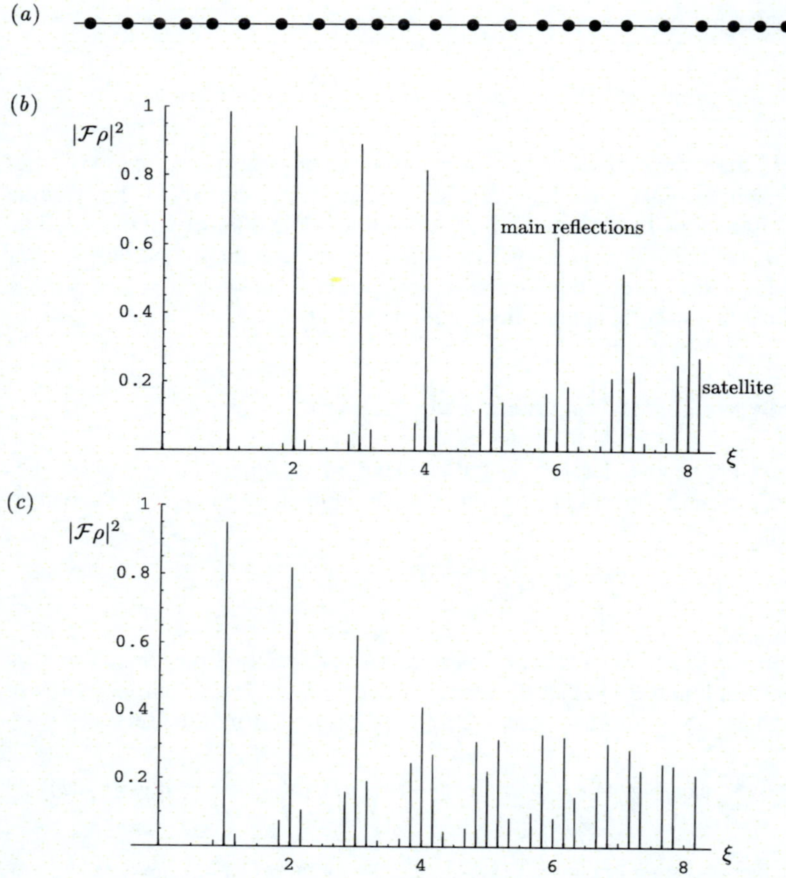

Fig. 10.5 Example of one-dimensional 'crystal' with sinusoidally modulated displacement for $M = 6$. (*a*) Distribution of scatterers with modulation amplitude $B = 1/5$. Spectra $|\mathcal{F}\rho|^2$ for amplitudes $B = 1/40$ (*b*) and $B = 1/20$ (*c*).

values of the order of $|m|$ or larger, relative intensities of satellites with respect to mains vary widely; see Fig. 10.5*c*.

10.4.2 Modulation Propagation Vector

The fundamental reason for having satellite reflections in diffraction patterns from three-dimensional crystals is the same as in the above one-dimensional cases; it is the periodicity of the modulation. The key difference is that besides its period, the modulation additionally has the attribute of propagation direction.

As was noted above, analysis of diffraction patterns showing main reflections and satellite reflections relies on making distinction between the 'mains' and the satellites. In the first approximation, satellites are neglected, and the average (undisturbed) structure is investigated. Knowing the average structure, the satellites are used to deduce the physical nature of the modulations.

The first step is to determine the modulation direction and modulation period. Let $\check{\boldsymbol{h}} = h_i \boldsymbol{a}^i$ be a vector of the reciprocal lattice Λ^* of the undisturbed crystal. It corresponds to a main reflection. Key for resolving a modulated structure is the vector indicating the position of the reflection's nearest satellite relative to $\check{\boldsymbol{h}}$. Let this vector be denoted by $\boldsymbol{q}$. It is easy to see, that the modulation direction is $\widehat{\boldsymbol{q}} = \boldsymbol{q}/|\boldsymbol{q}|$: Let $\boldsymbol{x}$ be a given vector in the direct space. For arbitrary $\boldsymbol{x}_\perp$ perpendicular to $\boldsymbol{q}$, the points $\boldsymbol{x} + \boldsymbol{x}_\perp$ satisfy $\boldsymbol{q} \cdot (\boldsymbol{x} + \boldsymbol{x}_\perp) = \boldsymbol{q} \cdot \boldsymbol{x}$, i.e., they all have the same phase. For $\boldsymbol{x}' = \boldsymbol{x} + \widehat{\boldsymbol{q}}/|\boldsymbol{q}|$, one has $\boldsymbol{q} \cdot \boldsymbol{x}' = \boldsymbol{q} \cdot \boldsymbol{x} + 1$, i.e., the period of the modulation is $1/|\boldsymbol{q}|$. Thus $\boldsymbol{q}$ is the modulation propagation vector.

The components of the modulation propagation vector in the basis $\boldsymbol{a}^i$ of the underlying reciprocal lattice Λ^* are $\sigma_i = \boldsymbol{q} \cdot \boldsymbol{a}_i$, and one has $\boldsymbol{q} = \sigma_i \boldsymbol{a}^i$. If all σ_i are rational, some multiples of $\boldsymbol{q}$ are nodes of Λ^*. Let $M > 1$ be the smallest integer such that $M\boldsymbol{q}$ is a node of that lattice, i.e.,

$$M\boldsymbol{q} = \check{\boldsymbol{h}} .$$

In this case, the modulation is metrically commensurate with the average crystal. It is truly commensurate if the modulated structure is a periodic crystal and its direct lattice Λ_s is a sublattice of the direct lattice Λ of the undistorted crystal; see, Fig. 10.6. The index of Λ_s with respect to Λ is M. The unit cell of the sublattice Λ_s is called a supercell. The basis of the lattice Λ_s^* reciprocal to Λ_s can be determined from the vectors $\boldsymbol{a}^i$ and $\boldsymbol{q}$ using the method described in Sect. 5.4.2.

10.4.3 Indexing

Position of the m-th satellite of $\check{\boldsymbol{h}}$ is

$$\boldsymbol{h}' = \check{\boldsymbol{h}} + m\boldsymbol{q} .$$

The integer m is zero for the main reflection. It is non-zero for satellites, and $|m|$ is the order of the satellite. Typically, only satellites of the lowest orders are visible, and they are observed near main reflections.

Let the indices of the main reflection $\check{\boldsymbol{h}}$ in the basis $\boldsymbol{a}^i$ of the underlying reciprocal lattice Λ^* of the undisturbed crystal be h_i, i.e. $\check{\boldsymbol{h}} = h_i \boldsymbol{a}^i$. The vector $\boldsymbol{h}' = \check{\boldsymbol{h}} + m\boldsymbol{q}$ representing the m-th satellite of $\check{\boldsymbol{h}}$ is identified by four indices. In the conventional notation with $\boldsymbol{h}' = h\mathbf{a}^* + k\mathbf{b}^* + l\mathbf{c}^* + m\boldsymbol{q}$, the indices are $hklm$. This identification of satellites can be seen as indexing in the frame $\boldsymbol{a}^\mu$ ($\mu = 1, 2, 3, 4$) with the first

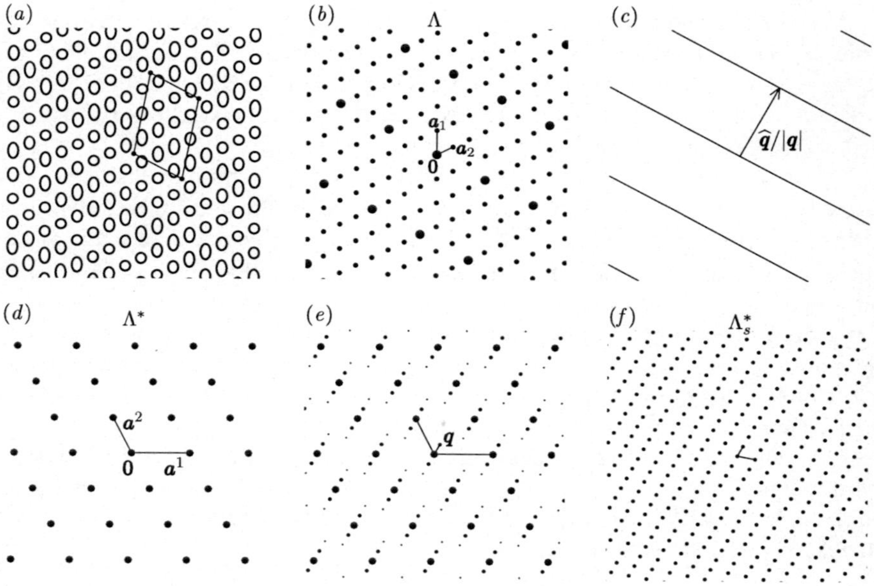

Fig. 10.6 Schematic two-dimensional illustration of a commensurately modulated crystal. (*a*) Model of a modulated structure. Elongation of an ellipse represents level of disturbance of the structure in a cell. The parallelogram marks a primitive cell of Λ_s (supercell). (*b*) Direct lattice Λ of the average structure. Large spots represent the direct lattice Λ_s of truly periodic structure. The index of Λ_s with respect Λ is 11. (*c*) Direction of the modulation propagation vector $\boldsymbol{q}$ and lines of constant phase of the modulation. The lines are separated by the period of the modulation $1/|\boldsymbol{q}|$. (*d*) The reciprocal lattice Λ* of the average structure, i.e., the lattice of main reflections. (*e*) Main reflections and low order satellites with $\boldsymbol{q} = (2\boldsymbol{a}^1 + 3\boldsymbol{a}^2)/11$. (*f*) The lattice Λ_s^* of main reflections and satellite reflections. None of its proper sublattices would cover all reflections. The lattice nodes are at integer combinations of $\boldsymbol{a}^1$, $\boldsymbol{a}^2$ and $\boldsymbol{q}$.

three vectors being the basis vectors $\boldsymbol{a}^i$ of Λ* and $\boldsymbol{a}^4 = \boldsymbol{q}$. The vector $\boldsymbol{h}' = h_\mu \boldsymbol{a}^\mu$ has indices h_μ ($\mu = 1, 2, 3, 4$) of which the fourth one is $h_4 = m$.

With the modulation extending over several primitive cells, its period exceeds the largest interplanar distances. Let the modulation period $1/|\boldsymbol{q}|$ be more than twice larger than the interplanar distance $d_{\check{\boldsymbol{h}}}$

$$1/|\boldsymbol{q}| > 2d_{\check{\boldsymbol{h}}} \,. \tag{10.3}$$

Let $\check{\boldsymbol{h}}$ be a short vector of the reciprocal lattice of the undisturbed crystal such that its direction is close to that of $\boldsymbol{q}$. With $\widehat{\boldsymbol{h}} = \check{\boldsymbol{h}}/|\check{\boldsymbol{h}}|$, one has $1/2 > |\boldsymbol{q}|\, d_{\check{\boldsymbol{h}}} = \boldsymbol{q} \cdot \widehat{\boldsymbol{q}} d_{\check{\boldsymbol{h}}} \geq \boldsymbol{q} \cdot \widehat{\boldsymbol{h}} d_{\check{\boldsymbol{h}}} = \boldsymbol{q} \cdot \widehat{\boldsymbol{h}}/|\check{\boldsymbol{h}}| = \boldsymbol{q} \cdot \check{\boldsymbol{h}}/|\check{\boldsymbol{h}}|^2$ or

$$\boldsymbol{q} \cdot \check{\boldsymbol{h}}/2 < (\check{\boldsymbol{h}}/2)^2 \,.$$

This inequality has the form analogous to conditions for points inside the Brillouin zone; cf. Sect. 2.2.1. With (10.3) satisfied for the vector $\check{\boldsymbol{h}}$ determining the boundary of the Brillouin zone in the direction $\widehat{\boldsymbol{q}}$, the modulation propagation vector is in the Brillouin zone. This makes the indexing in the four-vector frame unique if satellites of the order $|m| \geq M/2$ are ignored. However, in general, with both $\check{\boldsymbol{h}}_1$ and $\check{\boldsymbol{h}}_2 = \check{\boldsymbol{h}}_1 + M\boldsymbol{q}$ being vectors of Λ^*, one has $\check{\boldsymbol{h}}_1 + m\boldsymbol{q} = \check{\boldsymbol{h}}_2 + (m - M)\boldsymbol{q}$, i.e., the $(m - M)$-th satellite of $\check{\boldsymbol{h}}_2$ overlaps the m-th satellite of $\check{\boldsymbol{h}}_1$.

A single propagation vector characterizes a unidirectional modulation. A structure with multiple, say n, modulations is characterized by n distinct propagation vectors $\boldsymbol{q}_i$, $i = 1, \ldots, n$. Each propagation vector gives rise to an additional index. The diffraction pattern will show satellites located at multiples of $\boldsymbol{q}_i$ relative to main reflections.

As for tools for structure determination of modulated crystals based on single-crystal or powder diffraction data, one is usually referred to *Jana* [25], but this software does not perform indexing. In other programs, indexing of single-crystal diffraction patterns relies on prior discrimination of main reflections and satellites [26, 27].

10.4.4 Incommensurately Modulated Structures

Formally, a structure is incommensurately modulated if a component of the modulation propagation vector in the basis $\boldsymbol{a}^i$ of the main reciprocal lattice is irrational. Such structures are formally described by methods similar to those used for describing quasicrystals. The methods involve adding extra dimensions to the three-dimensional physical space. The diffraction reflections are projections of points of a lattice of higher dimension onto the three-dimensional space. The reflections form a 'quasi-lattice' with arbitrarily small distances between its nodes. See Chap. 13 for details. In practice, a structure is said to be incommensurate if it is not truly commensurate, in particular, if M large and not determinable.

10.5 Non-Conventional Structure Determination Methods

Some crystal structures cannot be solved by conventional methods. Data resulting from some dedicated experimental techniques require special methods of processing. This section presents briefly two prominent example cases in which special indexing methods are used.

The first example of a non-conventional method is the investigation of polycrystalline thin films by grazing-incidence X-ray diffraction (GIXD).

10.5.1 Indexing Grazing-Incidence X-ray Diffraction Data

Some substrate-induced phases exist only in the form of thin films, or thin-film phases differ structurally from their bulk counterparts. The flag example of such materials are molecular thin films for applications in organic electronics. Structural studies of such phases usually rely on GIXD.

With GIXD, the angle between sample surface and the incident beam is small and the X-ray illuminated area is large. The value of the angle is just above the critical angle for total external reflection for the film (and in some cases below the critical angle for the substrate). This allows for controlling the penetration depth. Moreover, the intensity of reflected beams is enhanced.

One of the steps in such GIXD-based studies is the determination of lattice parameters. Even if the film is polycrystalline, such data differ from powder data. When substrates mediate the crystallization process, new phases grow frequently with a specific crystal plane parallel to the substrate. The films are then strongly textured, i.e., preferred orientations are present. On isotropic substrates, the growth leads to axial textures, i.e, the distribution of orientations is symmetric with respect to rotations about the normal to the substrate.[1]

With an ideal single-component axial texture, the GIXD data are in a sense two-dimensional. The experiment provides two components of a scattering vector: the component normal to the substrate and the one along the substrate. The azimuthal angle of the vector (rotation about the normal) is undetermined, and therefore, special indexing methods are needed. For descriptions of programs indexing such GIXD data see, e.g., [28–30].[2] As in conventional methods, indexing is based on matching positions of observed and calculated reflections. The key difference is in calculation of reflection positions which must take into account the special geometry of GIXD and texture of the thin film. In some cases, advantage is taken of a specular scan [29], i.e., the scan with changing magnitude of the scattering vector and vector direction perpendicular to the thin film. Peaks on the scan correspond to (specular) reflections from crystal planes parallel to the substrate.

10.5.2 Serial Crystallography

The standard method of collecting data for structure determination is by rotating or oscillating a specimen illuminated by X-rays. It requires reasonably large single crystals. Growing sufficiently large well-diffracting crystals of molecules can be challenging. This is particularly true for macromolecular crystallography – the field of studying biological molecules such as proteins, nucleic acids or viruses. Moreover, analysis of biological specimens is severely limited by radiation damage. Finally, X-

[1] On the other hand, the growth can be epitaxial on monocrystalline substrates.

[2] The scheme described in [31] is based indexing thin-film reflections of known structures; elimination of these structures allows only for testing if any novel thin-film phases are present.

ray (and electron) scattering of biological specimens is weaker than that of inorganic materials as the former consist primarily of carbon, oxygen, nitrogen and hydrogen, i.e., elements of low atomic number.

One of the alternatives to conventional single-crystal methods is *serial crystallography*. This term refers to collecting 'imperfect' diffraction patterns from multiple randomly oriented crystals, and solving the structure based on a number of incomplete or low-resolution data sets. Serial crystallography is expected to become an important method when applied to data obtained from X-ray free-electron laser (XFEL) [32].

Free-electron laser uses relativistic electrons passing through periodic magnetic field. Emitted radiation interacts with the electrons, and a process known as self-amplified spontaneous emission (SASE) leads to intense coherent beam. The high-brilliance XFEL sources provide short (tens of femtoseconds) X-ray pulses. Specimens are delivered into the beam using substrate-free methods; typically, microcrystals suspended in a liquid are injected at a proper rate [33]. XFEL pulses are shorter than the time needed for radiation-induced specimen damage. In *serial femtosecond crystallography*, radiation destroys the structure, but that happens only after it is diffracted. A scattering event gives rise to a 'snapshot' diffraction pattern corresponding to the state for which radiation-induced atomic displacement can be neglected.

⊡ It is worth noting parenthetically that experiments based on serial crystallography but using a beamline of third-generation synchrotron instead of XFEL have been carried out. With the approach described in [34], crystals were exposed to X-rays as they passed through the beam. In order to get patterns originating from single crystals, the parameters of sample concentration and flow rate, and the detector frame rate need to be properly adjusted. (For another sample delivery method, see [35, 36].)

Indexing

Of the large number of collected 'snapshot' patterns, only some carry diffraction information. The first step of data processing is to extract those patterns. In the simplest case, after correction for extraneous and detector artifacts and background subtraction, quality of individual patterns is evaluated based on the number of detected Bragg peaks; the greater the number of peaks, the better the pattern. Generally, taking into account the volume of produced data, pattern filtering needs to be performed at various stages of pattern recording; see, [36, 37].

One way to index 'snapshot' diffraction patterns is by using indexing packages originally developed for conventional single-crystal data [38–40], but there is also programs dedicated to such patterns, e.g., [41]. The use of dedicated software is justified by specific features of 'snapshot' patterns. With very small crystals, besides the Bragg peaks, the pattern may contain shape-induced subsidiary maxima; if present, they may critically affect indexing. Moreover, since diffraction patterns are indexed independently, the resulting bases of the lattices differ not only by their orientations, but they may also span different reduced cells; one needs to determine transformations, so all correct bases are related by rotations only.

Having a large number of patterns originating from variously oriented crystals, one can combine them into a single ring pattern [42]. By integration over the rings, one obtains a powder diffraction pattern indexable by methods of Chap. 7.

Once reliable lattice parameters are obtained from ab initio indexing of some patterns, the remaining snapshots can be processed using software for end-indexing (orientation determination). See [43–45] for accounts on end-indexing software dedicated to data from serial crystallography experiments.

Indexing of 'snapshot' patterns is affected by ambiguities (due to quality of patterns or pseudosymmetry) and a procedure for resolving them is needed. One can deal with the ambiguities by taking the advantage of the large number of patterns [40, 46, 47]: Given a tentative indexing solution (set of lattice parameters), one can check how well intensities on a given pattern correlate with intensities of other patterns. The set of lattice parameters leading to the highest correlations between all pairs of the patterns is taken as the final indexing solution.

Finally, experimental conditions which improve time efficiency and reduce consumption of samples may lead to individual patterns originating from *multiple* crystals. Such patterns can still be of use. Knowing (or assuming) lattice parameters, orientations of crystals contributing to multi-crystal diffraction patterns can be determined [48, 49].

Peak Intensities

Further processing of XFEL data also relies on their high redundancy. A single snapshot corresponds to a fixed (still) random crystal orientation and most reflections are only partially recorded. As a result, intensities of diffraction peaks on individual patterns do not represent true intensities of Bragg peaks. Integrated intensities and structure factors are obtained by using multiple patterns. In the simplest approach, assuming random distribution of crystal orientations, integrated intensity of a (symmetrically unique) reflection is evaluated via averaging over multiple diffraction patterns on which the reflection is present [38, 50, 51]. The number of diffraction patterns needed depends on various factors, but generally it is of the order of 10^4.

References

1. H. Inui, A. Fujii, K. Tanaka, H. Sakamoto, K. Ishizuka, New electron diffraction method to identify the chirality of enantiomorphic crystals. Acta Cryst. B **59**, 802–810 (2003)
2. U. Burkhardt, H. Borrmann, P. Moll, M. Schmidt, Y. Grin, A. Winkelmann, Absolute structure from scanning electron microscopy. Sci. Rep. **10**, 4065 (2020)
3. A.L. Spek, Single-crystal structure validation with the program *PLATON*. J. Appl. Cryst. **36**, 7–13 (2003)

4. H.T. Stokes, D.M. Hatch, *FINDSYM*: program for identifying the space-group symmetry of a crystal. J. Appl. Cryst. **38**, 237–238 (2005)
5. A. Togo, I. Tanaka, *Spglib*: a software library for crystal symmetry search, 2018. https://arxiv.org/pdf/1808.01590.pdf. Accessed Aug 2022
6. D. Hicks, C. Oses, E. Gossett, G. Gomez, R.H. Taylor, C. Toher, M.J. Mehl, O. Levy, S. Curtarolo, *AFLOW-SYM*: platform for the complete, automatic and self-consistent symmetry analysis of crystals. Acta Cryst. A **74**, 184–203 (2018)
7. H. Li, X. Li, M. He, Y. Li, J. Liu, G. Shen, Z. Zhang, Indexing of multi-particle diffraction data in a high-pressure single-crystal diffraction experiment. J. Appl. Cryst. **46**, 387–390 (2013)
8. C. Wejdemann, H.F. Poulsen, Multigrain indexing of unknown multiphase materials. J. Appl. Cryst. **49**, 616–621 (2016)
9. L. Buts, M.H. Dao-Thi, L. Wyns, R. Loris, Untangle, a tool for filtering overlapping diffraction patterns from multicrystals. Acta Cryst. D **60**, 983–984 (2004)
10. N.K. Sauter, B.K. Poon, Autoindexing with outlier rejection and identification of superimposed lattices. J. Appl. Cryst. **43**, 611–616 (2010)
11. H.R. Powell, O. Johnson, A.G.W. Leslie, Autoindexing diffraction images with *iMosflm*. Acta Cryst. D **69**, 1195–1203 (2013)
12. G. Donnay, J.D.H. Donnay, Classification of triperiodic twins. Canad. Mineral. **12**, 422–425 (1974)
13. A. Santoro, Characterization of twinning. Acta Cryst. A **30**, 224–231 (1974)
14. M. Catti, G. Ferraris, Twinning by merohedry and X-ray structure determination. Acta Cryst. A **32**, 163–165 (1976)
15. A.J.M. Duisenberg, Indexing in single-crystal diffractometry with an obstinate list of reflections. J. Appl. Cryst. **25**, 92–96 (1992)
16. R.A. Sparks, *GEMINI* Twinning solution program suite. Technical report, Bruker AXS, Madison, Wisconsin, USA (1999)
17. K. Chen, M. Kunz, N. Tamura, H.R. Wenk, Evidence for high stress in quartz from the impact site of Vredefort. South Africa. Eur. J. Mineral. **23**, 169–178 (2011)
18. K. Chen, C. Dejoie, H.R. Wenk, Unambiguous indexing of trigonal crystals from white-beam Laue diffraction patterns: application to Dauphiné twinning and lattice stress mapping in deformed quartz. J. Appl. Cryst. **45**, 982–989 (2012)
19. S.R. Dey, A. Morawiec, E. Bouzy, A. Hazotte, J.J. Fundenberger, A technique for determination of γ/γ interface relationships in a $(\alpha_2 + \gamma)$ TiAl base alloy using TEM Kikuchi patterns. Mater. Lett. **60**, 646–650 (2006)
20. B. Jackson, D. Fullwood, J. Christensen, S. Wright, Resolving pseudosymmetry in γ-TiAl using cross-correlation electron backscatter diffraction with dynamically simulated reference patterns. J. Appl. Cryst. **51**, 655–669 (2018)
21. A. Morawiec, E. Bouzy, H. Paul, J.J. Fundenberger, Orientation precision of TEM-based orientation mapping techniques. Ultramicroscopy **136**, 107–118 (2014)
22. G.D. Preston, The diffraction of X-rays by an age-hardening aluminium alloys. Proc. Roy. Soc. Lond. **167**, 526–538 (1938)
23. V. Daniel, H. Lipson, An X-ray study of the dissociation of an alloy of copper, iron and nickel. Proc. Roy. Soc. A **181**, 368–378 (1943)
24. V. Daniel, H. Lipson, The dissociation of an alloy of copper, iron.and nickel. Further X-ray work. *Proc. Roy. Soc. A*, **182**, 378–387 (1944)
25. V. Petříček, M. Dušek, and L. Palatinus. Crystallographic computing system Jana, General features. Z. Kristallogr. **229**(345–352), 2014 (2006)
26. K. Pilz, M. Estermann, S. van Smaalen, Automatic indexing of area-detector data of periodic and aperiodic crystals. J. Appl. Cryst. **35**, 253–260 (2002)
27. J. Porta, J.J. Lovelace, A.M.M. Schreurs, L.M.J. Kroon-Batenburg, G.E.O. Borgstahl, Processing incommensurately modulated protein diffraction data with Eval15. Acta Cryst. D **67**, 628–638 (2011)

28. A.K. Hailey, A.M. Hiszpanski, D.M. Smilgies, Y.L. Loo, The Diffraction Pattern Calculator (DPC) toolkit: a user-friendly approach to unit-cell lattice parameter identification of two-dimensional grazing-incidence wide-angle X-ray scattering data. J. Appl. Cryst. **47**, 2090–2099 (2014)
29. J. Simbrunner, C. Simbrunner, B. Schrode, C. Röthel, N. Bedoya-Martinez, I. Salzmannd, R. Resel, Indexing of grazing-incidence X-ray diffraction patterns: the case of fibre-textured thin films. Acta Cryst. A **74**, 373–387 (2018)
30. V. Savikhin, H.G. Steinrück, R.Z. Liang, B.A. Collins, S.D. Oosterhout, P.M. Beaujuged, M.F. Toney, GIWAXS-SIIRkit: scattering intensity, indexing and refraction calculation toolkit for grazing-incidence wide-angle X-ray scattering of organic materials. J. Appl. Cryst. **53**, 1108–1129 (2020)
31. D.M. Smilgies, D.R. Blasini, Indexation scheme for oriented molecular thin films studied with grazing-incidence reciprocal-space mapping. J. Appl. Cryst. **40**, 716–718 (2007)
32. A. Förster, C. Schulze-Briese, A shared vision for macromolecular crystallography over the next five years. Struct. Dyn. **6**, 064302 (2019)
33. U. Weierstall, Liquid sample delivery techniques for serial femtosecond crystallography. Phil. Trans. R. Soc. B **369**, 20130337 (2014)
34. F. Stellato, D. Oberthür, M. Liang, R. Bean, C. Gati, O. Yefanov, A. Barty, A. Burkhardt, P. Fischer, L. Galli, R.A. Kirian, J. Meyer, S. Panneerselvam, C.H. Yoon, F. Chervinskii, E. Speller, T.A. White, C. Betzel, A. Meents, H.N. Chapman, Room-temperature macromolecular serial crystallography using synchrotron radiation. IUCrJ **1**, 204–212 (2014)
35. C. Gati, G. Bourenkov, M. Klinge, D. Rehders, F. Stellato, D. Oberthür, O. Yefanov, B.P. Sommer, S. Mogk, M. Duszenko, C. Betzel, T.R. Schneider, H.N. Chapman, L. Redecke, Serial crystallography on in vivo grown microcrystals using synchrotron radiation. IUCrJ **1**, 87–94 (2014)
36. N. Coquelle, A.S. Brewster, U. Kapp, A. Shilova, B. Weinhausen, M. Burghammer, J.P. Colletier, Raster-scanning serial protein crystallography using micro- and nano-focused synchrotron beams. Acta Cryst. D **71**, 1184–1196 (2015)
37. A. Barty, R.A. Kirian, F.R.N.C. Maia, M. Hantke, C.H. Yoon, T.A. White, H. Chapman, *Cheetah*: software for high-throughput reduction and analysis of serial femtosecond X-ray diffraction data. J. Appl. Cryst. **47**, 1118–1131 (2014)
38. T.A. White, R.A. Kirian, A.V. Martin, A. Aquila, K. Nass, A. Bartyand, H.N. Chapman, *CrystFEL*: a software suite for snapshot serial crystallography. J. Appl. Cryst. **45**, 335–341 (2012)
39. T.A. White, A. Barty, F. Stellato, J.M. Holton, R.A. Kirian, N.A. Zatsepin, H.N. Chapman, Crystallographic data processing for free-electron laser sources. Acta Cryst. D **69**, 1231–1240 (2013)
40. T.A. White, V. Mariani, W. Brehm, O. Yefanov, A. Barty, K.R. Beyerlein, F. Chervinskii, L. Galli, C. Gati, T. Nakane, A. Tolstikova, K. Yamashita, C.H. Yoon, K. Diederichs, H.N. Chapman, Recent developments in *CrystFEL*. J. Appl. Cryst. **49**, 680–689 (2016)
41. Y. Gevorkov, O. Yefanov, A. Barty, T.A. White, V. Mariani, W. Brehm, A. Tolstikova, R.R. Grigat, H.N. Chapman, *XGANDALF* - extended gradient descent algorithm for lattice finding. Acta Cryst. A **75**, 694–704 (2019)
42. A.S. Brewster, M.R. Sawaya, J. Rodriguez, J. Hattne, N. Echols, H.T. McFarlane, D. Cascio, P.D. Adams, D.S. Eisenberg, N.K. Sauter, Indexing amyloid peptide diffraction from serial femtosecond crystallography: new algorithms for sparse patterns. Acta Cryst. D **71**, 357–366 (2015)
43. H.M. Ginn, P. Roedig, Kuo A, G. Evans, N.K. Sauter, O.P. Ernst, A. Meents, H. Mueller-Werkmeister, R.J.D. Millerg, D.I. Stuart, *TakeTwo*: an indexing algorithm suited to still images with known crystal parameters. *Acta Cryst. D*, **72**, 956–965 (2016)
44. C. Li, X. Li, R. Kirian, J.C.H. Spence, H. Liu, N.A. Zatsepin, SPIND: a reference-based auto-indexing algorithm for sparse serial crystallography data. IUCrJ **6**, 72–84 (2019)
45. C. Dejoie, N. Tamura, Pattern-matching indexing of Laue and monochromatic serial crystallography data for applications in materials science. J. Appl. Cryst. **53**, 824–836 (2020)

46. W. Kabsch, Processing of X-ray snapshots from crystals in random orientations. Acta Cryst. D **70**, 2204–2216 (2014)
47. W. Brehm, K. Diederichs, Breaking the indexing ambiguity in serial crystallography. Acta Cryst. D **70**, 101–109 (2014)
48. K.R. Beyerlein, T.A. White, O. Yefanov, C. Gati, I.G. Kazantsev, N.F.G. Nielsen, P.M. Larsen, H.N. Chapman, S. Schmidt, *FELIX*: an algorithm for indexing multiple crystallites in X-ray free-electron laser snapshot diffraction images. J. Appl. Cryst. **50**, 1075–1083 (2017)
49. Q. Zhou, Z.Q. Gao, Z. Dong, Y.M. Jiang, Z. She, Z. Geng, Y.H. Dong, A reference-based multi-lattice indexing method integrating prior information correction and iterative refinement in protein crystallography. Acta Cryst. A **77**, 277–288 (2021)
50. R.A. Kirian, X. Wang, U. Weierstall, K.E. Schmidt, J.C.H. Spence, M. Hunter, P. Fromme, T. White, H.N. Chapman, J.M. Holton, Femtosecond protein nanocrystallography - data analysis methods. Optics Express **18**, 5713–5723 (2010)
51. R.A. Kirian, T.A. White, J.M. Holton, H.N. Chapman, P. Fromme, A. Barty, L. Lomb, A. Aquila, F.R.N.C. Maia, A.V. Martin, R. Fromme, X. Wang, M.S. Hunter, K.E. Schmidt, J.C.H. Spence, Structure-factor analysis of femtosecond microdiffraction patterns from protein nanocrystals. Acta Cryst. A **67**, 131–140 (2011)

Chapter 11
Multigrain Indexing

Determination of crystal orientations via end-indexing of diffraction data is a key element of characterization of polycrystalline materials by non-destructive *tomographic* methods. Such methods have been developed to reconstruct three-dimensional microstructures of polycrystals, to reveal spatial grain boundary networks, and to provide three-dimensional orientation and strain fields. For the most part, these are transmission-type methods relying on high-flux highly collimated X-ray beams from synchrotron sources.

The beam penetrating a polycrystalline sample diffracts in multiple crystallites. The task is to determine the spatial arrangement of grains and their orientations. This means that the software must extract information corresponding to a single grain or volume element from patterns affected by many grains, and use that information to get both the position of the origin of reflections and orientation of the diffracting crystallite. The result of processing such diffraction data may have the form of a three-dimensional orientation map of a voxelated[1] specimen volume or a list of parameters (position, volume, orientation) of individual grains. Since multiple grains contribute to diffraction patterns, the assignment of indices to reflections is referred to as multigrain indexing.

The methods of reconstructing three-dimensional microstructures are related to X-ray absorption tomography, in which the attenuation of the transmitted beam is used to get the optical density field. The absorption-based *diffraction* contrast tomography uses diffracted beams instead of the transmitted beam. Clearly, absorption tomography cannot provide orientations of crystallites, and, for most materials, it cannot distinguish neighboring grains of the same phase.

The methods of three-dimensional orientation or strain mapping are still in statu nascendi. Early experiments demonstrated their utility, but the work is at an exploratory stage. However, there is clearly a demand for experimental

[1] Voxel is an elementary volume element. It is an analogue of a pixel in a two-dimensional image. A three-dimensional image is composed of voxels.

A. Morawiec, *Indexing of Crystal Diffraction Patterns*, Springer Series in Materials Science 326, https://doi.org/10.1007/978-3-031-11077-1_11

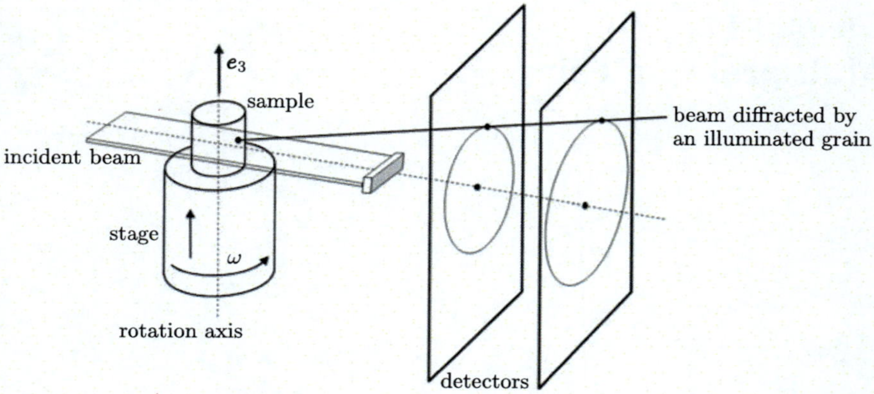

Fig. 11.1 Schematic of the experimental 3DXRD setup described in [5, 6].

three-dimensional data, and further progress in three-dimensional reconstruction is expected. Validations against EBSD-based two-dimensional orientation maps show that the three-dimensional orientation mapping techniques are promising.

The methods described shortly below can be broadly classified into methods in which a narrow beam illuminating a limited number of grains scans the investigated field, and full-field imaging with whole volume of interest illuminated. They differ in respect of specimen volumes which can be investigated, spatial and orientation resolutions, pattern acquisition times, times needed for data processing, et cetera. These methods are designed for investigating (mainly metallic) aggregates with relatively simple crystal structures. (On the other hand, multigrain indexing is related to indexing X-ray diffraction patterns from assemblies of macromolecular crystals of known structure [1, 2]. Programs intended for indexing serial-crystallography snapshots originating from multiple crystallites were presented in [3, 4]. Cf. Sect. 10.5.2.)

⊡ With synchrotron methods based on transmission of a monochromatic beam, the choice of wave energy plays an important role. Since the attenuation of X-rays decreases with the energy, the latter must be sufficiently high to penetrate through a polycrystalline specimen of a given thickness. On the other hand, with increasing energy (and decreasing wavelength), the Bragg angles decrease. With small Bragg angles, the probed regions have elongated (cigar-like) shapes, and this negatively influences spatial resolution in the elongated direction. Moreover, the Bragg angles are linked to the sample-to-detector distance; detector of a given size and resolution must be at an optimal distance allowing for recording all reflections of interest and the largest possible separation of peaks. In practice, X-rays with energies in the 15–100 keV range (corresponding to wavelengths in the range 0.83–0.12 Å) are used.

11.1 Three-Dimensional X-ray Diffraction

The three-dimensional X-ray diffraction (3DXRD) method relies on monochromatic parallel X-ray beam penetrating through a specimen mounted on a rotation stage with the axis perpendicular to the beam. Diffraction patterns are recorded on X-ray imaging detector(s) perpendicular to the beam. Over time, a number of experimental setups were proposed. Data obtained using these setups have been shown to have spatial resolution sufficient to provide the grain positions to within a few micrometers.

Clearly, if the number of illuminated grains is large, the diffraction spots originating from different grains overlap. The simplest way to alleviate the problem, and to make the determination of grain positions easier, is by reducing the volume contributing to diffraction patterns and scan the sample. In an early approach, absorbing conical slits were used. Figure 11.1 shows schematic of a more convenient and simpler approach with the beam narrow in one direction and wide in the other, so only a thin layer of the specimen is illuminated [5, 6]. Besides the rotation about $\boldsymbol{e}_3$, the stage allows for translation along this direction. Suitably oriented grains of the illuminated layer reflect, and traces of diffracted beams are recorded using multiple semi-transparent (high resolution, near-field, planar) detectors located at different distances from the rotation axis. To get three-dimensional results, such patterns are recorded for multiple specimen orientations (rotation about $\boldsymbol{e}_3$), and for multiple layers (translation along $\boldsymbol{e}_3$).

Focusing an X-ray beam to a narrow line is practically difficult. Focusing is avoided by the use of 'wide- (or box-) beam' geometry; instead of the 'line-beam', the whole specimen is illuminated, so only sample rotation is performed, and there in no need for translation [7, 8]. In effect, the data acquisition is faster compared to the setup with layer-by-layer illumination. This comes at the cost of increased density of spots and additional spot overlap. With yet another approach, the spot overlap is reduced by using point-focused beam scanned across the sample in the plane perpendicular to the rotation axis [9]; see also [10].

A single fixed (low resolution) detector can be located far from the specimen [11–15]. In such setups, the position of the diffracting grain in the specimen affects the positions of the diffraction spots on the distant detector but only with reduced spatial resolution. On the other hand, the increased angular resolution allows for determining average strains in individual grains [13].

Clearly, distant detectors provide more accurate angles (thus, more accurate orientations), while data from near-field detectors have better spatial resolution. A way of combining data from near-field and far-field detectors was described in [16].

11.2 X-ray Diffraction Contrast Tomography

Directly related to 3DXRD is the approach referred to as X-ray diffraction contrast tomography (DCT). It combines analysis of diffraction patterns with microtomographic imaging [17]. Again, the sample mounted on a rotation stage is illuminated

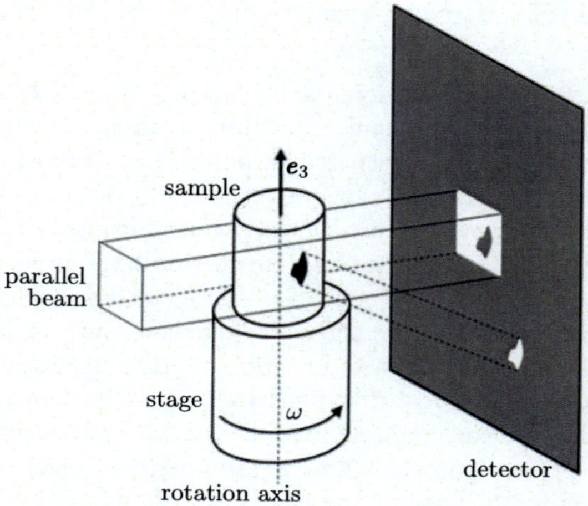

Fig. 11.2 Schematic of experimental DCT setup [17].

by parallel monochromatic X-ray beam. The axis of sample rotation is perpendicular to the beam. High-resolution detector is positioned close to the specimen. The grain structure is reconstructed from a series of patterns recorded as the sample rotates. In DCT, besides diffraction peaks, also intensities in the transmitted beam are recorded. When a crystallite embedded in a polycrystal is aligned for diffraction, there is a reduction of the transmitted intensity; see Fig. 11.2. An observed diffraction spot is expected to have a corresponding extinction spot in the transmitted beam. Such pairs are used for identification of grains and reconstruction of their shapes. Clearly, pure DCT-based reconstruction provides only grain arrangement, and additional measures are needed to get grain orientations.

There are various variants of the DCT method. The reconstruction can be based entirely on diffraction spots, and detectors can be in positions other than that shown in Fig. 11.2 [18]. The method is applicable to multiphase materials. It can be combined with other tomographic techniques to show inclusions of other phases. Particularly interesting are the cases with grain boundary networks decorated by second phase particles [19]. Results can be confronted by superimposing the tomographic images on the orientation-based grain maps.

11.3 Processing of Diffraction Data

The key point in data processing is to determine grain orientation and its location in the specimen. The methods of data processing partly depend on the experimental setup. The initial steps of spot detection and spot characterization are not discussed here. The reconstruction of three-dimensional microstructures is usually performed in stages or iteratively. There are two types of the reconstruction algorithms. In the parlance of tomography, they are based on 'forward' or 'backward' projections.

The forward projection (or modeling) means simulation of patterns. The reconstruction is performed by searching through the space of possible solutions and comparing all simulated and experimental patterns. The resulting solution corresponds to the best match between the patterns. The method can be seen as global optimization (over the product space of grain positions and orientations) by exhaustive trial-and-error. Thus, this approach is analogous to the direct matching of Sect. 8.9 for orientation determination.

The backward-projection relies on mathematical transforms of the experimental patterns. It is analogous to orientation determination via indexing of detected reflections. Clearly, the reconstruction based on forward-projection is significantly slower than that based on backward-projection, but it is less susceptible to peak overlaps, noise in experimental patterns, and to intra-granular orientation variations. As in the case of the two-dimensional orientation mapping, the latter feature is key for handling deformed specimens.

Before proceeding to indexing and reconstruction methods, three technical aspects are briefly described.

11.3.1 *Location of a Diffraction Spot as a Function of Grain Position*

To link the grain position to positions of indexed diffraction spots on a flat area detector, one needs to recall (2.11) of Sect. 2.3.2. The position $\boldsymbol{y}$ of a spot on a planar detector is a function of the wave vector $\boldsymbol{k}$ of the diffracted beam and the vector $\boldsymbol{L}$ relating position of the origin of the diffracted beam to the orthogonal projection of the origin on the detector; one has $\boldsymbol{y} = \boldsymbol{y}(\boldsymbol{L}, \boldsymbol{k}) = \boldsymbol{L} \times (\boldsymbol{k} \times \boldsymbol{L})/(\boldsymbol{k} \cdot \boldsymbol{L})$. Let $\boldsymbol{L}_0$ be a known vector from $\boldsymbol{0}$ of the laboratory reference frame to the orthogonal projection $\boldsymbol{0}_\perp$ of the point $\boldsymbol{0}$ on the detector (Fig. 11.3). If diffraction occurred at $\boldsymbol{x}$, then

$$\boldsymbol{x} + \boldsymbol{L} = \boldsymbol{x}_\perp + \boldsymbol{L}_0 \ ,$$

where $\boldsymbol{x}_\perp = \boldsymbol{x} - (\boldsymbol{x} \cdot \widehat{\boldsymbol{L}_0})\widehat{\boldsymbol{L}_0}$ is the orthogonal projection of $\boldsymbol{x}$ on the detector. Hence, one has $\boldsymbol{L} = \boldsymbol{L}_0 - (\boldsymbol{x} \cdot \widehat{\boldsymbol{L}_0})\widehat{\boldsymbol{L}_0}$, and the position $\boldsymbol{y}_n$ of the diffraction spot corresponding to reciprocal lattice vector $\check{\boldsymbol{h}}_n$ with respect to the point $\boldsymbol{0}_\perp$ is

$$\boldsymbol{y}_n = \boldsymbol{y}_n(\boldsymbol{x}, \boldsymbol{k}) = \boldsymbol{x}_\perp + \boldsymbol{y}(\boldsymbol{L}, \boldsymbol{k}) = \frac{\boldsymbol{L}_0 \times (\boldsymbol{k} \times (\boldsymbol{L}_0 - \boldsymbol{x}))}{\boldsymbol{k} \cdot \boldsymbol{L}_0} \ ,$$

where $\boldsymbol{k} = \boldsymbol{k}_0 + \check{\boldsymbol{h}}_n$, and $\boldsymbol{k}_0$ is the wave vector of the incident beam. Thus, the location $\boldsymbol{y}_n$ of the diffraction spot depends on $\boldsymbol{x}$ linearly. A shift of the grain position in a direction perpendicular to $\boldsymbol{L}_0$ corresponds to the same shift of the diffraction spot. A

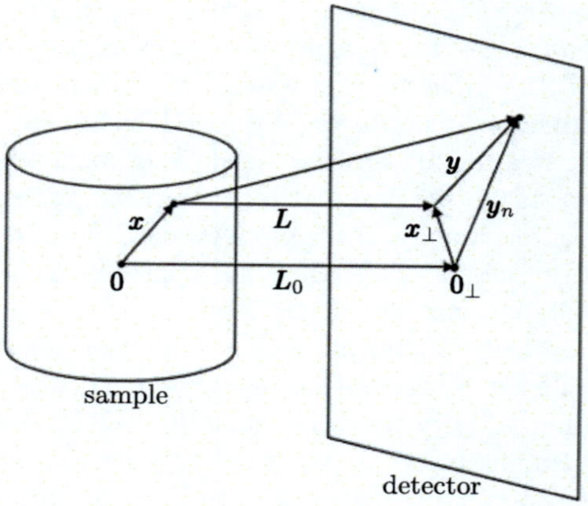

Fig. 11.3 Schematic illustrating the link between the position $\boldsymbol{x}$ of diffracting grain in the sample and the location of the diffraction spot on a planar detector. The point $\mathbf{0}$ is the origin of the sample reference frame, and $\mathbf{0}_\perp$ is the orthogonal projection of $\mathbf{0}$ on the detector. Both $\boldsymbol{L}_0$ and $\boldsymbol{L}$ are perpendicular to the detector, and the vectors $\boldsymbol{y}_n$, $\boldsymbol{y}$ and $\boldsymbol{x}_\perp$ are in the plane of the detector.

change of the grain position along $\boldsymbol{L}_0$ corresponds to a shift of the spot proportional to tangent of the angle between the diffracted beam and $\boldsymbol{L}_0$, i.e., with the typical geometry of 3DXRD and small Bragg angle, the latter shift is small.

Experiments provide positions $\boldsymbol{y}_n^{\text{exp}}$ $(n = 1, 2, \ldots, N)$ of indexed spots ascribed to a grain located at unknown $\boldsymbol{x}$. In principle, with $N \geq 3$, the grain position $\boldsymbol{x}$ can be determined by minimization of the $\boldsymbol{x}$-dependent function

$$\sum_n \left(\boldsymbol{y}_n^{\text{exp}} - \boldsymbol{y}_n(\boldsymbol{x}\,,\ \boldsymbol{k}_0 + \check{\boldsymbol{h}}_n) \right)^2 .$$

This problem is linear, but in practice, one needs to take into account that the spots are detected at different sample orientations. Moreover, the vectors $\check{\boldsymbol{h}}_n$ are known accurately only in the crystal reference frame, and the orientation of the latter is approximate (because it is determined based on approximate $\boldsymbol{x}$); thus, also components of $\check{\boldsymbol{h}}_n$ in the laboratory reference frame are approximate.

11.3.2 Algebraic Reconstruction Technique

As was already noted, grain structures are reconstructed in stages. Initial reconstruction results are crude. Additional refining steps can be used to resolve ambiguities (voids and grain overlaps) and to create 'smooth' three-dimensional orientation (or strain) maps. They rely on optimization and use results of initial reconstruction as input data. A number of methods have been used, e.g., [7, 11]. One of them is based on the standard method of discrete tomography: the algebraic reconstruction technique (ART) [20]. Generally, in ART, the area of interest is tessellated in voxels, and the optical density is sought for all voxels. The relationship between the unknown density x_j at jth voxel and the intensity b_i in the ith detector pixel is linear, i.e.,

$\sum_j A_{ij}x_j = b_i$, where the matrix A is determined by geometry of the experimental setup. Knowing the entries A_{ij} and b_i, the system of linear equations is solved (by one of the iterative methods) with respect to the non-negative densities x_j [8, 21]. In a very simple purely tomographic case, A_{ij} is the length of the segment of the ith ray inside the jth voxel. In reconstruction of a three-dimensional orientation map, x_j is an element of the product space of all possible voxel positions and orientations, so both positions and orientations need to be taken into account in constructing rows of the matrix A; see, e.g., [22].

11.3.3 Friedel Pairs

With the full turn of a crystal, a given plane in the crystal reflects four times. The reflections correspond to two Friedel pairs. Assuming that $\boldsymbol{k}_0$ is perpendicular to $\boldsymbol{e}_3$ and the rotation is about $\boldsymbol{e}_3$, if a spot corresponding to reciprocal lattice vector $\check{\boldsymbol{h}}$ appears on a pattern, then another one corresponding to the same vector shows up after a certain rotation when $\check{\boldsymbol{h}}$ intersects the Ewald sphere again, and there are also two analogously related spots corresponding to $-\check{\boldsymbol{h}}$; see, Fig. 11.4. If the reflection corresponding to $\check{\boldsymbol{h}}$ appears when the stage is at the angle ω, then the other three appear when the stage is at $\omega + \psi$, $\omega + \pi$ and $\omega + \pi + \psi$, where

$$\psi = 2 \arccos \left(-\widehat{\boldsymbol{k}}_0 \cdot \widehat{\boldsymbol{h}}_\perp \right) ,$$

$\boldsymbol{h}_\perp = \check{\boldsymbol{h}} - (\check{\boldsymbol{h}} \cdot \boldsymbol{e}_3)\, \boldsymbol{e}_3$ is the orthogonal projection of $\check{\boldsymbol{h}}$ on the plane perpendicular to $\boldsymbol{e}_3$, and $\widehat{\boldsymbol{h}}_\perp = \boldsymbol{h}_\perp / |\boldsymbol{h}_\perp|$. Friedel pairs are used to speed-up indexing and for refinement of grain orientations and positions [13, 18, 19, 23].

11.3.4 Indexing and Reconstruction

In the case of the setup with multiple detectors shown in Fig. 11.1, the reconstruction described in [5] is based on ray tracing: after spot detection, spots of the same reflection on patterns acquired at different detector positions are identified, and the position of the reflection origin in the specimen is computed. That position is at the intersection of the ray and the illuminated slice of the specimen. For choosing the best solution, the software (*GRAINDEX*) uses spatial criterion (reflections originate from approximately the same location) and crystallographic criterion (scattering vectors match reciprocal lattice vectors). The second criterion relies on grain orientations. In [5], the orientations were determined by the brute-force trial-and-error approach: scattering vectors are simulated for points of a grid in orientation space and compared to scattering vectors obtained from spot positions on the experimental patterns.

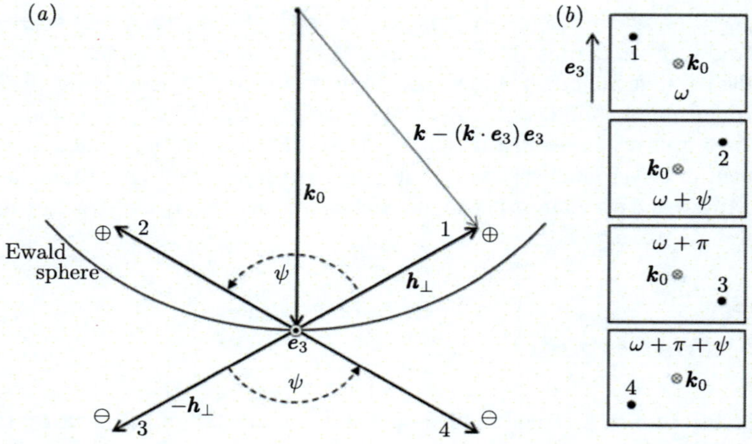

Fig. 11.4 (*a*) Projection of two Friedel pairs on the plane perpendicular to the sample rotation axis. Terminal point of the vector $\check{\boldsymbol{h}}$ is assumed to be on the Ewald sphere. Vector $\boldsymbol{h}_\perp$ is the orthogonal projection of $\check{\boldsymbol{h}}$ on the plane perpendicular to $\boldsymbol{e}_3$. The angle between $-\boldsymbol{k}_0$ and $\boldsymbol{h}_\perp$ is $\psi/2$. After rotation of the specimen about $\boldsymbol{e}_3$ by ψ, the terminal point of $\check{\boldsymbol{h}}$ is on the Ewald sphere again, and after rotation by π, the terminal point of $-\check{\boldsymbol{h}}$ is on the Ewald sphere. If the reciprocal lattice vectors corresponding to the projections marked by $\oplus$ are above the plane of the drawing, those corresponding to the projections marked by $\ominus$ are below that plane. (*b*) Positions of spots on the detector for particular orientations of the specimen.

Orientation for which the largest number of simulated vectors is sufficiently close to observed scattering vectors is ascribed to an individual grain [5]. Cf. Sect. 8.9.

Data from the setup of Fig. 11.1, i.e., with the line-focused beam, can be processed using the pure forward modeling. In [6, 24], the illuminated layer of the sample is meshed. For each area element, the code searches orientation space for crystallographic orientations that generate patterns matching experimental data. The reconstruction is performed layer-by-layer. That approach was modified in [25]; to speed-up the reconstruction, the modified method uses adaptive sampling of orientations and the fact that neighboring area elements are likely to have similar orientations.

In the case of setups with 'wide-beam' geometry, the number of mapped grains is limited by spot overlap. Clearly, the overlap is more likely with a larger number of grains and larger spread of spots. If the number of grains is large, individual grains can be indexed only if the tolerances on the differences between the measured and predicted scattering vectors are small. If the diffraction spots are sufficiently separated, at the first stage, tolerances for accepting an orientation can be relatively large, and the positions of grains in the specimen can be ignored as in *GrainSpotter* described in [12]. With larger number of grains, one needs to apply alternate orientation determination and refinement of remaining grain parameters until a satisfactory match is reached; see, e.g. [13]. Both *GrainSpotter* of [12] as well as the indexing program described in [13] rely on the same principles as accumulation along curves in the rotation space; cf. Sect. 8.6.2.

Reconstruction via "Surface Scanning"

A different approach to processing 3DXRD data was proposed in [15]. The processing is conducted in two stages. The first one comprises indexing and assignment of diffraction spots to individual grains. The second one is the refinement of the grain orientation, position of grain centroid and strain; it is based on the diffraction spots assigned to grains during indexing.

There are two modes of indexing. The first mode is applicable when the number of grains per sample cross-section is small. The positions of the diffracting grains in the sample are ignored. A version of direct pattern matching limited by a detected reflection is used; see Sect. 8.9.1.

If the number of grains is large, the second mode referred to as "surface scanning" is used. Given a diffraction spot, one needs to determine the intersection of the specimen with all rays which could be responsible for that spot. The intersection is a common of the specimen and a cone with its vertex at the diffraction spot, the axis parallel to the incident beam direction, and the opening angle of $2 \times 2\theta$ (Fig. 11.5). The intersection surface is divided into a grid, and as in the first mode, the method described in Sect. 8.9.1 is used to get the orientation. The position on the surface and orientation which lead to the best match are taken as the result. Clearly, since the position is assumed to be correct, the tolerances used for orientation determination in the second mode can be much narrower than those in the first one.

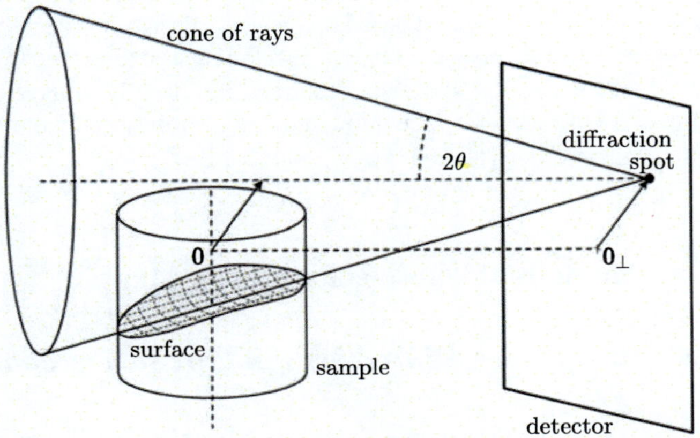

Fig. 11.5 Schematic illustration of the real-space surface searched for the position of the grain responsible for a given diffraction spot [15]. The surface is an intersection of the specimen with all potential rays through the spot.

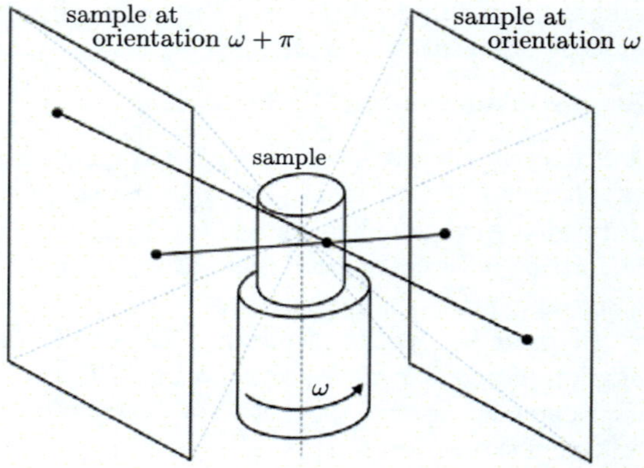

Fig. 11.6 Two Friedel pairs originating from the same grain determine its position in the sample.

Reconstruction via Friedel Pairs

In relation to DCT, another program (*Indexter*) combining determination of grain orientation and grain position was devised [18, 19]. It relies on Friedel pairs. In particular, a detected pair is used to determine diffracted beam path through the specimen. A set of Friedel pairs corresponds to a grain, and diffracted beam paths intersect (within a tolerance) at the same point of the specimen (Fig. 11.6). Moreover, lines in the orientation space determined by the experimental scattering vectors and vectors of crystal reciprocal lattice need to intersect at the same point; cf. Sect. 8.6.2. Thus, grain position and grain orientation are determined by identifying line intersections in real space and orientation space, respectively. I.e., grain position and grain orientation are determined based on the same principle. Clearly, the actual algorithm is more involved. Key are criteria and tolerances for accepting resulting positions and orientations. The approach described in [18, 19] is iterative with the tolerances gradually loosened, and the initial indexing is followed by the ART-based refinement.

11.4 Other Methods of Three-Dimensional Mapping

11.4.1 Laboratory X-ray Diffraction Contrast Tomography

The laboratory X-ray diffraction contrast tomography (labDCT) relies on a divergent polychromatic X-ray beam from a laboratory source [26]. As in the case of DCT, the sample is mounted on a rotation stage. The grain structure is reconstructed from a set of diffraction patterns recorded on a high-resolution detector at different sample orientations.

The beam geometry influences shapes of spots on diffraction patterns. With a point X-ray source and divergent incident beam, a diffracted beam is focused in the plane of diffraction at the distance equal to the distance between the point source and the sample [27]. Such configuration is known as Laue focusing geometry. The beam keeps its divergence in the plane perpendicular to the plane of diffraction; in geometric description, with the detector at the focusing plane, a diffraction spot from a grain is magnified in that direction by the factor of two compared to the original grain size. In effect diffraction spots from individual grains have elongated forms. Spatial resolution of labDCT can be slightly improved by using out-of-focus geometry with the specimen-to-detector distance larger than source-to-specimen distance which leads to magnification of diffraction spots (and some loss of pattern sharpness) [28].

A reconstruction of a shape of a grain via pure ART was demonstrated in [29], but usually the labDCT reconstructions are carried out using proprietary software package *GrainMapper3D* [30]. The labDCT setup can be easily modified for collection of absorptive contrast tomography projections by removing the beam-stop and the aperture used for forming the divergent beam.

11.4.2 Differential Aperture X-ray Microscopy

The approach known as 'differential aperture X-ray microscopy' (DAXM) uses a point-focused polychromatic X-ray beam from a synchrotron source; see, e.g., [31] and references therein. Experiments were carried out in reflection geometry with the specimen surface inclined with respect to the beam by about $\pi/4$ (Fig. 11.7). A three-dimensional orientation or strain map is obtained by moving the sample under the beam; at each position of the sample, a highly absorbing (tungsten) wire is moved along the specimen surface. The wire casts a sharp shadow on the diffraction patterns recorded at subsequent wire positions. Ray tracing is used to extract the position of the origin of each diffraction spot on the detector. The method has high spatial resolution, but due to small probe and long scanning times, it is limited to relatively small specimen volumes. Moreover, the information originating from regions buried at a distance from the surface is limited by exponential reduction of intensity due to absorption. DAXM can be seen as a depth profiling technique applicable to analysis of near surface layers. Clearly, depth-resolved diffraction information can also be collected by scanning absorbing structures other then wires. For similar depth-resolved experiments without a wire, see [32]. Micro-beam Laue diffraction setups using transmission geometry and high-energy (i.e., >50 keV) X-rays have also been described; see, e.g., [33, 34]. In most applications, the indexing and the rest of processing of three-dimensional diffraction data were carried out using the software package *XMAS* [35].

11.4.3 Three-Dimensional Orientation Mapping in TEM

A very high spatial resolution can be achieved using TEM. The first approach to three-dimensional orientation mapping in TEM was presented in [36]. The method

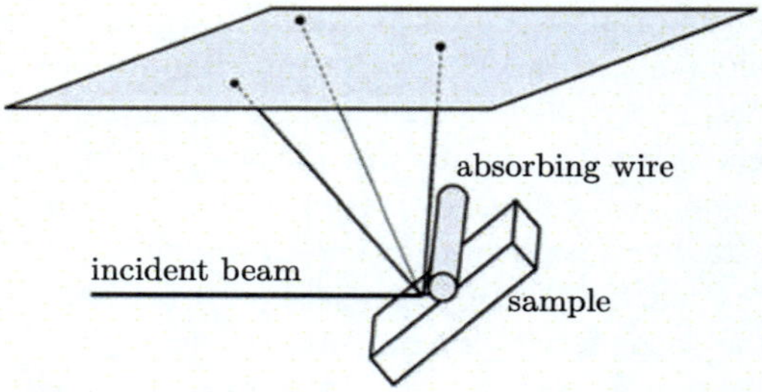

Fig. 11.7 Schematic of experimental DAXM setup [31].

is based on conical dark-field scanning (Sect. 8.1) at multiple sample tilts. Both the dark-field images and orientations determined based on the scattering vectors derived from the dark-field images were used for constructing a three-dimensional orientation map. In the case of the version described in [36], indexing and reconstruction were performed by tools developed for 3DXRD [5] and *GrainSpotter* [12].

A different approach was used in [37] to get three-dimensional morphological information from a small volume of Ni-based superalloy with precipitates. These measurements also relied on a tilt series, but instead of dark-field images, beam scanning was used. For each tilt, the beam was shifted step by step and precession diffraction patterns were acquired. These patterns, with components from both the matrix and the precipitates, were decomposed using non-negative matrix factorization[2]—a convenient tool for image classification and image decomposition. The resulting component patterns resembling conventional single-phase patterns were used to get orientations (by template matching) and for confirming the structures of the involved phases. The maps of spatial origins of component patterns (resembling virtual dark-field images) were used for reconstruction of grain locations in real space.

A similar method was used in [38] to determine arrangement of nanocrystalline grains. It also relied on a tilt series and step by step recording of nanodiffraction patterns (affected by multiple grains). Positions and intensities of spots on the patterns were determined, and these data were used to compute virtual dark-field images and hence 'image-filtered' diffraction patterns (i.e., decomposed patterns of the previous paragraph). To get orientations, the image-filtered patterns were matched to simulated patterns (with some pre-selection to speed up the process). The grain locations were obtained via tomographic ART based on the virtual dark-field images.

Also in more recent applications [39, 40], orientations are determined by direct matching of experimental patterns to patterns simulated for nodes of a grid in the orientation space.

[2] Formally, non-negative matrix factorization is a numerical decomposition of a given matrix F with non-negative entries into a product WH, where the matrices W and H have non-negative entries, and the number of columns in the weight matrix W is much smaller than in F.

11.4.4 *Three-Dimensional Mapping Using Neutron Diffraction*

In terms of spatial resolution, neutron techniques are on the side opposite to TEM. Due to high penetration of neutrons, only large (centimeter-sized) samples with large grains can be mapped. In the setup referred to as 'neutron diffraction contrast tomography' pink (narrow spectrum) neutron beam was used [41]. The patterns were recorded in back-reflection. Indexing and reconstruction were performed using the aforementioned *Indexter* [18, 19].

White neutron beam and two detectors (for back- and forward-diffraction) were used in experiments described in [42]. The reconstruction was performed via forward projection i.e., a search for the best possible match between the simulated and recorded Laue patterns.

In yet another approach, three-dimensional orientation maps were created using time-of-flight neutron diffraction[3] [43].

References

1. K.S. Paithankar, H.O. Sørensen, J.P. Wright, S. Schmidt, H.F. Poulsen, E.F. Garman, Simultaneous X-ray diffraction from multiple single crystals of macromolecules. Acta Cryst. D **67**, 608–618 (2011)
2. R.J. Gildea, D.G. Waterman, J.M. Parkhurst, D. Axford, G. Sutton, D.I. Stuart, N.K. Sauter, G. Evans, G. Winter, New methods for indexing multi-lattice diffraction data. Acta Cryst. D **70**, 2652–2666 (2014)
3. K.R. Beyerlein, T.A. White, O. Yefanov, C. Gati, I.G. Kazantsev, N.F.G. Nielsen, P.M. Larsen, H.N. Chapman, S. Schmidt, *FELIX*: an algorithm for indexing multiple crystallites in X-ray free-electron laser snapshot diffraction images. J. Appl. Cryst. **50**, 1075–1083 (2017)
4. Q. Zhou, Z.Q. Gao, Z. Dong, Y.M. Jiang, Z. She, Z. Geng, Y.H. Dong, A reference-based multi-lattice indexing method integrating prior information correction and iterative refinement in protein crystallography. Acta Cryst. A **77**, 277–288 (2021)
5. E.M. Lauridsen, S. Schmidt, R.M. Suter, H.F. Poulsen, Tracking: a method for structural characterization of grains in powders or polycrystals. J. Appl. Cryst. **34**, 744–750 (2001)
6. R.M. Suter, D. Hennessy, C. Xiao, U. Lienert, Forward modeling method for microstructure reconstruction using X-ray diffraction microscopy: single-crystal verification. Rev. Sci. Instrum. **77**, 123905 (2006)
7. H.F. Poulsen, S. Schmidt, Reconstruction of grain boundaries in polycrystals by filtered back-projection of diffraction spots. J. Appl. Cryst. **36**, 319–325 (2003)
8. T. Markussen, X. Fu, L. Margulies, E.M. Lauridsen, S.F. Nielsen, S. Schmidt, H.F. Poulsen, An algebraic algorithm for generation of three-dimensional grain maps based on diffraction with a wide beam of hard X-rays. J. Appl. Cryst. **37**, 96–102 (2004)
9. Y. Hayashi, Y. Hirose, Y. Seno, Polycrystal orientation mapping using scanning three-dimensional X-ray diffraction microscopy. J. Appl. Cryst. **48**, 1094–1101 (2015)
10. J. Hektor, S.A. Hall, N.A. Henningsson, J. Engqvist, M. Ristinmaa, F. Lenrick, J.P. Wright, Scanning 3DXRD measurement of grain growth, stress, and formation of Cu_6Sn_5 around a tin whisker during heat treatment. Materials **12**, 446 (2019)

[3] The energies and corresponding wavelengths of neutrons are determined based on the time it takes to fly a known distance.

11. J. Oddershede, S. Schmidt, H.F. Poulsen, H.O. Sørensen, J. Wright, W. Reimers, Determining grain resolved stresses in polycrystalline materials using three-dimensional X-ray diffraction. J. Appl. Cryst. **43**, 539–549 (2010)
12. S. Schmidt, *GrainSpotter*: a fast and robust polycrystalline indexing algorithm. J. Appl. Cryst. **47**, 276–284 (2014)
13. J.V. Bernier, N.R. Barton, U. Lienert, M.P. Miller, Far-field high-energy diffraction microscopy: a tool for intergranular orientation and strain analysis. J. Strain Anal. Eng. Des. **46**, 527–547 (2011)
14. H. Sharma, R.M. Huizenga, S.E. Offerman, A fast methodology to determine the characteristics of thousands of grains using three-dimensional X-ray diffraction. I. Overlapping diffraction peaks and parameters of the experimental setup. J. Appl. Cryst. **45**, 693–704 (2012)
15. H. Sharma, R.M. Huizenga, S.E. Offerman, A fast methodology to determine the characteristics of thousands of grains using three-dimensional X-ray diffraction. II. Volume, centre-of-mass position, crystallographic orientation and strain state of grains. J. Appl. Cryst. **45**, 705–718 (2012)
16. K.E. Nygren, D.C. Pagan, J.V. Bernier, M.P. Miller, An algorithm for resolving intragranular orientation fields using coupled far-field and near-field high energy X-ray diffraction microscopy. Mater. Charact. **165**, 110366 (2020)
17. W. Ludwig, E.M. Lauridsen, S. Schmidt, H.F. Poulsen, J. Baruchel, High-resolution three-dimensional mapping of individual grains in polycrystals by topotomography. J. Appl. Cryst. **40**, 905–911 (2007)
18. P. Reischig, A. King, L. Nervo, N. Vigano, Y. Guilhem, W.J. Palenstijn, K.J. Batenburg, M. Preuss, W. Ludwig, Advances in X-ray diffraction contrast tomography: flexibility in the setup geometry and application to multiphase materials. J. Appl. Cryst. **46**, 297–311 (2013)
19. W. Ludwig, P. Reischig, A. King, M. Herbig, E.M. Lauridsen, G. Johnson, J. Marrow, J.Y. Buffière, Three-dimensional grain mapping by X-ray diffraction contrast tomography and the use of Friedel pairs in diffraction data analysis. Rev. Sci. Instrum. **80**, 033905 (2009)
20. R. Gordon, R. Bender, G.T. Herman, Algebraic reconstruction techniques (ART) for three-dimensional electron microscopy and X-ray photography. J. Theor. Biol. **29**, 471–481 (1970)
21. H.F. Poulsen, X. Fun, Generation of grain maps by an algebraic reconstruction technique. J. Appl. Cryst. **36**, 1062–1068 (2003)
22. N. Viganò, W. Ludwig, K.J. Batenburg, Reconstruction of local orientation in grains using a discrete representation of orientation space. J. Appl. Cryst. **47**, 1826–1840 (2014)
23. M. Moscicki, P. Kenesei, J. Wright, H. Pinto, T. Lippmann, A. Borbély, A.R. Pyzalla, Friedel-pair based indexing method for characterization of single grains with hard X-rays. Mater. Sci. Eng. A **524**, 64–68 (2009)
24. C.M. Hefferan, S.F. Li, J. Lind, R.M. Suter, Tests of microstructure reconstruction by forward modeling of high energy X-ray diffraction microscopy data. Powder Diffr. **25**, 132–137 (2010)
25. S.F. Li, R.M. Suter, Adaptive reconstruction method for three-dimensional orientation imaging. J. Appl. Cryst. **46**, 512–524 (2013)
26. A. King, P. Reischig, J. Adrien, W. Ludwig, First laboratory X-ray diffraction contrast tomography for grain mapping of polycrystals. J. Appl. Cryst. **46**, 1734–1740 (2013)
27. A. Guinier, J. Tennevin, Sur deux variantes de la méthode de Laue et leurs applicationes. Acta Cryst. **2**, 133–138 (1949)
28. H. Fang, D. Juul Jensen, Y. Zhang, Improved grain mapping by laboratory X-ray diffraction contrast tomography. IUCrJ **8**, 559–573 (2021)
29. W. van Aarle, W. Ludwig, A. King, D. Penumadu, An accurate projection model for diffraction image formation and inversion using a polychromatic cone beam. J. Appl. Cryst. **48**, 334–343 (2015)
30. F. Bachmann, H. Bale, N. Gueninchault, C. Holzner, E.M. Lauridsen, 3D grain reconstruction from laboratory diffraction contrast tomography. J. Appl. Cryst. **52**, 643–651 (2019)
31. G.E. Ice, J.W.L. Pang, Tutorial on X-ray microLaue diffraction. Mater. Charact. **60**, 1191–1201 (2009)

32. D. Ferreira Sanchez, J. Villanova, J. Laurencin, J.S. Micha, A. Montani, P. Gergauda, P. Bleuet, X-ray micro Laue diffraction tomography analysis of a solid oxide fuel cell. J. Appl. Cryst. **48**, 357–364 (2015)
33. F. Hofmann, B. Abbey, L. Connor, N. Baimpas, X. Song, S. Keegan, A.M. Korsunsky, Imaging of grain-level orientation and strain in thicker metallic polycrystals by high energy transmission micro-Laue (HETL). Int. J. Mater. Res. **103**, 192–199 (2012)
34. F. Hofmann, X. Song, B. Abbey, T.S. Jund, A.M. Korsunsky, High-energy transmission Laue micro-beam X-ray diffraction: a probe for intra-granular lattice orientation and elastic strain in thicker samples. J. Synchrotron Rad. **19**, 307–318 (2012)
35. N. Tamura, *XMAS*: a versatile tool for analyzing synchrotron X-ray microdiffraction data, in *Strain and Dislocation Gradients from Diffraction*, ed. by R. Barabash, G. Ice (Imperial College Press, London, 2014), pp. 125–155
36. H.H. Liu, S. Schmidt, H.F. Poulsen, A. Godfrey, Z.Q. Liu, J.A. Sharon, X. Huang, Three-dimensional orientation mapping in the transmission electron microscope. Science **332**, 833–834 (2011)
37. A.S. Eggeman, R. Krakow, P.A. Midgley, Scanning precession electron tomography for three-dimensional nanoscale orientation imaging and crystallographic analysis. Nat. Commun. **6**, 7267 (2015)
38. Y. Meng, J.M. Zuo, Three-dimensional nanostructure determination from a large diffraction data set recorded using scanning electron nanodiffraction. IUCrJ **3**, 300–308 (2016)
39. G. Wu, W. Zhu, Q. He, Z. Feng, T. Huang, L. Zhang, S. Schmidt, A. Godfrey, X. Huang, 2D and 3D orientation mapping in nanostructured metals: a review. Nano Mater. Sci. **2**, 50–57 (2020)
40. S. Colding-Jørgensen, 3D electron microscopy of nanostructures in energy devices. Ph.D. thesis, Technical University of Denmark (DTU), Kongens Lyngby, Denmark (2020)
41. S. Peetermans, W. Ludwig, A. King, P. Reischig, E.H. Lehmann, Cold neutron diffraction contrast tomography of polycrystalline material. Analyst **139**, 5766–5772 (2014)
42. M. Raventós, M. Tovar, M. Medarde, T. Shang, M. Strobl, S. Samothrakitis, E. Pomjakushina, C. Grünzweig, S. Schmidt, Laue three dimensional neutron diffraction. Sci. Rep. **9**, 4798 (2019)
43. A. Cereser, M. Strobl, S. Hall, A. Steuwer, R. Kiyanagi, A. Tremsin, E. Bergbäck Knudsen, T. Shinohara, P. Willendrup, A. Bastos da Silva Fanta, S. Iyengar, P.M. Larsen, T. Hanashima, T. Moyoshi, P.M. Kadletz, P. Krooss, T. Niendorf, M. Sales, W.W. Schmahl, S. Schmidt, Time-of-flight three dimensional neutron diffraction in transmission mode for mapping crystal grain structures. Sci. Rep. **7**, 9561 (2017)

Chapter 12
An Excursion Beyond Diffraction by Periodic Crystals

The considerations above were focused on diffraction by crystalline materials with three-dimensionally periodic order of atoms. However, diffraction is also used for determining structures lacking a three-dimensional lattice. There is a large diversity of such structures. Only selected subjects will be considered below. These are the indexing of patterns resulting from diffraction on helical structures (this chapter) and on quasicrystals (next chapter). Before that, two topics related to characterization of arbitrary arrangements of atoms are briefly discussed: the Debye scattering formula and diffraction imaging of single molecules.

12.1 Debye Scattering Formula

Description of crystal diffraction relies on Fourier transformation. General arrangements of atoms can be dealt with in a similar way. The broad kinematic description of Fraunhofer diffraction can be summarized in the following way: First, the material is characterized by a position dependent 'transmittance' function (e.g., electron density) $\rho = \rho(\boldsymbol{x})$. Second, with $\boldsymbol{k}_0$ being the wave vector of the incident beam, the intensity $I = I(\boldsymbol{k} - \boldsymbol{k}_0)$ in the scattering direction determined by wave vector $\boldsymbol{k}$ is $I = |\mathcal{F}\rho|^2$, i.e., the intensity is the squared magnitude of Fourier transform of ρ at $\boldsymbol{k} - \boldsymbol{k}_0$. It follows from (1.76) that the intensity can also be expressed as the Fourier transform of autocorrelation of ρ: $I = \mathcal{F}(\rho \star \rho)$. Diffraction patterns consist of two components: 'discrete' and 'continuous'. The discrete component (pure point diffraction) is directly linked to some order in the scattering material, whereas the continuous one is an indicator of departures from the order.

Let n enumerate atoms in a general arrangement. The electron density at $\boldsymbol{x}$ is $\rho(\boldsymbol{x}) = \sum_n \rho_n(\boldsymbol{x} - \boldsymbol{x}_n)$, where ρ_n denotes the electron density of the n-th atom at the point $\boldsymbol{x}$ separated from the average atom position $\boldsymbol{x}_n$ by $\boldsymbol{x} - \boldsymbol{x}_n$. As in the derivation of (2.15) and (2.23), the expression for the structure factor contains the sum over the

A. Morawiec, *Indexing of Crystal Diffraction Patterns*, Springer Series in Materials Science 326, https://doi.org/10.1007/978-3-031-11077-1_12

scattering centers, but without the condition of periodicity, one is forced to use the sum over all, say, N, scattering sites. With

$$F = \mathcal{F}\rho = \sum_{n=1}^{N} \mathcal{F}_{\boldsymbol{x}} \left[\rho_n(\boldsymbol{x} - \boldsymbol{x}_n) \right] ,$$

by the shift property $\mathcal{F}_{\boldsymbol{x}} \left[\rho_n(\boldsymbol{x} - \boldsymbol{x}_n) \right] (\boldsymbol{\xi}) = \mathcal{F}_{\boldsymbol{x}} \left[\rho_n(\boldsymbol{x}) \right] (\boldsymbol{\xi}) \exp(-2\pi i \boldsymbol{\xi} \cdot \boldsymbol{x}_n)$ (cf. Sect. 1.11.1), one has

$$F(\boldsymbol{\xi}) = \sum_{n=1}^{N} f_n(\boldsymbol{\xi}) \exp(-2\pi i \boldsymbol{\xi} \cdot \boldsymbol{x}_n) ,$$

where $f_n = \mathcal{F}_{\boldsymbol{x}} \left[\rho_n(\boldsymbol{x}) \right]$ denotes the scattering factor of the n-th site. (Clearly, if ρ is not periodic, F is not a coefficient of Fourier series, and the scattering vector $\boldsymbol{\xi} = \boldsymbol{k} - \boldsymbol{k}_0$ has no interpretation as the reciprocal lattice vector.) In this general case, the formula for the intensity is

$$I(\boldsymbol{\xi}) \propto F^*(\boldsymbol{\xi}) F(\boldsymbol{\xi}) = \sum_{m=1}^{N} \sum_{n=1}^{N} f_m^*(\boldsymbol{\xi}) f_n(\boldsymbol{\xi}) \exp(2\pi i \boldsymbol{\xi} \cdot (\boldsymbol{x}_m - \boldsymbol{x}_n)) ; \qquad (12.1)$$

cf. (2.41). With an 'irregular' arrangement of atoms, the intensity is calculated by averaging $\exp(2\pi i \boldsymbol{\xi} \cdot \boldsymbol{x})$ over directions of $\boldsymbol{x} = \boldsymbol{x}_m - \boldsymbol{x}_n$. Generally, an orientation distribution needs to be taken into account [1], but frequently the distribution of directions is assumed to be random. In this case, the orientation average is equal to the integral of $\exp(2\pi i \boldsymbol{\xi} \cdot \boldsymbol{x})$ over the unit sphere ($\times (4\pi)^{-1}$). Let α denote the polar angle of $\boldsymbol{x}$ with respect to the direction of $\boldsymbol{\xi}$. Since the integrand is axially symmetric with respect to $\boldsymbol{\xi}$, the sought mean is

$$\frac{1}{4\pi} \int_0^{\pi} 2\pi \exp(2\pi i \, |\boldsymbol{\xi}| \, |\boldsymbol{x}| \cos\alpha) \sin\alpha \, \mathrm{d}\alpha = \mathrm{sinc}(2 \, |\boldsymbol{\xi}| \, |\boldsymbol{x}|) .$$

With the factors f_m depending only on the magnitude of the scattering vector, also the intensity is determined by $|\boldsymbol{\xi}|$ so one can write $I(|\boldsymbol{\xi}|)$ instead of $I(\boldsymbol{\xi})$, and

$$I(|\boldsymbol{\xi}|) \propto \sum_{m=1}^{N} \sum_{n=1}^{N} f_m^*(|\boldsymbol{\xi}|) \, f_n(|\boldsymbol{\xi}|) \, \mathrm{sinc}(2 \, |\boldsymbol{\xi}| \, d_{mn}) , \qquad (12.2)$$

where $d_{mn} = |\boldsymbol{x}_m - \boldsymbol{x}_n|$ is the distance between atoms m and n. This is the well known *Debye scattering formula* [2]. Formally, the Debye formula is applicable to any arrangement of atoms (crystalline or not) in any state (solid, liquid, gaseous). In fact, it is applied to crystalline nanopowders, polycrystals with ultrafine grains, suspensions of molecules, dilute solutions of proteins, et cetera.[1] Some 'particles' are modeled not as aggregates of atoms but as aggregates of spherical assemblies of atoms or globular subunits of the particle with given form factors F_m, and the formula $I(|\boldsymbol{\xi}|) \propto \sum\sum F_m^* F_n \operatorname{sinc}(2|\boldsymbol{\xi}|\,d_{mn})$ analogous to (12.2) is applicable; here, d_{mn} represents distances between centers of the assemblies. With particle sizes large compared to the radiation wavelength, the scattering angles become small. Consequently, the Debye scattering formula constitutes a basis of the research area known as small angle scattering [3].

The right side of the Debye formula is related (through Fourier transformation) to *radial* (or *pair*) *distribution*— a function (of the direct-space distance) describing the density of probability of having a pair of atoms separated by a given distance. Diffraction-based experimental radial distributions are compared to corresponding distributions obtained from models of atomic systems.

In practical calculations, the number N in the Debye formula must be reasonably small because the number of distances d_{mn} and, accordingly, the number of terms in (12.2) grow with the square of N. Numerous methods of speeding up the calculations have been devised to allow for larger N; see, e.g., [4, 5].

12.2 Single-Particle Diffraction Imaging

Some structures are difficult to crystallize. Many of them are studied using electron microscopy methods.[2] However, since electrons have low penetration depth, it is difficult to image structures of thick objects. Single-particle coherent X-ray diffraction imaging is an arising alternative to electron microscopy. The idea is to use X-rays to image individual structures of nanoparticles or biomolecules without crystallization. There are two key impediments to the development of such a method. First, the diffraction signal originating from a single molecule is very weak. Second, even if an experiment provides distribution of diffracted intensities, one still faces the fundamental 'phase problem'; cf. Sect. 2.6.

12.2.1 Phase Problem

Let the function ρ represent the electron density in a particle, and let $\rho(\boldsymbol{x})$ be equal to zero for any point $\boldsymbol{x}$ outside the particle. The function ρ extended beyond the particle has compact support. The intensities on a diffraction pattern represent squared moduli

[1] As before, one must take into account that the actual intensity involves factors specific to a particular experiment (e.g., polarization, Debye-Waller, geometric absorption factors).

[2] Electron microscopy methods can be directly applied to observation of inorganic nanoparticles. Biomolecules are investigated using cryogenic electron microscopy: to alleviate radiation damage, the molecules are frozen in ice. Three-dimensional images of investigated structures are reconstructed from micrographs (real space two-dimensional projections) of a number of single particles in random orientations. Alternatively, with cryogenic electron *tomography*, a tilt-series of micrographs is used to reconstruct three-dimensional structures.

$|\mathcal{F}\rho|$ of the Fourier transform $F(\boldsymbol{\xi}) = \mathcal{F}\rho(\boldsymbol{\xi}) = \int \rho(\boldsymbol{x}) \exp(-2\pi i \boldsymbol{x} \cdot \boldsymbol{\xi})\, \mathrm{d}^3\boldsymbol{x}$. Let $\boldsymbol{x}_l$ be nodes of a a cubic grid covering the support of ρ with the voxel of volume Δ^3, and let $\rho(\boldsymbol{x}_l) = \varrho_l$. In other words, the values ϱ_l constitute a discrete approximation of ρ on the regular grid with Δ being the interval between closest nodes. The discrete three-dimensional Fourier transform with values given on a grid in the Fourier space has the form $F_k = \rho(\boldsymbol{\xi}_k) = \Delta^3 \sum_l \varrho_l \exp(-2\pi i \boldsymbol{x}_l \cdot \boldsymbol{\xi}_k)$. Knowing the moduli $|F_k|$ (i.e., square roots of diffraction intensities), one has the system of equations

$$\Delta^3 \left| \sum_l \varrho_l \exp(-2\pi i \boldsymbol{x}_l \cdot \boldsymbol{\xi}_k) \right| = |F_k| \tag{12.3}$$

with respect to ϱ_l. Clearly, the solution of this system is not unique. With constant $\boldsymbol{x}_0$ and ϕ_0, the moduli of Fourier transforms of the functions $\rho(\boldsymbol{x})$, $\rho^*(-\boldsymbol{x})$, $\rho(\boldsymbol{x} + \boldsymbol{x}_0)$ and $\rho(\boldsymbol{x})e^{-2\pi i \phi_0}$ are all equal, i.e., the modulus is invariant with respect to inversion, translation and constant phase shift. In practice, this invariance can be ignored because the interest is in parameters describing the shape of the function irrespective of its absolute orientation, position or phase. Thus, the actual question is about the determinability of these other parameters. It turns out that having $|F|$ sampled on sufficiently dense grid and some additional constraints on ρ, these other characteristics of the function ρ can be computed using phase retrieval algorithms.

12.2.2 Iterative Phase Retrieval Algorithms

It was first observed that with sufficiently densely sampled $|F|$ and $|\rho|$, good approximations of the complete complex functions F and ρ can be recovered using a simple algorithm. The (Gerchberg-Saxton) algorithm [6] starts with known $|F_k|$ and $|\varrho_l|$ and random phases. It performs cycling computation of forward and inverse Fourier transforms, and in each cycle, the computed moduli are replaced by the actual $|F_k|$ and $|\varrho_l|$. These steps are repeated until an error criterion is satisfied.

This approach was modified by Fienup [7, 8] so it became applicable to diffraction experiments in which $|\varrho_l|$ are not known; that requirement was replaced by other constraints. As in the Gerchberg-Saxton algorithm, the method is iterative; in each step the computed moduli are replaced by the actual $|F_k|$, and $|\varrho_l|$ are modified to satisfy the constraints. Versions of Fienup's 'hybrid input-output' algorithm [8] perform very well in many practical cases even with quite noisy data.

Bounds on the support of the function ρ can be derived from the support of its autocorrelation; the former is contained in the latter. The support of the autocorrelation of ρ is accessible as the autocorrelation is equal to the inverse Fourier transform of $|F|^2$ (i.e., of the diffraction intensities; cf. Sect. 2.6.) Using this information and modified 'hybrid input-output' algorithm, phases can be retrieved without prior knowledge on the object's compact support [9].[3]

Application of iterative phase retrieval algorithms allowed for demonstrating (soft) X-ray imaging of two-dimensional micrometre-sized non-periodic objects [10]. This experiment demonstrated the feasibility of single-molecule imaging if sufficiently intense diffraction data are collected.

12.2.3 Single-Particle Imaging With XFEL

X-ray free-electron lasers open the opportunity to collect such data. The principles of the collection are similar to those used in the case of microcrystals. (See, Sect. 10.5.2). The event of diffraction in a particle precedes its (radiation-caused) disintegration. With a proper sample delivery method, one can collect a large number of patterns. However, even with a very intense pulse, the diffraction pattern from a single particle is very faint. This is especially true of weakly scattering bioparticles. The idea is to process such data sets based on their high redundancy. One needs to take the advantage of the large number of diffraction patterns to change the signal-to-noise ratio: from a very low signal-to-noise ratio of individual patterns to a sufficiently high ratio of the final assembled set of consistent diffraction data.

As in the case of patterns from microcrystals, the first step of processing is to get rid of blanks, patterns with artifacts due to extraneous scattering, patterns from aggregates *et cetera*. Assuming homogeneity of the particle population i.e., that particles are similar in shapes and sizes, and that they are fully illuminated, a snapshot diffraction pattern depends on orientation of the diffracting particle. After recording a large number of two-dimensional patterns, one needs to assemble the patterns into a single dataset in three-dimensional Fourier space, i.e., into a distribution of intensity of diffracted radiation, and hence get the moduli $|F_k| = |F(\boldsymbol{\xi}_k)|$.

A number of methods for retrieving three-dimensional information from large sets of images have been suggested; see e.g. [11–13]. Roughly, these methods rely on the fact that snapshots originating from particles in close orientations are expected to be similar. Therefore, one needs to identify diffraction data originating from particles in nearly the same orientations. Knowing mutual orientations of particles corresponding to individual snapshots, one can reconstruct their orientations with respect to external reference frame and then average intensities at corresponding locations of the patterns. Having sufficiently reliable three-dimensional distribution of diffracted intensities, and hence the moduli $|F_k|$, the structure of the particle can be determined using the phase retrieval algorithms.

---

[3] One might consider solving the system (12.3) by defining a cost function (e.g., as the sum of squared residuals, i.e., differences between right and computed left sides of (12.3)) and using continuous optimization techniques. This approach, however, frequently fails because of local minima of the cost function.

A number of cases have been studied at low resolution and based on relatively small sets of patterns; see, e.g., [14, 15]. The low resolution is an effect of low intensities at higher scattering angles. The relatively intense more central parts of diffraction patterns correspond to short scattering vectors and represent crude features of the particles.

12.3 Indexing of Diffraction Patterns of Helical Structures

This section is devoted to the important case of indexing diffraction patterns originating from helical structures. Such structures are abundant and play significant roles in biological systems. Theory of X-ray diffraction by helical molecules was first described by Cochran, Crick and Vand [16]. It is known as CCV theory. CCV is applicable to other filamentous or tubular structures, in particular it is used for describing electron diffraction by carbon nanotubes [17].

12.3.1 Helix

With the right-handed Cartesian coordinate system, points $\boldsymbol{x} = x^i \boldsymbol{e}_i$ of a right-handed helix of radius r wound around the $\boldsymbol{e}_3$ axis have coordinates

$$x^1(\varphi) = r \cos\varphi \,, \quad x^2(\varphi) = r \sin\varphi \,, \quad x^3(\varphi) = P\,\varphi/(2\pi) \,,$$

where φ is the curve parameter and P (> 0) is the axial spacing (pitch) of the helix, i.e., the axial displacement after one full turn along the helix. In cylindrical coordinates (r, φ, z), the helix is given simply by $r =$ const and $z = P\,\varphi/(2\pi)$. The mathematical helix is invariant with respect to screw displacement: simultaneous right-hand rotation about the axis of the helix by an arbitrary angle φ and translation along the axis by $P\,\varphi/(2\pi)$. It is also invariant of translation by multiples of P along the axis.

12.3.2 Helical Structure

A helical structure is a set of atoms decorating a helix in periodic way, i.e., the structure is invariant with respect to simultaneous translation along the axis by a certain step and rotation about the axis of the helix by a certain angle. An ideal helical structure is infinite in the direction of its axis. Although helices are chiral, helical structures can be achiral.

The principal symmetry of a helical structure is the aforementioned screw displacement: the structure is invariant with respect to simultaneous translation along the axis by the step p and rotation about the axis of the helix by the angle $\varphi_1 = 2\pi\ p/P$. Hence, the electron density $\rho(\boldsymbol{x})$ satisfies the relationship

$$\rho(\boldsymbol{x}) = \rho(R_{\boldsymbol{e}_3}(\varphi_k)\boldsymbol{x} + kp\,\boldsymbol{e}_3)\ , \tag{12.4}$$

where $\varphi_k = k\varphi_1 = 2\pi\ kp/P$ and $R_{\boldsymbol{e}_3}(\varphi_k)$ is the rotation by φ_k about $\boldsymbol{e}_3$. With the first structural unit at the node $\boldsymbol{x}_0$, the other symmetrically equivalent units are at

$$\boldsymbol{x}_k = R_{\boldsymbol{e}_3}(\varphi_k)\boldsymbol{x}_0 + kp\,\boldsymbol{e}_3\ .$$

Let individual atoms be at the points $\boldsymbol{x}_{k,j}$, where k enumerates nodes on the helix (and corresponding structural units or motifs repeating along the helix), and j enumerates atoms of the structural unit ascribed the k-th node. In the independent atom approximation, the density function can be expressed as the sum of densities ρ_j of individual atoms

$$\rho(\boldsymbol{x}) = \sum_{k=-\infty}^{\infty} \sum_{j} \rho_j(R_{\boldsymbol{e}_3}(-\varphi_k)(\boldsymbol{x} - \boldsymbol{x}_{k,j}))\ . \tag{12.5}$$

Using $\boldsymbol{x}_{k+k',j} = R_{\boldsymbol{e}_3}(\varphi_{k'})\boldsymbol{x}_{k,j} + k'p\,\boldsymbol{e}_3$, one can show that the function defined by (12.5) has the principal symmetry (12.4).

12.3.3 Structure Factor

The parameters of a helical structure (in particular, P and p) are to be determined from the geometry of a diffraction pattern. The general relationship between the electron density $\rho(\boldsymbol{x})$ and the diffraction pattern is applicable. With $\boldsymbol{k}_0$ being the wave vector of incident beam, the intensity of the scattered beam with the wave vector $\boldsymbol{k} = \boldsymbol{k}_0 + \boldsymbol{\xi}$ is $|F|^2$, where $F(\boldsymbol{\xi}) = \mathcal{F}_{\boldsymbol{x}}[\rho(\boldsymbol{x})](\boldsymbol{\xi})$.

Based on (12.5), the structure factor $F(\boldsymbol{\xi}) = \mathcal{F}_{\boldsymbol{x}}[\rho(\boldsymbol{x})](\boldsymbol{\xi}) = \int \rho(\boldsymbol{x}) \exp(-2\pi i\boldsymbol{\xi} \cdot \boldsymbol{x})\, \mathrm{d}_3\boldsymbol{x}$ can be expressed as

$$F(\boldsymbol{\xi}) = \int \sum_{k,\,j} \rho_j(R_{\boldsymbol{e}_3}(-\varphi_k)(\boldsymbol{x} - \boldsymbol{x}_{k,j}))\, \exp(-2\pi i\boldsymbol{\xi} \cdot \boldsymbol{x})\, \mathrm{d}_3\boldsymbol{x}\ .$$

Using the substitution $\boldsymbol{x} = R(\varphi_k)\boldsymbol{y} + \boldsymbol{x}_{k,j}$, one obtains

$$\begin{aligned} F(\boldsymbol{\xi}) &= \sum_{k,\,j} \int \rho_j(\boldsymbol{y}) \exp(-2\pi i\boldsymbol{\xi} \cdot (R(\varphi_k)\boldsymbol{y} + \boldsymbol{x}_{k,j}))\, \mathrm{d}_3\boldsymbol{y} \\ &= \sum_{k,\,j} f_j(R_{\boldsymbol{e}_3}(-\varphi_k)\boldsymbol{\xi})\, \exp(-2\pi i\boldsymbol{\xi} \cdot \boldsymbol{x}_{k,j})\ , \end{aligned}$$

where $f_j(\boldsymbol{\xi}) = \int \rho_j(\boldsymbol{y}) \exp(-2\pi i \boldsymbol{\xi} \cdot \boldsymbol{y})\, d_3\boldsymbol{y}$ is the scattering factor of the j-th atom. Assuming spherical symmetry of ρ_j, f_j depends only on the magnitude $|\boldsymbol{\xi}|$. With an abuse of notation, one has $f_j(R(\varphi)\boldsymbol{\xi}) = f_j(|\boldsymbol{\xi}|) = f_j(\boldsymbol{\xi})$, and the structure factor takes the simple form

$$F(\boldsymbol{\xi}) = \sum_j f_j(|\boldsymbol{\xi}|) \sum_k \exp(-2\pi i \boldsymbol{\xi} \cdot \boldsymbol{x}_{k,j}) .$$

This expression is valid for arbitrary helical structures.

The positions $\boldsymbol{x}_{k,j}$ of atoms can be expressed as $\boldsymbol{x}_{k,j} = r_j \cos\phi_{k,j}\, \boldsymbol{e}_1 + r_j \sin\phi_{k,j}\, \boldsymbol{e}_2 + z_{k,j}\boldsymbol{e}_3$, where r_j is the distance of the j-th atom from the axis,

$$\phi_{k,j} = \varphi_k + \phi_{0,j} = 2\pi kp/P + \phi_{0,j} \ \text{ and } \ z_{k,j} = kp + z_{0,j} \tag{12.6}$$

Figure 12.1, let the cylindrical coordinates of $\boldsymbol{\xi}$ be R, ψ and $\xi_3 = \zeta$, i.e., $\boldsymbol{\xi} = R\cos\psi\, \boldsymbol{e}^1 + R\sin\psi\, \boldsymbol{e}^2 + \zeta\, \boldsymbol{e}^3$. With this notation, one has

$$\boldsymbol{\xi} \cdot \boldsymbol{x}_{k,j} = Rr_j \cos(\psi - \phi_{k,j}) + \zeta z_{k,j} ,$$

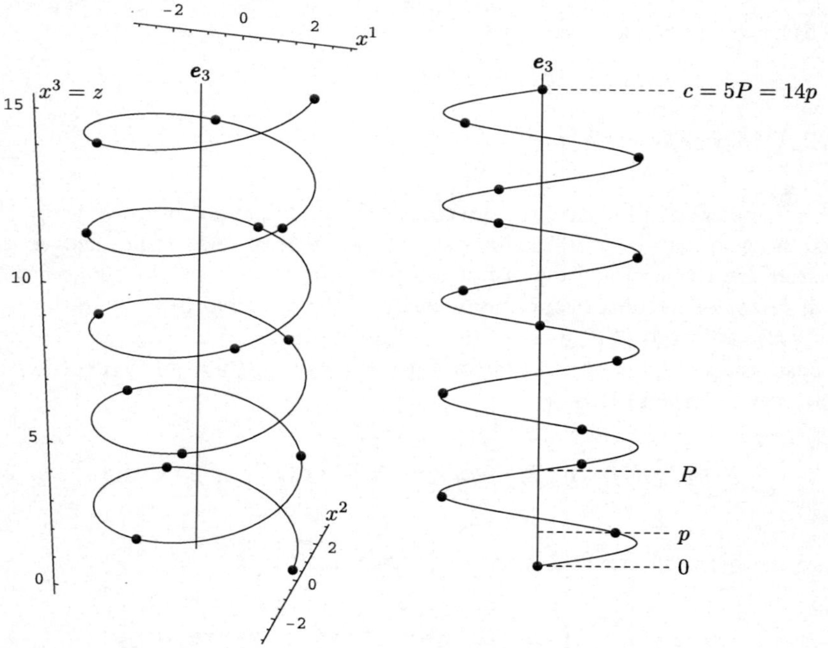

Fig. 12.1 Example helix with point scatterers distributed at equal intervals and its projection along $\boldsymbol{e}_1$. The parameters are $r = 3$, $\phi_0 = 0$, $P = 15/5 = 3$, $p = 15/14$. The structure is periodic along $\boldsymbol{e}_3$ with the period $c = 15$.

and the (conjugate) structure factor is given by

$$F^*(\boldsymbol{\xi}) = \sum_j f_j^*(|\boldsymbol{\xi}|) \sum_{k=-\infty}^{\infty} \exp(2\pi i R r_j \cos(\psi - \phi_{k,j})) \exp(2\pi i \zeta z_{k,j}) .$$

This expression is simplified using the Jacobi-Anger expansion (1.65). With J_n denoting the n-th Bessel function of the first kind, one has

$$F^*(\boldsymbol{\xi}) = \sum_j f_j^*(|\boldsymbol{\xi}|) \sum_{k=-\infty}^{\infty} \exp(2\pi i \zeta z_{k,j}) \sum_{n=-\infty}^{\infty} i^n J_n(2\pi R r_j) \exp(in(\psi - \phi_{k,j})) .$$

By substituting $z_{k,j}$ and $\phi_{k,j}$ with (12.6), one obtains

$$F^*(\boldsymbol{\xi}) = \sum_j f_j^*(|\boldsymbol{\xi}|) \sum_{n=-\infty}^{\infty} i^n J_n(2\pi R r_j) \exp(in(\psi - \phi_{0,j})) \exp(2\pi i \zeta z_{0,j}) \,\text{III}(\zeta p - np/P) ,$$

where the enumerated by n Dirac combs $\text{III}(\zeta p - np/P)$ stand for the sums $\sum_k \exp(2\pi i k(\zeta p - np/P))$ based on the identity (1.80). With this formula, one has

$$\text{III}(\zeta p - np/P) = \sum_{m=-\infty}^{\infty} \delta(\zeta p - np/P - m) = \frac{1}{p} \sum_{m=-\infty}^{\infty} \delta(\zeta - n/P - m/p) . \tag{12.7}$$

Thus, the principal symmetry of a helical structure is reflected in discrete character of diffraction pattern in the $\boldsymbol{e}^3$ direction. With the constant $1/p$ in (12.7) ignored, the (conjugate) structure factor at

$$\zeta = \frac{n}{P} + \frac{m}{p} \tag{12.8}$$

is

$$F^*(\boldsymbol{\xi}) = \sum_{(n)=-\infty}^{\infty} \exp(in(\psi + \pi/2)) \sum_j J_n(2\pi R r_j)\, f_j^*(|\boldsymbol{\xi}|) \exp(2\pi i \zeta z_{0,j} - in\phi_{0,j}) , \tag{12.9}$$

and it is zero everywhere else. The part of the pattern for fixed ζ given by (12.8) is known as a layer line.

12.3.4 Selection Rule

If P/p is rational, the structure is periodic along its axis, i.e., repeats after some distance c along the axis and the density ρ satisfies the relationship $\rho(\boldsymbol{x} + kc\,\boldsymbol{e}_3) = \rho(\boldsymbol{x})$, where k is an integer (Fig. 12.1). E.g., in ideal single-wall carbon tubes, the ratio P/p is rational, i.e., these structures are periodic along their axes. In general, if not exactly, the periodicity can be achieved with an arbitrarily good approximation, but this may require large c.

The period c is the smallest common multiple of P and p, i.e.,

$$c = u_P P \quad \text{and}\; c = u_p p\,,$$

with co-prime integers u_P and u_p. Hence, based on (12.8),

$$\zeta = \frac{n}{P} + \frac{m}{p} = \frac{nu_P + mu_p}{c} = \frac{l}{c}\,,$$

where l takes arbitrary integer values. The layer lines in the diffraction pattern are numbered by l with the line at $\zeta = 0$ corresponding to $l = 0$. The relationship

$$\frac{l}{c} = \frac{n}{P} + \frac{m}{p} \qquad (\text{or } l = nu_P + mu_p) \tag{12.10}$$

is known as the *selection rule*. It reflects the link between the screw-displacement symmetry of the structure and its translational symmetry.

The selection rule determines which terms of the series (12.9) contribute to a given layer line. For a fixed layer line l, the series consists only of terms $i^n J_n e^{in\psi} T_n$ with n such that $lp/c - np/P = (l - nu_P)/u_p$ is an integer. To emphasize this rule, the index n under summation sign in (12.9) is put in parentheses.

12.3.5 Single-Wall Tubes

To explain diffraction patterns it is convenient to limit further considerations to tubular structures with all atoms at the same distance r from the axis. (A real example of such structure is a single-wall carbon nanotube.) Under this assumption, the structure factor is

$$F^*(\boldsymbol{\xi}) = \sum_{(n)=-\infty}^{\infty} J_n(2\pi Rr)\, \exp(in(\psi + \pi/2))\, T_n(\boldsymbol{\xi})\,, \tag{12.11}$$

where

$$T_n(\boldsymbol{\xi}) = \sum_j f_j^*(|\boldsymbol{\xi}|) \exp(2\pi i \zeta z_{0,j} - in\phi_{0,j}) , \tag{12.12}$$

i.e., it is given by the series with the n-term being a product of the Bessel function $J_n(2\pi Rr)$, the phase factor $\exp(in(\psi + \pi/2))$ and the $\boldsymbol{\xi}$-dependent sum over atoms of the structural unit denoted by T_n.

Typical fiber diffraction patterns originate from samples being aggregates of helical structures with parallel axes. In some cases (e.g., with two, three, four or six nodes per helical turn), the structures may form a crystal, and peaks on the pattern correspond to discrete values of ψ and nodes of a three-dimensional lattice. On the other hand, when the orientations in the plane normal to the axes are not correlated, one needs to average the intensities over rotations about z, i.e., to take the average of $|F(\boldsymbol{\xi})|^2 = F(\boldsymbol{\xi})F^*(\boldsymbol{\xi})$ over ψ. With $F(\boldsymbol{\xi})$ given by (12.11), one has

$$\int_0^{2\pi} |F|^2 \frac{d\psi}{2\pi} = \int_0^{2\pi} \sum_{(n),\,(n')} J_n(2\pi Rr) J_{n'}(2\pi Rr) \exp(i(n'-n)(\psi + \pi/2))\, T_n(\boldsymbol{\xi}) T_{n'}^*(\boldsymbol{\xi}) \frac{d\psi}{2\pi} .$$

Since $T_n(\boldsymbol{\xi})$ is axially symmetric, i.e., independent of ψ and $\int_0^{2\pi} \exp(i(n'-n)(\psi + \pi/2))d\psi = 2\pi\delta_{nn'}$, the expression for the intensity has the form

$$\sum_{(n),\,(n')} \delta_{nn'} J_n(2\pi Rr)\, T_n(\boldsymbol{\xi}) J_{n'}(2\pi Rr)\, T_{n'}^*(\boldsymbol{\xi}) = \sum_{(n)=-\infty}^{\infty} J_n^2(2\pi Rr)\, |T_n(\boldsymbol{\xi})|^2 . \tag{12.13}$$

In the simplistic artificial case with structural unit consisting of a single atom, one has $|T_n(\boldsymbol{\xi})|^2 = |f_1(|\boldsymbol{\xi}|)|^2$, and the intensity is determined by $I \propto \sum_{(n)} J_n^2(2\pi Rr)$. See Fig. 12.2. For a formula corresponding to (12.13) but applicable in the case of helical structures with multiple radii r_j, see [18].

12.3.6 Intensities in Layer Lines

In the standard arrangement, a diffraction pattern is recorded with the incident beam direction perpendicular to axes of helices and to the detector. For convenience, the coordinate frame is taken with $\boldsymbol{e}_1$ along these vectors, i.e., $\widehat{\boldsymbol{k}}_0 = \boldsymbol{e}_1 = \widehat{\boldsymbol{L}}$. On a diffraction pattern, the line parallel (perpendicular) to the axes and passing through the incident beam direction is referred to as the meridian (the equator). By convention, the meridian is taken to be vertical.

The distribution of intensities in a layer line is determined by the character of the Bessel J-functions. The key point for the distribution is that $J_n(x)$ it is nearly zero for $|x|$ appreciably smaller than $|n|$; see Fig. 12.3. In effect, the peaks closest to the axis are linked to the terms with the smallest $|n|$.

Fig. 12.2 Simulated helical diffraction pattern on a flat detector. It is based on the formula $I \propto \sum_{(n)} J_n^2(2\pi Rr)$ and the structure shown in Fig. 12.1. The wave vector of the incident beam was taken to be $\boldsymbol{k}_0 = 1\,\boldsymbol{e}_1$, and also $\boldsymbol{L}$ was directed along $\boldsymbol{e}_1$.

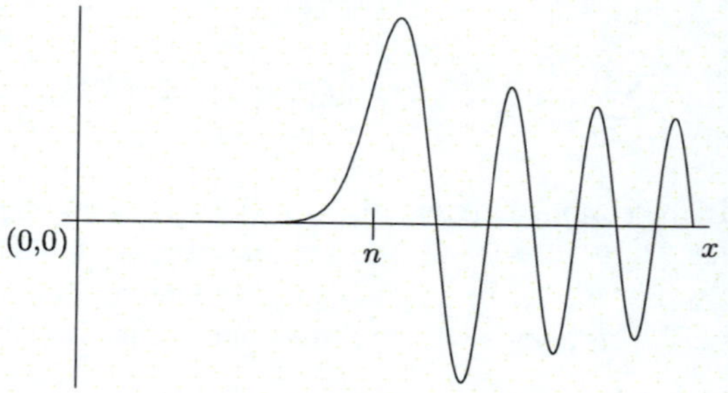

Fig. 12.3 Example Bessel function J_n for non-negative arguments. For small even (odd) $|n| \geq 1$, the maximum (minimum) of the function is at the argument close to $|n| + 1$, and it is at the argument close to $|n| + 2$ when $|n|$ is near 15.

With (12.9) and (12.11), the distance of a point on a layer line from the meridian is determined by R and ψ. A major contribution to a layer line is made by only a single (lowest $|n|$) Bessel function. Higher-order terms occur at larger distances from the meridian and frequently can be ignored. In such cases, with a single term, intensities along the layer line do not depend on ψ; otherwise, the phases of different terms affect (via $e^{in\psi}$) the intensities along the layer line. To give an example, for the first layer line ($l = 1$) of the case with nodes shown in Fig. 12.1, the leading terms allowed by the selection rule $1 = l = 5n + 14m$ are $n = 3, -11, 17, \ldots$; they correspond to $m = -1, 4, -6, \ldots$, respectively. The peak closest to the axis arises due to the term $n = 3$. The impact of the term $n = -11$ is affected by the tail of J_3^2 and the term $-2J_3J_{-11}\cos(14\psi)$ due to the phase factor $e^{in\psi}$ in (12.11). The leading

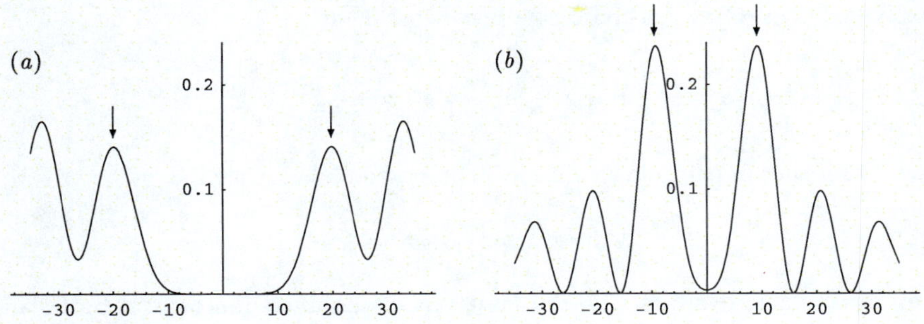

Fig. 12.4 Intensities (in arbitrary units) along layer lines $l = 3$ (*a*) and $l = 4$ (*b*) of the pattern shown in Fig. 12.2. The numbers on the abscissa represent the azimuth angle (in degrees) with respect to the meridian. The leading term for the third layer line has the index $n = -5$, so the nearest-to-meridian peaks (indicated by arrows) are separated more than in the case of fourth layer line with the leading term having the index $n = -2$.

terms for other layer lines given in the form (n, l) are $(0, 0)$, $(6, 2)$, $(-5, 3)$, $(-2, 4)$, $(1, 5)$, $(4, 6)$, $(7, 7)$ et cetera.

The interpretation of patterns is simpler when intensities are averaged over ψ. In this case, the distance of a peak from the meridian is determined only by R. Example intensity profiles are shown in Fig. 12.4.

12.3.7 Indices of Helical Reflections

The pairs (n, l) are conventionally used as indices of reflections in patterns of helical structures [19]. This is justified by an analogy to reflection indices in crystal diffraction: the factor $\exp(2\pi i \zeta z_{0,j} - in\phi_{0,j})$ in (12.9) and (12.12) can be expressed in the form $\exp(2\pi i \,\check{\boldsymbol{\xi}} \cdot \boldsymbol{y}_j)$, where $\check{\boldsymbol{\xi}}$ is a vector of a two dimensional lattice. The lattice is referred to as helical net. With the helical net basis given by

$$\boldsymbol{a}_1 = -a\,\boldsymbol{e}_2 \ \text{ and } \ \boldsymbol{a}_2 = \frac{a\,p}{P}\,\boldsymbol{e}_2 + p\,\boldsymbol{e}_3 \,, \tag{12.14}$$

where $a = 2\pi r$, the reciprocal net is based on the vectors

$$\boldsymbol{a}^1 = -\frac{1}{a}\boldsymbol{e}^2 + \frac{1}{P}\boldsymbol{e}^3 \ \text{ and } \ \boldsymbol{a}^2 = \frac{1}{p}\boldsymbol{e}^3 \,. \tag{12.15}$$

The vector of the reciprocal net

$$\check{\boldsymbol{\xi}} = n\boldsymbol{a}^1 + m\boldsymbol{a}^2$$

expressed in the Cartesian coordinate system has the form

$$\check{\boldsymbol{\xi}} = -\frac{n}{a}\boldsymbol{e}^2 + \left(\frac{n}{P} + \frac{m}{p}\right)\boldsymbol{e}^3 = -\frac{n}{a}\boldsymbol{e}^2 + \frac{l}{c}\boldsymbol{e}^3 .$$

Let coordinates of points in an abstract helix-based 'direct space' be

$$y^2 = \phi r \text{ and } y^3 = z = x^3 ,$$

i.e., the ordinate overlaps with the third axis of the three-dimensional Cartesian system and the axis of the helix, and the ϕ angle scaled by r is on the abscissa. For the point

$$\boldsymbol{y}_{k,j} = y^2_{k,j}\,\boldsymbol{e}_2 + y^3_{k,j}\,\boldsymbol{e}_3 = \phi_{k,j} r\,\boldsymbol{e}_2 + z_{k,j}\,\boldsymbol{e}_3$$

having coordinates $\phi_{k,j}$ and $z_{k,j}$ given by (12.6), the expression

$$\exp(2\pi i\,\check{\boldsymbol{\xi}}\cdot\boldsymbol{y}_{k,j}) = \exp(2\pi i k m)\exp(2\pi i l z_{0,j}/c - i n\phi_{0,j}) = \exp(2\pi i\,\check{\boldsymbol{\xi}}\cdot\boldsymbol{y}_{0,j})$$

is independent of k enumerating the nodes on the helix. With $\boldsymbol{y}_j = \boldsymbol{y}_{0,j}$, the sum $T_n(\boldsymbol{\xi})$ in (12.12) at $\boldsymbol{\xi} = \check{\boldsymbol{\xi}}$ can be written in the form

$$T_n(\check{\boldsymbol{\xi}}) = \sum_j f_j^*(|\check{\boldsymbol{\xi}}|)\exp(2\pi i\,\check{\boldsymbol{\xi}}\cdot\boldsymbol{y}_j) .$$

Thus, as in the case of crystal diffraction, reflections correspond to net vectors $\check{\boldsymbol{\xi}} = n\boldsymbol{a}^1 + m\boldsymbol{a}^2$ identified by arbitrary integer indices (n, m). Since the interpretation of l is simpler than that of m, reflections are usually identified by the indices (n, l) with l satisfying the selection rule (12.10).

Transforming a helix to flat two-dimensional space with coordinates $y^2 = \varphi r$ and $y^3 = z = x^3$ is like cutting the cylindrical surface coaxial with the helix along a line parallel to the axis at $\phi = \phi_0$ (i.e., at $\varphi = 0$) and unfolding it.[4] Figure 12.5 shows the net and basis vectors $\boldsymbol{a}_1$ and $\boldsymbol{a}_2$ for the helix of Fig. 12.1. The net reciprocal to the direct net is shown in Fig. 12.6. The symmetrized (reflected with respect to the vertical axis through net node) figure is directly related to the diffraction pattern. This is illustrated in Fig. 12.7. The near-meridian peaks on the pattern are distributed in a manner corresponding to the distribution of the near-meridian spots of the symmetrized reciprocal net.

---

[4] Projection of a helical structure on that surface along lines intersecting the axis at the right angle is called "radial projection" [19].

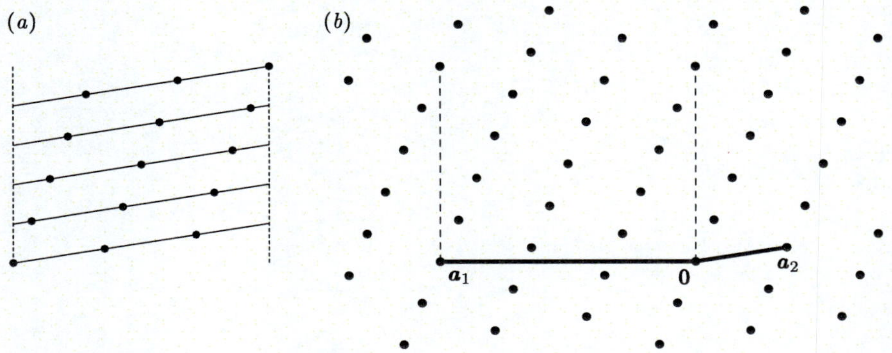

Fig. 12.5 (*a*) Unfolded (and reduced in size) cylindrical surface with the helix of Fig. 12.1 and points forming part of the helical net. (*b*) A larger part of the direct helical net with basis vectors $\boldsymbol{a}_1$ and $\boldsymbol{a}_2$.

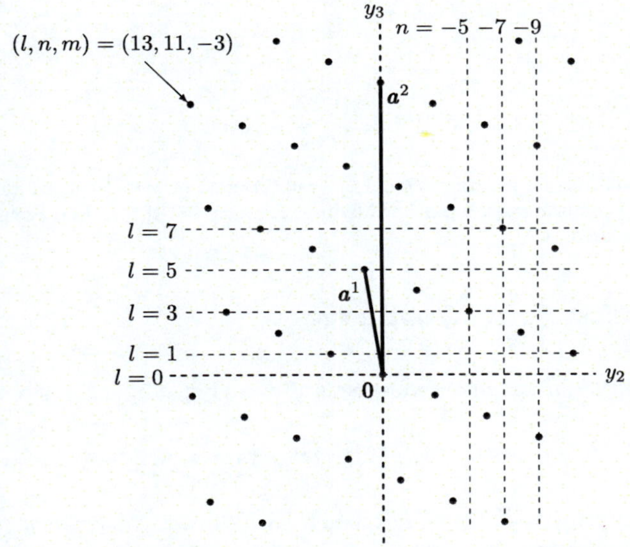

Fig. 12.6 Net reciprocal to the helix net of Fig. 12.5*b*. The net nodes can be identified by pairs (n, l) satisfying the selection rule (12.10), or simply by integers (n, m). In the latter case, the node (n, m) is at $n\boldsymbol{a}^1 + m\boldsymbol{a}^2$. In the Cartesian coordinate system, the nodes are at $(y_2, y_3) = (-n/a, l/c)$ with l satisfying the selection rule.

Clearly, the pair of bases (12.14) and (12.15) can be replaced by other bases [20]. For instance, one might consider reduced bases, but then one would need extra formulae for layer-line number l and leading term number n.

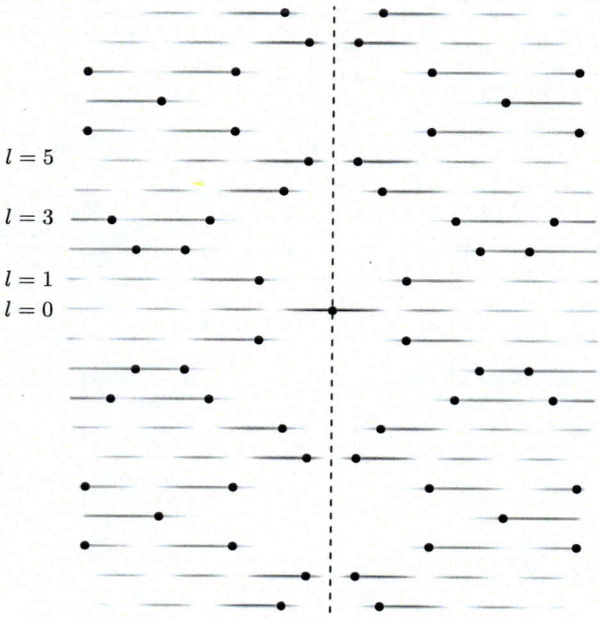

Fig. 12.7 The diffraction pattern of Fig. 12.1 (with inverted gray level) transformed from flat detector to the reciprocal space overlapped by the reciprocal net of Fig. 12.6 symmetrized with respect to the meridian.

12.3.8 *Indices (l, n, m) and a Frame*

All three indices l, n and m are most naturally combined by using the direct space frame[5]

$$\boldsymbol{a}_1 = -a\,\boldsymbol{e}_2\,, \quad \boldsymbol{a}_2 = \frac{ap}{P}\,\boldsymbol{e}_2 + p\,\boldsymbol{e}_3 \text{ and } \boldsymbol{a}_3 = c\,\boldsymbol{e}_3\,,$$

i.e., the helical net basis (12.14) is appended with linearly dependent vector $\boldsymbol{a}_3 = u_P\,\boldsymbol{a}_1 + u_p\,\boldsymbol{a}_2$. The vectors $\boldsymbol{a}_1$ and $\boldsymbol{a}_3$ are orthogonal, and they can be easily identified on the unfolded cylindrical surface. The frame $\boldsymbol{a}^\mu$ dual to $\boldsymbol{a}_\mu$ ($\mu = 1, 2, 3$) is given by

$$\boldsymbol{a}^1 = -\frac{\alpha}{a}\left(\frac{1}{c^2} + \frac{1}{p^2}\right)\boldsymbol{e}^2 + \frac{\alpha}{Pc^2}\boldsymbol{e}^3,\ \boldsymbol{a}^2 = \frac{\alpha}{aPp}\boldsymbol{e}^2 + \frac{\alpha}{c^2 p}\boldsymbol{e}^3,\ \boldsymbol{a}^3 = -\frac{\alpha}{aPc}\boldsymbol{e}^2 + \frac{\alpha}{c}\left(\frac{1}{P^2} + \frac{1}{p^2}\right)\boldsymbol{e}^3,$$

[5] For an account on frames, see Sect. 1.5.2.

where $\alpha = \delta_{\mu\nu} v^\mu v^\nu$ and $(v^1,\ v^2,\ v^3) = (1/P,\ \ 1/p,\ \ -1/c)$. The vectors of the dual frame satisfy the relationship $\boldsymbol{a}^3 = u_P\, \boldsymbol{a}^1 + u_p\, \boldsymbol{a}^2$. The products $\boldsymbol{a}_\mu \cdot \boldsymbol{a}^\nu = g_\mu^{\ \nu}$ are given by

$$g_\mu^{\ \nu} = \delta_\mu^{\ \nu} - \delta_{\mu\kappa} v^\kappa v^\nu \ ,$$

and the conditions (1.26) have the form of the selection rule (12.10). Each integer combination of the vectors $\boldsymbol{a}_\mu$ ($\mu = 1, 2, 3$) is a node of the direct helix net. Nodes of the reciprocal net are combinations

$$n\boldsymbol{a}^1 + m\boldsymbol{a}^2 + l\boldsymbol{a}^3 = l_\mu \boldsymbol{a}^\mu$$

with integer coefficients $(n, m, l) = (l_1, l_2, l_3)$ satisfying the selection rule (12.10) or $g_\mu^{\ \nu} l_\nu = l_\mu$.

12.3.9 Helical Structures Spanning a Range of Radial Values

Real helical structures range from filaments—thread-like elongated assemblies of molecules, to tubular (cylindrical) structures with a hollow center. In general, they can be seen as collections of coaxial helices of different radii but the same spacing p. The same spacing implies (exact or approximate) periodicity along the axis and validity of the selection rule (12.10). The diffraction pattern consists of layer lines, but because of different radii, instead of (12.11), structure factors are given by the more general formula (12.9).

12.3.10 Procedures for Indexing Helical Diffraction Patterns

As in the case of crystals, indexing is an early stage of reconstruction of three-dimensional helical structure from diffraction patterns [21, 22]. Formally, having positions of local maxima on particular layer lines of a pattern, the point is to determine the basis vectors $\boldsymbol{a}^1$ and $\boldsymbol{a}^2$ of (12.15) such that the corresponding helical reciprocal net best fits the positions of the maxima. Then, knowing $\boldsymbol{a}^k$ and hence $\boldsymbol{a}_k$ ($k = 1, 2$), the relationships (12.14) determine the helix parameters: $a = 2\pi r = (\boldsymbol{a}_1 \cdot \boldsymbol{a}_1)^{1/2} = -\boldsymbol{a}_1 \cdot \boldsymbol{e}_2$, the step $p = \boldsymbol{a}_2 \cdot \boldsymbol{e}_3$, the pitch $P = ap/(\boldsymbol{a}_2 \cdot \boldsymbol{e}_2)$, and the period c as the shortest multiple of p and P.

Indexing of patterns originating from helical structures can be seen as the assignment of the Bessel order to intensity maxima on particular layer lines. With a diffraction pattern showing sharp layer lines, the layer-line numbers can be easily assigned, and the key problem is to get the radial positions of diffraction peaks. The positions are affected by the spread caused by the Bessel functions, the phase factors, and the fact that usually structural units span a range of radii. The initial step of indexing is to get the position of the first (closest to the meridian) maximum on each layer line. One should start with examining the layer lines having first peaks closest to the meridian. For each layer line, one determines n from the position of the peak. Ascribing n to first maxima distant from the axis (i.e., those with large $|n|$) is usually ambiguous but the problem can be resolved using higher-order layer lines with maxima close to the axis. Indexing is relatively easy when atoms are at a single radius r. It is more difficult when there are a number of radii; with a range of radii, there is a range of possible peaks along the layer line. When ambiguities in indexing arise, one needs to test a number of alternative values of n. Typically, indexing is performed using dedicated software; see, e.g., [20, 23, 24].

In practice, real helical structures are rarely perfectly straight, and the bending affects the geometry of patterns. The peak positions on layer lines can also be shifted due to a tilt of helical axis with respect to the detector and incident beam direction. In such cases, the larger the layer number $|l|$ the larger the shift, i.e., least affected and most reliable are positions of peaks in layers closest to the equator. Moreover, diffraction patterns can be affected by interferences between neighboring helices. With complicated helical structures, e.g., proteins, the low signal-to-noise ratio and low accuracy of peak positions make ascribing the right values of n to a peak difficult, and alternative more sophisticated indexing schemes are needed; e.g., [25].

Solving helical structures via Fourier methods is not always easy. As was noted above, even indexing can be ambiguous. Therefore other methods are also used. An alternative to X-ray methods are real-space reconstruction methods relying on processing of electron microscopy images of helical structures and methods combining real-space reconstruction with Fourier-based analysis [26–28]; cf. footnote 2.

References

1. T. Wieder, A generalized Debye scattering formula and the Hankel transform. Z. Naturforsch. **54a**, 124–130 (1999)
2. A. Guinier, *X-Ray Diffraction (Imperfect Crystals, and Amorphous Bodies* (Freeman, San Francisco, In Crystals, 1963)
3. O. Glatter, O. Kratky, *Small-Angle X-ray Scattering* (Academic Press, London, 1982)
4. L.H. Germer, A.H. White, Electron diffraction studies of thin films. II. Anomalous powder patterns produced by small crystals. Phys. Rev. **60**, 447–455 (1941)
5. L. Gelisio, C.L.A. Ricardo, M. Leoni, P. Scardi, Real-space calculation of powder diffraction patterns on graphics processing units. J. Appl. Cryst. **43**, 647–653 (2010)
6. R.W. Gerchberg, W.O. Saxton, A practical algorithm for the determination of the phase from image and diffraction plane pictures. Optik **35**, 237–246 (1972)
7. J.R. Fienup, Reconstruction of an object from the modulus of its Fourier transform. Opt. Lett. **3**, 27–29 (1978)
8. J.R. Fienup, Phase retrieval algorithms: a comparison. Appl. Opt. **21**, 2758–2769 (1982)
9. S. Marchesini, H. He, H.N. Chapman, S.P. Hau-Riege, A. Noy, M.R. Howells, U. Weierstall, J.C.H. Spence, X-ray image reconstruction from a diffraction pattern alone. Phys. Rev. B **68**, 140101 (2003)
10. J. Miao, P. Charalambous, J. Kirz, D. Sayre, Extending the methodology of X-ray crystallography to allow imaging of micrometre-sized non-crystalline specimens. Nature **400**, 342–344 (1999)

11. N.T.D. Loh, V. Elser, A reconstruction algorithm for single-particle diffraction imaging experiments. Phys. Rev. E. **80**, 026705 (2009)
12. M. Tegze, G. Bortel, Atomic structure of a single large biomolecule from diffraction patterns of random orientations. J. Struct. Biol. **179**, 41–45 (2012)
13. S. Kassemeyer, A. Jafarpour, L. Lomb, J. Steinbrener, A.V. Martin, I. Schlichting, Optimal mapping of X-ray laser diffraction patterns into three dimensions using routing algorithms. Phys. Rev. E **88**, 042710 (2013)
14. T. Kimura, Y. Joti, Shibuya, A., C. Song, S. Kim, K. Tono, M. Yabashi, M. Tamakoshi, T. Moriya, T. Oshima, T. Ishikawa, Y. Bessho, Y. Nishino, Imaging live cell in micro-liquid enclosure by X-ray laser diffraction. Nat. Commun. **5**, 3052 (2014)
15. A. Kobayashi, Y. Takayama, T. Hirakawa, K. Okajima, M. Oide, T. Oroguchi, Y. Inui, M. Yamamoto, S. Matsunaga, M. Nakasako, Common architectures in cyanobacteria Prochlorococcus cells visualized by X-ray diffraction imaging using X-ray free electron laser. Sci. Rep. **11**, 3877 (2021)
16. W. Cochran, F.H.C. Crick, V. Vand, The structure of synthetic polypeptides. I. The transform of atoms on a helix. *Acta Cryst.* **5**, 581–586 (1952)
17. S. Lee, P.C. Doerschuk, J.E. Johnson, Electron diffraction from cylindrical nanotubes. J. Mater. Res. **9**, 2450–2456 (1994)
18. R.E. Franklin, A. Klug, The splitting of layer lines in X-ray fibre diagrams of helical structures: Application to tobacco mosaic virus. Acta Cryst. **8**, 777–780 (1955)
19. A. Klug, F.H.C. Crick, H.W. Wyckoff, Diffraction by helical structures. Acta Cryst. **11**, 199–212 (1958)
20. C. Toyoshima. Structure determination of tubular crystals of membrane proteins. 1. Indexing of diffraction patterns. Ultramicroscopy **84**, 1–14 (2000)
21. D.J. DeRosier, P.B. Moore, Reconstruction of three-dimensional images from electron micrographs of structures with helical symmetry. J. Mol. Biol. **52**, 355–369 (1970)
22. R. Diaz, W.J. Rice, D.L. Stokes, Fourier-Bessel reconstruction of helical assemblies. Methods Enzymol. **482**, 131–165 (2010)
23. A. Ward, M.F. Moody, B. Sheehan, R.A. Milligan, B. Carragher, *Windex*: a toolset for indexing helices. J. Struct. Biol. **144**, 172–183 (2003)
24. X. Zhang, Python-based *Helix Indexer*: A graphical user interface program for finding symmetry of helical assembly through Fourier-Bessel indexing of electron microscopic data. Protein Sci. **31**, 107–117 (2022)
25. N. Coudray, R. Lasala, Z. Zhang, K.M. Clark, M.E. Dumont, D.L. Stokes, Deducing the symmetry of helical assemblies: Applications to membrane protein. J. Struct. Biol. **195**, 167–178 (2016)
26. E.H. Egelman, The iterative helical real space reconstruction method: Surmounting the problems posed by real polymers. J. Struct. Biol. **157**, 83–94 (2007)
27. A. Desfosses, R. Ciuffa, I. Gutsche, C. Sachse, *SPRING* - an image processing package for single-particle based helical reconstruction from electron cryomicrographs. J. Struct. Biol. **185**, 15–26 (2014)
28. E.H. Egelman, Three-dimensional reconstruction of helical polymers. Arch. Biochem. Biophys. **581**, 54–58 (2015)

Chapter 13
Indexing of Quasicrystal Diffraction Patterns

The conventional crystallography relies on the notion of periodicity. Quasicrystallography allows for aperiodicity; for an aperiodic structure, there is no non-zero translation which would leave it invariant, and there are no arbitrarily long periodic sequences. A quasicrystal is characterized by a kind of repetitiveness, usually referred to as quasiperiodicity, which leads to sharp peaks in its diffraction patterns. To account for such structures, a crystal is defined as "any solid having an essentially discrete diffraction diagram". This definition clearly incorporates periodic crystals. Like periodic crystals, quasicrystals and their diffraction patterns have well defined discrete point symmetries. However, for a quasicrystal, some of these symmetries are incompatible with periodic translational order. A quasicrystal is a crystal with a point symmetry violating the crystallographic restriction of Sect. 1.3.3. Known quasicrystals and their diffraction patterns show icosahedral, octagonal, decagonal and dodecagonal symmetries (Table 13.1).

Standard crystallographic narrative begins with physical-space lattice which is used to define the reciprocal lattice, and the latter is linked to diffraction peaks. With the new definition of the crystal, the direction in a sense is reversed. One begins with a discrete diffraction pattern, and Bragg peaks of the pattern are associated with a 'quasilattice' in the reciprocal space. It is assumed that the corresponding scattering vectors can be expressed as integer linear combinations of a finite number of vectors. This assumption considerably limits the class of quasicrystals compared to all structures leading to discrete diffraction diagrams, but the class covers cases of physical interest.

Quasiperiodic ordering is usually associated with intermetallic compounds.[1] In practice, research on quasicrystals is a small niche of materials science (because of lack of any important applications), but quasicrystals are an interesting theoretical subject. The theory of quasicrystals involves incommensurate parameters, i.e.,

[1] It is also observed in mesoscopic systems known as *soft quasicrystals* in which quasicrystalline arrangements are built of molecular assemblies [1].

A. Morawiec, *Indexing of Crystal Diffraction Patterns*, Springer Series in Materials Science 326, https://doi.org/10.1007/978-3-031-11077-1_13

Table 13.1 Pentagonal, octagonal, decagonal, dodecagonal and icosahedral point groups, i.e., groups of interest in research on quasicrystals. 'Holohedries' are in the line marked by H, hemihedries (groups with only half the number of operations in H) by h. Laue groups are marked by L, and enantiomorphic groups (i.e., groups of the first kind) are marked by e.

	Pentagonal	Octagonal	Decagonal	Dodecagonal	Icosahedral
H, L	D_{5d}	D_{8h}	D_{10h}	D_{12h}	I_h
h, e	D_5	D_8	D_{10}	D_{12}	I
h	C_{5v}	C_{8v}	C_{10v}	C_{12v}	
h		D_{4d}	D_{5h}	D_{6d}	
h, L	S_{10}	C_{8h}	C_{10h}	C_{12h}	
e	C_5	C_8	C_{10}	C_{12}	
		S_8	C_{5h}	S_{12}	

parameters with irrational ratios. Clearly, no experiment can distinguish between rational and irrational numbers. From the experimental viewpoint, the borderline between periodic crystals on one side and quasicrystals on the other is not sharp; intermediate states range from defects, mosaicity, particular types of twinning, composite crystals, incommensurately modulated structures, to periodic approximants of quasicrystals.

The problem of quasicrystal structure determination is much more complicated than in the case of periodic crystals. Standard techniques are used to get quasicrystal diffraction patterns, from selected area electron diffraction, through X-ray single-crystal and powder diffraction methods, to electron Kikuchi diffraction and EBSD.

With a periodic crystal lattice, one can select a single primitive cell. Quasicrystal is a repetition of two or more structural units packed in space.[2] Differently than in periodic crystals, the units are *not* repeated periodically. The units are referred to as tiles, and their arrangement is called a tiling. Below, the units are also called segments, tiles and cells in one-, two- and three-dimensional cases, respectively. The three-dimensional analogue of two-dimensional 'tiling' is 'packing'.

Quasicrystal structure determination comprises the determination of metric parameters of the tiles, the arrangement of the tiles, and the ultimate goal is to unravel the (short-range) atomic arrangement, i.e., atomic decoration of the tiles. Even the first step of structure determination, i.e., the indexing of quasicrystal diffraction patterns, is not trivial.

The issue of indexing quasicrystal diffraction patterns is complicated by the so-called scaling 'symmetry'. In classical crystallography, if a pattern is correctly indexable with a given basis of the reciprocal lattice, then it is not indexable or indexing is considered to be incorrect if the basis is scaled by a nontrivial factor. A quasicrystalline diffraction pattern can equally well be indexed in a given frame and in its properly scaled counterpart.

[2] There is also a quasi-unit-cell approach based on a single motif, with the motifs covering the structure permitted to overlap in specific ways; see, e.g., [2].

Additionally, several indexing schemes are used to index patterns of three-dimensional quasicrystals. The indexing scheme depends on the choice of a frame (overcomplete set of vectors) determining positions of peaks in diffraction patterns and its dual which is linked to the space tiling. The choice of the frame corresponds to the choice of the lattice basis in conventional crystallography. In quasicrystallography, the choice of a proper frame is not obvious [3].

In order to explain indexing of quasicrystal diffraction patterns, one needs an introduction to quasicrystallography in general. The subject is only sketched. The text below is limited to primitive (or 'pure-point') quasicrystals with identical point-like atoms at the vertices of cells. This approach is sufficient for presenting indexing schemes applicable to real quasicrystals.

For simplicity, positions in direct space will be given in terms of a unit length a. This implies that positions in reciprocal space are in units of a^{-1}. More generally, all quantities involving length in m-th dimension need to be multiplied by a^m. There are exceptions from this notation in subsections in which determination of a is considered; there is a warning wherever a is explicitly used. In practice, the actual value of the unit a must be sufficiently large in order to allow for a meaningful atomic decoration of tiles, but on the other hand, it should be as small as possible (to have a reasonably small number of atoms in the tiles).

Like many texts on quasicrystallography, this one also begins with a tinker-toy model—a one-dimensional quasicrystal. Numerous features of the one-dimensional case are shared by quasicrystalline structures of higher dimension. The model is helpful in understanding the more demanding formalism needed for describing real quasicrystals.

13.1 Example One-Dimensional Quasicrystal

Our first one-dimensional example is a 'pure-point crystal' built of points with coordinates

$$x_n = n + \frac{1}{q} \left\lfloor \frac{n}{q} \right\rfloor ; \tag{13.1}$$

$q > 1$ is assumed to be a real number and the points are indexed by integer n. Since

$$x_{n+1} - x_n = 1 + \iota/q \ ,$$

where ι equals 0 or 1, the points tessellate the line into a chain of segments of two lengths: 1 and $1 + 1/q$ (Fig. 13.1). If q is rational, i.e., $q = k_1/k_2$ with co-prime integers k_1 and k_2, the segments constitute a periodic structure; the period of length $(k_1^2 + k_2^2)/k_1$ consists of $k_1 - k_2$ short segments and k_2 long segments. If q is irrational, the structure governed by (13.1) is not periodic, but it has a long-range

Fig. 13.1 A fragment of a chain of segments separated by points at x_n given by (13.1).

order as the distance between points separated by m segments takes one of just two values

$$x_{n+m} - x_n = x_m + \iota/q \ .$$

The positions x_n of the nodes are correlated, and the density supported by the nodes has a discrete Fourier transform.

13.1.1 Fourier Transform

To show that the set of points with coordinates x_n is a quasicrystal, one needs to get Fourier transform of the density

$$\rho(x) = Ш(x) = \sum_{n=-\infty}^{\infty} \delta(x - x_n) \ . \tag{13.2}$$

This can be done using the following observation. In two-dimensional Cartesian frame, if one takes the points with integer coordinates $(n, \ \lfloor n/q \rfloor)$ and rotates them by the angle of $\phi = \arctan(1/q)$ about the origin of the coordinate system, one obtains

$$\begin{bmatrix} \cos(\phi) & \sin(\phi) \\ -\sin(\phi) & \cos(\phi) \end{bmatrix} \begin{bmatrix} n \\ \lfloor n/q \rfloor \end{bmatrix} = \cos(\phi) \begin{bmatrix} n + (1/q) \lfloor n/q \rfloor \\ \lfloor n/q \rfloor - n/q \end{bmatrix} = \cos(\phi) \begin{bmatrix} x_n^1 \\ x_n^2 \end{bmatrix} ,$$

where $x_n^1 = x_n$ of (13.1), and $x_n^2 = \lfloor n/q \rfloor - n/q$ satisfies the conditions $-1 < x_n^2 \leq 0$. This can be re-written as

$$\mathcal{T}^T \begin{bmatrix} n \\ \lfloor n/q \rfloor \end{bmatrix} = \begin{bmatrix} x_n^1 \\ x_n^2 \end{bmatrix} ,$$

with the transformation matrix $\mathcal{T}$ of the form

$$\mathcal{T} = [\mathcal{T}_\mu^{\ \nu}] = \frac{1}{\cos(\phi)} \begin{bmatrix} \cos(\phi) & -\sin(\phi) \\ \sin(\phi) & \cos(\phi) \end{bmatrix} = \begin{bmatrix} 1 & -\tan(\phi) \\ \tan(\phi) & 1 \end{bmatrix} = \begin{bmatrix} 1 & -1/q \\ 1/q & 1 \end{bmatrix} . \tag{13.3}$$

Clearly, the points with integer coordinates $(n, \ \lfloor n/q \rfloor)$ are a part of the $\boldsymbol{e}_i$-based square lattice. The transformation of this lattice by $\mathcal{T}$ gives a lattice, say, Λ, based on $\boldsymbol{b}_\mu = \mathcal{T}_\mu^{\ \nu} \boldsymbol{e}_\nu$ or

$$\boldsymbol{b}_1 = \boldsymbol{e}_1 - \boldsymbol{e}_2/q \ , \qquad \boldsymbol{b}_2 = \boldsymbol{e}_1/q + \boldsymbol{e}_2 \ ,$$

with the cell volume $V = \det(\mathcal{T}) = 1 + 1/q^2$ (Fig. 13.2).

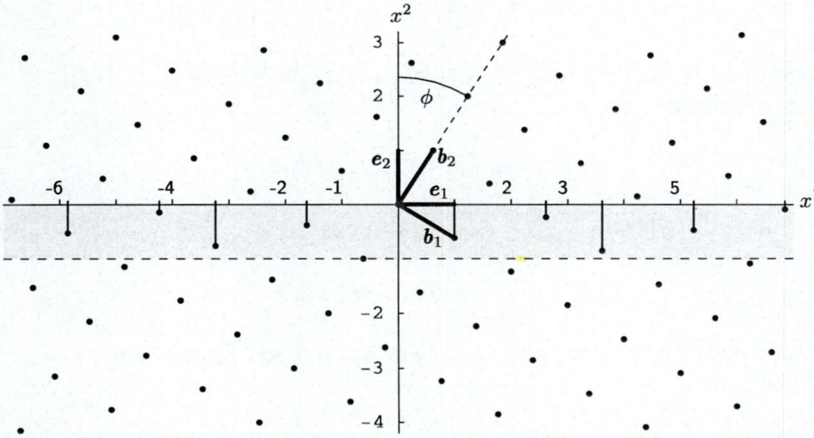

Fig. 13.2 The square lattice $m^i e_i$ transformed by $\mathcal{T}$ and the projection of the nodes located in the strip $-1 \leq x^2 < 0$ onto abscissa. The x^1 coordinates of these nodes are x_n of (13.1). A segment on the abscissa is a projection of either a square edge or a square diagonal. Square edges have two inclinations with respect to the abscissa, but only one type is projected, and the same applies to diagonals. The figure is drawn for $q = \tau =$ golden ratio.

The above means that by transforming the square lattice by $\mathcal{T}$ (rotation by ϕ and scaling by $1/\cos(\phi)$), and by multiplying it by the window function[3]

$$w(x^1, x^2) = \Pi(x^2 + 1/2) \tag{13.4}$$

to account for the conditions $-1 < x_n^2 \leq 0$, one gets points with the first components being elements of the sequence x_n of (13.1). The density function (13.2) is a projection (on the axis directed along e_1) of the product of the transformed Shah 'function' and the window function, i.e., with

$$\psi(x^1, x^2) = \text{III}\left(\mathcal{T}^{-T}\begin{bmatrix} x^1 \\ x^2 \end{bmatrix}\right) w(x^1, x^2) , \tag{13.5}$$

one has $\rho(x^1) = \int \psi(x^1, x^2)\, dx^2$. With $\mathcal{F}[\psi](\xi_1, \xi_2)$ being the Fourier transform of ψ at (ξ_1, ξ_2), by the projection-slice theorem (Sect. 1.11.5), the Fourier transform of ρ is

$$\mathcal{F}[\rho](\xi_1) = \mathcal{F}[\psi](\xi_1, 0) . \tag{13.6}$$

Thus, one needs to get the explicit formula for $\mathcal{F}[\psi](\xi_1, \xi_2)$.

To match the notation of previous Sects. (1.2, 1.11.3), the arguments (x^1, x^2) will be denoted by the vector $\boldsymbol{x}$ $(= x^\mu \boldsymbol{e}_\mu)$. In this notation, (13.5) takes the form

[3] As in Sect. 1.11.1, Π denotes the rectangular function.

$$\psi(\boldsymbol{x}) = \text{III}\left(\mathcal{T}^{-T}\boldsymbol{x}\right) w(\boldsymbol{x}) = \det(\mathcal{T})\, Ш_\Lambda\,(\boldsymbol{x})\, w(\boldsymbol{x})\,.$$

The last equality follows from (1.86). By the convolution theorem (Sect. 1.11.3), the Fourier transform of ψ equals

$$\mathcal{F}[\psi] = \det(\mathcal{T})\,\mathcal{F}[Ш_\Lambda] * \mathcal{F}[w]\,.$$

Based on $\det(\mathcal{T})\mathcal{F}[Ш_\Lambda] = Ш_{\Lambda^\star}$ (see, (1.87)) one has

$$\mathcal{F}[\psi] = Ш_{\Lambda^\star} * \mathcal{F}[w]\,,$$

where $\Lambda^\star$ denotes the lattice reciprocal to Λ. $\Lambda^\star$ is based on vectors $\boldsymbol{b}^\mu = \mathcal{T}^\mu_{\ \nu}\boldsymbol{e}^\nu = (\mathcal{T}^{-1})_\nu^{\ \mu}\boldsymbol{e}^\nu$ or

$$\boldsymbol{b}^1 = \det(\mathcal{T})^{-1}\left(\boldsymbol{e}^1 - \boldsymbol{e}^2/q\right)\,, \qquad \boldsymbol{b}^2 = \det(\mathcal{T})^{-1}\left(\boldsymbol{e}^1/q + \boldsymbol{e}^2\right)\,. \tag{13.7}$$

Because of (1.85), $\mathcal{F}[\psi]$ at $\boldsymbol{\xi} = \xi_\mu \boldsymbol{e}^\mu$ can be expressed as

$$\mathcal{F}[\psi](\boldsymbol{\xi}) = (Ш_{\Lambda^\star} * \mathcal{F}[w])\,(\boldsymbol{\xi}) = \sum\nolimits_{\check{\boldsymbol{l}}} \mathcal{F}[w](\boldsymbol{\xi} - \check{\boldsymbol{l}})\,,$$

where $\check{\boldsymbol{l}} = l_\mu \boldsymbol{b}^\mu$ is a node of the reciprocal lattice $\Lambda^\star$ (i.e., l_μ are integers). Thus, all one needs to complete the calculations is an explicit expression for the transform $\mathcal{F}[w]$ of the window function (13.4). Using $\mathcal{F}[1](\xi) = \delta(\xi)$ and $\mathcal{F}_x[\Pi(x + 1/2)](\xi) = e^{i\pi\xi}\,\text{sinc}(\xi)$ (cf. (1.63)), one obtains $\mathcal{F}[w](\xi_1, \xi_2) = \delta(\xi_1)\, e^{i\pi\xi_2}\,\text{sinc}(\xi_2) = \delta(\xi_1)\, S(-\xi_2)$, where $S(\xi) = e^{-i\pi\xi}\,\text{sinc}(\xi)$. In vector notation, the transform of the window function is

$$\mathcal{F}[w](\boldsymbol{\xi}) = \delta(\boldsymbol{\xi}\cdot\boldsymbol{e}_1)\, S(-\boldsymbol{\xi}\cdot\boldsymbol{e}_2)\,.$$

With this formula, the transform of ψ is

$$\mathcal{F}[\psi](\boldsymbol{\xi}) = \sum\nolimits_{\check{\boldsymbol{l}}} \delta\left((\boldsymbol{\xi} - \check{\boldsymbol{l}})\cdot\boldsymbol{e}_1\right)\, S\left(-(\boldsymbol{\xi} - \check{\boldsymbol{l}})\cdot\boldsymbol{e}_2\right)\,.$$

Hence, based on (13.6) and $\xi_\mu = \boldsymbol{\xi}\cdot\boldsymbol{e}_\mu$, one finally gets

$$\mathcal{F}[\rho](\xi) = \sum\nolimits_{\check{\boldsymbol{l}}} \delta(\xi - \check{\boldsymbol{l}}\cdot\boldsymbol{e}_1)\, S(\check{\boldsymbol{l}}\cdot\boldsymbol{e}_2)\,, \tag{13.8}$$

where $\check{\boldsymbol{l}} = l_\mu \boldsymbol{b}^\mu$ with integer coordinates l_μ, and as above $\sum_{\check{\boldsymbol{l}}}$ denotes summation over nodes of the reciprocal lattice $\Lambda^\star$. Thus, the transform of the density function ρ is the sum of S-weighted Dirac δ 'functions', i.e., the structure has a discrete diffraction diagram. It satisfies the general definition of a crystal. The function S, sometimes referred to as geometric form (or structure) factor [4], has no counterpart in description of diffraction by periodic crystals.

As was already mentioned, if $q = k_1/k_2$ with co-prime integers k_1 and k_2, the segments constitute a structure with the period $(k_1^2 + k_2^2)/k_1$. As expected, the peaks are at $\xi = \check{\boldsymbol{l}} \cdot \boldsymbol{e}_1 = l_\mu \boldsymbol{b}^\mu \cdot \boldsymbol{e}_1 = (l_1 + l_2/q)/\det(\mathcal{T}) = (k_1 l_1 + k_2 l_2)k_1/(k_1^2 + k_2^2) = (k_1 l_1 + k_2 l_2)/\text{period}$, i.e., at multiples of the inverse of the period of the structure. The weight of each spectral peak is the same $|S(\check{\boldsymbol{l}} \cdot \boldsymbol{e}_2)|^2 = 1$. If an irrational number q is approximated by the rational $k_1/k_2 \approx q$, the resulting structure is a periodic *approximant* of the q-based quasicrystal. The closer the approximation, the larger the integers k_1 and k_2, the larger the period $(k_1^2 + k_2^2)/k_1$ of the approximant.

Of our interest here is the case with irrational q when the points constitute an non-periodic structure. It is easy to see that when q is irrational, the distribution of peaks is dense and their weights vary. The peaks are separated in the sense that a sufficiently small neighborhood of a peak contains only much weaker peaks. Example spectra for irrational $q = 2\cos(\pi/5) = \tau$ and $q = \pi$ are shown in Fig. 13.3. They were computed based on

$$\mathcal{F}_x[\rho(x)](\xi) = \int \sum_n \delta(x - x_n) \exp(-2\pi i \xi x) \mathrm{d}x = \sum_n \exp(-2\pi i \xi x_n)$$

using only a finite number of terms.

13.1.2 Fibonacci Chain

The Fibonacci chain consists of segments of two types denoted by L and S with the ratio of the lengths of L and S equal to the golden ratio τ.[4] The chain is conventionally constructed by iteration. The initial configuration consists of just one segment L. In each iteration step, each segment L is replaced by LS and each segment S is replaced by L.[5] The first five steps are

L → LS → LSL → LSLLS → LSLLSLSL → LSLLSLSLLSLLS → ... .

The type of the n-th ($n = 1, 2, \ldots$) segment of the chain can be determined algebraically using the function

$$\iota(n) = \lfloor (n+1)/\tau \rfloor - \lfloor n/\tau \rfloor\ ;$$

the n-th segment of the Fibonacci chain is L if $\iota(n) = 1$, and it is S if $\iota(n) = 0$.

[4] The golden ratio $\tau = 2\cos(\pi/5) = \left(1 + \sqrt{5}\right)/2 \approx 1.618034$ is the positive root of the algebraic equation $x^2 - x - 1 = 0$.

[5] The number F_m of segments in m-th iteration step satisfies the relationship $F_{m+1} = F_{m-1} + F_m$, i.e., that number is given by the m-th element of the Fibonacci sequence $1, 2, 3, 5, 8, 13, \ldots$.

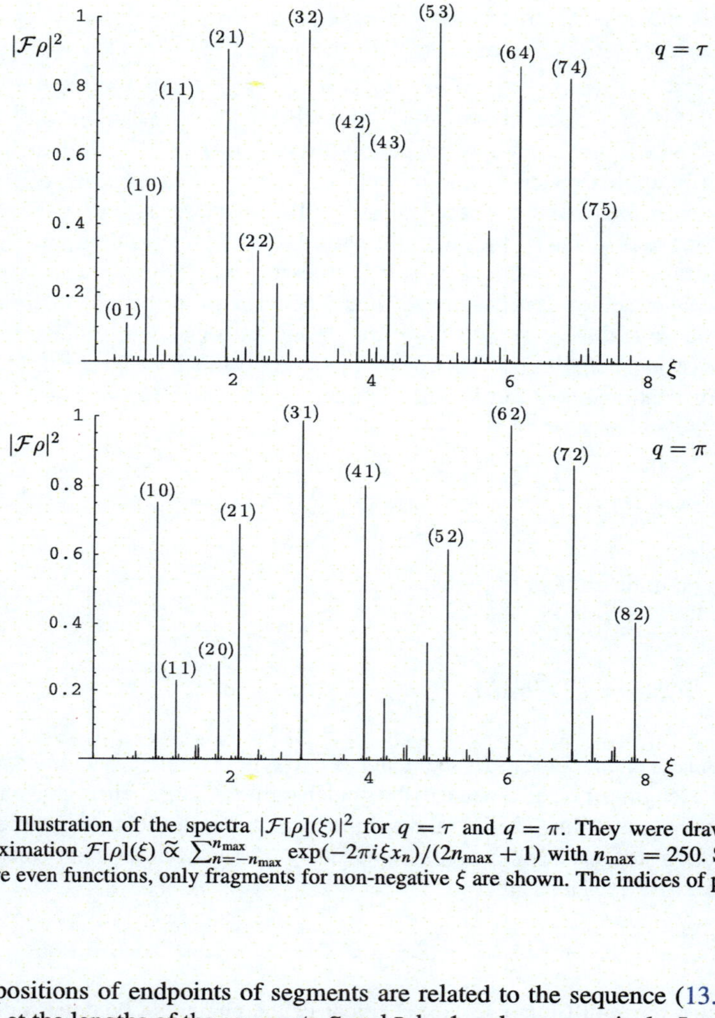

Fig. 13.3 Illustration of the spectra $|\mathcal{F}[\rho](\xi)|^2$ for $q = \tau$ and $q = \pi$. They were drawn using the approximation $\mathcal{F}[\rho](\xi) \mathrel{\widetilde{\propto}} \sum_{n=-n_{\max}}^{n_{\max}} \exp(-2\pi i \xi x_n)/(2n_{\max}+1)$ with $n_{\max} = 250$. Since the spectra are even functions, only fragments for non-negative ξ are shown. The indices of peaks are $(l_1\ l_2)$.

The positions of endpoints of segments are related to the sequence (13.1) with $q = \tau$. Let the lengths of the segments S and L be 1 and τ, respectively. Let $f_0 = 0$ be the coordinate of the left endpoint of the first segment, and let f_n denote the coordinate of right endpoint of the n-th segment. The number of L-type segments in the chain consisting of n segments is $n_{\mathrm{L}} = \sum_{k=1}^{n} \iota(k) = \lfloor (n+1)/\tau \rfloor$. Hence, one has

$$f_n = n_{\mathrm{L}} \times \tau + (n - n_{\mathrm{L}}) \times 1 = n + n_{\mathrm{L}}/\tau = n + \lfloor (n+1)/\tau \rfloor/\tau ,$$

and these coordinates are related those given by (13.1) for $q = \tau$ via $f_n = x_{n+1} - 1$. Therefore (13.1) for $q = \tau$, or the more general sequence $a(n + \alpha + \lfloor n/\tau + \beta \rfloor/\tau)$ with $0 \leq \alpha, \beta < 1$, are seen as endpoints of segments in Fibonacci chains.

The Fibonacci chain is convenient for explaining the structural property known as self-similarity, i.e., an invariance with respect to some discrete scaling transformations. For the Fibonacci chain, the transformation (deflation) rule is the same as the substitution rule used for constructing the Fibonacci chain: S → L and L → LS. Deflation shortens the segments by the factor of τ. Clearly, the process of deflation can be repeated again and again.

Analogous principle applies to self-similar tilings of higher dimension. The tiles can be divided into smaller tiles with the same shapes and using the same rule, and the division can be performed again and again. Given an initial configuration of a small number of tiles and a deflation rule, repeated deflations can be used for constructing a tiling consisting of arbitrarily large number of tiles.

13.1.3 Four-Segment Quasicrystal

Clearly, the Fibonacci chain or other two-segment chains are not the only possible one-dimensional quasicrystals. Let q_2, q_3 (each > 1) and q_2/q_3 be all irrational, and let

$$x_n = n + \frac{1}{q_2}\left\lfloor\frac{n}{q_2}\right\rfloor + \frac{1}{q_3}\left\lfloor\frac{n}{q_3}\right\rfloor . \tag{13.9}$$

This time, the segments have four lengths, and the distance between $x_{n+m} - x_n$ takes one of four values

$$x_{n+m} - x_n = x_m + \iota_1/q_2 + \iota_2/q_3 ,$$

where the coefficients ι_1 and ι_2 are 0 or 1. The Fourier transform of $\rho(x) = \sum_n \delta(x - x_n)$ can be derived by method analogous to that used in the two-segment case.

⊡ With

$$\mathcal{T} = \begin{bmatrix} 1 & -1/q_2 & -1/q_3 \\ 1/q_2 & 1 & 0 \\ 1/q_3 & 0 & 1 \end{bmatrix} \tag{13.10}$$

the sequence (13.9) is the first coordinate $x_n^1 = x_n$ of the triple of the integers $[n,\ \lfloor n/q_2\rfloor,\ \lfloor n/q_2\rfloor]$ transformed by $\mathcal{T}^T$

$$\mathcal{T}^T \begin{bmatrix} n \\ \lfloor n/q_2\rfloor \\ \lfloor n/q_3\rfloor \end{bmatrix} = \begin{bmatrix} x_n^1 \\ x_n^2 \\ x_n^3 \end{bmatrix} ,$$

where $x_n^k = \lfloor n/q_k\rfloor - n/q_k$ $(k = 2, 3)$ satisfy the conditions $-1 < x_n^k \le 0$. Thus, the function characterizing the acceptance domain is $w(x^1, x^2, x^3) = \Pi(x^2 + 1/2)\ \Pi(x^3 + 1/2)$, and with $\mathcal{F}_x[\Pi(x + 1/2)](\xi) = e^{i\pi\xi}\,\mathrm{sinc}(\xi) = S(-\xi)$, the Fourier transform of the window function has the form $\mathcal{F}[w]\,(\xi_1, \xi_2, \xi_3) = \delta(\xi_1)\ S(-\xi_2)\ S(-\xi_3)$. Proceeding as in the two-segment case, one gets the Fourier transform of ρ

$$\mathcal{F}[\rho](\xi) = \sum\nolimits_{\check{l}} \delta(\xi - \check{\boldsymbol{l}} \cdot \boldsymbol{e}_1)\ S(\check{\boldsymbol{l}} \cdot \boldsymbol{e}_2)\ S(\check{\boldsymbol{l}} \cdot \boldsymbol{e}_3) ,$$

where $\sum_{\check{l}}$ denotes summation over nodes of the reciprocal lattice based on the vectors $\boldsymbol{b}^\mu = (\mathcal{T}^{-1})_\nu{}^\mu \boldsymbol{e}^\nu = \mathcal{T}^\mu{}_\nu \boldsymbol{e}^\nu$, and $\check{\boldsymbol{l}} = l_\mu \boldsymbol{b}^\mu$ with integer coefficients l_μ. For an example spectrum, see Fig. 13.4.

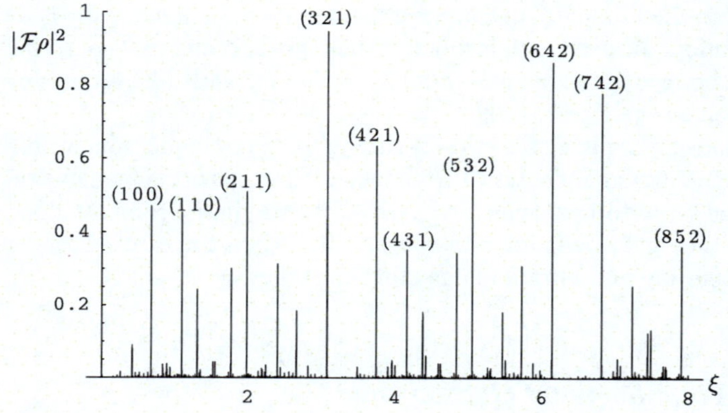

Fig. 13.4 The spectrum $|\mathcal{F}[\rho](\xi)|^2$ for the four-segment chain with $q_2 = \tau$ and $q_3 = \pi$. It was drawn using the approximation $\mathcal{F}[\rho](\xi) \mathrel{\tilde{\propto}} \sum_{n=-n_{\max}}^{n_{\max}} \exp(-2\pi i \xi x_n)/(2n_{\max}+1)$ with $n_{\max} = 250$. The indices of peaks are $(l_1\, l_2\, l_3)$.

This example slightly generalizes features of the chain defined by (13.1). The segments of the chain have four lengths, but there are only three numbers ($1, q_2$ and q_3) defining the sequence (13.9). The dimension of the lattice used to derive the Fourier transform is three, i.e., it is equal to the number of sequence-defining numbers. In the four-segment case, the lattice based on $\boldsymbol{b}^\mu$ and the one based on $\boldsymbol{b}_\mu = \mathcal{T}_\mu{}^\nu \boldsymbol{e}_\nu$ are not cubic; they were square lattices in the two-segment case.

13.1.4 The Strip Projection in One-Dimensional Cases

The derivation of the Fourier transform in Sect. 13.1.1 illustrates the projection method. Briefly, using a window (acceptance domain), a part of lattice Λ in space of higher dimension, say N, is selected, and projection of the selected part on the physical space of lower dimension gives the quasiperiodic structure of interest. The Fourier transform $\mathcal{F}[\rho]$ of the structure is obtained from the Fourier transform of Λ by projection-slice theorem.

With $\Lambda^\star$ denoting the reciprocal of Λ and $\boldsymbol{b}^\mu$ being a vector of the basis of $\Lambda^\star$, the resulting transform $\mathcal{F}[\rho]$ at $\boldsymbol{\xi}$ in the space reciprocal to the physical space has the form

$$\mathcal{F}[\rho](\boldsymbol{\xi}) = \sum_{\check{\boldsymbol{l}}} \delta(\boldsymbol{\xi} - P_\parallel(\check{\boldsymbol{l}}))\, \mathsf{S}(P_\perp(\check{\boldsymbol{l}})) = \sum_{l_1,\ldots,l_N} \delta(\boldsymbol{\xi} - l_\mu \boldsymbol{a}_\parallel^\mu)\, \mathsf{S}(l_\mu \boldsymbol{a}_\perp^\mu)\,, \qquad (13.11)$$

where l_μ are integers, $P_\parallel$ denotes the orthogonal projection on the physical space, $P_\perp$ is the orthogonal projection on its orthogonal complement, and

$$\boldsymbol{a}_{\parallel}^{\mu} = P_{\parallel}(\boldsymbol{b}^{\mu}) \,, \quad \boldsymbol{a}_{\perp}^{\mu} = P_{\perp}(\boldsymbol{b}^{\mu}) \,.$$

The projection of $\check{\boldsymbol{l}} = l_{\mu}\boldsymbol{b}^{\mu}$ on the physical space determines the position of the peak, i.e., the vector $\boldsymbol{\xi} = P_{\parallel}(\check{\boldsymbol{l}})$ plays the same role as a reciprocal lattice vector in conventional crystallography. The projection $P_{\perp}(\check{\boldsymbol{l}})$ has no conventional counterpart. It determines the weight $\mathsf{S}(P_{\perp}(\check{\boldsymbol{l}}))$ ascribed to the peak located at $P_{\parallel}(\check{\boldsymbol{l}})$ and influences the spectral intensity of the peak.

In the case of two-segment chain based on the sequence (13.1), the physical space has dimension 1, N equals 2, the acceptance domain is the segment $-1 < x^2 \leq 0$ of the orthogonal complement to the physical space, and

$$\boldsymbol{a}_{\parallel}^{\mu} = P_{\parallel}(\boldsymbol{b}^{\mu}) = (\boldsymbol{b}^{\mu} \cdot \boldsymbol{e}_1)\boldsymbol{e}^1 \,, \quad \boldsymbol{a}_{\perp}^{\mu} = P_{\perp}(\boldsymbol{b}^{\mu}) = (\boldsymbol{b}^{\mu} \cdot \boldsymbol{e}_2)\boldsymbol{e}^2$$

with basis vectors $\boldsymbol{b}^{\mu}$ given by (13.7), i.e.,

$$\boldsymbol{a}_{\parallel}^{1} = \frac{1}{\det(\mathcal{T})}\boldsymbol{e}^1 \,, \; \boldsymbol{a}_{\parallel}^{2} = \frac{1}{\det(\mathcal{T})\, q}\boldsymbol{e}^1 \,, \tag{13.12}$$

and similarly $\boldsymbol{a}_{\perp}^{1} = -\boldsymbol{e}^2/(q\det(\mathcal{T}))$ and $\boldsymbol{a}_{\perp}^{2} = \boldsymbol{e}^2/\det(\mathcal{T})$, where $\det(\mathcal{T}) = 1 + 1/q^2$. The function S determining the peak intensity is $\mathsf{S}(P_{\perp}(\check{\boldsymbol{l}})) = S(\check{\boldsymbol{l}} \cdot \boldsymbol{e}_2)$.

The strip projection is also used for tessellating the direct physical space. The subject will be described in a more general framework in Sect. 13.2 below, but it worth sketching the method using the one-dimensional Fibonacci chain. In Fig. 13.2, segments of the chain were projections of square edges and diagonals, but construction can be slightly modified so segments of the chain are projections of square edges only. This is illustrated in Fig. 13.5. The strip has the width equal to the length of the projection of the square cell of the lattice on the ordinate (or the orthogonal complement of the physical space). The edges (one-dimensional facets of the square cells) entirely contained in it constitute a polygonal chain (one-dimensional surface) which projects on the abscissa (i.e., on the physical space) without any gaps or overlaps. The chain is a union of the facets entirely contained in the strip, and the projection of the chain on the abscissa is one-to-one. Clearly, the projection of the edges constituting the chain gives a segmentation of the abscissa.

Similar constructions can be made for other values of the angle ϕ. The irrationality of $\tan\phi$ implies that any finite configuration of chain segments appears in it infinitely many times. (This feature is known as local isomorphism property; see Sect. 13.2 below.) Moreover, it is easy to see that one can move the strip along the ordinate to a new location, and get a new segmentation of the abscissa. However, any finite configuration of old chain segments appears in the new one infinitely many times and vice versa.

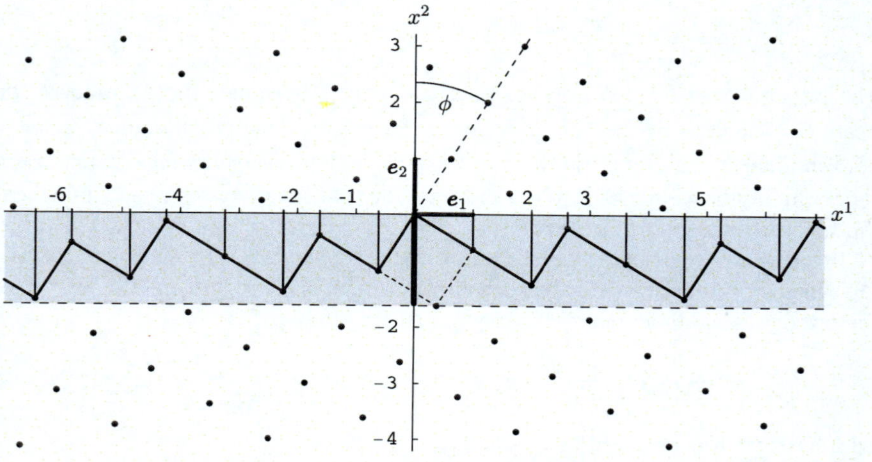

Fig. 13.5 The rotated squared lattice and the projection of the nodes in the strip onto abscissa. Segments of the chain are projections of edges (one-dimensional facets) of the square cells of the lattice. This is a modified version of Fig. 13.2 for the Fibonacci chain, i.e., for $\phi = \arctan \tau^{-1}$. Segments of the chain are scaled by τ compared to those of Fig. 13.2. Each long segment of Fig. 13.2 is replaced by a pair of a long segment and a short segment, and each short segment of Fig. 13.2 is replaced by a long segment. Thus, the resulting sequence is also a Fibonacci chain.

⊡ It has been argued that the physical space should be sufficient to describe quasicrystals, and there is no need to refer to spaces of higher dimension; e.g., [5]. However, at the current stage, the projection method is strongly entrenched in the field of quasicrystallography and communication would be difficult without referring to it.

13.1.5 Indexing

Similarly to diffraction by periodic crystals, a quasicrystalline pattern consists of peaks at

$$l_\mu \boldsymbol{a}_\|^\mu ,$$

where $\boldsymbol{a}_\|^\mu$ are vectors of a frame (overcomplete set of vectors) in reciprocal space. Vectors in (13.12) are an example of such a frame. The point in indexing a diffraction pattern is to assign indices l_μ to diffraction peaks, and to determine the frame (including the structure parameter a used above as a unit), and organization of 'unit cells' or chain segments, i.e., to get restrictions on positions of endpoint of the segments. Having the (reciprocal space) frame $\boldsymbol{a}_\|^\mu$, one can get its dual $\boldsymbol{a}_\mu^\|$ in the direct space. The frame $\boldsymbol{a}_\mu^\|$ in the direct space conforms to 'unit cells' of the quasicrystal.

For brevity, starting from here, the vectors $\boldsymbol{a}_\mu^\|$ and $\boldsymbol{a}_\|^\mu$ in the physical space will be denoted by $\boldsymbol{a}_\mu$ and $\boldsymbol{a}^\mu$, respectively. Only if the context of the projection method is needed, the complete symbols $\boldsymbol{a}_\mu^\|$ and $\boldsymbol{a}_\|^\mu$ will be used.

It is instructive to consider the case of the quasicrystal described by (13.1). The two vectors $\boldsymbol{a}^\mu$ of (13.12) determine a frame in the one-dimensional (reciprocal) space. The direct frame $\boldsymbol{a}_\mu$ dual to $\boldsymbol{a}^\mu$ differs from the latter by the factor $\det(\mathcal{T})$; one has $\boldsymbol{a}_\mu = \det(\mathcal{T})\boldsymbol{a}^\mu$ ($\mu = 1, 2$) or

$$\boldsymbol{a}_1 = \boldsymbol{e}_1 \, , \quad \boldsymbol{a}_2 = \boldsymbol{e}_1/q \, .$$

With irrational q and arbitrary integer coefficients m^μ, the direct-space points $m^\mu \boldsymbol{a}_\mu$ would be distributed densely, but the actual distribution of the nodes described by (13.1) is restricted by the selection rule $m^2 = \lfloor m^1/q \rfloor$. (Clearly, the same pattern is obtained when $m^2 = \lfloor m^1/q + \beta \rfloor$, where β is a real number.)

Let a one-dimensional diffraction pattern be given, and one needs to solve it under the assumption is that it is described by the model (13.1). One needs to get the irrational parameter q, the lattice constant a and the indices l_μ of the diffraction peaks in the reciprocal space frame (13.12). They can be determined based on the positions x of peaks in the diffraction pattern. With the constant a taken into account, one has

$$\boldsymbol{x} = x\boldsymbol{e}^1 = l_\mu \boldsymbol{a}^\mu = \frac{1}{a\,(1 + 1/q^2)}\,(l_1 + l_2/q)\,\boldsymbol{e}^1 \, . \tag{13.13}$$

It is easy to see that for a given a and a given irrational q, the pair of indices of a peak is unique: If there were two pairs $(l_1\, l_2)$ and $(l_1'\, l_2')$ leading to the same x, one would have $l_1 + l_2/q = l_1' + l_2'/q$ or $(l_1 - l_1')q = -(l_2 - l_2')$, i.e., assuming $l_1 \neq l_1'$, $q = -(l_2 - l_2')/(l_1 - l_1')$ would be rational; thus, one must have $l_1 = l_1'$ and $l_2 = l_2'$.

The above feature can be expressed in a more formal way: A set of N vectors $\boldsymbol{a}^\mu$ ($\mu = 1, \ldots, N$) is *rationally* linearly independent if $r_\mu \boldsymbol{a}^\mu = \boldsymbol{0}$ implies that all *rational* numbers r_μ are zero. In the one-dimensional example, with $N = 2$ and irrational q, the vectors $\boldsymbol{a}^\mu$ of the considered frame are rationally independent, and this implies that the pair $(l_1\, l_2)$ representing a given $\boldsymbol{x} = l_\mu \boldsymbol{a}^\mu$ is unique.[6]

One can formally define the (reciprocal) *quasilattice* as a set of integer linear combinations $l_\mu \boldsymbol{a}^\mu$ of the vectors $\boldsymbol{a}^\mu$ ($\mu = 1, \ldots, N$). Clearly, this set is closed under addition and subtraction. Compared to the lattices defined in Sect. 1.2 for describing periodic crystals, there is no requirement that the vectors $\boldsymbol{a}^\mu$ are linearly independent. Unlike reciprocal lattice of an ordinary crystal, reciprocal quasilattice has no minimum distance between its nodes; the distance between nodes can be arbitrarily small.

⊡ One needs to note that the term 'quasilattice' is frequently used in reference to vertices of the tiles, i.e., it has nodes in direct space. In the case of one-dimensional quasicrystals, a direct-space quasilattice consists of endpoints of non-overlapping line segments filling the line. Positions of the vertices can be expressed as integer combinations of the vectors $\boldsymbol{a}_\mu$, but only some combinations are allowed. There is a selection rule that filters out the set of tiling vertices from the complete direct-space quasilattice.

---

[6] As will be discussed below (Sect. 13.3.1), to account for symmetry, some quasicrystal frames are chosen to rely on vectors which are not rationally independent.

If the assumed model of the one-dimensional quasicrystal is correct, indexing can be relatively easily done by trial and error. Assuming that a peak at $x = x_a$ corresponds to certain $(l_1, l_2) = (l_1^a, l_2^a)$ and another one at $x = x_b$ corresponds to (l_1^b, l_2^b), one has

$$x_a\, a\,(1 + 1/q^2) = (l_1^a + l_2^a/q) \quad \text{and} \quad x_b\, a\,(1 + 1/q^2) = (l_1^b + l_2^b/q)\,.$$

These equations are solved by

$$a = \left(l_1^a\, l_2^b - l_1^b\, l_2^a\right) u_2 \,/\, \left(u_1^2 + u_2^2\right) \quad \text{and} \quad q = -u_2/u_1\,,$$

where $u_i = x_a l_i^b - x_b l_i^a$ $(i = 1, 2)$. To check correctness of the assumption and the obtained parameters a and q, one needs to compare the pattern simulated based on these parameters to the original pattern. Clearly, one needs to start with high peaks at small x and indices with low (absolute) values like $(l_1, l_2) = (1, 0), (0, 1), (1, 1)$ et cetera.

But will this approach lead to a unique result? If intensities are ignored, the choice of the tested peaks is generally ambiguous. This can be illustrated using the Fibonacci chain. From (13.13) and $\tau = 1 + 1/\tau$, one has

$$x\,(1 + 1/\tau^2) = \frac{l_1 + l_2/\tau}{a} = \frac{(l_1 + l_2) + l_1/\tau}{a\tau} = \frac{l_1' + l_2'/\tau}{a'}\,.$$

Thus, based on the peak positions alone, the pattern can be indexed using the parameter a and indices $(l_1\, l_2)$ or the parameter $a' = a\tau$ and indices $(l_1'\, l_2')$ such that $l_1' = l_1 + l_2$ and $l_2' = l_1$.

For future use, it is worth to reformulate the above paragraph in a slightly more formal way. The ambiguity is related to the inflation of self-similar Fibonacci chain. The inflation in direct space corresponds to deflation in the reciprocal space. With the frame (13.12) and $q = \tau$, the vectors $\boldsymbol{a}^\mu$ deflated by τ are $\boldsymbol{a}'^\mu = \boldsymbol{a}^\mu/\tau$. Based on $\tau = 1 + 1/\tau$, one has $\boldsymbol{a}^1 = \tau\boldsymbol{a}'^1 = \boldsymbol{a}'^1 + \boldsymbol{a}'^1/\tau = \boldsymbol{a}'^1 + \boldsymbol{a}'^2$ and $\boldsymbol{a}^2 = \tau\boldsymbol{a}'^2 = \boldsymbol{a}'^1$, or briefly,

$$\boldsymbol{a}^\mu = \tau\boldsymbol{a}'^\mu = M^\mu{}_\nu \boldsymbol{a}'^\nu\,,$$

where the (inflation) matrix M is

$$[M^\mu{}_\nu] = \begin{bmatrix} 1 & 1 \\ 1 & 0 \end{bmatrix}.$$

Since $l_\mu \boldsymbol{a}^\mu = l_\mu \tau\boldsymbol{a}'^\mu = l_\mu M^\mu{}_\nu \boldsymbol{a}'^\nu = l_\nu' \boldsymbol{a}'^\nu$, the indices l_μ and l_μ' of the same reflection in the frames $\boldsymbol{a}^\mu$ and $\boldsymbol{a}'^\mu$, respectively, are related by

$$l_\nu' = l_\mu M^\mu{}_\nu\,,$$

i.e., by the relationship listed above.[7]

Clearly, the ambiguity in indexing is related to the choice of the fundamental vector indexed as $(l_1\, l_2) = (0\,1)$. Unlike in conventional crystallography, here, there is no shortest lattice vector, and formally, the vectors $\boldsymbol{a}^\mu$ can be arbitrarily short. The choice of the fundamental lengths needs to be resolved by comparing positions and intensities of highest peaks. With a long reciprocal space vector chosen as $(l_1\, l_2) = (0\,1)$, it is difficult to ascribe indices to shorter vectors.

Finally, it must be stressed again that key for the above approach was to assume the right model. The described indexing method would work for the patterns shown in Fig. 13.3, but it would fail if applied to that of Fig. 13.4.

13.2 The Strip Projection Method

Formally, a tiling (or a packing of unit cells) is defined as a covering of the space by polyhedra with no holes or tile overlaps such that the number of different types of tiles is finite. As in the simple one-dimensional case considered in Sect. 13.1.4, tilings of n_d-dimensional direct physical space can be generated using the strip projection. The physical space is seen as a subspace of an N-dimensional space. A cell of N-dimensional lattice has a number of distinct n_d-dimensional facets. One needs a part of the N-dimensional space (a strip), such that the n_d-dimensional facets entirely contained in the strip constitute a surface which projects on the physical subspace without any gaps or overlaps. In other words, the n_d-dimensional surface is a union of the facets entirely contained in the strip, and it can be projected one-to-one on the subspace. Clearly, the projection of the facets constituting the surface gives a tessellation of the physical space into tiles.

There is a plethora of theoretically possible quasiperiodic two-dimensional tilings and three-dimensional packings. Two tilings (packings) are called locally isomorphic if any finite configuration of cells present in one also appears in the other in the same orientation [6, 7]. The diffraction patterns of locally isomorphic tilings (packings) are the same. The notion of local isomorphism is convenient for defining tiling (packing) symmetry: A tiling (packing) is considered to have a given point group symmetry if transformation by any group element gives a locally isomorphic tiling (packing).

In the context of the strip projection method, the required symmetry of the tiling (packing) is the key aspect for the choice of the N-dimensional lattice and its n_d-dimensional section. The lattice is chosen so it has the needed symmetry, and it is projected in such a way that the symmetry is preserved in the projection. The above principles are best illustrated using the two-dimensional case with fivefold symmetry.

[7] According to the crystallographic convention, the notation for reflection indices is hkl without parentheses. In the case of quasicrystals, the number of indices in a set changes with the type of quasicrystal and indexing scheme. Here, for clarity, the reflection indices are inside parentheses.

13.3 Two-Dimensional Pentagonal Case

Similarly to the conventional cube having threefold symmetry along the direction of its diagonal, the five-dimensional cube with vertices at $\iota^\mu \boldsymbol{e}_\mu$, where $\mu = 1, 2, \ldots, 5$ and ι^μ equals 0 or 1, has the fivefold symmetry along the $\sum_{\mu=1}^{5} \boldsymbol{e}_\mu$ direction. The projection of the five-dimensional hypercubic lattice[8] on the two-dimensional plane perpendicular to $\sum_{\mu=1}^{5} \boldsymbol{e}_\mu$ also has this symmetry. One will ultimately project only nodes contained in a strip, so locally the symmetry will be broken, but it will remain as the symmetry of the resulting quasiperiodic tiling.

Cyclic permutations of the basis vectors $\boldsymbol{e}_\mu$ are represented by $Q_1 = I_5$ and matrices Q_2, Q_3, Q_4, Q_5 equal to

$$\begin{bmatrix} 0&0&0&0&1 \\ 1&0&0&0&0 \\ 0&1&0&0&0 \\ 0&0&1&0&0 \\ 0&0&0&1&0 \end{bmatrix}, \quad \begin{bmatrix} 0&0&0&1&0 \\ 0&0&0&0&1 \\ 1&0&0&0&0 \\ 0&1&0&0&0 \\ 0&0&1&0&0 \end{bmatrix}, \quad \begin{bmatrix} 0&0&1&0&0 \\ 0&0&0&1&0 \\ 0&0&0&0&1 \\ 1&0&0&0&0 \\ 0&1&0&0&0 \end{bmatrix}, \quad \begin{bmatrix} 0&1&0&0&0 \\ 0&0&1&0&0 \\ 0&0&0&1&0 \\ 0&0&0&0&1 \\ 1&0&0&0&0 \end{bmatrix},$$

respectively. Let the matrix $\mathcal{T} = [\mathcal{T}_\mu{}^\nu]$ have the form

$$[\mathcal{T}_\mu{}^\nu] = \begin{bmatrix} c_1 & s_1 & C_1 & S_1 & 1/\sqrt{2} \\ c_2 & s_2 & C_2 & S_2 & 1/\sqrt{2} \\ c_3 & s_3 & C_3 & S_3 & 1/\sqrt{2} \\ c_4 & s_4 & C_4 & S_4 & 1/\sqrt{2} \\ c_5 & s_5 & C_5 & S_5 & 1/\sqrt{2} \end{bmatrix}, \tag{13.14}$$

where $c_\mu = \cos(2\pi(\mu - 1)/5)$, $s_\mu = \sin(2\pi(\mu - 1)/5)$, $C_\mu = \cos(4\pi(\mu - 1)/5)$ and $S_\mu = \sin(4\pi(\mu - 1)/5)$. The matrix $R = \sqrt{2/5}[\mathcal{T}_\mu{}^\nu]$ is special orthogonal. It is easy to verify that

$$R^T Q_\mu R = \begin{bmatrix} c_\mu & -s_\mu & 0 & 0 & 0 \\ s_\mu & c_\mu & 0 & 0 & 0 \\ 0 & 0 & C_\mu & -S_\mu & 0 \\ 0 & 0 & S_\mu & C_\mu & 0 \\ 0 & 0 & 0 & 0 & 1 \end{bmatrix}.$$

Thus, the decomposition of Q_μ into irreducible representations gives three invariant subspaces. One of the two-dimensional subspaces serves as the physical space, and the other two subspaces constitute the orthogonal complement of the physical space.

Take the five-dimensional hypercubic lattice $m^\mu \boldsymbol{b}_\mu$ based on the vectors

$$\boldsymbol{b}_\mu = \mathcal{T}_\mu{}^\nu \boldsymbol{e}_\nu \, .$$

---

[8] A hypercubic lattice is the higher-dimensional analogue of the square and cubic lattices in two- and three-dimensional spaces, respectively.

The $\boldsymbol{b}_\mu$-based lattice is obtained from the $\boldsymbol{e}_\mu$-based lattice via rotation by R and scaling by $\sqrt{5/2}$. With $m'^\mu = m^\mu + m$, where m is an integer, one has

$$m'^\mu \boldsymbol{b}_\mu = m^\mu \boldsymbol{b}_\mu + m \sum_\mu \boldsymbol{b}_\mu = m^\mu \boldsymbol{b}_\mu + m(5/\sqrt{2})\boldsymbol{e}^5$$

i.e., the lattice is invariant with respect to translations by multiples of $(5/\sqrt{2})\boldsymbol{e}^5$. Since

$$m^\mu \boldsymbol{b}_\mu \cdot \boldsymbol{e}^5 = m^\mu \mathcal{T}_\mu{}^\nu \boldsymbol{e}_\nu \cdot \boldsymbol{e}^5 = m^\mu \mathcal{T}_\mu{}^\nu \delta_\nu^5 = m^\mu \mathcal{T}_\mu{}^5 = \sum_\mu m^\mu/\sqrt{2}\,,$$

all lattice nodes are on planes normal to $\boldsymbol{e}_5$ with the inter-planar distance of $1/\sqrt{2}$. Thus, there are five such planes along one period $(5/\sqrt{2})\boldsymbol{e}^5$.

The projections of the lattice basis vectors $\boldsymbol{b}_\mu$ on the physical space spanned by $\boldsymbol{e}_1$ and $\boldsymbol{e}_2$ and on the orthogonal component spanned by $\boldsymbol{e}_3$, $\boldsymbol{e}_4$ and $\boldsymbol{e}_5$ are

$$\boldsymbol{a}_\mu^\parallel = P_\parallel(\boldsymbol{b}_\mu) = (\boldsymbol{b}_\mu \cdot \boldsymbol{e}^1)\boldsymbol{e}_1 + (\boldsymbol{b}_\mu \cdot \boldsymbol{e}^2)\boldsymbol{e}_2 = \mathcal{T}_\mu{}^1 \boldsymbol{e}_1 + \mathcal{T}_\mu{}^2 \boldsymbol{e}_2 \tag{13.15}$$

and

$$\boldsymbol{a}_\mu^\perp = P_\perp(\boldsymbol{b}_\mu) = (\boldsymbol{b}_\mu \cdot \boldsymbol{e}^3)\boldsymbol{e}_3 + (\boldsymbol{b}_\mu \cdot \boldsymbol{e}^4)\boldsymbol{e}_4 + (\boldsymbol{b}_\mu \cdot \boldsymbol{e}^5)\boldsymbol{e}_5 = \mathcal{T}_\mu{}^3 \boldsymbol{e}_3 + \mathcal{T}_\mu{}^4 \boldsymbol{e}_4 + \mathcal{T}_\mu{}^5 \boldsymbol{e}_5\,, \tag{13.16}$$

respectively. The five unit vectors (13.15)

$$\boldsymbol{a}_\mu = \boldsymbol{a}_\mu^\parallel = c_\mu \boldsymbol{e}_1 + s_\mu \boldsymbol{e}_2\,, \quad \mu = 1, \ldots, 5\,, \tag{13.17}$$

point to vertices of a regular pentagon (Fig. 13.6). This frame reflects the desired five-fold symmetry, but the vectors $\boldsymbol{a}_\mu$ are not rationally linearly independent as $\sum_{\mu=1}^5 \boldsymbol{a}_\mu = \boldsymbol{0}$. With integer m^μ and m, if $m'^\mu = m^\mu + m$, then

$$m'^\mu \boldsymbol{a}_\mu = m^\mu \boldsymbol{a}_\mu + m \sum_\mu \boldsymbol{a}_\mu = m^\mu \boldsymbol{a}_\mu + m\boldsymbol{0} = m^\mu \boldsymbol{a}_\mu\,,$$

and this is the most general formulation of non-uniqueness of the indices m^μ.

□ The last statement can be justified in the following way. If $m'^\mu \boldsymbol{a}_\mu = m^\mu \boldsymbol{a}_\mu$, and the indices are related via μ-dependent difference $n^\mu = m'^\mu - m^\mu$, then $n^\mu \boldsymbol{a}_\mu = \boldsymbol{0}$. Scalar multiplication by the dual $\boldsymbol{a}^\nu$ gives the homogeneous system of equations $n^\mu g_\mu{}^\nu = 0$, where $g_\mu{}^\nu = \boldsymbol{a}_\mu \cdot \boldsymbol{a}^\nu = \mathcal{T}_\mu{}^j \mathcal{T}^\nu{}_j$ or explicitly

$$\left[g_\mu^\nu\right] = \frac{1}{5}\begin{bmatrix} 2 & \tau^{-1} & -\tau & -\tau & \tau^{-1} \\ \tau^{-1} & 2 & \tau^{-1} & -\tau & -\tau \\ -\tau & \tau^{-1} & 2 & \tau^{-1} & -\tau \\ -\tau & -\tau & \tau^{-1} & 2 & \tau^{-1} \\ \tau^{-1} & -\tau & -\tau & \tau^{-1} & 2 \end{bmatrix}. \tag{13.18}$$

Fig. 13.6 The vectors (13.17) pointing to vertices of a regular pentagon.

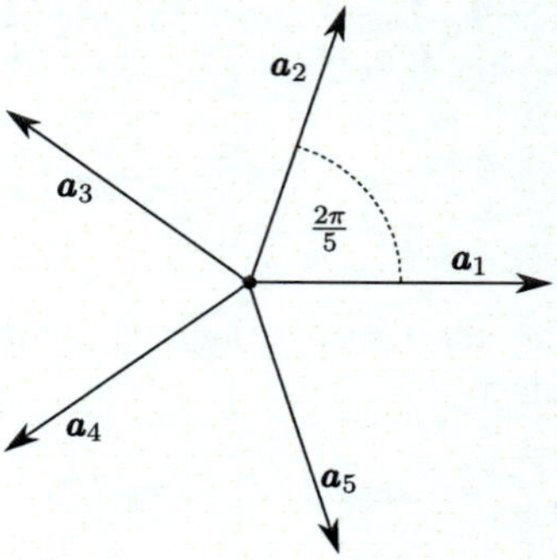

Integer solutions of the system $n^\mu g_\mu{}^\nu = 0$ satisfy four conditions $n^1 = n^2 = n^3 = n^4 = n^5$. A single additional condition (like $m^1 = 0$) eliminates the non-uniqueness. Thus, any four of the $\boldsymbol{a}_\mu$ vectors are rationally linearly independent.[9]

The integer combination $m^\mu \boldsymbol{a}_\mu^\parallel$ is a node of the tiling of the physical space if $m^\mu \boldsymbol{a}_\mu^\perp$ falls in the window being the projection of the lattice cell (rotated hypercube) on the orthogonal complement of the physical space. So what is the shape of this projection? The vertices of the original cube with vertices at $\mathbf{0}$ and at $\sum_{\mu=1}^5 \boldsymbol{b}_\mu$ are $\iota^\mu \boldsymbol{b}_\mu$, where ι^μ equals 0 or 1. In the basis $\boldsymbol{e}_\mu$, they are $\iota^\mu T_\mu{}^\nu \boldsymbol{e}_\nu$. The projection of these $2^5 = 32$ points on the three-dimensional space spanned by $\boldsymbol{e}_3, \boldsymbol{e}_4$ and $\boldsymbol{e}_5$ gives 22 points which are vertices of the a figure known as rhombic icosahedron (Fig. 13.7) and 10 points which are inside this figure.

The procedure of determining the window for pentagonal quasicrystals is affected by an additional complication compared to that applicable the one-dimensional quasicrystal. In the latter case (and also in the case of icosahedral quasicrystal), the integer combination $m^\mu \boldsymbol{a}_\mu^\parallel$ of vectors $\boldsymbol{a}_\mu^\parallel$ corresponds to exactly one vector $m^\mu \boldsymbol{a}_\mu^\perp$, and the window is densely filled with the points $m^\mu \boldsymbol{a}_\mu^\perp$. With the frames (13.15) and (13.16), the situation is different. First, since $m^\mu \boldsymbol{a}_\mu^\perp \cdot \boldsymbol{e}^5 = \sum_\mu m^\mu/\sqrt{2}$, the window (rhombic icosahedron) is not densely filled with points $m^\mu \boldsymbol{a}_\mu^\perp$. Moreover, as was noted above, the vectors $\boldsymbol{a}_\mu^\parallel$ are not rationally independent. With $m'^\mu = m^\mu + m$, the vectors $m'^\mu \boldsymbol{a}_\mu^\parallel$ and $m^\mu \boldsymbol{a}_\mu^\parallel$ are equal, but $m^\mu \boldsymbol{a}_\mu^\perp$ and $m'^\mu \boldsymbol{a}_\mu^\perp \left(= m^\mu \boldsymbol{a}_\mu^\perp + m \sum_\mu \boldsymbol{a}_\mu^\perp = m^\mu \boldsymbol{a}_\mu^\perp + m(5/\sqrt{2})\boldsymbol{e}^5\right)$ are generally different, and multiple vectors $m^\mu \boldsymbol{a}_\mu^\perp$ separated by $(5/\sqrt{2})\boldsymbol{e}^5$ correspond to a single vector $m^\mu \boldsymbol{a}_\mu^\parallel$. If, however, $m^\mu \boldsymbol{a}_\mu^\parallel$ has a counterpart $m^\mu \boldsymbol{a}_\mu^\perp$ inside the rhombic icosahedron, the point is unique because the length of the polyhedron's diagonal along $\boldsymbol{e}_5$ is $5/\sqrt{2}$. For a given position of the rhombic icosahedron with respect to the lattice, all these

[9] It is known that a minimum number of rationally linearly independent vectors required to span a planar quasilattice with n-fold symmetry is $E(n)$, where $E(n)$ is the Euler's totient function [8]. For $n = 5, 8, 10, 12$, $E(n)$ equals 4.

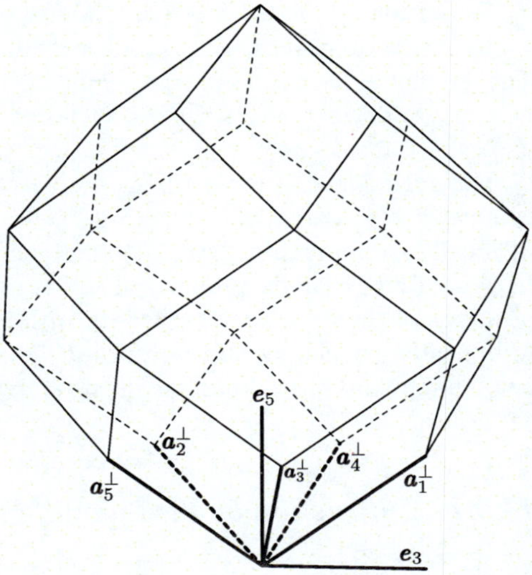

Fig. 13.7 Rhombic icosahedron and the frame $\boldsymbol{a}_\mu^\perp$. The omitted vector $\boldsymbol{e}_4$ forms right-handed Cartesian frame with $\boldsymbol{e}_3$ and $\boldsymbol{e}_5$.

counterparts are densely distributed on five planes perpendicular to $\boldsymbol{e}_5$ and separated by the distance of $1/\sqrt{2}$ (Fig. 13.8).

Some properties of the tilings resulting from the strip projection depend on the position of the rhombic icosahedron with respect to these planes, but their general features are common. The integer combinations $m^\mu \boldsymbol{a}_\mu^\parallel$ such that $m^\mu \boldsymbol{a}_\mu^\perp$ are inside the rhombic icosahedron are nodes of a *generalized* Penrose tiling. The five-dimensional hypercubic cell has 2-facets in ten orientations. At a given vertex, the five-dimensional hypercubic cell has ten 2-facets, each spanned by a pair of vectors

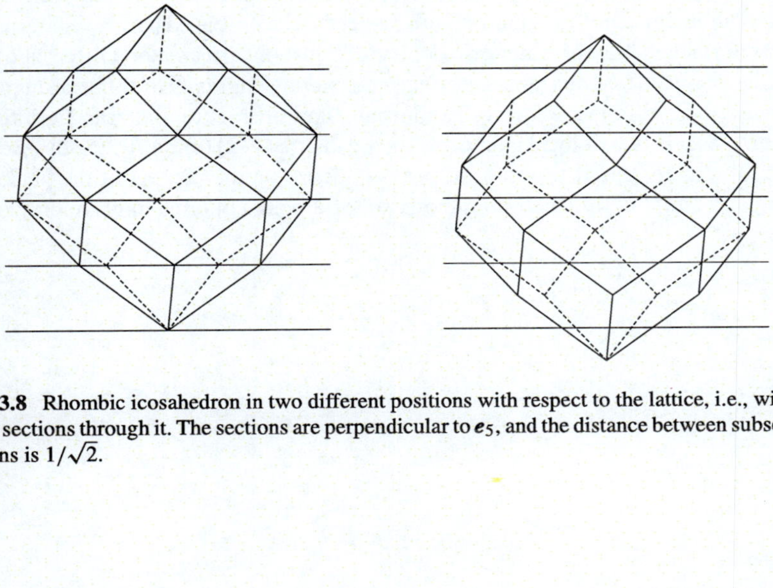

Fig. 13.8 Rhombic icosahedron in two different positions with respect to the lattice, i.e., with different sections through it. The sections are perpendicular to $\boldsymbol{e}_5$, and the distance between subsequent sections is $1/\sqrt{2}$.

$\boldsymbol{e}_\mu$. Projections of these 2-facets are rhombuses spanned by pairs of vectors $\boldsymbol{a}_\mu$; there are two types of the rhombuses (thick and thin), and each appears in five orientations. The rhombuses have sides of equal length and angles being multiples of $\pi/5$. In units of $\pi/5$, the thick rhombus has the angles of 2 and 3, and the thin rhombus has the angles of 1 and 4.

The generalized Penrose tilings are aperiodic. They have macroscopic tenfold symmetry with ten mirror lines (Fig. 13.9). The whole family consists of classes of locally isomorphic patterns. Among generalized Penrose tilings, only the true Penrose tilings have the property of self-similarity (e.g., [9]). They arise for special positions of the rhombic icosahedron with respect to the lattice. Original Penrose tilings satisfy matching rules, which are violated by generalized Penrose tilings not belonging to this class. For more details, see [10, 11].

13.3.1 Diffraction by Pentagonal Primitive Quasicrystal

The frames dual to $\boldsymbol{a}^{\parallel}_{\mu}$ and $\boldsymbol{a}^{\perp}_{\mu}$ of (13.15) and (13.16) are

$$\boldsymbol{a}_{\parallel}^{\mu} = \frac{2}{5}\boldsymbol{a}^{\parallel}_{\mu} \quad \text{and} \quad \boldsymbol{a}_{\perp}^{\mu} = \frac{2}{5}\boldsymbol{a}^{\perp}_{\mu} \ . \tag{13.19}$$

They are projections of the basis $\boldsymbol{b}^{\mu}$ $(= (2/5)\boldsymbol{b}_{\mu})$ dual to $\boldsymbol{b}_{\mu}$. As in the one-dimensional case (see (13.11)), the geometric structure factor is given by

$$\mathcal{F}[\rho](\boldsymbol{\xi}) = \sum \delta(\boldsymbol{\xi} - l_\mu \boldsymbol{a}_{\parallel}^{\mu})\ \mathsf{S}(l_\mu \boldsymbol{a}_{\perp}^{\mu}) \ ,$$

with the summation over integer five-tuples l_μ and $\mathsf{S}(\boldsymbol{\xi})$ linked to the Fourier transform of the characteristic function of sections through the rhombic icosahedron. The position of the rhombic icosahedron with respect to the $\boldsymbol{b}_\mu$-based lattice is determined by five parameters, but the transform depends on just one parameter. Different tilings with the same value of this parameter have the same weights S of diffraction peaks. Moreover, these tilings are locally isomorphic, and diffraction patterns from tilings from different classes of local isomorphism are distinct [12] (although the differences are small). Again, the distribution of peaks is dense on the plane, but the peaks are separated: a sufficiently small neighborhood of a peak contains only much weaker peaks.

(a)

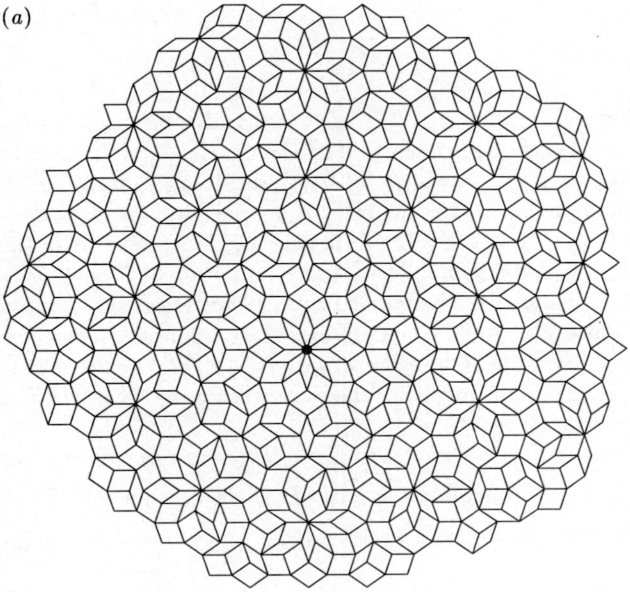

(b)

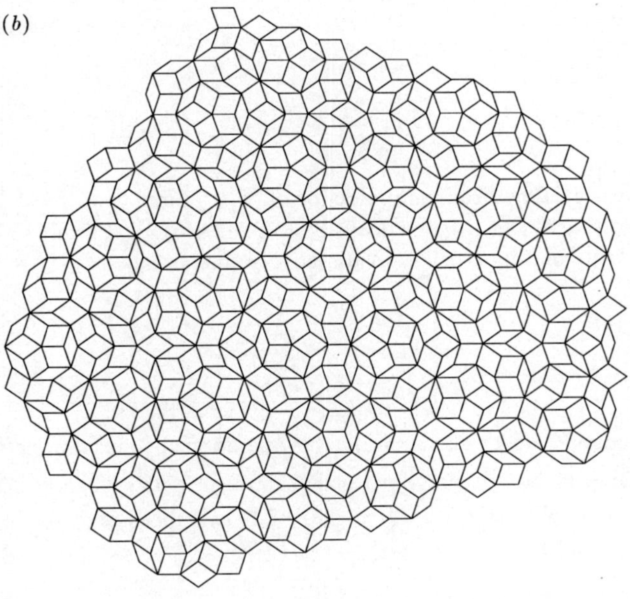

Fig. 13.9 Fragments of a generalized Penrose tiling (*a*) and a Penrose tiling (*b*). Some tile configurations present in (*a*) (like ten thin rhombuses with a common vertex marked by a disk) are not allowed in the original Penrose tilings.

13.3.2 Ambiguity Due to Rational Linear Dependence of Vectors $\boldsymbol{a}_\mu$

Analogously to the vectors $\boldsymbol{a}_\mu = \boldsymbol{a}^{\|}_\mu$ of direct space, the vectors $\boldsymbol{a}^\mu = \boldsymbol{a}^\mu_{\|}$ of the reciprocal space are rationally linearly dependent; if $l'_\mu = l_\mu + l$, where l is an integer, then $l'_\mu \boldsymbol{a}^\mu = l_\mu \boldsymbol{a}^\mu + l \sum_\mu \boldsymbol{a}^\mu = l_\mu \boldsymbol{a}^\mu + l\,\boldsymbol{0} = l_\mu \boldsymbol{a}^\mu$. To make the indices l_μ unique, an additional condition needs to be imposed. A simple option is to set the value of one of the indices, say, l_1, to 0. This is equivalent to using just four-vector frame (as, e.g., in [13]) which does not reflect the five-fold symmetry.

Other schemes preserve the symmetry. With the "least path" criterion [14], the sum $\sum_{\mu=1}^5 |l_\mu|$ takes the minimal value, and this is achieved by subtracting from all indices their median value. For instance, with $(l'_1\, l'_2\, l'_3\, l'_4\, l'_5) = (6\,0\,4\,2\,1)$, the median value is $l'_4 = 2$, and the equivalent indices satisfying the "least path" criterion are $(l_1\, l_2\, l_3\, l_4\, l_5) = (4\,\bar{2}\,2\,0\,\bar{1})$.

In a manner similar to the "least path" criterion, one can use indices which minimize $\left|\sum_{\mu=1}^5 l_\mu\right|$. To get them, one needs to subtract $\left\lfloor \sum_\mu l'_\mu/5 \right\rceil$ from each l'_μ. E.g., with $(l'_1\, l'_2\, l'_3\, l'_4\, l'_5) = (6\,0\,4\,2\,1)$, $\left\lfloor \sum_\mu l'_\mu/5 \right\rceil = \lfloor 13/5 \rceil = 3$, and the equivalent indices minimizing $\left|\sum_{\mu=1}^5 l_\mu\right|$ are $(l_1\, l_2\, l_3\, l_4\, l_5) = (3\,\bar{3}\,1\,\bar{1}\,\bar{2})$.

Finally, in analogy to Bravais–Miller indices in the hexagonal case, the condition

$$\sum_{\mu=1}^{5} l_\mu = 0 \tag{13.20}$$

can be imposed. It follows from minimization of $\sum_\mu l_\mu^2$. Since $\sum_\mu l'_\mu = \sum_\mu (l_\mu + l) = 5l + \sum_\mu l_\mu$, to satisfy (13.20), one needs $l_\mu = l'_\mu - (1/5) \sum_\mu l'_\mu$, and with integer l'_μ, the numbers l_μ are multiples of $1/5$. Alternatively, integer indices

$$l_\mu = 5 l'_\mu - \sum_{\mu=1}^{5} l'_\mu\,,$$

can be used, but one must remember that this operation changes the vector $\check{\boldsymbol{h}} = l'_\mu \boldsymbol{a}^\mu$ to $l_\mu \boldsymbol{a}^\mu = 5\check{\boldsymbol{h}}$, i.e., the vector direction remains the same but its magnitude is increased by the factor of 5. With $(l'_1\, l'_2\, l'_3\, l'_4\, l'_5) = (6\,0\,4\,2\,1)$ as indices of $\check{\boldsymbol{h}}$, the indices of $5\check{\boldsymbol{h}}$ which satisfy (13.20) are $(l_1\, l_2\, l_3\, l_4\, l_5) = (17\,\overline{13}\,7\,\bar{3}\,\bar{8})$.

13.3.3 Ambiguity in Assignment of Indices Due to Scaling 'Symmetry'

In the considered two-dimensional case, indexing is affected by the ambiguity related to scaling via inflation or deflation (analogous to that described in Sect. 13.1.5 for the one-dimensional Fibonacci chain). For the vector $\boldsymbol{a}_1$ of the frame (13.17), one has $\boldsymbol{a}_1\tau = -\boldsymbol{a}_3 - \boldsymbol{a}_4$ and $\boldsymbol{a}_1/\tau = \boldsymbol{a}_1(\tau - 1) = \boldsymbol{a}_1\tau - \boldsymbol{a}_1 = -\boldsymbol{a}_3 - \boldsymbol{a}_4 - \boldsymbol{a}_1$. By symmetry, there are analogous relationships for the remaining vectors. Generally, one can write

$$\boldsymbol{a}_\mu\tau = \boldsymbol{a}_\nu M^\nu{}_\mu\,, \qquad \boldsymbol{a}_\mu/\tau = \boldsymbol{a}_\nu N^\nu{}_\mu = \boldsymbol{a}_\nu\left(M^\nu{}_\mu - \delta^\nu{}_\mu\right)\,,$$

where

$$[M^\mu{}_\nu] = -\begin{bmatrix} 0\,0\,1\,1\,0 \\ 0\,0\,0\,1\,1 \\ 1\,0\,0\,0\,1 \\ 1\,1\,0\,0\,0 \\ 0\,1\,1\,0\,0 \end{bmatrix}, \quad [N^\mu{}_\nu] = -\begin{bmatrix} 1\,0\,1\,1\,0 \\ 0\,1\,0\,1\,1 \\ 1\,0\,1\,0\,1 \\ 1\,1\,0\,1\,0 \\ 0\,1\,1\,0\,1 \end{bmatrix}.$$

Inflation annihilates deflation and vice versa. The inverse of M is not equal to N, but $M^{-1} = N + O$, where $O^\mu_\nu = 1/2$ for all entries of the matrix, and $\boldsymbol{a}_\nu O^\nu{}_\mu = \boldsymbol{0}$. Thus, in agreement with $(\boldsymbol{a}_\mu/\tau)\tau = \boldsymbol{a}_\mu = (\boldsymbol{a}_\mu\tau)/\tau$, one has

$$\boldsymbol{a}_\kappa N^\kappa{}_\nu M^\nu{}_\mu = \boldsymbol{a}_\mu = \boldsymbol{a}_\kappa M^\kappa{}_\nu N^\nu{}_\mu\,.$$

Inflation $\boldsymbol{a}'_\mu = \boldsymbol{a}_\mu\tau$ in the direct space corresponds to deflation $\boldsymbol{a}'^\mu = \boldsymbol{a}^\mu/\tau$ in the reciprocal space. With $\boldsymbol{a}^\mu = \tau\boldsymbol{a}'^\mu = M^\mu{}_\nu\boldsymbol{a}'^\nu$, the quasilattice vector $l_\mu\boldsymbol{a}^\mu$ can be expressed as $l_\mu\boldsymbol{a}^\mu = l_\mu\tau\boldsymbol{a}'^\mu = l_\mu M^\mu{}_\nu\boldsymbol{a}'^\nu = l'_\nu\boldsymbol{a}'^\nu$, and hence, the indices in the frame $\boldsymbol{a}'^\nu$ are

$$l'_\nu = l_\mu M^\mu{}_\nu\,. \tag{13.21}$$

Similarly, deflation $\boldsymbol{a}'_\mu = \boldsymbol{a}_\mu/\tau$ in the direct space corresponds to inflation $\boldsymbol{a}'^\mu = \boldsymbol{a}^\mu\tau$ in the reciprocal space, and as above, sets of indices l_μ and l'_μ in the frames $\boldsymbol{a}^\mu$ and $\boldsymbol{a}'^\mu$, respectively, are related by

$$l'_\nu = l_\mu N^\mu{}_\nu\,. \tag{13.22}$$

The ambiguity is resolved by the choice of the fundamental vector indexed as $(l_1\,l_2\,l_3\,l_4\,l_5) = (1\,0\,0\,0\,0)$, and for that, one needs to account for peak intensities.

13.4 Frame-Based Tilings

As was already noted in relation to the one-dimensional examples and the two-dimensional pentagonal model, one can see the primitive quasicrystals as point sets for which scattering vectors corresponding to peaks on a discrete diffraction diagram can be expressed as integer linear combinations of a finite number of vectors. This is valid for all idealized quasicrystals. Peak positions are at nodes of a quasilattice, i.e., at integer linear combinations of some N vectors $\boldsymbol{a}^\mu$ ($\mu = 1, 2, \ldots, N$). For a structure to be a proper quasicrystal, the number N must exceed the space dimension. The set of vectors $\boldsymbol{a}^\mu$ is frequently referred to as 'a star of vectors'. It is convenient to see it as a finite frame over real numbers. (See Sect. 1.1.2.)

If a tiling gives a diffraction pattern with reflections corresponding to nodes of the quasilattice based on the frame $\boldsymbol{a}^\mu$, then the dual frame $\boldsymbol{a}_\mu$ in the direct space is expected to be linked to the geometry of the tiling. In conventional crystallography, *all* integer combinations of the vectors $\boldsymbol{a}_\mu$ point to nodes of the crystal lattice or vertices of the cells. In the case of quasicrystals, with a suitable choice of $\boldsymbol{a}_\mu$, *some* integer combinations of these vectors point to vertices of the cells. A way of constructing a tiling based on assumed vectors $\boldsymbol{a}_\mu$ is known as the grid method of de Bruijn.

13.4.1 Grid Method of De Bruijn

The method was originally developed to deal with Penrose tilings [15]. De Bruijn devised an algebraic procedure for getting vertices of these tilings. For that, he used a planar tessellation by five grids of straight lines (or a pentagrid as it was called in [15]) and showed that the pentagrid determines a dual tessellation in Penrose tiles. The original method can be easily generalized to generate other quasiperiodic two-dimensional tilings or three-dimensional packings [7, 16].

Before describing the essence of the actual de Bruijn's approach, some related aspects of the strip projection method will be considered. Given the N-dimensional lattice based on vectors $\boldsymbol{b}_\mu$ and a certain vector $\boldsymbol{\gamma}$, with vectors $\boldsymbol{b}^\mu$ dual to $\boldsymbol{b}_\mu$ and K^μ taking integer values, the equation

$$(\boldsymbol{x} - \boldsymbol{\gamma}) \cdot \boldsymbol{b}^\mu = K^\mu \,, \tag{13.23}$$

determines a family of equispaced hyperplanes. Clearly, there are N such families, one for each μ. If the lattice origin is at $\boldsymbol{\gamma}$, the hyperplanes pass through lattice nodes. One needs to consider the intersection of a hyperplane with the n_d-dimensional physical space. Let $\boldsymbol{x}$ be in the physical space, i.e., $\boldsymbol{x} \cdot \boldsymbol{a}^\mu_\perp = 0$ for all μ. The points $\boldsymbol{x}$ of the physical space belonging also to the μ-th hyperplane satisfy

$$\boldsymbol{x} \cdot \boldsymbol{a}^\mu = \boldsymbol{x} \cdot \boldsymbol{a}^\mu_\parallel = \boldsymbol{x} \cdot (\boldsymbol{a}^\mu_\parallel + \boldsymbol{a}^\mu_\perp) = \boldsymbol{x} \cdot \boldsymbol{b}^\mu = K^\mu + \boldsymbol{\gamma} \cdot \boldsymbol{b}^\mu = K^\mu + \gamma^\mu \,,$$

where $\gamma^\mu = \boldsymbol{\gamma} \cdot \boldsymbol{b}^\mu$. The dimension of the intersection of a hyperplane with the physical space is $n_d - 1$, i.e., the intersection of a μ-th family of hyperplanes (referred to as a grid) consists of points in the one-dimensional case (Fig. 13.10), parallel lines in the two-dimensional case and parallel planes in the three-dimensional case. With N vectors $\boldsymbol{a}^\mu$, the intersections constitute N grids collectively referred to as an N-grid.

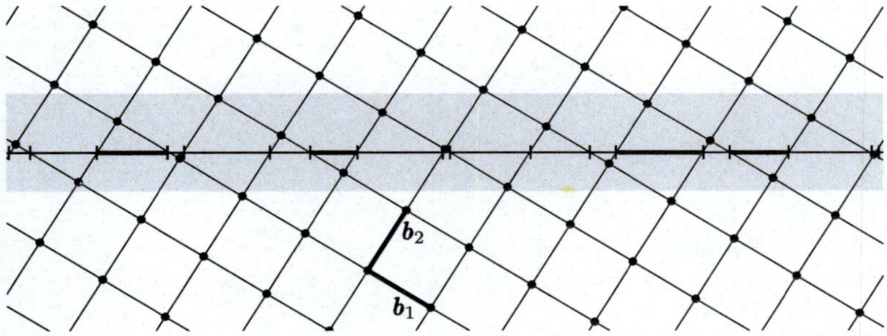

Fig. 13.10 The lines through nodes of $\boldsymbol{b}_\mu$-based lattice represent the (one-dimensional) hyperplanes described by (13.23). The lines of the μ-th family are perpendicular to $\boldsymbol{b}^\mu$. As in previous figures, the horizontal line represents the physical space. Thick segments mark four arbitrarily selected regions of the physical space. The 1-grid in the physical space is marked by vertical ticks.

With the $\boldsymbol{b}_\mu$-based lattice, one can select a strip such that n_d-dimensional facets of lattice cells entirely contained in the strip constitute a surface with a one-to-one orthogonal projection on the physical space. The intersection of a cell with the physical space will be called *region*. Now, the following observations are in place: First, a region is bounded by the intersections of the hyperplanes (13.23) with the physical space, i.e., it is a single mesh of the N-grid. Second, as was noted by de Bruijn, there are ways to ascribe each region to a unique vertex of the cell in the strip, i.e., one can have a one-to-one correspondence between the regions and vertices in the strip, and hence, (via arguments used to justify the strip projection method) a one-to-one correspondence between the regions and vertices of the tiling in the physical space.

The method can be most easily explained in the two-dimensional case (i.e., $n_d = 2$). To generate a tiling, the method relies on a frame of vectors $\boldsymbol{a}^\mu$ and the parameters γ^μ. With integer K^μ, given vector $\boldsymbol{a}^\mu$ and parameter γ^μ determine the grid of lines $\boldsymbol{x} \cdot \boldsymbol{a}^\mu = K^\mu + \gamma^\mu$ separated by the distance of $|\boldsymbol{a}^\mu|^{-1}$. All N vectors and parameters determine an N-grid. It is assumed that γ^μ are such that no more than two lines of the grid intersect at one point; such N-grid is called *regular*. For illustration, one can take the vectors pointing to vertices of a regular pentagon (13.17). Figure 13.11 shows an example 5-grid (= pentagrid) for some γ^μ satisfying the condition $\sum_\mu \gamma^\mu = 0$.

Now comes the aforementioned de Bruijn's assignment of lattice nodes in the stripe to regions in the physical space. For a given point $\boldsymbol{x}$, one defines the integers

$$K^\mu(\boldsymbol{x}) = \lceil \boldsymbol{x} \cdot \boldsymbol{a}^\mu - \gamma^\mu \rceil \ .$$

Interior of each region bounded by the lines of the N-grid is ascribed unique integers $K^\mu(\boldsymbol{x})$. Two regions sharing a common edge have one index different and the difference is ± 1. (For illustration, see Fig. 13.12.) In the considered case of two-dimensional physical space, each intersection point of the N-grid is at the boundary of four regions. With the region containing the point $\boldsymbol{x}$ ascribed to the vertex

$$K^\mu(\boldsymbol{x})\, \boldsymbol{a}_\mu$$

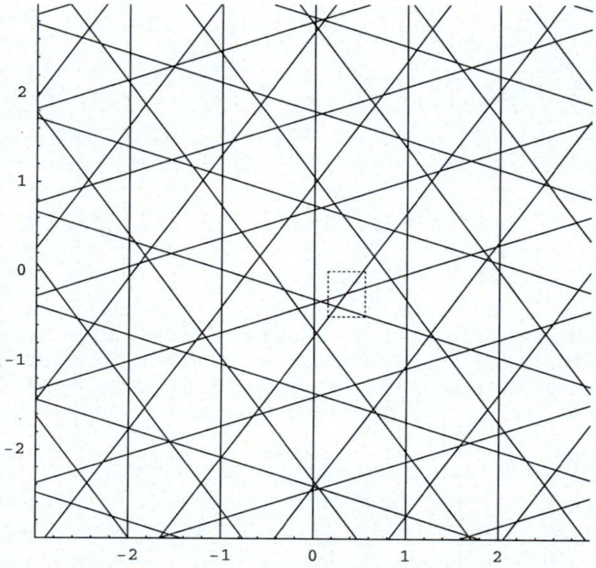

Fig. 13.11 An example de Bruijn's pentagrid. Dashed rectangle marks the fragment shown in Fig. 13.12*a*.

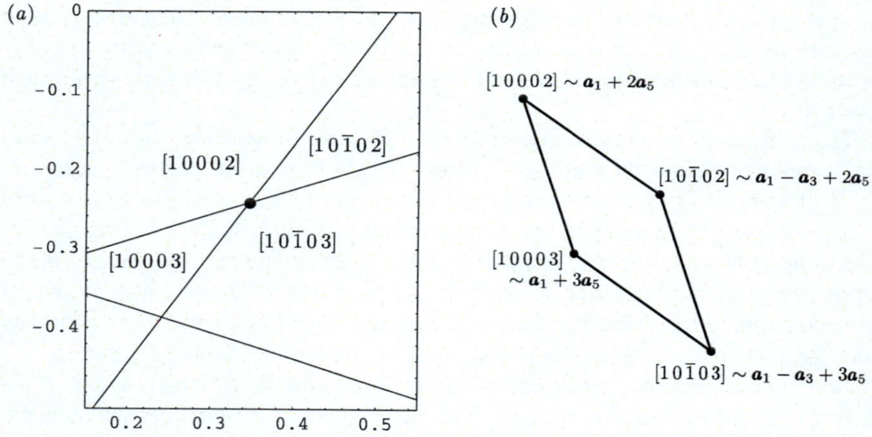

Fig. 13.12 (*a*) Fragment of the pentagrid shown in Fig. 13.11 with indices K^μ of regions meeting at the marked line intersection. (*b*) The corresponding tile with edges along single $\boldsymbol{a}_3$ and $\boldsymbol{a}_5$ vectors.

of the tiling, the four regions meeting at an N-grid intersection point are ascribed to four vertices of a single tile. It is easy to see this by noting that the sums $K^\mu \boldsymbol{a}_\mu$ corresponding to two regions sharing a common edge differ by one of the vectors $\pm\boldsymbol{a}_\mu$. Two regions having a common face correspond to two vertices connected by a tiling edge perpendicular to the face. Generally, each k-gonal region of the grid corresponds to a vertex of k tiles. Summarizing, one has the mapping of the tessellation by N-grid to tessellation into tiles with no gaps or overlaps. Each edge of the resulting tiling is parallel to one of the vectors $\boldsymbol{a}_\mu$, and the edge length is equal to the vector magnitude.

An example (Penrose) tiling corresponding to the vectors (13.17) is shown in Fig. 13.13*a*. Clearly, different sets of vectors $\boldsymbol{a}_\mu$ lead to different tilings. For illustration, let $\boldsymbol{a}_1$ be removed from the set (13.17). With the frame shown in Fig. 13.13*b*, the vector $-(\boldsymbol{a}_1 + \boldsymbol{a}_2 + \boldsymbol{a}_3 + \boldsymbol{a}_4)$ equals to the missing vector, but the tilings of Figs. 13.13*a* and *b* are different: the Penrose tiling has five-fold symmetry, while the pattern shown in Fig. 13.13*b* does not have it.

The same scheme is applicable in three-dimensions ($n_d = 3$) with lines of grids replaced by planes. With a regular N-grid (i.e., such that no more than three planes of the N-grid intersect at one point), each intersection point is linked to eight regions which are mapped on eight vertices of a parallelepiped of a packing.

The important and convenient features of de Bruijn's method for generating a tiling are that it relies solely on the vectors $\boldsymbol{a}_\mu$, and that the tiling inherits the symmetry of the figure built of these vectors.

13.5 Indices of Reflections

In indexing, it is important to know sets of indices representing symmetrically equivalent reflections. It is convenient to determine these sets algebraically without referring to geometry and high-dimensional spaces. A procedure for that, and procedures for transformation of indices between frames (Sect. 13.5.2 below) and for determination of zone axes (Sect. 13.5.3), rely on the rational linear independence of vectors of quasilattice frames.

13.5.1 Indices of Symmetrically Equivalent Reflections

Let $R_i^{\ j}$ be entries of an orthogonal matrix representing an element of the point group in the physical space in the Cartesian system. The $N \times N$ matrix obtained via $R_\mu^{\ \nu} = \mathcal{T}_\mu^{\ i} \mathcal{T}^\nu_{\ j} R_i^{\ j}$ represents this element in the frame $\boldsymbol{a}_\mu = \mathcal{T}_\mu^{\ i} \boldsymbol{e}_i$; see (1.27). The matrix $[\mathcal{T}^\mu_{\ i}]$ is the transposed pseudoinverse of $[\mathcal{T}_\mu^{\ i}]$. Given the indices l_μ, one seeks indices l'_μ ($\mu = 1, \ldots, N$) equivalent to l_μ due to R. The numbers

$$\lambda_\nu = R_\mu^{\ \nu} l_\nu$$

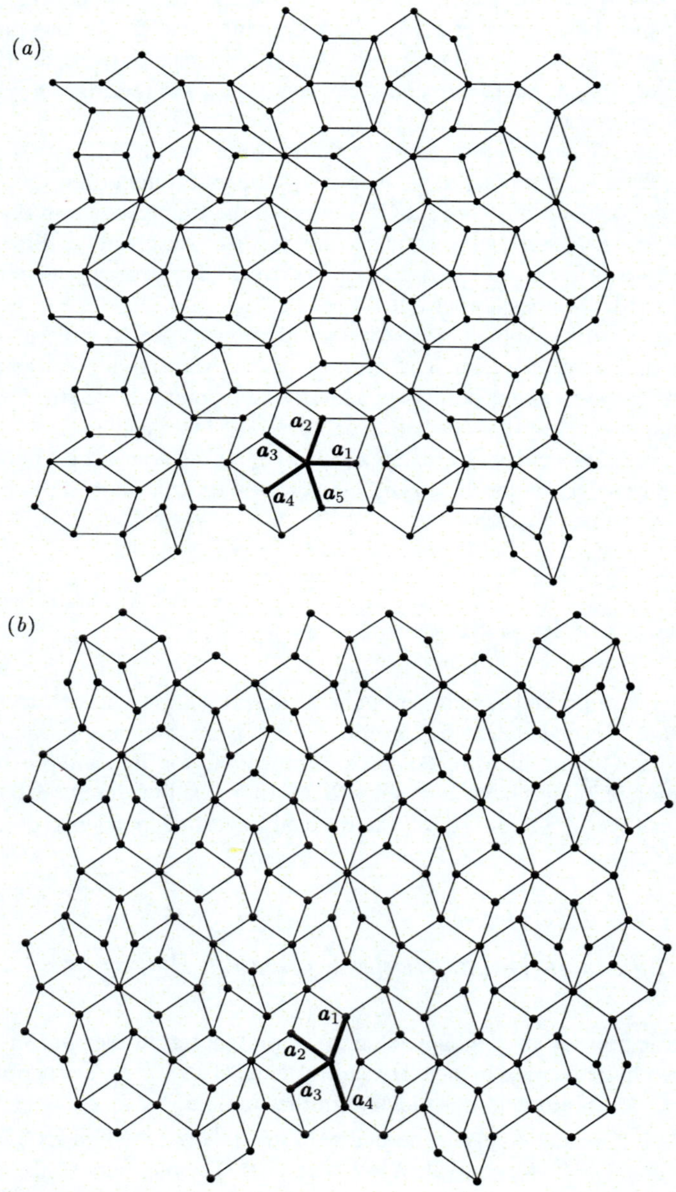

Fig. 13.13 (*a*) Fragment of Penrose tiling (generated from the frame (13.17) and γ^μ such that $\sum_\mu \gamma^\mu = 0$) based on the pentagrid shown in Fig. 13.11. (*b*) Fragment of the tiling generated from the four vectors $\boldsymbol{a}_\mu$ drawn at the lower part of the figure. Compared to the frame (13.17), the horizontally oriented vector is absent. The tiling (*b*) can be obtained from the tiling (*a*) by removing all tiles having horizontally oriented edges.

are components of the projection of the rotated vector indexed by l'_μ, i.e.,

$$g_\mu{}^\nu l'_\nu = \lambda_\mu \,. \tag{13.24}$$

By imposing these conditions, the requirement that irrational terms of l'_μ are zero, and any extra conditions which may arise due to rational linear dependence of vectors of the frame, one obtains the sought set of equivalent indices.

It is instructive to illustrate the above procedure by considering the particular two-dimensional pentagonal case. The only convenient form of the extra condition is $\sum_\mu l'_\mu = 0 = \sum_\mu l_\mu$, i.e., that given in (13.20). Take the frame (13.17), and the rotation by $2\pi/5$. In the Cartesian fame, the rotation is represented by

$$\left[R_i{}^j\right] = \frac{1}{2}\begin{bmatrix} \tau^{-1} & -\sqrt{\tau+2} \\ \sqrt{\tau+2} & \tau^{-1} \end{bmatrix} .$$

With the $\mathcal{T}_\mu{}^i$ $(i = 1, 2)$ given by (13.14), and $\mathcal{T}^\mu{}_i = (2/5)\mathcal{T}_\mu{}^i$, the 5×5 representation of the rotation is

$$\left[R_\mu{}^\nu\right] = \left[\mathcal{T}_\mu{}^i \mathcal{T}^\nu{}_j R_i{}^j\right] = \frac{1}{5}\begin{bmatrix} \tau^{-1} & -\tau & -\tau & \tau^{-1} & 2 \\ 2 & \tau^{-1} & -\tau & -\tau & \tau^{-1} \\ \tau^{-1} & 2 & \tau^{-1} & -\tau & -\tau \\ -\tau & \tau^{-1} & 2 & \tau^{-1} & -\tau \\ -\tau & -\tau & \tau^{-1} & 2 & \tau^{-1} \end{bmatrix} ,$$

and the rotated components $\lambda_\mu = R_\mu{}^\nu l_\nu$ are

$$\begin{aligned}(\lambda_1\, \lambda_2\, \lambda_3\, \lambda_4\, \lambda_5) = (\, & \tau^{-1}l_1 - \tau l_2 - \tau l_3 + \tau^{-1}l_4 + 2l_5 \quad 2l_1 + \tau^{-1}l_2 - \tau l_3 - \tau l_4 + \tau^{-1}l_5 \\ & \tau^{-1}l_1 + 2l_2 + \tau^{-1}l_3 - \tau l_4 - \tau l_5 \quad -\tau l_1 + \tau^{-1}l_2 + 2l_3 + \tau^{-1}l_4 - \tau l_5 \\ & -\tau l_1 - \tau l_2 + \tau^{-1}l_3 + 2l_4 + \tau^{-1}l_5\,)/5 \,.\end{aligned}$$

Now, with $g_\mu{}^\nu$ given by (13.18) the conditions $g_\mu{}^\nu l'_\nu = \lambda_\mu$ together with $\sum_\mu l'_\mu = 0$ lead to

$$\begin{aligned} l'_1 &= \left(-5l_2 - 5l_3 + 10l_5 + 5l'_3 + 5l'_4 + \sqrt{5}(2l_1 - 3l_2 - 3l_3 + 2l_4 + 2l_5 + 5l'_3 + 5l'_4)\right)/10 \,, \\ l'_2 &= \left(5l_1 + 5l_3 - 5l_4 - 5l_5 - 5l'_3 - 10l'_4 + \sqrt{5}(-l_1 + 4l_2 - l_3 - l_4 - l_5 - 5l'_3)\right)/10 \,. \end{aligned}$$

The requirement that the coefficients of $\sqrt{5}$ are zero and $l'_5 = -(l'_1 + l'_2 + l'_3 + l'_4)$ give

$$\begin{aligned}\left(l'_1\, l'_2\, l'_3\, l'_4\, l'_5\right) = (\, & -l_1 - l_2 - l_3 - l_4 + 4l_5 \quad 4l_1 - l_2 - l_3 - l_4 - l_5 \\ & -l_1 + 4l_2 - l_3 - l_4 - l_5 \quad -l_1 - l_2 + 4l_3 - l_4 - l_5 \\ & -l_1 - l_2 - l_3 + 4l_4 - l_5\,)/5 \,.\end{aligned}$$

Table 13.2 Indices symmetrically equivalent to $(l_1\, l_2\, l_3\, l_4\, l_5)$ for the five-element cyclic group in the frame (13.17).

$(l_1\, l_2\, l_3\, l_4\, l_5)$	$(l_2\, l_3\, l_4\, l_5\, l_1)$	$(l_3\, l_4\, l_5\, l_1\, l_2)$	$(l_4\, l_5\, l_1\, l_2\, l_3)$	$(l_5\, l_1\, l_2\, l_3\, l_4)$

By adding $0 = (l_1 + l_2 + l_3 + l_4 + l_5)/5$, one ultimately gets the expected result

$$\left(l'_1\, l'_2\, l'_3\, l'_4\, l'_5\right) = (l_5\, l_1\, l_2\, l_3\, l_4)\ .$$

All sets of equivalent indices for the five-element cyclic group of rotations about the origin of the frame are listed in Table 13.2. The considered example is trivial, but the described approach is general, and it is used below to create analogous tables for other frames and symmetries.

13.5.2 Transformation of Indices Between Frames

Given indices l_μ of a reflection in the frame $\boldsymbol{a}^\mu$, what are its indices in a different frame $\boldsymbol{a}'^\mu$? The reflection corresponds to a unique quasilattice vector, i.e.,

$$l'_\nu \boldsymbol{a}'^\nu = l_\nu \boldsymbol{a}^\nu\ .$$

Scalar multiplication of both sides by the dual $\boldsymbol{a}'_\mu$ gives

$$g'^{\ \nu}_{\mu} l'_\nu = \lambda_\mu\ ,$$

where $g'^{\ \nu}_{\mu} = \boldsymbol{a}'^\mu \cdot \boldsymbol{a}'_\mu$ and $\lambda_\mu = l_\nu\, \boldsymbol{a}^\nu \cdot \boldsymbol{a}'_\mu$. The above equation with respect to l'_ν has the same form as (13.24), and it can be solved in the same way.

13.5.3 Zone Law

Reflection with indices l_μ belongs to the zone specified by indices m^μ if the vector $\check{\boldsymbol{l}} = l_\mu \boldsymbol{a}^\mu$ is perpendicular to $\check{\boldsymbol{m}} = m^\mu \boldsymbol{a}_\mu$ i.e., if $\check{\boldsymbol{l}} \cdot \check{\boldsymbol{m}} = 0$. Explicitly, one has

$$\check{\boldsymbol{l}} \cdot \check{\boldsymbol{m}} = l_\mu m^\nu\, \boldsymbol{a}^\mu \cdot \boldsymbol{a}_\nu = g^\mu_{\ \nu}\, l_\mu m^\nu = 0\ .$$

The relationship $g^\mu_{\ \nu}\, l_\mu m^\nu = 0$ involves integer l_μ and m^μ, and the coefficients $g^\mu_{\ \nu}$ having both rational and irrational terms. Thus, the single scalar relationship $g^\mu_{\ \nu}\, l_\mu m^\nu = 0$ generally provides multiple conditions on l_μ.

Determination of zone axis direction based on sets of indices l^a_μ and l^b_ν of two reflections belonging to the zone is also straightforward. With (non-collinear) $\check{\boldsymbol{l}}^a = l^a_\mu \boldsymbol{a}^\mu$ and $\check{\boldsymbol{l}}^b = l^b_\mu \boldsymbol{a}^\mu$, the zone axis is parallel to

$$\check{\boldsymbol{m}} = m^\mu \boldsymbol{a}_\mu \propto \check{\boldsymbol{l}}^a \times \check{\boldsymbol{l}}^b = l^a_\mu l^b_\nu \boldsymbol{a}^\mu \times \boldsymbol{a}^\nu = l^a_\mu l^b_\nu T^\mu{}_i T^\nu{}_j \boldsymbol{e}^i \times \boldsymbol{e}^j = l^a_\mu l^b_\nu T^\mu{}_i T^\nu{}_j \varepsilon^{ijk} \boldsymbol{e}_k \,.$$

Scalar multiplication of both sides by $\boldsymbol{a}^\rho$ leads to

$$m^\mu g_\mu^{\ \rho} \propto \lambda^\rho \,,$$

where $\lambda^\rho = l^a_\mu l^b_\nu T^\mu{}_i T^\nu{}_j T^\rho{}_k \varepsilon^{ijk}$. The above equation with known $g_\mu^{\ \nu}$ and λ^ρ is similar to (13.24), and it can be solved with respect to m^μ (up to a scale or magnitude of $\check{\boldsymbol{m}}$) by the method described in Sect. 13.5.1.

13.5.4 Indices of Peaks in Powder Diffraction Diagrams

The principles of powder diffraction by quasicrystal are the same as in convectional crystallography. With fixed wavelength λ, positions of diffraction peaks are determined at various scattering angles θ or at various $(2\sin\theta/\lambda)^2$. The latter quantity is equal to the squared magnitude of a certain scattering vector or of a certain vector $\check{\boldsymbol{l}}$ of the quasilattice, i.e.,

$$(2\sin\theta/\lambda)^2 = |\check{\boldsymbol{l}}|^2 = g^{\mu\nu} l_\mu l_\nu \,,$$

where $g^{\mu\nu} = \boldsymbol{a}^\mu \cdot \boldsymbol{a}^\nu$. The point is to determine a suitable frame $\boldsymbol{a}^\mu$ and the sets of indices l_μ corresponding to the measured peak positions. Clearly, the indexing of powder diffraction patterns is also affected by the scaling ambiguity; with $l'_\nu = l_\mu M^\mu{}_\nu$, $\boldsymbol{a}^\mu = M^\mu{}_\nu \boldsymbol{a}'^\mu$, $g'^{\mu\nu} = \boldsymbol{a}'^\mu \cdot \boldsymbol{a}'^\nu$ and $g^{\mu\nu} = M^\mu{}_\kappa M^\nu{}_\rho g'^{\kappa\rho}$, one has $g^{\mu\nu} l_\mu l_\nu = g'^{\mu\nu} l'_\mu l'_\nu$.

13.6 The Decagonal and Other Axial Quasicrystals

Armed with the experience based on one- and two-dimensional quasicrystals, the strip projection method and de Bruijn's grid method, one can proceed to considering three-dimensional cases.

The decagonal, octagonal and dodecagonal quasicrystals belong to the class with periodicity in one direction and quasiperiodicity in the perpendicular planes. An example diffraction pattern of a decagonal phase is shown in Fig. 13.14. The standard decagonal indexing scheme [17] relies on a vector along the direction in which quasicrystals are periodic (assumed to be parallel to $\boldsymbol{e}_3$) and five unit vectors perpendicular to the first one pointing to vertices of a regular pentagon. I.e., the frame is built of the vectors defined by (13.17) plus the vector $\boldsymbol{a}_6 = (c/a)\boldsymbol{e}_3$ with $\boldsymbol{a}_\mu \cdot \boldsymbol{a}_6 = (c/a)^2\delta_{\mu 6}$ (Fig. 13.15*a*).

With this frame, the description of the decagonal quasicrystals is directly related to that of the two-dimensional pentagonal case, and features of decagonal quasicrystals can be deduced from that case. For instance, it is easy to see that the packings generated by the frame are the two-dimensional generalized Penrose tilings in the quasicrystalline plane perpendicular to $\boldsymbol{a}_6 \propto \boldsymbol{e}_3$, extended in the third direction along $\boldsymbol{a}_6$. Thus, the cells are two types of right prisms with Penorse thick and thin rhombuses as their base faces. Each type appears in five orientations, and the symmetry of the packing is D_{10h}. The 10-fold axis is along the vectors with indices $[0\,0\,0\,0\,0\,m]$ ($m \neq 0$), and there are ten two-fold axes parallel to $[m\,0\,0\,0\,0\,0]$ and $[m\,0\,0\,\overline{m}\,0\,0] \parallel [0\,m\,\overline{m}\,0\,0\,0]$.

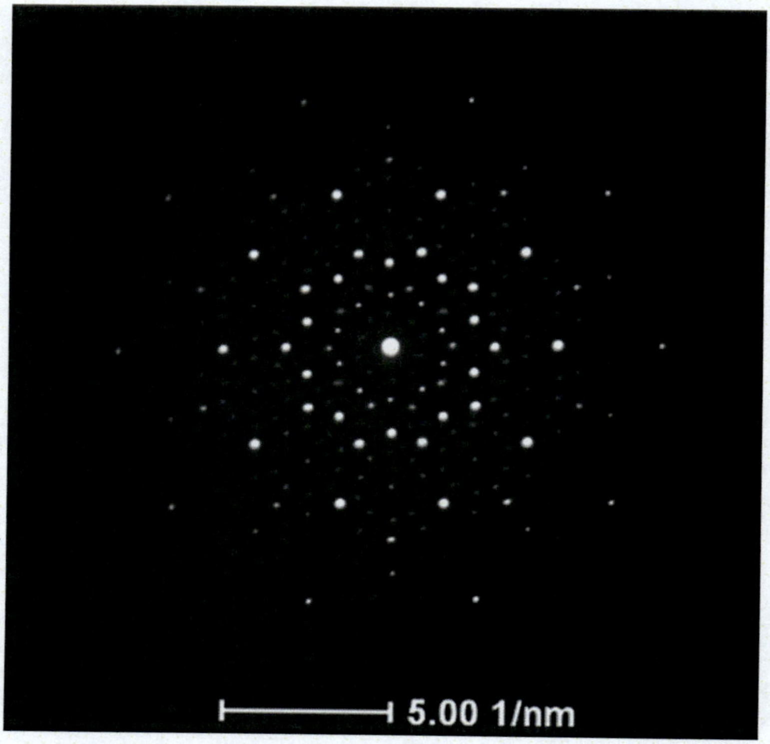

Fig. 13.14 Selected area electron diffraction pattern of decagonal phase of rapidly cooled Al-Mn-Fe alloy. Courtesy of K. Stan-Glowinska.

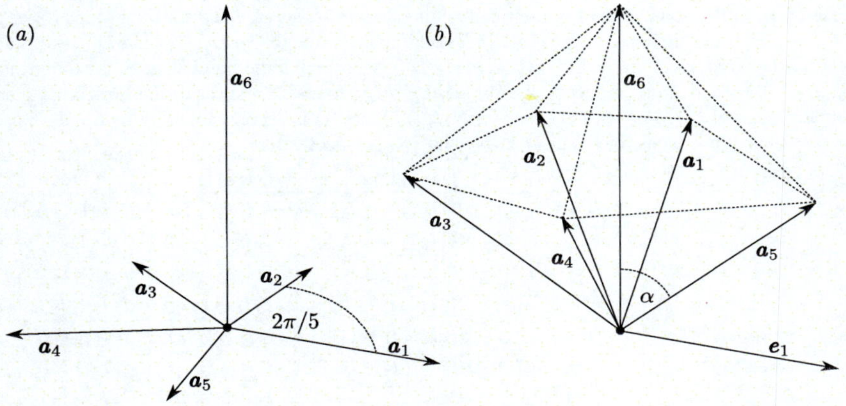

Fig. 13.15 (*a*) The vectors of the standard frame used in the decagonal case. Vector $\boldsymbol{a}_6$ is perpendicular to the remaining five vectors. (*b*) Vectors of the Ho's frame (13.26).

The frame dual to that of Fig. 13.15*a* is

$$\boldsymbol{a}^{\mu} = \frac{2}{5}\boldsymbol{a}_{\mu}\,,\ \mu = 1, 2, \ldots, 5 \ \text{ and } \boldsymbol{a}^6 = \frac{a}{c}\boldsymbol{e}^3\,. \tag{13.25}$$

These vectors are not rationally linearly independent (as $\sum_{\mu=1}^{5}\boldsymbol{a}^{\mu} = \boldsymbol{0}$). Therefore, for uniqueness of reflection indexes, a condition analogous to one of those described in Sect. 13.3.2 is needed. For simplicity, it is assumed below that $\sum_{\mu=1}^{5} l_{\mu} = 0$.

The zone rule is analogous to 13.5.3. Reflection with indices l_{μ} belongs to the zone specified by indices m^{μ} if

$$\check{\boldsymbol{h}} \cdot \check{\boldsymbol{m}} = l_{\mu} m^{\nu}\, \boldsymbol{a}^{\mu} \cdot \boldsymbol{a}_{\nu} = g^{\mu}_{\ \nu}\, l_{\mu} m^{\nu} = 0\,,$$

where $[g^{\mu}_{\ \nu}]$ is obtained from (13.18) by appending the sixth column and the sixth row with all entries equal to 0 except $g^{6}_{\ 6} = 1$.

Moreover, one faces the ambiguity in assignment of indices to diffraction reflections due to scaling. The quasilattice can be inflated or deflated in the quasicrystalline plane, and the rules of Sect. 13.3.3 are applicable to the first five indices l_{μ}.

Sets of equivalent reflection indices for pentagonal and decagonal point groups and the frame (13.25) are listed in Table 13.3. Although this is the simplest way of expressing the equivalences, the indices need to be handled with care. Take for instance the case of highest decagonal symmetry D_{10h}; one can check that the set $(1\,\bar{1}\,0\,0\,0\,0)$ is equivalent to $(1\,0\,0\,0\,\bar{1}\,0)$, but it is not equivalent to $(1\,0\,\bar{1}\,0\,0\,0)$.

The Fourier transform of decagonal (and other axial) 'pure-point' quasicrystal is a product of one-dimensional Shah 'function' reflecting periodicity along the principal axis and the geometric structure factor for a two-dimensional quasicrystal. Since these structures are periodic in the direction perpendicular to quasiperiodic layers, there is

Table 13.3 Reflection indices symmetrically equivalent to $(l_1\, l_2\, l_3\, l_4\, l_5\, l_6)$ for the groups C_5, C_{5v}, C_{10}, C_{10v}, D_5 and D_{10} in the frame (13.17). By adding inversion to C_5, D_5, C_{10} and D_{10}, (i.e., each $(l_1\, l_2\, l_3\, l_4\, l_5\, l_6)$ is accompanied by $(\bar{l}_1\, \bar{l}_2\, \bar{l}_3\, \bar{l}_4\, \bar{l}_5\, \bar{l}_6)$) one obtains the equivalences for the groups S_{10}, D_{5d}, C_{10h} and D_{10h}, respectively. By adding reflection with respect to the horizontal plane (perpendicular to $\boldsymbol{e}^3$) to C_5 and D_5 (i.e., each $(l_1\, l_2\, l_3\, l_4\, l_5\, l_6)$ is accompanied by $(l_1\, l_2\, l_3\, l_4\, l_5\, \bar{l}_6)$), one gets the equivalences for the groups C_{5h} and D_{5h}, respectively.

C_5	$(l_1\, l_2\, l_3\, l_4\, l_5\, l_6)$	$(l_2\, l_3\, l_4\, l_5\, l_1\, l_6)$	$(l_3\, l_4\, l_5\, l_1\, l_2\, l_6)$	$(l_4\, l_5\, l_1\, l_2\, l_3\, l_6)$	$(l_5\, l_1\, l_2\, l_3\, l_4\, l_6)$
C_{5v}	$(l_1\, l_2\, l_3\, l_4\, l_5\, l_6)$	$(l_2\, l_3\, l_4\, l_5\, l_1\, l_6)$	$(l_3\, l_4\, l_5\, l_1\, l_2\, l_6)$	$(l_4\, l_5\, l_1\, l_2\, l_3\, l_6)$	$(l_5\, l_1\, l_2\, l_3\, l_4\, l_6)$
	$(\bar{l}_5\, \bar{l}_4\, \bar{l}_3\, \bar{l}_2\, \bar{l}_1\, l_6)$	$(\bar{l}_4\, \bar{l}_3\, \bar{l}_2\, \bar{l}_1\, \bar{l}_5\, l_6)$	$(\bar{l}_3\, \bar{l}_2\, \bar{l}_1\, \bar{l}_5\, \bar{l}_4\, l_6)$	$(\bar{l}_2\, \bar{l}_1\, \bar{l}_5\, \bar{l}_4\, \bar{l}_3\, l_6)$	$(\bar{l}_1\, \bar{l}_5\, \bar{l}_4\, \bar{l}_3\, \bar{l}_2\, l_6)$
C_{10}	$(l_1\, l_2\, l_3\, l_4\, l_5\, l_6)$	$(l_2\, l_3\, l_4\, l_5\, l_1\, l_6)$	$(l_3\, l_4\, l_5\, l_1\, l_2\, l_6)$	$(l_4\, l_5\, l_1\, l_2\, l_3\, l_6)$	$(l_5\, l_1\, l_2\, l_3\, l_4\, l_6)$
	$(\bar{l}_1\, \bar{l}_2\, \bar{l}_3\, \bar{l}_4\, \bar{l}_5\, l_6)$	$(\bar{l}_2\, \bar{l}_3\, \bar{l}_4\, \bar{l}_5\, \bar{l}_1\, l_6)$	$(\bar{l}_3\, \bar{l}_4\, \bar{l}_5\, \bar{l}_1\, \bar{l}_2\, l_6)$	$(\bar{l}_4\, \bar{l}_5\, \bar{l}_1\, \bar{l}_2\, \bar{l}_3\, l_6)$	$(\bar{l}_5\, \bar{l}_1\, \bar{l}_2\, \bar{l}_3\, \bar{l}_4\, l_6)$
C_{10v}	$(l_1\, l_2\, l_3\, l_4\, l_5\, l_6)$	$(l_2\, l_3\, l_4\, l_5\, l_1\, l_6)$	$(l_3\, l_4\, l_5\, l_1\, l_2\, l_6)$	$(l_4\, l_5\, l_1\, l_2\, l_3\, l_6)$	$(l_5\, l_1\, l_2\, l_3\, l_4\, l_6)$
	$(l_5\, l_4\, l_3\, l_2\, l_1\, l_6)$	$(l_4\, l_3\, l_2\, l_1\, l_5\, l_6)$	$(l_3\, l_2\, l_1\, l_5\, l_4\, l_6)$	$(l_2\, l_1\, l_5\, l_4\, l_3\, l_6)$	$(l_1\, l_5\, l_4\, l_3\, l_2\, l_6)$
	$(\bar{l}_1\, \bar{l}_2\, \bar{l}_3\, \bar{l}_4\, \bar{l}_5\, l_6)$	$(\bar{l}_2\, \bar{l}_3\, \bar{l}_4\, \bar{l}_5\, \bar{l}_1\, l_6)$	$(\bar{l}_3\, \bar{l}_4\, \bar{l}_5\, \bar{l}_1\, \bar{l}_2\, l_6)$	$(\bar{l}_4\, \bar{l}_5\, \bar{l}_1\, \bar{l}_2\, \bar{l}_3\, l_6)$	$(\bar{l}_5\, \bar{l}_1\, \bar{l}_2\, \bar{l}_3\, \bar{l}_4\, l_6)$
	$(\bar{l}_5\, \bar{l}_4\, \bar{l}_3\, \bar{l}_2\, \bar{l}_1\, l_6)$	$(\bar{l}_4\, \bar{l}_3\, \bar{l}_2\, \bar{l}_1\, \bar{l}_5\, l_6)$	$(\bar{l}_3\, \bar{l}_2\, \bar{l}_1\, \bar{l}_5\, \bar{l}_4\, l_6)$	$(\bar{l}_2\, \bar{l}_1\, \bar{l}_5\, \bar{l}_4\, \bar{l}_3\, l_6)$	$(\bar{l}_1\, \bar{l}_5\, \bar{l}_4\, \bar{l}_3\, \bar{l}_2\, l_6)$
D_5	$(l_1\, l_2\, l_3\, l_4\, l_5\, l_6)$	$(l_2\, l_3\, l_4\, l_5\, l_1\, l_6)$	$(l_3\, l_4\, l_5\, l_1\, l_2\, l_6)$	$(l_4\, l_5\, l_1\, l_2\, l_3\, l_6)$	$(l_5\, l_1\, l_2\, l_3\, l_4\, l_6)$
	$(l_5\, l_4\, l_3\, l_2\, l_1\, \bar{l}_6)$	$(l_4\, l_3\, l_2\, l_1\, l_5\, \bar{l}_6)$	$(l_3\, l_2\, l_1\, l_5\, l_4\, \bar{l}_6)$	$(l_2\, l_1\, l_5\, l_4\, l_3\, \bar{l}_6)$	$(l_1\, l_5\, l_4\, l_3\, l_2\, \bar{l}_6)$
D_{10}	$(l_1\, l_2\, l_3\, l_4\, l_5\, l_6)$	$(l_2\, l_3\, l_4\, l_5\, l_1\, l_6)$	$(l_3\, l_4\, l_5\, l_1\, l_2\, l_6)$	$(l_4\, l_5\, l_1\, l_2\, l_3\, l_6)$	$(l_5\, l_1\, l_2\, l_3\, l_4\, l_6)$
	$(\bar{l}_1\, \bar{l}_2\, \bar{l}_3\, \bar{l}_4\, \bar{l}_5\, l_6)$	$(\bar{l}_2\, \bar{l}_3\, \bar{l}_4\, \bar{l}_5\, \bar{l}_1\, l_6)$	$(\bar{l}_3\, \bar{l}_4\, \bar{l}_5\, \bar{l}_1\, \bar{l}_2\, l_6)$	$(\bar{l}_4\, \bar{l}_5\, \bar{l}_1\, \bar{l}_2\, \bar{l}_3\, l_6)$	$(\bar{l}_5\, \bar{l}_1\, \bar{l}_2\, \bar{l}_3\, \bar{l}_4\, l_6)$
	$(l_5\, l_4\, l_3\, l_2\, l_1\, \bar{l}_6)$	$(l_4\, l_3\, l_2\, l_1\, l_5\, \bar{l}_6)$	$(l_3\, l_2\, l_1\, l_5\, l_4\, \bar{l}_6)$	$(l_2\, l_1\, l_5\, l_4\, l_3\, \bar{l}_6)$	$(l_1\, l_5\, l_4\, l_3\, l_2\, \bar{l}_6)$
	$(\bar{l}_5\, \bar{l}_4\, \bar{l}_3\, \bar{l}_2\, \bar{l}_1\, \bar{l}_6)$	$(\bar{l}_4\, \bar{l}_3\, \bar{l}_2\, \bar{l}_1\, \bar{l}_5\, \bar{l}_6)$	$(\bar{l}_3\, \bar{l}_2\, \bar{l}_1\, \bar{l}_5\, \bar{l}_4\, \bar{l}_6)$	$(\bar{l}_2\, \bar{l}_1\, \bar{l}_5\, \bar{l}_4\, \bar{l}_3\, \bar{l}_6)$	$(\bar{l}_1\, \bar{l}_5\, \bar{l}_4\, \bar{l}_3\, \bar{l}_2\, \bar{l}_6)$

periodicity along the same direction in the reciprocal space. (In diffraction patterns of real decagonal phases, intensities of reflections along the principal axis vary due to presence of quasicrystalline layers. Typically, the number of the layers is 2, 4, 6 or 8.)

13.6.1 Other Frames

For completeness, one needs to mention other indexing schemes. As was already explained, the vectors of the standard frame are not rationally independent. Reflections originating from the decagonal phase can be indexed using four vectors in the quasicrystalline plane and a vector along the periodic direction, i.e., five vectors in total. Such five-vector indexing scheme is preferred by some authors [13]. With this frame, however, sets of indices of symmetry-related vectors look different, and it is difficult to identify symmetry-related reflections. Moreover, it is clear that symmetry of the packing generated by such frame does not match the symmetry of the decagonal quasicrystal; c.f. Fig. 13.13.

Some studies (e.g., [18]) relied on the 'pentagonal-bipyramid' indexing scheme devised by Ho [19]. (A similar scheme was considered in [20].) The frame in the reciprocal space is

$$a^\mu = \cos\alpha\, e^3 + \sin\alpha\, R^\mu_{e^3} e^1\,, \quad \mu = 1, 2, \ldots, 5 \quad \text{and} \quad a^6 = 2\cos\alpha\, e^3\,, \qquad (13.26)$$

where $R^\mu_{e^3}$ is the rotation about e^3 by the angle $2\mu\pi/5$. (With $\mu = 1, 2, \ldots, 5$, the vectors a^μ and $a^\mu - a^6$ are parallel to edges of a pentagonal bipyramid; Fig. 13.15*b*.) The set of integer combinations $l_\mu a^\mu$ is periodic along e^3 with the period $\cos\alpha$. The Ho's frame is similar to a frame applicable to the icosahedral case; see (13.32) below. The motivation behind suggesting the frame (13.26) is that decagonal structure can be considered as a distorted icosahedral structure. In a sense, decagonal quasicrystals can be seen as an approximants of the icosahedral phase. Electron diffraction patterns from decagonal and icosahedral phases exhibit similar features; see, e.g. [21]. With the frame (13.26), however, the vectors a^μ ($\mu = 1, 2, \ldots, 5$) have components along both periodic and quasiperiodic directions, and this makes the Ho's indexing scheme cumbersome and impractical.

13.6.2 Other Axial Quasicrystals

Two other types of quasicrystals with one periodic direction are rare octagonal and dodecagonal phases. The indexing methods are similar to those used for the decagonal phase. For illustration, the octagonal case is described briefly.

The natural frame for two-dimensional octagonal tilings and quasilattices consists of the four unit vectors

$$a_\mu = \cos(2\pi(\mu - 1)/8)\, e_1 + \sin(2\pi(\mu - 1)/8)\, e_2\,, \quad \mu = 1, \ldots, 4. \qquad (13.27)$$

These vectors are rationally linearly independent. Using de Bruijn's method (with pentagrid replaced by a tetragrid), the vectors generate tilings with eight-fold symmetry. The tiles are square and rhombus with the angle of $\pi/4$, and all sides of the tiles have the same length; see Fig. 13.16.

In the octagonal case, the ratio $q = 1 + \sqrt{2}$ plays the same role as the golden ratio in the case of Penrose patterns. It solves the equation $q^2 - 2q - 1 = 0$, i.e., one has $1/q = q - 2 = \sqrt{2} - 1$. The scaling law for two-dimensional octagonal case is

$$a_\mu q = a_\nu M^\nu{}_\mu\,, \qquad a_\mu/q = a_\nu N^\nu{}_\mu\,,$$

where

$$[M^\mu{}_\nu] = \begin{bmatrix} 1 & 1 & 0 & -1 \\ 1 & 1 & 1 & 0 \\ 0 & 1 & 1 & 1 \\ -1 & 0 & 1 & 1 \end{bmatrix}, \quad [N^\mu{}_\nu] = \begin{bmatrix} -1 & 1 & 0 & -1 \\ 1 & -1 & 1 & 0 \\ 0 & 1 & -1 & 1 \\ -1 & 0 & 1 & -1 \end{bmatrix},$$

and $N = M^{-1} = M - 2I_4$. The expressions for the transformation of indices of reflections are analogous to (13.21) and (13.22).

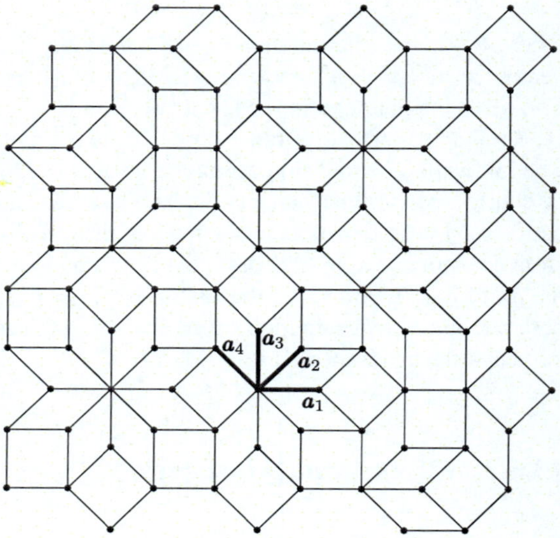

Fig. 13.16 A fragment of a tiling with eight-fold symmetry. It was generated using de Bruin's method based on the four vectors $\boldsymbol{a}_\mu$. The canonical octagonal tiling (introduced by R. Ammann and first described by Beenker [22]) is known as Ammann-Beenker tiling.

The complete direct space frame suitable for describing three-dimensional quasicrystal consist of the four vectors (13.27) and the fifth along the eightfold axis

$$\boldsymbol{a}_5 = (c/a)\boldsymbol{e}_3 \ .$$

The dual frame for indexing of reflections in diffraction patterns is

$$\boldsymbol{a}^\mu = \boldsymbol{a}_\mu/2 \ \text{for} \ \mu = 1, 2, 3, 4 \ \text{and} \ \boldsymbol{a}^5 = (a/c)\,\boldsymbol{e}_3 \ . \tag{13.28}$$

Lists of symmetrically equivalent reflection indices for octagonal groups are collected in Table 13.4.

13.7 The Icosahedral Quasicrystal

The icosahedral point group is a subgroup of the symmetry group of six-dimensional hypercubic lattice. The decomposition of the representation of the icosahedral group which permutes the orthonormal basis $\boldsymbol{e}_\mu$ ($\mu = 1, 2, \ldots, 6$) into irreducible representations gives two invariant three-dimensional subspaces with two non-equivalent three-dimensional representations of the group. By properly orienting such a lattice and projecting nodes located in a stripe on the three-dimensional space, one gets vertices of an aperiodic packing showing icosahedral symmetry.

Table 13.4 Reflection indices symmetrically equivalent to $(l_1\,l_2\,l_3\,l_4,\,l_5)$ for the octagonal groups C_8, C_{8v}, D_8, D_{4d} and S_8 in the frame (13.28). By adding inversion to C_8 and D_8, (i.e., each $(l_1\,l_2\,l_3\,l_4\,l_5)$ is accompanied by $(\bar{l}_1\,\bar{l}_2\,\bar{l}_3\,\bar{l}_4\,\bar{l}_5)$) or by adding reflection with respect to the horizontal plane (perpendicular to $\boldsymbol{e}^3$) to C_8 and D_8 (i.e., each $(l_1\,l_2\,l_3\,l_4\,l_5)$ is accompanied by $(l_1\,l_2\,l_3\,l_4\,\bar{l}_5)$), one obtains the equivalences for the groups C_{8h} and D_{8h}, respectively.

C_8	$(l_1\,l_2\,l_3\,l_4\,l_5)$	$(l_2\,l_3\,l_4\,\bar{l}_1\,l_5)$	$(l_3\,l_4\,\bar{l}_1\,\bar{l}_2\,l_5)$	$(l_4\,\bar{l}_1\,\bar{l}_2\,\bar{l}_3\,l_5)$
	$(\bar{l}_1\,\bar{l}_2\,\bar{l}_3\,\bar{l}_4\,l_5)$	$(\bar{l}_2\,\bar{l}_3\,\bar{l}_4\,l_1\,l_5)$	$(\bar{l}_3\,\bar{l}_4\,l_1\,l_2\,l_5)$	$(\bar{l}_4\,l_1\,l_2\,l_3\,l_5)$
C_{8v}	$(l_1\,l_2\,l_3\,l_4\,l_5)$	$(l_2\,l_3\,l_4\,\bar{l}_1\,l_5)$	$(l_3\,l_4\,\bar{l}_1\,\bar{l}_2\,l_5)$	$(l_4\,\bar{l}_1\,\bar{l}_2\,\bar{l}_3\,l_5)$
	$(\bar{l}_1\,\bar{l}_2\,\bar{l}_3\,\bar{l}_4\,l_5)$	$(\bar{l}_2\,\bar{l}_3\,\bar{l}_4\,l_1\,l_5)$	$(\bar{l}_3\,\bar{l}_4\,l_1\,l_2\,l_5)$	$(\bar{l}_4\,l_1\,l_2\,l_3\,l_5)$
	$(l_4\,l_3\,l_2\,l_1\,l_5)$	$(l_3\,l_2\,l_1\,\bar{l}_4\,l_5)$	$(l_2\,l_1\,\bar{l}_4\,\bar{l}_3\,l_5)$	$(l_1\,\bar{l}_4\,\bar{l}_3\,\bar{l}_2\,l_5)$
	$(\bar{l}_4\,\bar{l}_3\,\bar{l}_2\,\bar{l}_1\,l_5)$	$(\bar{l}_3\,\bar{l}_2\,\bar{l}_1\,l_4\,l_5)$	$(\bar{l}_2\,\bar{l}_1\,l_4\,l_3\,l_5)$	$(\bar{l}_1\,l_4\,l_3\,l_2\,l_5)$
D_8	$(l_1\,l_2\,l_3\,l_4\,l_5)$	$(l_2\,l_3\,l_4\,\bar{l}_1\,l_5)$	$(l_3\,l_4\,\bar{l}_1\,\bar{l}_2\,l_5)$	$(l_4\,\bar{l}_1\,\bar{l}_2\,\bar{l}_3\,l_5)$
	$(\bar{l}_1\,\bar{l}_2\,\bar{l}_3\,\bar{l}_4\,l_5)$	$(\bar{l}_2\,\bar{l}_3\,\bar{l}_4\,l_1\,l_5)$	$(\bar{l}_3\,\bar{l}_4\,l_1\,l_2\,l_5)$	$(\bar{l}_4\,l_1\,l_2\,l_3\,l_5)$
	$(l_4\,l_3\,l_2\,l_1\,\bar{l}_5)$	$(l_3\,l_2\,l_1\,\bar{l}_4\,\bar{l}_5)$	$(l_2\,l_1\,\bar{l}_4\,\bar{l}_3\,\bar{l}_5)$	$(l_1\,\bar{l}_4\,\bar{l}_3\,\bar{l}_2\,\bar{l}_5)$
	$(\bar{l}_4\,\bar{l}_3\,\bar{l}_2\,\bar{l}_1\,\bar{l}_5)$	$(\bar{l}_3\,\bar{l}_2\,\bar{l}_1\,l_4\,\bar{l}_5)$	$(\bar{l}_2\,\bar{l}_1\,l_4\,l_3\,\bar{l}_5)$	$(\bar{l}_1\,l_4\,l_3\,l_2\,\bar{l}_5)$
D_{4d}	$(l_1\,l_2\,l_3\,l_4\,l_5)$	$(l_2\,l_3\,l_4\,\bar{l}_1\,\bar{l}_5)$	$(l_3\,l_4\,\bar{l}_1\,\bar{l}_2\,l_5)$	$(l_4\,\bar{l}_1\,\bar{l}_2\,\bar{l}_3\,\bar{l}_5)$
	$(\bar{l}_1\,\bar{l}_2\,\bar{l}_3\,\bar{l}_4\,l_5)$	$(\bar{l}_2\,\bar{l}_3\,\bar{l}_4\,l_1\,\bar{l}_5)$	$(\bar{l}_3\,\bar{l}_4\,l_1\,l_2\,l_5)$	$(\bar{l}_4\,l_1\,l_2\,l_3\,\bar{l}_5)$
	$(l_4\,l_3\,l_2\,l_1\,l_5)$	$(l_3\,l_2\,l_1\,\bar{l}_4\,\bar{l}_5)$	$(l_2\,l_1\,\bar{l}_4\,\bar{l}_3\,l_5)$	$(l_1\,\bar{l}_4\,\bar{l}_3\,\bar{l}_2\,\bar{l}_5)$
	$(\bar{l}_4\,\bar{l}_3\,\bar{l}_2\,\bar{l}_1\,l_5)$	$(\bar{l}_3\,\bar{l}_2\,\bar{l}_1\,l_4\,\bar{l}_5)$	$(\bar{l}_2\,\bar{l}_1\,l_4\,l_3\,l_5)$	$(\bar{l}_1\,l_4\,l_3\,l_2\,\bar{l}_5)$
S_8	$(l_1\,l_2\,l_3\,l_4\,l_5)$	$(l_2\,l_3\,l_4\,\bar{l}_1\,\bar{l}_5)$	$(l_3\,l_4\,\bar{l}_1\,\bar{l}_2\,l_5)$	$(l_4\,\bar{l}_1\,\bar{l}_2\,\bar{l}_3\,\bar{l}_5)$
	$(\bar{l}_1\,\bar{l}_2\,\bar{l}_3\,\bar{l}_4\,l_5)$	$(\bar{l}_2\,\bar{l}_3\,\bar{l}_4\,l_1\,\bar{l}_5)$	$(\bar{l}_3\,\bar{l}_4\,l_1\,l_2\,l_5)$	$(\bar{l}_4\,l_1\,l_2\,l_3\,\bar{l}_5)$

It is convenient to use six-dimensional primitive[10] hypercubic lattice $m^\mu \boldsymbol{b}_\mu$ based on the vectors

$$\boldsymbol{b}_\mu = \mathcal{T}_\mu^{\ \nu} \boldsymbol{e}_\nu \,,$$

where

$$[\mathcal{T}_\mu^{\ \nu}] = \frac{3-\tau}{10} \begin{bmatrix} 1 & \tau & 0 & \tau & -1 & 0 \\ 1 & -\tau & 0 & \tau & 1 & 0 \\ 0 & 1 & \tau & 0 & \tau & -1 \\ 0 & 1 & -\tau & 0 & \tau & 1 \\ \tau & 0 & 1 & -1 & 0 & \tau \\ -\tau & 0 & 1 & 1 & 0 & \tau \end{bmatrix} \tag{13.29}$$

[10] As was noted by Kalugin *et al.* [23], besides the $\boldsymbol{e}_\mu$-based primitive six-dimensional cubic lattice, there are other six-dimensional lattices with three-dimensional projections having icosahedral symmetry. With $\boldsymbol{e}_6$ replaced by $\sum_{\mu=1}^{6} \boldsymbol{e}_\mu/2$, one has 'body' centered lattice. Centering of two-dimensional facets implies centering of four-dimensional facets. Centering of three- or five-dimensional facets leads to the primitive lattice based on the vectors $\boldsymbol{e}_\mu/2$. So there are three types of 'cubic' lattices. In analogy to the conventional case, they are referred to as fcc and bcc lattices; the former consist of points $m^\mu \boldsymbol{e}_\mu/2$ with even $\sum_\mu m^\mu$, and for the latter, all m^μ have the same parity. Only the primitive case will be considered here.

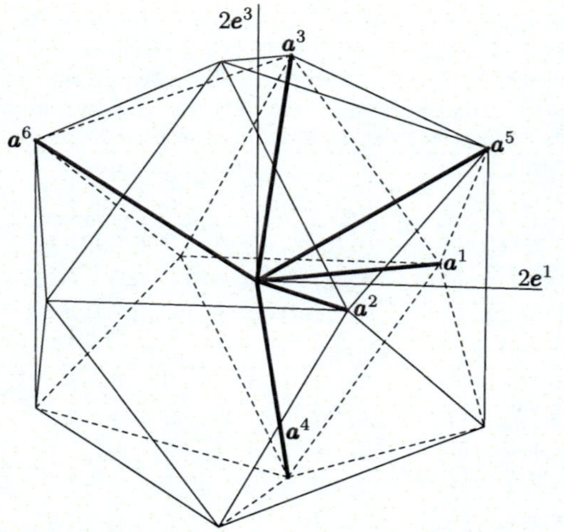

Fig. 13.17 Vectors of the Bancel's frame (13.31) for icosahedral quasicrystals.

The matrix $R = \sqrt{2(\tau+2)}\,[T_\mu^{\ \nu}]$ is special orthogonal. Thus, the considered $\boldsymbol{b}_\mu$-based lattice is obtained from the $\boldsymbol{e}_\mu$-based lattice via rotation by R and scaling by $\sqrt{2(\tau+2)}$. The projections of the lattice basis $\boldsymbol{b}_\mu$ on the physical space spanned by $\boldsymbol{e}_1, \boldsymbol{e}_2$ and $\boldsymbol{e}_3$, and on the orthogonal component spanned by $\boldsymbol{e}_4, \boldsymbol{e}_5$ and $\boldsymbol{e}_6$ are

$$\boldsymbol{a}_\mu = \boldsymbol{a}_\mu^\| = P_\|(\boldsymbol{b}_\mu) = (\boldsymbol{b}_\mu \cdot \boldsymbol{e}^1)\boldsymbol{e}_1 + (\boldsymbol{b}_\mu \cdot \boldsymbol{e}^2)\boldsymbol{e}_2 + (\boldsymbol{b}_\mu \cdot \boldsymbol{e}^3)\boldsymbol{e}_3 = T_\mu^{\ 1}\boldsymbol{e}_1 + T_\mu^{\ 2}\boldsymbol{e}_2 + T_\mu^{\ 3}\boldsymbol{e}_3 \quad (13.30)$$

and

$$\boldsymbol{a}_\mu^\perp = P_\perp(\boldsymbol{b}_\mu) = (\boldsymbol{b}_\mu \cdot \boldsymbol{e}^4)\boldsymbol{e}_4 + (\boldsymbol{b}_\mu \cdot \boldsymbol{e}^5)\boldsymbol{e}_5 + (\boldsymbol{b}_\mu \cdot \boldsymbol{e}^6)\boldsymbol{e}_6 = T_\mu^{\ 4}\boldsymbol{e}_4 + T_\mu^{\ 5}\boldsymbol{e}_5 + T_\mu^{\ 6}\boldsymbol{e}_6 \,,$$

respectively; c.f. Sect. 13.3.

The vectors $\boldsymbol{a}_\mu$ of (13.30) point to six vertices of icosahedron having two-fold symmetry axes along $\boldsymbol{e}_1$, $\boldsymbol{e}_2$ and $\boldsymbol{e}_3$; see Fig. 13.17. Cartesian coordinates of the 12 vertices of this icosahedron are $(\pm 1, \pm\tau, 0)$, $(0, \pm 1, \pm\tau,)$, $(\pm\tau, 0, \pm 1)$. The vectors $\boldsymbol{a}_\mu$ are parallel to the five-fold symmetry axes. They form a frame reflecting the icosahedral symmetry. With integer m^μ, the equality $m^\mu \boldsymbol{a}_\mu = \boldsymbol{0}$ implies $m^\mu = 0$, i.e., the vectors $\boldsymbol{a}_\mu$ are rationally linearly independent.

The 'icosahedral' packing can be obtained by the strip projection [10] or generalized de Bruijn's grid method based on the $\boldsymbol{a}_\mu$ vectors. Again, each edge of the packing is parallel to one of the vectors $\boldsymbol{a}_\mu$, and the edge length is equal to the vector magnitude. The angles between vectors $\boldsymbol{a}_1$, $\boldsymbol{a}_3$, $\boldsymbol{a}_5$ are $\arctan(2)$, and those between $\boldsymbol{a}_2, \boldsymbol{a}_4, \boldsymbol{a}_6$ are $\pi - \arctan(2)$, and generally, the angle between any distinct $\boldsymbol{a}_\mu$

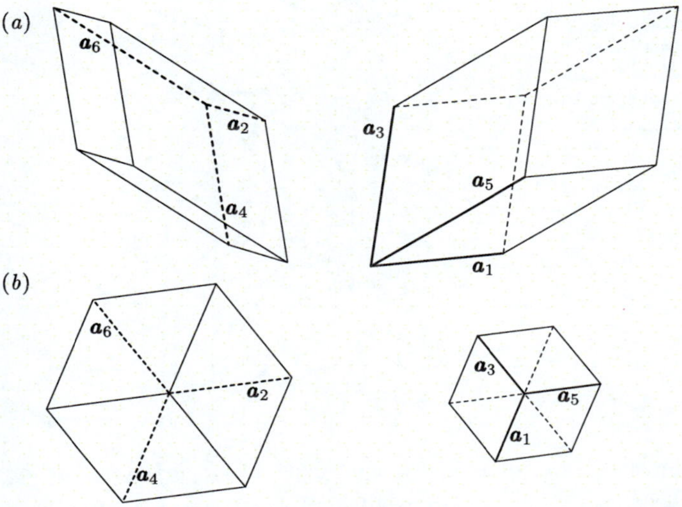

Fig. 13.18 (*a*) Three-dimensional Ammann cells, i.e., oblate (left) and prolate (right) rhombohedra for icosahedral packing. To stress their relationship to the vectors of the frame (13.30), they are shown from the same perspective as in Fig. 13.17. (*b*) The same rhombohedra projected along their three-fold symmetry axes.

vectors takes one of these two values. Therefore, the packing is composed of cells with identical rhombic facets. The cells are two Ammann rhombohedra (Fig. 13.18).[11]

Analogously to the previously considered cases, the integer combination $m^\mu \boldsymbol{a}^{\parallel}_\mu$ is a node of the packing of the physical space if $m^\mu \boldsymbol{a}^{\perp}_\mu$ falls inside the projection of the lattice cell (rotated hypercube) on the orthogonal complement of the physical space. The vertices of the original hypercube with vertices at $\sum_{\mu=1}^{6} \boldsymbol{b}_\mu$ are $\iota^\mu \boldsymbol{b}_\mu$, where ι^μ equals 0 or 1. In the basis $\boldsymbol{e}_\mu$, they are $\iota^\mu \mathcal{T}_\mu{}^\nu \boldsymbol{e}_\nu$. The projection of these $2^6 = 64$ points on the three-dimensional space spanned by $\boldsymbol{e}_3$, $\boldsymbol{e}_4$ and $\boldsymbol{e}_5$ gives 32 points which are vertices of the a figure known as rhombic triacontahedron (Fig. 13.19) and 32 points which are inside this figure.

Vectors $\boldsymbol{a}^\mu$ of the frame dual to (13.30) are $\boldsymbol{a}^\mu = 2(\tau + 2)\, \boldsymbol{a}_\mu$ or

$$\begin{aligned} \boldsymbol{a}^1 &= \boldsymbol{e}^1 + \tau \boldsymbol{e}^2 , \quad & \boldsymbol{a}^2 &= \boldsymbol{e}^1 - \tau \boldsymbol{e}^2 , \\ \boldsymbol{a}^3 &= \boldsymbol{e}^2 + \tau \boldsymbol{e}^3 , \quad & \boldsymbol{a}^4 &= \boldsymbol{e}^2 - \tau \boldsymbol{e}^3 , \\ \boldsymbol{a}^5 &= \boldsymbol{e}^3 + \tau \boldsymbol{e}^1 , \quad & \boldsymbol{a}^6 &= \boldsymbol{e}^3 - \tau \boldsymbol{e}^1 . \end{aligned} \tag{13.31}$$

The matrix coefficient in (13.29) was chosen to get the above expressions. This frame was introduced by Bancel et al. [26]. The same frame is sometimes used with a different enumeration of vectors, e.g. in [27].

[11] These two cells constituting icosahedral quasicrystals were devised by R.Ammann, first described by Mackay [24] and derived using the de Bruijn's approach by Kramer and Neri [25].

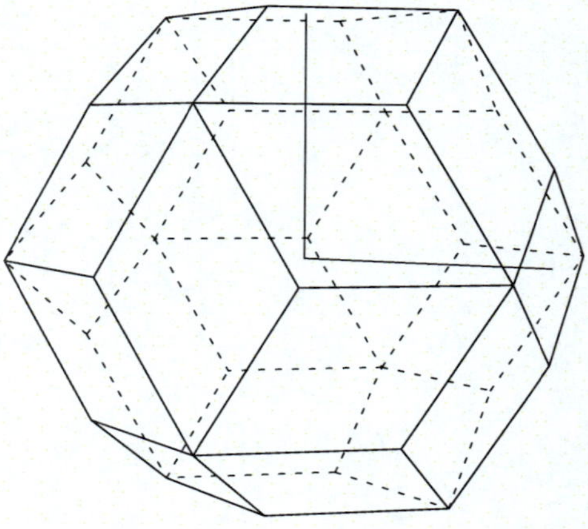

Fig. 13.19 Rhombic triacontahedron—the acceptance domain in the orthogonal complement of the physical space.

Reflection indices symmetrically equivalent to $(l_1\, l_2\, l_3\, l_4\, l_5\, l_6)$ for the Bancel's frame are listed in Table 13.5. The scaling law is

$$\boldsymbol{a}_\mu \tau^3 = \boldsymbol{a}_\nu M^\nu{}_\mu\,, \qquad \boldsymbol{a}_\mu/\tau^3 = \boldsymbol{a}_\nu N^\nu{}_\mu\,,$$

where

$$[M^\mu{}_\nu] = \begin{bmatrix} 2 & -1 & 1 & 1 & 1 & -1 \\ -1 & 2 & -1 & -1 & 1 & -1 \\ 1 & -1 & 2 & -1 & 1 & 1 \\ 1 & -1 & -1 & 2 & -1 & -1 \\ 1 & 1 & 1 & -1 & 2 & -1 \\ -1 & -1 & 1 & -1 & -1 & 2 \end{bmatrix}$$

Table 13.5 Reflection indices symmetrically equivalent to $(l_1\, l_2\, l_3\, l_4\, l_5\, l_6)$ for the group Y in the frame of Bancel et al. [26]. By adding inversion (i.e., each $(l_1\, l_2\, l_3\, l_4\, l_5\, l_6)$ is accompanied by $(\bar{l}_1\, \bar{l}_2\, \bar{l}_3\, \bar{l}_4\, \bar{l}_5\, \bar{l}_6)$) one obtains the equivalences for the group Y_h.

$(l_1\, l_2\, l_3\, l_4\, l_5\, l_6)$	$(l_1\, \bar{l}_3\, l_5\, \bar{l}_2\, \bar{l}_6\, \bar{l}_4)$	$(l_1\, \bar{l}_4\, \bar{l}_2\, \bar{l}_6\, l_3\, \bar{l}_5)$	$(l_1\, \bar{l}_5\, \bar{l}_6\, l_3\, l_4\, l_2)$	$(l_1\, l_6\, l_4\, l_5\, \bar{l}_2\, \bar{l}_3)$
$(\bar{l}_1\, \bar{l}_2\, \bar{l}_4\, \bar{l}_3\, l_6\, l_5)$	$(\bar{l}_1\, l_3\, l_2\, \bar{l}_5\, \bar{l}_4\, \bar{l}_6)$	$(\bar{l}_1\, l_4\, l_6\, l_2\, \bar{l}_5\, l_3)$	$(\bar{l}_1\, l_5\, \bar{l}_3\, l_6\, l_2\, l_4)$	$(\bar{l}_1\, \bar{l}_6\, \bar{l}_5\, \bar{l}_4\, \bar{l}_3\, \bar{l}_2)$
$(l_2\, l_1\, \bar{l}_3\, \bar{l}_4\, \bar{l}_6\, \bar{l}_5)$	$(l_2\, l_3\, \bar{l}_6\, \bar{l}_1\, l_5\, l_4)$	$(l_2\, l_4\, \bar{l}_1\, l_5\, \bar{l}_3\, l_6)$	$(l_2\, \bar{l}_5\, \bar{l}_4\, \bar{l}_6\, \bar{l}_1\, l_3)$	$(l_2\, l_6\, l_5\, \bar{l}_3\, \bar{l}_4\, l_1)$
$(\bar{l}_2\, \bar{l}_1\, l_4\, l_3\, \bar{l}_5\, \bar{l}_6)$	$(\bar{l}_2\, \bar{l}_3\, l_1\, l_6\, l_4\, l_5)$	$(\bar{l}_2\, \bar{l}_4\, \bar{l}_5\, l_1\, l_6\, \bar{l}_3)$	$(\bar{l}_2\, l_5\, l_6\, l_4\, l_3\, \bar{l}_1)$	$(\bar{l}_2\, \bar{l}_6\, l_3\, \bar{l}_5\, l_1\, \bar{l}_4)$
$(l_3\, \bar{l}_1\, \bar{l}_2\, l_5\, l_6\, l_4)$	$(l_3\, l_2\, l_6\, l_1\, \bar{l}_4\, \bar{l}_5)$	$(l_3\, l_4\, l_5\, l_6\, l_1\, l_2)$	$(l_3\, \bar{l}_5\, l_1\, \bar{l}_4\, \bar{l}_2\, \bar{l}_6)$	$(l_3\, \bar{l}_6\, \bar{l}_4\, \bar{l}_2\, l_5\, \bar{l}_1)$
$(\bar{l}_3\, l_1\, \bar{l}_5\, l_2\, l_4\, l_6)$	$(\bar{l}_3\, \bar{l}_2\, \bar{l}_1\, \bar{l}_6\, \bar{l}_5\, \bar{l}_4)$	$(\bar{l}_3\, \bar{l}_4\, \bar{l}_6\, \bar{l}_5\, l_2\, l_1)$	$(\bar{l}_3\, l_5\, l_4\, \bar{l}_1\, \bar{l}_6\, \bar{l}_2)$	$(\bar{l}_3\, l_6\, l_2\, l_4\, \bar{l}_1\, l_5)$
$(l_4\, \bar{l}_1\, \bar{l}_6\, \bar{l}_2\, \bar{l}_3\, l_5)$	$(l_4\, l_2\, l_1\, \bar{l}_5\, \bar{l}_6\, l_3)$	$(l_4\, l_3\, \bar{l}_5\, \bar{l}_6\, \bar{l}_2\, \bar{l}_1)$	$(l_4\, l_5\, \bar{l}_2\, \bar{l}_3\, l_1\, l_6)$	$(l_4\, l_6\, \bar{l}_3\, l_1\, \bar{l}_5\, l_2)$
$(\bar{l}_4\, l_1\, l_2\, l_6\, l_5\, \bar{l}_3)$	$(\bar{l}_4\, \bar{l}_2\, l_5\, \bar{l}_1\, l_3\, \bar{l}_6)$	$(\bar{l}_4\, \bar{l}_3\, l_6\, l_5\, \bar{l}_1\, \bar{l}_2)$	$(\bar{l}_4\, \bar{l}_5\, l_3\, l_2\, l_6\, l_1)$	$(\bar{l}_4\, \bar{l}_6\, \bar{l}_1\, l_3\, l_2\, \bar{l}_5)$
$(l_5\, \bar{l}_1\, l_3\, \bar{l}_6\, \bar{l}_4\, \bar{l}_2)$	$(l_5\, \bar{l}_2\, \bar{l}_6\, \bar{l}_4\, l_1\, \bar{l}_3)$	$(l_5\, \bar{l}_3\, \bar{l}_4\, l_1\, l_2\, l_6)$	$(l_5\, l_4\, l_2\, l_3\, \bar{l}_6\, \bar{l}_1)$	$(l_5\, l_6\, l_1\, l_2\, l_3\, l_4)$
$(\bar{l}_5\, l_1\, l_6\, \bar{l}_3\, \bar{l}_2\, \bar{l}_4)$	$(\bar{l}_5\, l_2\, l_4\, l_6\, \bar{l}_3\, l_1)$	$(\bar{l}_5\, l_3\, \bar{l}_1\, l_4\, l_6\, l_2)$	$(\bar{l}_5\, \bar{l}_4\, \bar{l}_3\, \bar{l}_2\, \bar{l}_1\, \bar{l}_6)$	$(\bar{l}_5\, \bar{l}_6\, \bar{l}_2\, \bar{l}_1\, l_4\, l_3)$
$(l_6\, l_1\, \bar{l}_4\, \bar{l}_5\, l_3\, l_2)$	$(l_6\, l_2\, \bar{l}_5\, l_3\, \bar{l}_1\, l_4)$	$(l_6\, \bar{l}_3\, \bar{l}_2\, \bar{l}_4\, \bar{l}_5\, l_1)$	$(l_6\, l_4\, l_3\, \bar{l}_1\, \bar{l}_2\, l_5)$	$(l_6\, l_5\, \bar{l}_1\, \bar{l}_2\, \bar{l}_4\, \bar{l}_3)$
$(\bar{l}_6\, \bar{l}_1\, l_5\, l_4\, l_2\, l_3)$	$(\bar{l}_6\, \bar{l}_2\, \bar{l}_3\, l_5\, l_4\, \bar{l}_1)$	$(\bar{l}_6\, l_3\, l_4\, l_2\, l_1\, \bar{l}_5)$	$(\bar{l}_6\, \bar{l}_4\, l_1\, \bar{l}_3\, l_5\, \bar{l}_2)$	$(\bar{l}_6\, \bar{l}_5\, l_2\, l_1\, \bar{l}_3\, \bar{l}_4)$

and $N = M^{-1} = M - 4I_6$.[12] Formulas for the transformation of indices of reflections are analogous to (13.21) and (13.22). With $\boldsymbol{a}^\mu = \tau^3 \boldsymbol{a}'^\mu = M^\mu{}_\nu \boldsymbol{a}'^\nu$ and $l_\mu \boldsymbol{a}^\mu = l_\mu \tau^3 \boldsymbol{a}'^\mu = l_\mu M^\mu{}_\nu \boldsymbol{a}'^\nu = l'_\nu \boldsymbol{a}'^\nu$, sets of indices l_μ and l'_μ in the frames $\boldsymbol{a}^\nu$ and $\boldsymbol{a}'^\nu$, respectively, are related via $l'_\nu = l_\mu M^\mu{}_\nu$. Similarly, with $\boldsymbol{a}^\mu = \boldsymbol{a}'^\mu / \tau^3$, one has $l'_\nu = l_\mu N^\mu{}_\nu$.

The zone rule is analogous to that given in Sect. 13.5.3. Reflection with indices l_μ belongs to the zone specified by indices m^μ if

$$\check{\boldsymbol{h}} \cdot \check{\boldsymbol{m}} = l_\mu m^\nu \boldsymbol{a}^\mu \cdot \boldsymbol{a}_\nu = g^\mu{}_\nu l_\mu m^\nu = 0 \,,$$

where

$$[g^\mu{}_\nu] = \frac{1}{2\sqrt{5}} \begin{bmatrix} \sqrt{5} & -1 & 1 & 1 & 1 & -1 \\ -1 & \sqrt{5} & -1 & -1 & 1 & -1 \\ 1 & -1 & \sqrt{5} & -1 & 1 & 1 \\ 1 & -1 & -1 & \sqrt{5} & -1 & -1 \\ 1 & 1 & 1 & -1 & \sqrt{5} & -1 \\ -1 & -1 & 1 & -1 & -1 & \sqrt{5} \end{bmatrix} .$$

Six five-fold symmetry axes of the icosahedron are along directions equivalent to $[m^1\,0\,0\,0\,0\,0]$, ten three-fold axes are along directions equivalent to $[m^1\,m^2\,m^1\,m^2\,m^1\,m^2]$, and fifteen two-fold axes are along directions equivalent to $[m^1\,m^1\,0\,0\,m^5\,\overline{m}^5]$; in each case, at least one index is assumed to be non-zero. Example diffraction patterns of an icosahedral phase are shown in Fig. 13.20.

⊡ *Related frame*
For completeness one needs to mention another frame built of vectors lined up with five-fold symmetry axes of the icosahedron; see, e.g. [28–30]. In this respect it is similar to the frame (13.30). The difference is that the first of the vectors is surrounded in cyclic order by the remaining five, i.e., one has unit vectors

$$\boldsymbol{a}_1 = \boldsymbol{e}_3 \quad \text{and} \quad \boldsymbol{a}_\mu = \cos\alpha\, \boldsymbol{e}_3 + \sin\alpha\, R_{\boldsymbol{e}_3}^{\mu-2} \boldsymbol{e}_1 \quad \text{for} \quad \mu = 2, 3, \ldots, 6 \,, \tag{13.32}$$

where $\alpha = \arctan(2) = \arccos\left(1/\sqrt{5}\right)$, and $R_{\boldsymbol{e}_3}^\mu$ is the rotation about $\boldsymbol{e}_3$ by the angle $2\mu\pi/5$. The frame is self-dual up to the factor of 1/2, i.e., $\boldsymbol{a}^\mu = \boldsymbol{a}_\mu/2$.

13.7.1 Alternative Icosahedral Indexing Scheme

The scheme proposed by Cahn *et al.* [31] relies on the frame with all vectors along three axes perpendicular two-fold symmetry axes of icosahedron. The vectors of the reciprocal space frame are

$$\begin{aligned} \boldsymbol{a}^1 &= \boldsymbol{e}^1 \,, & \boldsymbol{a}^2 &= \boldsymbol{e}^2 \,, & \boldsymbol{a}^3 &= \boldsymbol{e}^3 \,, \\ \boldsymbol{a}^4 &= \tau \boldsymbol{e}^1 \,, & \boldsymbol{a}^5 &= \tau \boldsymbol{e}^2 \,, & \boldsymbol{a}^6 &= \tau \boldsymbol{e}^3 \,. \end{aligned} \tag{13.33}$$

[12] This is related to the fact that $\tau^3 = 2 + \sqrt{5}$ solves $x^2 - 4x - 1 = 0$, i.e., $1/\tau^3 = \tau^3 - 4$.

(a)

(b)

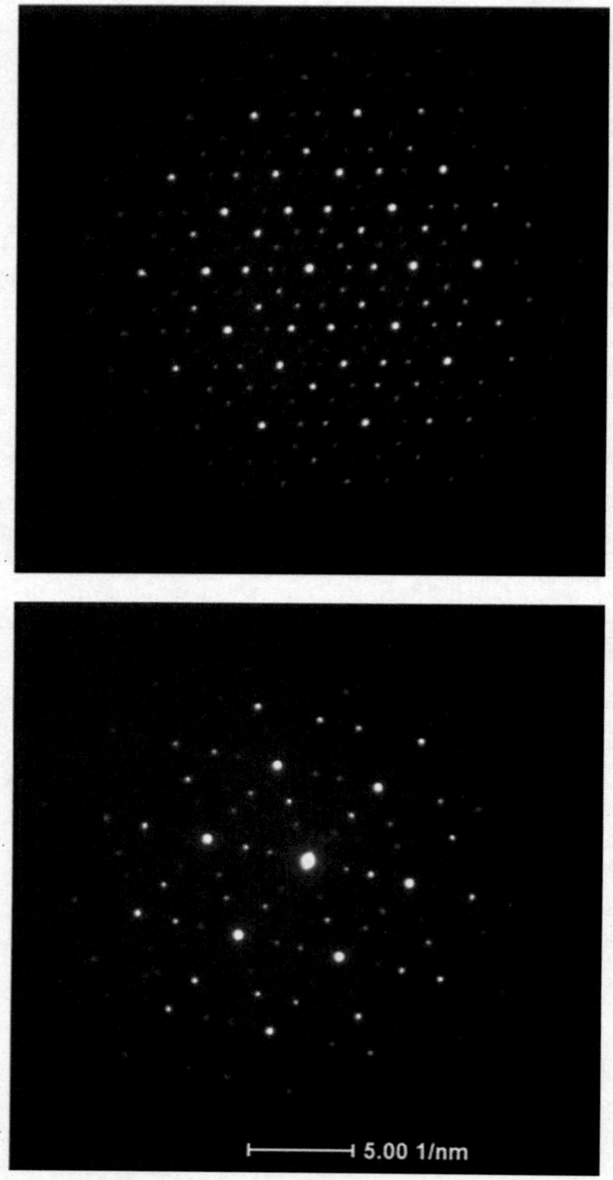

Fig. 13.20 Selected area electron diffraction patterns of icosahedral phase of rapidly cooled Al-Mn-Fe alloy. The zone axis was parallel to a five-fold symmetry axis in (*a*) and to a three-fold symmetry axis in (*b*). Courtesy of K. Stan-Glowinska.

The frame in self-dual up to a factor; one has $\boldsymbol{a}_\mu = \boldsymbol{a}^\mu/(\tau+2)$. The notation used in [31] is specific. Indices $(h\,k\,l,\, h'\,k'\,l')$ of the vector

$$\check{\boldsymbol{l}} = h\boldsymbol{a}^1 + k\boldsymbol{a}^2 + l\boldsymbol{a}^3 + h'\boldsymbol{a}^4 + k'\boldsymbol{a}^5 + l'\boldsymbol{a}^6$$

are written in the form

$$(h/h'\ k/k'\ l/l')\,.$$

These indices must satisfy the 'parity rule': the sums $h+k'$, $k+l'$ and $l+h'$ must be even. With this rule, indices of each reflection equivalent to $(h/h'\ k/k'\ l/l')$ are also integers; otherwise, some would be odd multiples of $1/2$. One can easily express vectors of the Bancel's frame (13.31) via the vectors $\boldsymbol{a}^\mu$ of (13.33) and the Bancel's reflection indices l_μ via h, k, l, h', k', l'; one has

$$\begin{array}{ll} l_1 = (h+k')/2\,, & l_2 = (h-k')/2\,, \\ l_3 = (k+l')/2\,, & l_4 = (k-l')/2\,, \\ l_5 = (l+h')/2\,, & l_6 = (l-h')/2\,. \end{array}$$

All involved numbers are integers thanks to the 'parity rule'. The scaling law is

$$\boldsymbol{a}_\mu\tau^3 = \boldsymbol{a}_\nu M^\nu{}_\mu\,, \qquad \boldsymbol{a}_\mu/\tau^3 = \boldsymbol{a}_\nu N^\nu{}_\mu\,,$$

where

$$[M^\mu{}_\nu] = \begin{bmatrix} 1\,0\,0\,2\,0\,0 \\ 0\,1\,0\,0\,2\,0 \\ 0\,0\,1\,0\,0\,2 \\ 2\,0\,0\,3\,0\,0 \\ 0\,2\,0\,0\,3\,0 \\ 0\,0\,2\,0\,0\,3 \end{bmatrix}$$

and $N = M^{-1} = M - 4I_6$.

The key drawback of the frame (13.33) is that equivalent reflections have dissimilar sets of indices. For instance, it is not immediately apparent that the reflection $(h/h'\ k/k'\ l/l') = (1/2\ \bar{3}/1\ 2/\bar{1})$ is equivalent to $(4/0\ 2/0\ 0/0)$.

13.8 Practical Aspects of Indexing

Indexing for determination of orientation of an a priori known structure (e.g. [32]) relies on the same principles as that for periodic crystals; see Chap. 8. As in the periodic case, to get the orientation, one needs to know the point symmetry, the frame spanning the reciprocal quasilattice, and the indices of intense reflections.

Conventional ab initio indexing comprises two tasks: assignment of indices to diffraction peaks, and determination of the dimensions and angles of repeating units.

In classical formulation of the indexing problem, these two aspects are closely related. In the case of quasicrystals, identification of repeating units in direct space is a separate intricate issue.

Like in the case of periodic crystals, the scattering vectors obtained from a quasicrystal diffraction pattern are integer linear combinations of properly oriented vectors of a frame of reciprocal quasilattice, and the pattern is indexed if (reasonably small, i.e., with low absolute value) integers are ascribed to observed peaks. However, the task of indexing quasicrystal patterns is complicated by the scaling symmetry. The guiding principle is to have smallest indices ascribed to the strongest peaks. Thus, to index a pattern obtained from a quasicrystal, peak intensities need to be taken into account, whereas indexing of patterns from periodic crystals is usually based on pattern geometry, with intensities used only for dividing reflections into 'on' and 'off' types.

The indexing process is simplified by symmetry. The symmetries can be determined from diffraction patterns. If diffraction patterns show that the quasicrystal has a given (say, icosahedral) symmetry, then the number of possible solutions is substantially limited.

Having a satisfactory result of indexing of a pattern from a quasicrystal, to communicate it, one needs to specify the frame (as different frames correspond to different indexing schemes). Then, one needs to specify the scale. Frequently, to compare the indexing results with published data, one is forced to apply rescaling formulas. Finally, if vectors of the frame are rationally linearly dependent (as in the pentagonal case), one needs to specify the conditions removing the corresponding ambiguity.

Ab initio indexing of a pattern from a quasicrystal is a basis for determination of its atomic structure. In conventional crystallography, all peaks in idealized diffraction by a pure-point crystal have the same weight; real peak intensities are determined by the structure factor based on the arrangement of atoms in the unit cell. In quasicrystallography, various weights are ascribed to peaks of pure-point quasicrystal. Peak intensities are influenced by both these weights and the arrangement of atoms, and determining atomic positions based on diffraction data is a difficult task. It is approached by assuming models and testing them against the experimental patterns. Two schemes are used. One is via decoration of periodic structures of higher dimension and cuts through it to get atomic arrangement in physical space. The second scheme is via specifying the cells of the packing, and subsequent cell decoration. Even if a quasicrystal is assumed to be perfect (no defects), the same atomic sites in cells of the same shape are not exactly equivalent as the surroundings of the cells are different.

References

1. T. Dotera, Quasicrystals in soft matter. Isr. J. Chem. **51**, 1197–1205 (2011)
2. P.J. Steinhardt, H.C. Jeong, A simpler approach to Penrose tiling with implications for quasicrystal formation. Nature **382**, 431–433 (1996)

3. F. Lançon, L. Billard, S. Burkov, M. de Boissieu, On choosing a proper basis for determining structures of quasicrystals. J. Phys. I France **4**, 283–301 (1994)
4. W. Steurer, Twenty years of structure research on quasicrystals. Part I. Pentagonal, octagonal, decagonal and dodecagonal quasicrystals. Z. Kristallogr. **219**, 391–446 (2004)
5. N.D. Mermin, Copernican crystallography. Phys. Rev. Lett. **68**, 1172–1175 (1992)
6. M. Gardner, Extraordinary nonperiodic tiling that enriches the theory of tiles. Sci. Am. **236**(1), 110–121 (1977)
7. J.E.S. Socolar, P.J. Steinhardt, D. Levine, Quasicrystals with arbitrary orientationtal symmetry. Phys. Rev B **32**, 5547–5550 (1985)
8. S. Ostlund, D. Wright, Scale invariance and the group structure of quasicrystals. Phys. Rev. Lett. **56**, 2068–2071 (1986)
9. A. Pavlovitch, M. Kléman, Generalised 2D Penrose tilings: structural properties. J. Phys. A: Math. Gen. **20**, 687–702 (1987)
10. M. Duneau, A. Katz, Quasiperiodic patterns. Phys. Rev. Lett. **54**, 2688–2891 (1985)
11. M.V. Jarić, Diffraction from quasicrystals: Geometric structure factor. Phys. Rev. B **34**, 4685–4698 (1986)
12. D. Levine and P.J. Steinhardt. Quasicrystals. I. Definition and structure. Phys. Rev. B, 34:596–616, 1996
13. W. Steurer, T. Haibach, B. Zhang, S. Kek, R. Lück, The structure of decagonal $Al_{70}Ni_{15}Co_{15}$. Acta Cryst. B **49**, 661–675 (1993)
14. N.K. Mukhopadhyay, E.A. Lord, Least path criterion (LPC) for unique indexing in a two-dimensional decagonal quasilattice. Acta Cryst. A **58**, 424–428 (2002)
15. N. de Bruijn, Algebraic theory of Penrose's non-periodic tilings of the plane. Ned. Akad. Wetensch. Proc. Ser. A **84**, 39–66 (1981)
16. F. Gähler, J. Rhyner, Equivalence of the generalised grid and projection methods for the construction of quasiperiodic tilings. J. Phys. A: Math. Gen. **19**, 267–277 (1986)
17. T.C. Choy, J.D. Fitz Gerald, A.C. Kalloniatis, The Al-Mn decagonal phase. Part 3. Diffraction pattern simulation based on a new index scheme. Philos. Mag. B. **58**, 35–46 (1988)
18. B. Koopmans, P.J. Schurer, F. van der Wounde, P. Bronsveld, X-ray diffraction and Mossbauer-effect study o the decagonal $Al_7(Mn_{1-x}Fe_x)_2$ alloy. Phys. Rev. B **35**, 3005–3008 (1987)
19. T.L. Ho, Periodic quasicrystal. Phys. Rev. Lett. **56**, 468–471 (1986)
20. R.K. Mandal, S. Lele, On the six-dimensional lattice of the decagonal phase. Philos. Mag. B **63**, 513–527 (1991)
21. T.L. Daulton, K.F. Kelton, P.C. Gibbonsi, Decagonal and related phases in Al-Mn alloys: electron diffraction and microstructure. Philos. Mag. B **63**, 687–714 (1991)
22. F.P.M. Beenker, Algebraic theory of non-periodic tilings by two simple building blocks: a square and a rhombus. Technical Report 82-WSK-04, Eindhoven University of Technology, 1982
23. P.A. Kalugin, A.Y. Kitayev, L.S. Levitov, 6-dimensional properties of $Al_{0.86}Mn_{0.14}$ alloy. J. Phys. Lett. **46**, L601–607 (1985)
24. A.L. Mackay, De Nive Quinquangula: On pentagonal snowflake. Sov. Phys. Crystallogr. **26**, 517–522 (1981)
25. P. Kramer, R. Neri, On periodic and non-periodic space fillings of $\mathbb{E}^m$ obtained by projection. Acta Cryst. A **40**, 580–587 (1984)
26. P.A. Bancel, P.A. Heiney, P.W. Stephens, A.I. Goldman, P.M. Horn, Structure of rapidly quenched Al-Mn. Phys. Rev. Lett. **54**, 2422–2425 (1985)
27. P.J. Lu, K. Deffeyes, P.J. Steinhardt, N. Yao, Identifying and indexing icosahedral quasicrystals from powder diffraction patterns. Phys. Rev. Lett. **87**, 275507 (2001)
28. V. Elser, Indexing problems in quasicrystal diffraction. Phys. Rev. B **32**, 4892–4898 (1985)
29. A. Katz, M. Duneau, Quasiperiodic patterns and icosahedral symmetry. J. Physique **47**, 181–196 (1986)
30. M. Duneau, A. Katz, The projection method: Fourier transforms and projected patterns. J. Microscopy **146**, 225–232 (1987)

31. J.W. Cahn, D. Shechtman, D. Gratias, Indexing of icosahedral quasiperiodic crystals. J. Mater. Res. **1**, 13–26 (1986)
32. R. Tanaka, S. Ohhashi, N. Fujita, M. Demura, A. Yamamoto, A. Kato, A.P. Tsai, Application of electron backscatter diffraction (EBSD) to quasicrystal-containing microstructures in the Mg-Cd-Yb system. Acta Mater. **119**, 193–202 (2016)

Chapter 14
Refinement of Lattice Parameters and Determination of Local Elastic Strains

Indexing of crystal diffraction patterns provides parameters of the crystal lattice. Indexing is usually seen as a step in crystal structure determination, but determination of lattice parameters is seen from another perspective when lattice parameters of *known* structures are measured with high accuracy. Such measurements are performed for studying phase transformations, in high-pressure experiments, to learn about thermal expansion or crystal chemical composition, and in numerous other circumstances. In most of these cases, the goal is to get parameters more accurate than their known rough estimates. Thus, from the computational point of view, the problem is analogous to the refinement of lattice parameters in the process of structure determination.

⊡ It is important to note, that refinements are used at various stages of crystal stricture determination, and various sets of parameters are refined. A good example are refinement procedures used in analysis of powder diffraction patterns. In the simplest case, one determines optimal lattice parameters and instrumental zero error by matching magnitudes of reciprocal lattice vectors to positions of lines in a pattern. More advanced are the Powley and Le Bail refinements in which peak intensities are used, and besides lattice parameters and instrumental zero error, also peak shape and background parameters are fitted. This is done without knowing the crystal structure. In Rietveld refinement, with a (known or assumed) crystal structure model, both experimental and structural parameters are fitted. The list includes lattice parameters, atomic positions, occupancies and displacements, peak broadening due to grain size or residual strain, crystallographic texture, instrumental parameters, et cetera. The Rietveld refinement is done by minimization of discrepancies between intensities on the measured and the computed powder diffraction profiles, and the latter are computed using structure factors of Bragg reflections and peak profile functions. Finally, one can apply methods based on Debye scattering formula (12.2) to describe total scattering, i.e., both Bragg and diffuse components of the pattern. With the latter approach, besides the crystalline structure and the above-listed factors, also various structural defects and distortions can be taken into account.

There are numerous experimental methods of lattice parameter determination; see part 5 of *ITC-C* or [1]. Methods used in refinement of structural parameters are described in part 8 of *ITC-C*.

A. Morawiec, *Indexing of Crystal Diffraction Patterns*, Springer Series in Materials Science 326, https://doi.org/10.1007/978-3-031-11077-1_14

Refinement of lattice parameters (or equivalently, the refinement of independent entries of the metric tensor) is closely related to the determination of local elastic strains. To give the reader a taste of procedures used in this type of refinement and issues which may arise, this chapter describes computational aspects of two example methods for diffraction-based determination of micro-strains.

External, or pure lattice strains, alter unit cell parameters. Accompanying adjustments of fractional atomic coordinates are referred to as internal strains [2]. In the standard approximation, the deformation of the crystal lattice is assumed to follow the overall deformation of the crystal: external strains imply accompanying internal strains, but the latter do not appear explicitly in any considerations.

Knowing strains present in crystallites is crucial for explaining deformation processes in (poly)crystalline materials. In general, understanding elastic strains plays a significant role in explaining properties of materials, and the determination of strains is important for selecting optimal processing and service conditions. There is a particular interest in strains linked to residual stresses, i.e., stresses present in materials despite absence of any external load. In polycrystalline materials, residual stresses are grouped based on the characteristic distances in which they are equilibrated. Our considerations are limited to stresses of the second and third kind (collectively referred to as micro-stresses) which vary, respectively, over distances comparable to or smaller than the grain size. Stresses of the first kind (macro-stresses) vary gradually only over large distances.

With the limitation to local strains, one can assume that the strain is homogeneous in the region covered by the probe. Clearly, at larger scales, deformation may vary from point to point. Non-uniform strains cause peak broadening,[1] whereas uniform strains cause shifts of peaks but no broadening. In methods of elastic strain determination considered below, with a uniform strain assumed, particular strain tensor components are computed from changes in the geometry of diffraction patterns.

Sensitivity of peak position to strain

Let ϵ_d denote a measure of the strain perpendicular to the reflecting planes separated by the distance d. By differentiating Bragg's law $d = \lambda/(2\sin\theta)$ with fixed λ, one obtains $|\epsilon_d| = |\delta d|/d = \delta\theta \cot\theta$, and

$$\delta\theta = |\epsilon_d| \tan\theta \ ,$$

i.e., for given strain magnitude $|\epsilon_d|$, the change $\delta\theta$ of the Bragg angle θ is proportional to $\tan\theta$. In other words, the larger the Bragg angle, the larger the strain-caused shift of the reflection. Since $|\check{\boldsymbol{h}}| = 1/d \propto \sin\theta$, the longer the reciprocal lattice vector $\check{\boldsymbol{h}}$ corresponding to the reflection, the larger the strain-caused shift of the reflection. In practice, mutual shifts of multiple reflections corresponding to different vectors of the reciprocal lattice are used, but the conclusion is generally the same: the larger

[1] Broadening is also caused by crystalline defects, small crystallite sizes (in the case of X-rays) and by instrumental effects.

the magnitudes of the involved reciprocal lattice vectors, the larger the sensitivity of the diffraction patterns to strain.

14.1 Methods of Local Strain Determination

Basically, there are two ways of approaching strain determination. The first one involves refinement of absolute values of lattice parameters, i.e., strains are lattice distortions measured in reference to some established parameters. Another option, applicable when the parameters themselves cannot be determined, is to obtain their *relative* variation at different points of the same material.

For the determination of micro-strains (and corresponding micro-stresses) a high spatial resolution is needed. There are a number of methods of local strain determination using X-ray and electron diffraction. Resolution of the order of a micron is achievable by synchrotron based methods. Very high resolution (in nanometers) can be reached only by methods based on electron diffraction.

Particularly useful are K-line electron diffraction patterns [3]. Lattice distortions cause pattern distortions which can be seen as shifts of K-lines and zone axes. Depending on the type of the pattern, one or the other type of shifts is determinable. If the analysis is based on the lines, the latter must be clearly visible and of sufficiently high reflection indices. These requirements are satisfied by setups for recording CBED patterns. On the other hand, if zone axes are used, the axes must be easily detectable and angles between them must be sufficiently large. These requirements are satisfied by EBSD patterns.

Below, methods utilizing central disks of CBED patterns will be described. With the potential strain sensitivity of two parts per ten thousand (see, e.g., [4, 5]) and the spatial resolution of a few nanometers, CBED is the method of choice for investigation of strains at the smallest scale, e.g. in microelectronic devices [6–9]. The technique has also been used for analysis of strain in superalloys [10–12], and other materials (e.g., around isolated precipitates [13] or near interfaces in metal matrix composites [14]).

The applicability of the CBED method is impaired to certain extent by issues inherent to measurements utilizing TEM. The main negative aspect of TEM-based strain measurements is stress relaxation due to sample thinning and difficulties with quantitative assessment of its influence. Plane stress conditions may be assumed,[2] but practice shows that considerable variations of lattice parameters or strain gradients are frequently present in materials [15, 16]. They are manifested in the form of 'line splitting' degrading the quality of CBED patterns.

At the end of this chapter, the related method of lattice parameter refinement and strain determination based on X-ray Kossel patterns will be briefly discussed. The lateral resolution of this technique is of the order of 1 micrometer, i.e., it is lower than that of CBED approach, but the technique avoids most of the difficulties of

[2] The assumption of plane strain is often made, but it is not justified in the presence of anisotropy.

CBED (like complicated specimen preparation, stress relaxation, limited number of determinable strain components, et cetera). The method relies on patterns covering relatively large solid angle. Different than lines in CBED patters, Kossel lines are manifestly conic sections. Otherwise, the geometry of the problem is as that faced in CBED-based calculations.

As for the aforementioned EBSD-based method, it allows for getting relative variation of lattice parameters at different points of the same crystallite [17–20]. Lattice distortions are calculated from relative shifts of zone axes in diffraction patterns originating from different points of the crystal. A shift is obtained via image analysis by locating the extremum of the correlation coefficient between corresponding fragments of the patterns. In another approach, the shift of diffraction bands in EBSD patterns was calculated from the shift of maxima on the Hough transforms of the patterns [21]. With zone axes involved, the resolution of the method increases with the solid angle covered by the diffraction patterns.

14.2 CBED-Based Determination of Micro-Strains

In the CBED method, strains are determined via refinement of lattice parameters. The strategy is straightforward: the strain with respect to the reference lattice is calculated by matching strain dependent simulated patterns to experimental patterns.[3] In principle, structural parameters can be obtained by fitting experimental and theoretical intensities [22–24], but that is difficult due to dynamical effects. In matching, simulated diffraction patterns must be calculated numerous times. Since computation times needed for dynamical simulations are long, the matchings usually rely on kinematic calculations. In the simplest approaches, they are based on geometry of deficiency lines.[4] Two such procedures are briefly described below. In the first case, the matching is done by minimization of deviations of experimentally obtained positions of lines from the positions obtained from formal equations of the lines [25]. In the second one, distances between intersections of the lines are fitted [26]. Another method was to match areas of triangles or polygons bounded by deficiency lines [27]. All these matchings are implemented as optimization procedures with properly defined objective functions [25–27]; for other approaches based on line intersections see [8, 28].

Additional complications arise due to the limited accuracy of the voltage and the camera length. The former needs to be calibrated using a reference sample. As for the camera length, it can be either fitted or eliminated by matching *ratios* of distances between intersections of lines (or *ratios* of areas of polygons).

[3] Here and below, it is assumed that there are no distortions of the pattern recording medium or device.

[4] If the pattern is registered with a large camera length and a low-index zone axis parallel to the optical axis of the microscope, the lines corresponding to reciprocal lattice nodes high on the Ewald sphere are referred to as high order Laue zone (HOLZ) lines.

14.2.1 K-Line Equation Based Scheme

A simple procedure for computation of lattice parameters is based directly on the underlying algebraic equation of K-lines of which deficiency lines in CBED patterns are a particular case. For brevity, this 'K-line equation based scheme' of refining lattice parameters is referred to as KLEBS. In the kinematic framework, the location of a deficiency line is described by (2.5)

$$2\check{\boldsymbol{h}} \cdot \boldsymbol{k} - \check{\boldsymbol{h}} \cdot \check{\boldsymbol{h}} = 0 \,, \quad \boldsymbol{k} \cdot \boldsymbol{k} = 1/\lambda^2 \,, \tag{14.1}$$

where $\check{\boldsymbol{h}}$ is the reciprocal lattice vector corresponding to the line, and $\boldsymbol{k}$ denotes the wave vector of the reflected beam.

In brief, the main idea of KLEBS is to determine the lattice parameters minimizing the sum of the expressions $(2\check{\boldsymbol{h}} \cdot \boldsymbol{k} - \check{\boldsymbol{h}} \cdot \check{\boldsymbol{h}})^2$ over lattice dependent $\check{\boldsymbol{h}}$ vectors and experimentally determined wave vectors. The point it to express the $\check{\boldsymbol{h}}$ and $\boldsymbol{k}$ vectors via the desired parameters (strain, orientation, camera length, wavelength). This concept can be put into practice in a number of ways. In particular, the scalar products of (14.1) can be calculated in a Cartesian coordinate system linked to the crystal. For the reference lattice, with the matrix $\mathcal{T}$ defined in Sect. 1.1 ($\mathcal{T}_j{}^i$ is the i-th coordinate of the j-th lattice basis vector in a Cartesian coordinate system linked to the crystal lattice; see also Sect. 1.8), the Cartesian components of the reciprocal lattice vector corresponding to the reflection $(h\,k\,l)$ are elements of the array

$$\mathcal{T}^{-T} \begin{bmatrix} h \\ k \\ l \end{bmatrix} = \check{\boldsymbol{h}}^0 \,.$$

Further, all coordinates refer to Cartesian systems. Based on (1.56) of Sect. 1.9, a homogeneous displacement $I + \epsilon$ alters the reciprocal lattice vector $\check{\boldsymbol{h}}^0$ to

$$\check{\boldsymbol{h}} = (I + \epsilon)^{-1} \check{\boldsymbol{h}}^0 \approx (I - \epsilon) \check{\boldsymbol{h}}^0 \,. \tag{14.2}$$

Let L^0 and λ^0 be the initial assessments of the camera length and the wavelength, respectively. The symbols L and λ denote their unknown actual values. The corresponding parameters X^L and X^λ are defined by the expressions

$$L = L^0(1 + X^L) \quad \text{and} \quad \lambda = \lambda^0(1 + X^\lambda) \,;$$

both X^L and X^λ are dimensionless and close to zero.

The strain of the crystal lattice is best interpreted in the crystal coordinate system, whereas the data originating from TEM are given in the microscope coordinate system. Assuming that entries of ϵ and $\boldsymbol{k}$ are in the Cartesian crystal reference frame, and the matrix $O_{c|m}$ represents the orientation of the crystal with respect to the microscope (see (4.1)), the strain tensor in the microscope coordinate system is $\epsilon^{\uparrow} = O^T_{c|m} \epsilon O_{c|m}$, and the wave vector in the microscope coordinate system has the form

$$\boldsymbol{k}^{\uparrow} = O^T_{c|m} \boldsymbol{k} \ .$$

The wave vector is obtained from the deficiency line location and the radiation wavelength via

$$\boldsymbol{k}^{\uparrow} = \boldsymbol{v}/(\lambda\, |\boldsymbol{v}|) \ , \tag{14.3}$$

where $\boldsymbol{v}$ is a vector from the source of diffraction to a point on the diffraction line. The algebraic steps for getting $\boldsymbol{v}$ depend on the way the lines are described.

⊡ Let the line parameters be the distance of the line from the pattern center ρ, and the directed angle between the outward-pointing normal to the line and the 1-axis ϕ. In the microscope system with the third axis along the optical axis, the vector $\boldsymbol{v}$ to a point on the diffraction line can be expressed as

$$\boldsymbol{v} = \boldsymbol{w} + s\, \boldsymbol{w}_p \ ,$$

where $\boldsymbol{w} = [\rho \cos\phi, \rho \sin\phi, -L]^T$ is a vector to the point closest to the pattern center, $\boldsymbol{w}_p = [-\sin\phi, \cos\phi, 0]^T$ is a vector perpendicular to $\boldsymbol{w}$ in the plane of the pattern, and s is a number parameterizing the line (Fig. 14.1).

Initially, the orientation is known based on the approximate reference lattice parameters; let that orientation be given by an orthogonal matrix O^0. The actual orientation $O_{c|m}$ deviates slightly from O^0. It is a product

$$O_{c|m} = O^0 R^X \ ,$$

where R^X is an orthogonal matrix corresponding to a small unknown orientation correction. The explicit rotation R^X (interpretable in terms of tilt angles of specimen holder) is used instead of the term r_{ij} of (1.56). The matrix R^X depends on three parameters. One can conveniently choose them to be the Rodrigues parameters X^R_i $(i = 1, 2, 3)$ so

$$R^X_{ij} = ((1 - X^R_k X^R_k)\delta_{ij} + 2X^R_i X^R_j - 2\varepsilon_{ijk} X^R_k)/(1 + X^R_l X^R_l) \ ;$$

see, e.g. [29].

Based on the $\check{\boldsymbol{h}}$ vectors (14.2) for a number of reflections and $\boldsymbol{k}$ vectors (14.3) determined for some points on the corresponding lines, one can define

$$\psi_1 = \sum (2\boldsymbol{k} \cdot \check{\boldsymbol{h}} - \check{\boldsymbol{h}} \cdot \check{\boldsymbol{h}})^2 \ , \tag{14.4}$$

where the sum is over the reflections and points used, and ψ_1 depends on the unknown ϵ, X^R_i, X^L and X^λ: $\psi_1 = \psi_1(\epsilon, X^R_i, X^L, X^\lambda)$. The parameters for which the ψ_1 function takes a minimal value are calculated numerically. Some of the parameters

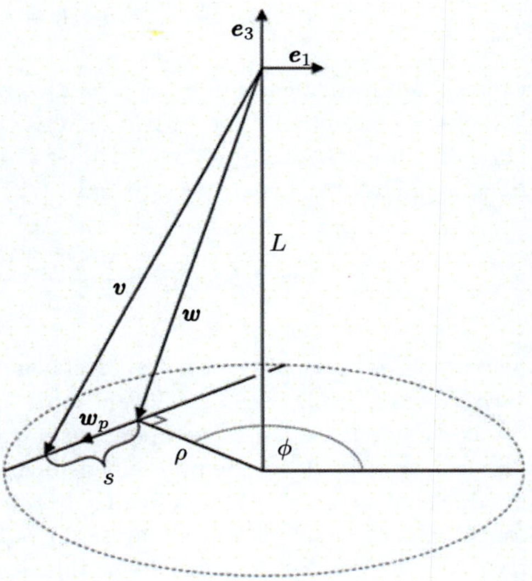

Fig. 14.1 Geometric characteristics of a HOLZ line in central disk of a CBED pattern.

can be fixed if the corresponding quantities are known; in particular, if the wavelength is assumed to be known exactly, X^λ is set to zero.

⊡ The above optimization problem is non-linear. The expression $2\boldsymbol{k} \cdot \check{\boldsymbol{h}} - \check{\boldsymbol{h}} \cdot \check{\boldsymbol{h}}$ linearized with respect to small ϵ, X_i^R, X^L and X^λ takes the form

$$2\boldsymbol{k} \cdot \check{\boldsymbol{h}} - \check{\boldsymbol{h}} \cdot \check{\boldsymbol{h}} \approx 2\check{\boldsymbol{h}}^0 \cdot \left((O^0 R^X - \epsilon\, O^0)\boldsymbol{k}^0 + O^0 (X^L \left[0, 0, k_3^0\right]^T - \xi\, \boldsymbol{k}^0) + \epsilon\, \check{\boldsymbol{h}}^0 - \check{\boldsymbol{h}}^0/2 \right), \tag{14.5}$$

where the matrix R^X depends on the parameters X_i^R through $R_{ij}^X = \delta_{ij} - 2\varepsilon_{ijk} X_k^R$, $\boldsymbol{k}^0$ is given by $\boldsymbol{k}^0 = \boldsymbol{v}^0/(\lambda^0\, |\boldsymbol{v}^0|)$, $\boldsymbol{v}^0 = \boldsymbol{w}^0 + s\boldsymbol{w}_p$, $\boldsymbol{w}^0 = \left[\rho \cos\phi, \rho \sin\phi, -L^0\right]^T$, and $\xi = X^L\, (k_3^0)^2 + X^\lambda$. Inserting the above expression (14.5) into (14.4) leads to a convex objective function.

If multiple patterns are used, a sum over all patterns needs to be taken in (14.4). This, however, has a significant influence on the number of fitted parameters because besides the sought-after components of ϵ, three orientation parameters (X_i^R) per pattern, one camera length correction (X^L) per pattern (and possibly wavelength correction X^λ) must be fitted.

14.2.2 Fitting Distances Between Line Intersections

Also in this approach, strain is calculated by fitting geometric elements of line patterns; this time distances between intersections of the lines are matched. Based on (14.1), the relationships

$$2\check{\boldsymbol{h}}_1 \cdot \boldsymbol{k} = \check{\boldsymbol{h}}_1 \cdot \check{\boldsymbol{h}}_1 \,, \quad 2\check{\boldsymbol{h}}_2 \cdot \boldsymbol{k} = \check{\boldsymbol{h}}_2 \cdot \check{\boldsymbol{h}}_2 \;\; \text{and} \;\; \boldsymbol{k} \cdot \boldsymbol{k} = 1/\lambda^2 \tag{14.6}$$

corresponding to reflections $\check{\boldsymbol{h}}_1 = (I - \epsilon)\check{\boldsymbol{h}}_1^0$ and $\check{\boldsymbol{h}}_2 = (I - \epsilon)\check{\boldsymbol{h}}_2^0$ determine the wave vector $\boldsymbol{k}$ of the intersection of two deficiency lines in a pattern originating from a crystallite transformed by strain ϵ. The solution of the system (14.6) can be obtained in an explicit form; (14.6) are satisfied by

$$\boldsymbol{k} = \boldsymbol{k}(\check{\boldsymbol{h}}_1, \check{\boldsymbol{h}}_2) = \frac{\boldsymbol{y} \times \boldsymbol{z} \pm \boldsymbol{y}\sqrt{4\boldsymbol{y} \cdot \boldsymbol{y}/\lambda^2 - \boldsymbol{z} \cdot \boldsymbol{z}}}{2\boldsymbol{y} \cdot \boldsymbol{y}} \,, \tag{14.7}$$

where $\boldsymbol{y} = \check{\boldsymbol{h}}_1 \times \check{\boldsymbol{h}}_2$ and $\boldsymbol{z} = \check{\boldsymbol{h}}_1(\check{\boldsymbol{h}}_2 \cdot \check{\boldsymbol{h}}_2) - \check{\boldsymbol{h}}_2(\check{\boldsymbol{h}}_1 \cdot \check{\boldsymbol{h}}_1)$. (For details see Sect. 14.4.)

A vector $\boldsymbol{k}^\uparrow = [k_1^\uparrow \; k_2^\uparrow \; k_3^\uparrow]^T$ given in the microscope coordinate system with the '3' axis along the microscope optical axis is projected on a flat detector perpendicular to $\boldsymbol{L} = [0\,0\,L]^T$ as the two-dimensional vector $\mathcal{P}(\boldsymbol{k}^\uparrow) = (L/k_3^\uparrow)[k_1^\uparrow \; k_2^\uparrow]^T$. The goodness-of-fit functions are defined based on locations of the line intersection points for the experimental pattern and those for $\mathcal{P}(\boldsymbol{k}^\uparrow)$ which involve strain. The standard method is to match distances between the intersections. The theoretical distance d between intersection points (i) and (j) is

$$d = \mid \mathcal{P}(\boldsymbol{k}^\uparrow_{(i)}) - \mathcal{P}(\boldsymbol{k}^\uparrow_{(j)}) \mid \,. \tag{14.8}$$

A simple goodness-of-fit function has the form

$$\psi_2 \propto \sum_n \left(d_{(n)} - c d_{(n)}^{exp} \right)^2 \,, \tag{14.9}$$

where n enumerates the measured distances $d_{(n)}^{exp}$, $d_{(n)}$ is given by (14.8), and c is a variable magnification factor linked to L; see [26, 30]. The factor is eliminated by fitting distance ratios; e.g., instead of (14.9), one can minimize $\sum_{m,n} \left(d_{(m)}/d_{(n)} - d_{(m)}^{exp}/d_{(n)}^{exp} \right)^2$. In all these cases, the goodness-of-fit function is constructed using the projection $\mathcal{P}(\boldsymbol{k}^\uparrow)$ to determine the distances, and therefore, it depends on strain ϵ via (14.7)

$$\boldsymbol{k}^\uparrow = O_{c|m}^T \, \boldsymbol{k}\left(\check{\boldsymbol{h}}_1, \check{\boldsymbol{h}}_2\right) = O_{c|m}^T \, \boldsymbol{k}\left((I - \epsilon)\check{\boldsymbol{h}}_1^0, \, (I - \epsilon)\check{\boldsymbol{h}}_2^0 \right)$$

with $\check{\boldsymbol{h}}_i = (I - \epsilon)\check{\boldsymbol{h}}_i^0$. The dependence of ψ_2 defined in (14.9) on the orientation $O_{c|m}$ is weak, and the approximation with fixed $O_{c|m}$ can be used. More precise orientation parameters can be obtained in an additional step by finding the 'absolute orientation' of Sect. 8.3.2, i.e., the best rotation relating two sets of vectors. The first set consists of wave vectors pointing to line intersections in the experimental pattern, and the second one contains the wave vectors pointing to corresponding intersections in patterns simulated with the lattice parameters and camera length obtained from the optimization.

One needs to note that a displacement of a line intersecting another line causes a large shift of the intersection point if the intersection angle is small. Consequently, small angle intersections have a much larger effect on ψ_2 than those with angles close to 90°. On the other hand, the locations of points of small angle intersections do not only depend strongly on lattice parameters but are also very sensitive to errors in determination of positions of the intersecting lines.

14.2.3 Ambiguities

As was mentioned above, besides strain, also orientation, camera length and voltage are known only approximately. Not all these parameters are determinable simultaneously.

Voltage and isotropic strain

In the kinematic framework, deviations from the actual voltage are directly linked to the trace of the strain tensor. With a change of the wavelength from λ to $(1-\zeta)\lambda$, (14.1) take the form $2\check{\boldsymbol{h}} \cdot \boldsymbol{k} = \check{\boldsymbol{h}} \cdot \check{\boldsymbol{h}}$ and $\boldsymbol{k} \cdot \boldsymbol{k} = 1/((1-\zeta)\lambda)^2$, and they can be written as

$$2\check{\boldsymbol{h}}' \cdot \boldsymbol{k}' = \check{\boldsymbol{h}}' \cdot \check{\boldsymbol{h}}' \;\text{ and }\; \boldsymbol{k}' \cdot \boldsymbol{k}' = 1/\lambda^2 ,$$

where $\boldsymbol{k}' = (1-\zeta)\boldsymbol{k}$ and $\check{\boldsymbol{h}}' = (1-\zeta)\check{\boldsymbol{h}}$. By comparing $\check{\boldsymbol{h}}' = (1-\zeta)\check{\boldsymbol{h}}$ with (14.2), one can see that the change of the wavelength $\lambda \to (1-\zeta)\lambda$ has the same effect on the geometry of lines as the isotropic strain

$$\epsilon = \zeta I \; ; \tag{14.10}$$

see Fig. 14.2. Therefore, the voltage cannot be subject to fitting at the same time as the diagonal of the strain tensor, and the trace of the strain tensor can be determined only if the voltage is known. Since the above tensor (14.10) is independent of the coordinate system, these conclusions apply to all possible foil orientations, i.e., using multiple patterns does not resolve the ambiguity.

Sensitivity of line geometry to particular strain components

It is clear from the definitions (14.4) and (14.9) that ψ_1 and ψ_2 are smooth functions of strain, and the considered minimization problem is well-posed (i.e. it has a unique solution depending continuously on the parameters). On the other hand,

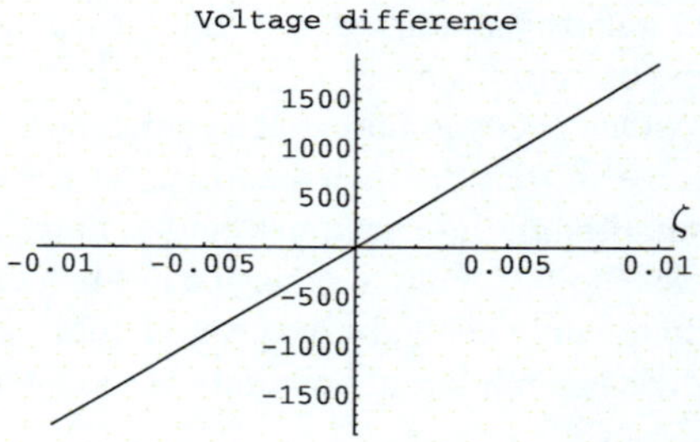

Fig. 14.2 Deviation of voltage (in [V]) from the value of 100 kV corresponding to strain $\epsilon = \zeta I$ versus ζ. With small ζ, the function is only weakly nonlinear: the deviation $\approx \zeta \times 183633$ V.

due to experimental conditions and dynamical effects, the objective functions can be noisy. Moreover, the problem of lattice parameter determination is ill-conditioned with barely detectable dependence on some parameters when compared to dependence on other parameters [31, 32]. Thus, the issue arises of separating determinable strain components from those which cannot be reliably evaluated.

It is frequently underlined that due to the presence of HOLZ lines, the CBED patterns carry 'three-dimensional information'. However, the region of the reciprocal space from which that data comes is still relatively flat; see Fig. 14.3. It is spread around the plane perpendicular to the microscope axis. Therefore, in terms of the sensitivity to strain, this particular direction is different than the directions perpendicular to it. Hence, it is rightful to expect that the issue will become clearer when the patterns are simulated for components of the strain tensor $\epsilon^{\uparrow}$ $(= O_{c|m}^T \epsilon O_{c|m})$ given in the microscope coordinate system. In fact, as it is shown below, the sensitivity of pattern geometry to different combinations of $\epsilon_{ij}^{\uparrow}$ varies drastically. This leads to a dichotomous discrimination of the parameters; those with negligible influence (smaller than the impact caused by random experimental errors) are referred to as indeterminable.

Since small (linear) strains are additive, an arbitrary strain can be represented as a composition of *elemental* strains with *one* non–vanishing $\epsilon_{ij}^{\uparrow}$ and other strain components fixed at zero. Due to the suitable choice of the coordinate system, some of the elemental strains can be identified as indeterminable. Simple analysis shows that the dependence on $\epsilon_{11}^{\uparrow}$, $\epsilon_{22}^{\uparrow}$, $\epsilon_{33}^{\uparrow}$ and $\epsilon_{12}^{\uparrow}$ is much stronger than the dependence on $\epsilon_{13}^{\uparrow}$ and $\epsilon_{23}^{\uparrow}$. In the case of non–zero $\epsilon_{11}^{\uparrow}$, $\epsilon_{22}^{\uparrow}$, $\epsilon_{33}^{\uparrow}$ and $\epsilon_{12}^{\uparrow}$, the strain caused shifts of the intersection points are small but they occur in different directions. On the other hand, large shifts caused by non–zero $\epsilon_{13}^{\uparrow}$ or $\epsilon_{23}^{\uparrow}$ are parallel. See Fig. 14.4. Generally, the strains which are combinations of only $\epsilon_{13}^{\uparrow}$ and $\epsilon_{23}^{\uparrow}$ components cause changes observable as parallel shifts of the patterns. Since this effect cannot be distinguished from a rotation of the sample, such strains have little influence on the strain-based goodness–of–fit. Figure 14.5a illustrates the typical dependence of ψ_2 on strain with varying determinable $\epsilon_{12}^{\uparrow}$ and indeterminable $\epsilon_{13}^{\uparrow}$ and other independent components equal to zero.

The strains $\epsilon_{13}^{\uparrow}$ and $\epsilon_{23}^{\uparrow}$ are the only indeterminable elemental strains but they do not exhaust the list of indeterminable parameters. Other indeterminable parameters can be linear combinations of $\epsilon_{11}^{\uparrow}$, $\epsilon_{22}^{\uparrow}$, $\epsilon_{33}^{\uparrow}$ and $\epsilon_{12}^{\uparrow}$. Due to the axial symmetry of the

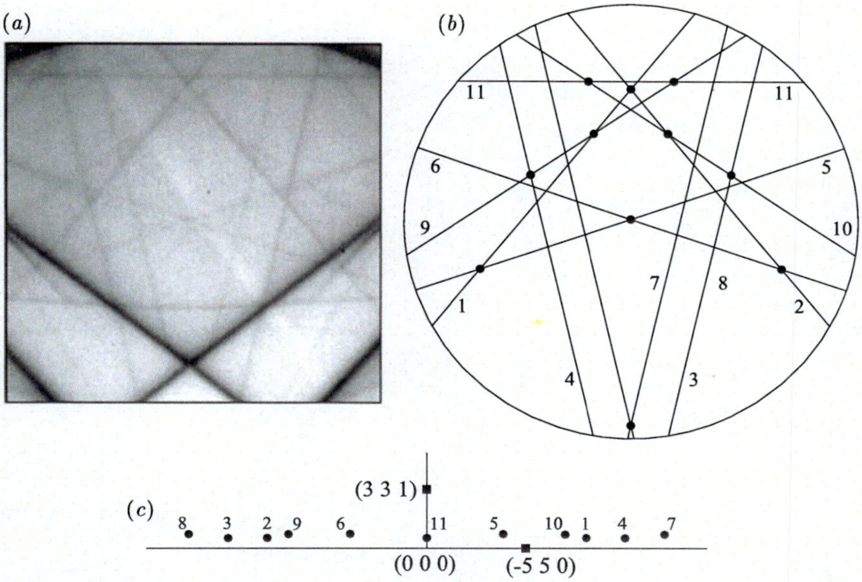

Fig. 14.3 (*a*) The [331] diffraction pattern of Si obtained for 119.2 kV. Geometry of this pattern is used as an example. (*b*) Schematic of the pattern shown in (*a*). Cubic lattice with the constant of $a = 0.54307$ nm was used to simulate the pattern. (*c*) Projection of the reciprocal lattice nodes corresponding to lines shown in (*b*) on the plane perpendicular to the $[\bar{1}\,\bar{1}\,6]$ direction. The nodes are marked by disks. For reference, the locations of (3 3 1) and $(\bar{5}\;5\;0)$ are marked as squares.

measurement (with particular choice of lines disregarded), the combinations must be symmetric with respect to indices 1 and 2. The most general strain of this kind is based on $\epsilon^{\uparrow}_{11} = \epsilon^{\uparrow}_{22}$, $\epsilon^{\uparrow}_{12}$ and $\epsilon^{\uparrow}_{33}$.

Formally, indeterminable strain components can be found by analyzing eigenvalues of the matrix of the second derivatives (Hessian) of the objective function with respect to parameters. Analytical approach would be too complicated for the function obtained from distances between line intersections (14.9). In the case of KLEBS, the calculations are manageable for the objective function depending only on the components of ϵ, with fixed experimental $\boldsymbol{k}$ vectors, camera length, wavelength and orientation. The objective function $\psi_1(\epsilon) = \sum \left(2\check{\boldsymbol{h}} \cdot \boldsymbol{k} - \check{\boldsymbol{h}} \cdot \check{\boldsymbol{h}}\right)^2$ with $\check{\boldsymbol{h}} = (I - \epsilon)\,\check{\boldsymbol{h}}^0$ is insensitive to changes along eigenvectors corresponding to the smallest eigenvalues of the Hessian $\partial^2\psi_1/\partial\epsilon_{ij}^2$ at $\epsilon = 0$. In the microscope coordinate system having the third axis along the optical axis of the microscope ($\boldsymbol{k} = \boldsymbol{k}^{\uparrow}$), one has $|k^{\uparrow}_3| \gg |k^{\uparrow}_1|$, $|k^{\uparrow}_2|$ and $|h^0_3| \ll |h^0_1|$, $|h^0_2|$. With such data, by far the smallest is the eigenvalue corresponding to

$$\epsilon^{\uparrow} \propto \mathrm{diag}(1, 1, \alpha) \tag{14.11}$$

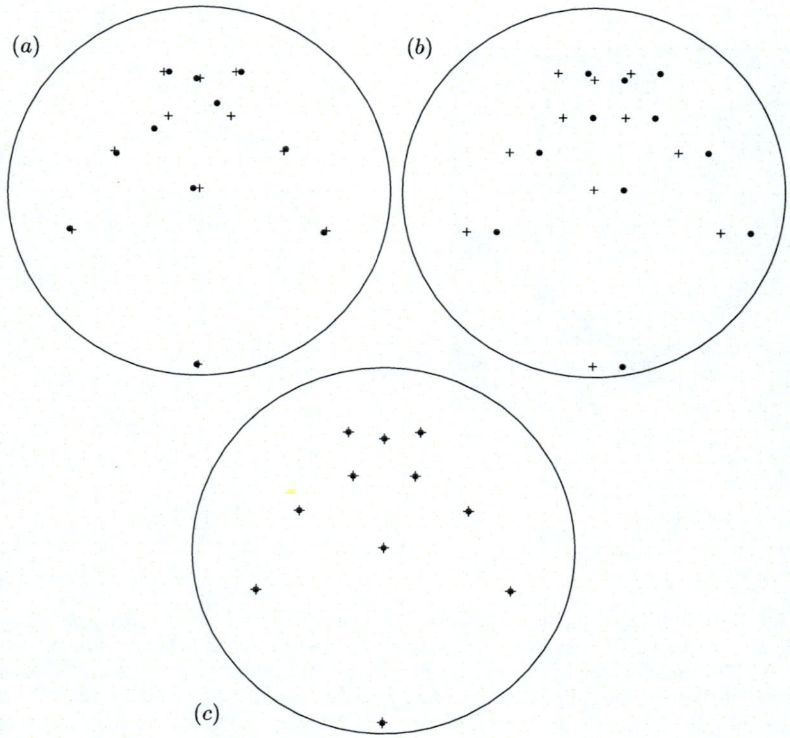

Fig. 14.4 The shift of the intersection points (dots) shown in Fig. 14.3*b* for the strain with $\epsilon^{\uparrow}_{12} = 0.001$ (*a*), $\epsilon^{\uparrow}_{13} = 0.001$ (*b*) and $\epsilon^{\uparrow} = 0.001 \times \text{diag}(1, 1, 2)$ (*c*). The other independent strain components are zero. For reference, the locations of the points corresponding to the material without strain are marked by crosses.

with α around 2. (In the considered example $\alpha \approx 2.010$.) This means that isotropic strain in the plane perpendicular to the beam has the same impact on a pattern as some extension along the beam [30], and—in effect—the combination (14.11) is highly indeterminable.

The coefficient α varies with the zone axis, selection of lines and the ratio of wavelength to the magnitude of lattice parameters. This relationship is relatively simple. An example plot of α versus the ratio of wavelength and the parameter a of cubic lattice is shown in Fig. 14.6. For the wavelengths used in practice (100–200 kV), the dependence is almost linear.

The above applies to a single pattern. One way to ease the ambiguity problem is to perform the calculation based on multiple patterns originating from one location. This is, however, at the price of additional experimental complications and a decrease in spatial resolution.

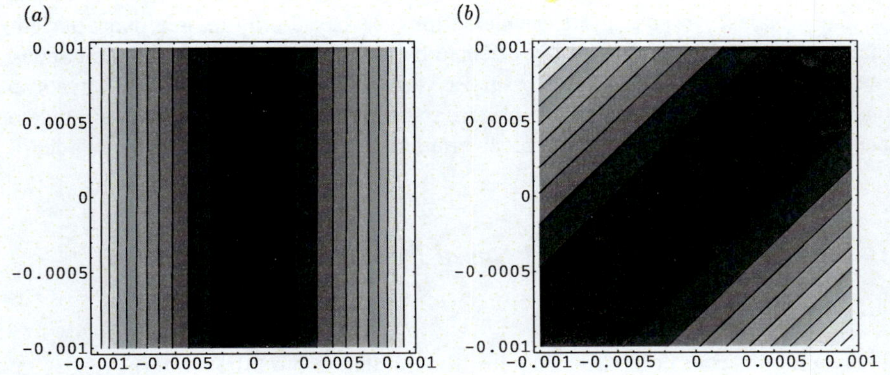

Fig. 14.5 (*a*) The dependence of ψ_2 on $\epsilon^{\uparrow}_{12}$ (horizontal axis) and $\epsilon^{\uparrow}_{13}$ (vertical axis). The other strain components are zero. The values of isolines are omitted for clarity; to make the point, it is sufficient to indicate that the ratio between the values at ($\epsilon^{\uparrow}_{12} = 0.001$, $\epsilon^{\uparrow}_{13} = 0$) and ($\epsilon^{\uparrow}_{12} = 0$, $\epsilon^{\uparrow}_{13} = 0.001$) is $\sim 1.2 \times 10^4$. (*b*) The dependence of ψ_2 on $\epsilon^{\uparrow}_{33}$ (horizontal axis) and $2\epsilon^{\uparrow}_{11}$ (vertical axis), for the strain tensor given by $\epsilon^{\uparrow} = \text{diag}(\epsilon^{\uparrow}_{11}, \epsilon^{\uparrow}_{11}, \epsilon^{\uparrow}_{33})$. The ratio between the values at $2\epsilon^{\uparrow}_{11} = 0.001 = -\epsilon^{\uparrow}_{33}$ and $2\epsilon^{\uparrow}_{11} = 0.001 = +\epsilon^{\uparrow}_{33}$ is $\sim 2.0 \times 10^4$.

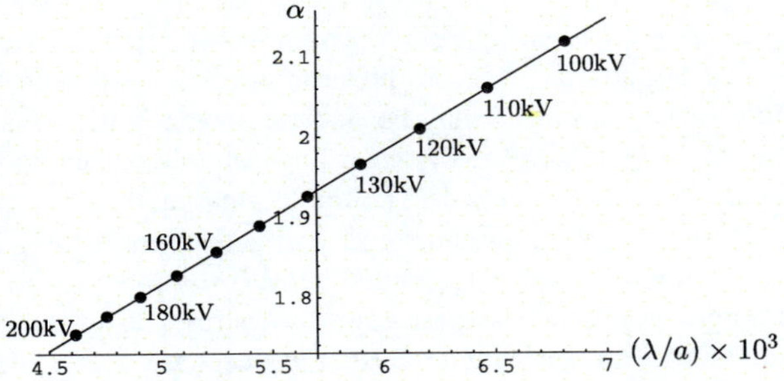

Fig. 14.6 The example dependence of α on the ratio wavelength/lattice parameter $= \lambda/a$ for the [331] pattern of Fig. 14.3. The numbers near points indicate the corresponding acceleration voltages for a corresponding to Si. The line fitted to the points is $\alpha(\lambda/a) = 165.9\lambda/a + 1.0$.

Impact of dynamical effects

The advancement of the CBED-based method is additionally hindered by the presence of dynamical effects in the diffraction patterns. In principle, the effects can be accounted for by fitting experimental and dynamically simulated patterns [33, 34]. Such fitting, however, is challenging in practice due to long computation times, and discrepancies between ideal simulated patterns and experimental patterns which are affected by factors ignored in the simulation. Moreover, it becomes necessary to input additional parameters (foil thickness, tilt angles).

The simplest lattice parameter calculations are based only on geometry of deficiency lines. Only lines negligibly affected by dynamical effects can be included in these calculations.[5] Thus, it is necessary to verify the impact of dynamical effects on each line used by comparing the geometric simulation with a dynamically simulated pattern, and this means that dynamical simulations cannot be completely avoided.

14.2.4 Software for CBED-Based Refinement of Lattice Parameters

There are numerous computer programs for simulation of CBED patterns (e.g., [35, 36]). However, the programs determining lattice parameters (or strain) from the patterns belong to a different category. Besides the simulation capability, they must contain routines for matching the simulated and experimental patterns. Usually, presentations of such procedures are subsidiary parts of accounts on applications of the CBED technique (e.g., [30]), and only a few publications are focused on the software itself [37, 38]. Worth noting is the system described in [5], which—according to the paper—is capable of automatic strain mapping for Si samples in some specific orientations.

Generally, it is not easy to compare the programs for getting lattice parameters or strain from CBED patterns because the problem has been seen from various perspectives. In effect, procedures may be disparate even if their main objectives are similar.[6] Additionally, it is not really clear how reliable particular programs are because in most accounts on the subject, a computational procedure is presented and then applied to a specific problem without convincing tests confirming its reliability.

Below is a description of a package called TEM*Strain* intended to facilitate the CBED-based refinement process. The package allows for tests of reliability by running it on simulated patterns for which the right answers are known. Moreover, it is available for an independent evaluation.

TEM*Strain*

TEM*Strain* is a Windows application allowing for simultaneous matching of multiple (up to 10) patterns. It is general, i.e., it is not limited to any particular material or structure. The software is capable of semiautomatic detection and indexing of lines. It allows for fitting user specified strain components, camera lengths and voltage. Both KLEBS and the standard strategies based on distances between intersections of lines are implemented. The software is capable of dynamical simulation of central

[5] Such lines are located far from low-index zone axes.

[6] Take for instance procedures for strain determination; they could be limited to calculation of the strain tensor components with the assumption of plane stress conditions (e.g., [39]), in the plane strain approximation (e.g., [40]), or some other selected components are determined (e.g., [9]).

disks of CBED patterns. Last not least, the program has a user-friendly interface. The most essential features and functions of TEM*Strain* 1.2 are described below. More details are in [41].

□ *Input and output*
Besides diffraction patterns, other mandatory input data are the reference lattice parameters, approximate crystal orientation for each pattern, approximate camera lengths, and voltage.

Relative deviations from reference values of lattice parameters, orientations, camera lengths, and voltages can be subject to fitting. If multiple patterns are used, the fitting concerns one set of lattice parameters (one strain tensor) and separate orientations, camera lengths and voltages for each pattern. Most of the optimization parameters can also be fixed; this does not apply to orientation parameters.

As for the output, one obtains new values of free optimization parameters. Strain can be obtained in three forms. The standard approach is to calculate (some or all) strain tensor components in a coordinate system linked to the crystal lattice. Another method is to calculate strain under the assumption of plane stress conditions. In both above cases, multiple patterns are allowed. Finally, with the ambiguity issue in mind, the strain components determinable from a single pattern can be calculated; they are given in the microscope coordinate system.

Orientation, camera length and voltage
The software allows for arbitrary sample orientations. Since the patterns suitable for the refinement are recorded with large camera lengths, relatively good approximations of orientations are needed. Orientations can be determined automatically using patterns recorded at smaller camera lengths; then the orientations for large camera-length patterns are fine-tunned manually. Also camera lengths and voltage can be manually adjusted.

Line detection and indexing
Lines in a CBED pattern can be marked manually with a computer mouse. Besides that, the program allows for automatic line detection, but its results may be unsatisfactory, and human inspection is needed. Corrections can be made using a tool displaying local profiles of intensities in directions perpendicular to the lines.

For the patterns recorded with large camera lengths, the indexing procedure relies on the known orientations. In order to index a given marking line, say, l, the program performs a search through reflections in a kinematically simulated diagram. If there is a reflection with the simulated line closer to l than a user specified limit (and it is the closest of such simulated lines), then indices of this reflection are ascribed to l.

Pattern simulation
The main procedures of TEM*Strain* rely on the geometry of diffraction lines. For calculating locations of the lines, the program essentially makes use of geometric theory of diffraction. Kinematic theory is needed only to decide whether a given line should be taken into account or not. For this purpose, structure factors are computed using the table of Waasmaier and Kirfel [42] for X-ray relativistic form factors and the Mott formula (Sect. 2.4.3). Debye-Waller temperature factors for some materials are collected in TEM*Strain* database. As for the dynamical simulations, they are carried out using the Bloch wave scheme (Chap. 3). Fourier coefficients of the superposition of atomic potentials are calculated from the structure factors of the kinematical theory.

Computational strategies
The strain or lattice parameters are calculated by fitting features of experimental patterns to corresponding features of simulated patterns. The patterns are matched by minimization of pre-defined objective functions. The principal tasks of the program are performed by using kinematically simulated patterns. In this kinematical framework, two variations of two strategies described above are available. The first strategy is based on matching distances between intersections of lines. One variant of this approach relies on matching the ratios of these distances to an average distance [26]; the ratios do not depend on camera lengths. In the second variant, the camera lengths are subject to

fitting [43]. The second strategy is KLEBS. A variant of this strategy uses a linearized form (14.5) of the 'K-line equation'. If the fitting converges, results of variants of a given strategy are similar. In order to improve the reliability of results, the above procedures should be combined.

TEM*Strain* is a computational tool intended to facilitate the CBED based measurements. It can be seen as a benchmark for future more comprehensive and more robust programs analyzing CBED patterns. The package has some limitations. With its relative generality allowing for diffraction patterns with diverse features, full automation of the lattice parameter determination and strain mapping are not achievable, and the refinement by TEM*Strain* is still a time consuming and tedious process. Moreover, the program is applicable only to simple patterns, in which a reflection leads to a line without splitting. On the other hand, TEM*Strain* combines capabilities of a number of earlier programs, from an easy adjustment of orientations, through a straightforward detection and indexing of lines, to a direct accessibility of dynamically simulated patterns. The most important feature of this software is that the reliability of results can be verified by comparing the output from a number of different strategies available in the package.

14.3 Kossel Micro-Diffraction

Kossel diffraction [44] or divergent beam X-ray diffraction [45] are other techniques suitable for orienting crystals, refinement of lattice parameters or determination of micro-strains. There is a variety of experimental setups and pattern acquisition geometries utilizing divergent X-rays. In particular, in the conventional Kossel diffraction, divergent X-rays are excited by an electron beam [44] or an X-ray beam in the presence of fluorescing atoms within the crystal or near its surface [45]. Kossel patterns are also acquired using synchrotron radiation [46–49]. With a planar detector, Kossel diffraction leads to a pattern of intersecting conic sections or fragments of such conics as shown in Figs. 2.9 and 8.7.

The determination of crystal orientations and lattice parameters by Kossel technique was advanced in the sixties with the use of electron probe microanalyzers (see, e.g., [50, 51]). As was noted in Chap. 8, the Kossel technique played a role in the development of the systems for orientation mappings. With the progress in digital cameras and the possibility of direct computer analysis of diffraction patterns, the use of the Kossel technique for strain determination is being considered again. In particular, the combination of scanning electron microscope equipped with a CCD camera and tensile loading device seems to be an attractive tool for in situ investigation of strains and stresses [52, 53].

The Kossel patterns usually cover a relatively large solid angle. Therefore, with known positions of Kossel lines, the end-indexing of Kossel patterns is relatively simple. With such large angles, there is also no problem with ambiguities in determination of lattice parameters. Furthermore, there is only one free surface of the investigated specimen and the relaxation problem faced by TEM-based techniques od strain determination is alleviated. On the other hand, spatial resolution of SEM-based Kossel micro-diffraction is only of the order of one micrometer. Moreover, the location of the pattern center must be determined.

Apart from line curvature, Kossel diffraction patterns—as K-line patterns—have geometry similar to that of CBED, Kikuchi or EBSD patterns. Therefore, some methods of analysis of these electron diffraction patterns are also applicable to Kossel patterns.

There are numerous programs for simulation of Kossel patterns (e.g., [54, 55]). They are usually based on the geometric diffraction theory [56] with kinematically calculated intensities used only to decide about the presence or absence of a given reflection. The problem of strain determination form Kossel patterns is more involved. Numerous procedures for solving it have been devised.[7] However, many of them are not really applicable to measurements in polycrystals because they rely on particular orientations and/or on some specific features of Kossel patterns (e.g., presence of 'lenses' or particular 'lenses'). More practical is to allow for arbitrary orientations, to consider the pattern as a whole, and to match all accessible features of experimental and simulated patterns. Since fitting intensities is difficult, the known procedures are based on pattern geometry.

With locations of points on indexed Kossel lines available, both KLEBS and the procedure based on distances between line intersections are applicable. The only modification of the algorithms of Sects. 14.2.1 and 14.2.2 needed to get lattice parameters from Kossel patterns is the calculation of the $\boldsymbol{k}^{\uparrow}$ vectors.

14.3.1 KSLStrain

KSL*Strain* [59, 60] is a package of computer programs facilitating the determination of orientations and strains based on the Kossel diffraction patterns similar to TEM*Strain* presented in Sect. 14.2.4. The package is capable of geometric simulation of patterns, orientation determination and lattice parameter refinement (or strain determination). Orientation determination and the refinement of lattice parameters are based on manually marked conics. Parameters are refined by matching geometric features (Kossel conics) of experimental and simulated patterns. To ensure good reliability of results, the program is capable of simultaneous matching of multiple patterns originating from the same crystallite. The software does not require special orientations, and it is not limited to any particular material. Operation of the package is controlled via a Windows user interface. As for limitations of KSL*Strain*, the package does not make any use of the fine structure of the Kossel reflections, and the refraction (as the beam comes from a crystal into the vacuum) is disregarded.

⊡ The operation of KSL*Strain* starts with loading (up to ten) Kossel patterns of arbitrary dimensions. Then conics on the patterns must be marked using a computer mouse. Based on the locations of the marking points, crystal orientations are calculated automatically. They are determined with the

[7] The older methods are reviewed in [57, 58].

help of suitably adapted *KiKoCh2*; cf. Sect. 8.5.2. The same marking points are used for lattice parameter refinement.

Crucial for the strain determination are values of the radiation wavelength, the sample-to-detector distance and the location of the pattern center. Since the wavelength is known with a relatively high accuracy, the user provided data are assumed to be exact. As for the pattern center and the sample-to-detector distance, the situation is different. In normal experimental conditions, the estimated values of these parameters are not sufficiently accurate for reliable strain determination. Getting better estimates based on direct measurements requires much effort. Therefore, an additional procedure is used to fit these parameters (with fixed reference values of lattice parameters). The fitting is based on the geometry of the pattern; the sample-to-detector distance is linked directly to pattern magnification, and the pattern center is placed at the intersection of major axes of the conics. Once the orientation is determined and specimen-to-detector distance and pattern center are tuned, one can proceed with the refinement of lattice parameters. Together with the strain components, the crystal orientation parameters, pattern center coordinates and the specimen-to-detector distance are fitted.

14.4 Appendix: Intersections of K-Lines

It can be verified by substitution that (14.6) for a wave vector along intersection of two Kossel cones are satisfied by

$$\boldsymbol{k} = \boldsymbol{k}(\check{\boldsymbol{h}}_1, \check{\boldsymbol{h}}_2) = \frac{\boldsymbol{y} \times \boldsymbol{z} \pm \boldsymbol{y}\sqrt{4\boldsymbol{y}\cdot\boldsymbol{y}/\lambda^2 - \boldsymbol{z}\cdot\boldsymbol{z}}}{2\boldsymbol{y}\cdot\boldsymbol{y}} ,$$

with $\boldsymbol{y} = \check{\boldsymbol{h}}_1 \times \check{\boldsymbol{h}}_2$ and $\boldsymbol{z} = \check{\boldsymbol{h}}_1(\check{\boldsymbol{h}}_2 \cdot \check{\boldsymbol{h}}_2) - \check{\boldsymbol{h}}_2(\check{\boldsymbol{h}}_1 \cdot \check{\boldsymbol{h}}_1)$; since $\check{\boldsymbol{h}}_1 \cdot \boldsymbol{y} = 0$ and $\check{\boldsymbol{h}}_1 \cdot (\boldsymbol{y} \times \boldsymbol{z}) = (\boldsymbol{y} \cdot \boldsymbol{y})(\check{\boldsymbol{h}}_1 \cdot \check{\boldsymbol{h}}_1)$, one obtains that $2\check{\boldsymbol{h}}_1 \cdot \boldsymbol{k}$ equals $\check{\boldsymbol{h}}_1 \cdot \check{\boldsymbol{h}}_1$, and an analogous relationship occurs for $\check{\boldsymbol{h}}_2$. An intersection occurs if $\boldsymbol{y} \neq \boldsymbol{0}$ and the argument $4\boldsymbol{y}\cdot\boldsymbol{y}/\lambda^2 - \boldsymbol{z}\cdot\boldsymbol{z}$ of the square root function is positive, i.e., there must occur

$$\frac{\boldsymbol{z}\cdot\boldsymbol{z}}{4\boldsymbol{y}\cdot\boldsymbol{y}} < 1/\lambda^2 . \tag{14.12}$$

This condition can be expressed in a more direct way via the angle between the reflecting planes β and semi-apex angles of the cones β_1 and β_2: the intersection occurs if

$$|\beta_1 - \beta_2| < \beta < \beta_1 + \beta_2 . \tag{14.13}$$

To show the equivalence of (14.12) to (14.13), one can write the latter as $\cos(\beta_1 - \beta_2) > \cos\beta > \cos(\beta_1 + \beta_2)$ and use $\cos\beta = \check{\boldsymbol{h}}_1 \cdot \check{\boldsymbol{h}}_2/(|\check{\boldsymbol{h}}_1|\,|\check{\boldsymbol{h}}_2|)$ and $\cos\beta_i = \check{\boldsymbol{h}}_i \cdot \boldsymbol{k}/(|\check{\boldsymbol{h}}_i||\boldsymbol{k}|) = |\check{\boldsymbol{h}}_i|/(2|\boldsymbol{k}|)$. Simple trigonometric and arithmetic steps lead from inequalities (14.13) to $4|\boldsymbol{k}|^4\left(|\check{\boldsymbol{h}}_1|^2\,|\check{\boldsymbol{h}}_2|^2 - \left(\check{\boldsymbol{h}}_1 \cdot \check{\boldsymbol{h}}_2\right)^2\right) > |\boldsymbol{k}|^2\,|\check{\boldsymbol{h}}_1 - \check{\boldsymbol{h}}_2|^2\,|\check{\boldsymbol{h}}_1|^2\,|\check{\boldsymbol{h}}_2|^2$. Hence, with $|\boldsymbol{k}|^2 > 0$ and $|\check{\boldsymbol{h}}_1|^2|\check{\boldsymbol{h}}_2|^2 > \left(\check{\boldsymbol{h}}_1 \cdot \check{\boldsymbol{h}}_2\right)^2$, one gets the inequality (14.12).

Points of intersection of three Kossel lines
Useful for lattice parameter determination are also intersections of multiple Kossel lines [61, 62]. Three lines corresponding to reciprocal lattice vectors $\check{\boldsymbol{h}}_i$ $(i = 1, 2, 3)$ may intersect each other at one point given by $\boldsymbol{k}$ if

$$2\check{\boldsymbol{h}}_i \cdot \boldsymbol{k} = \check{\boldsymbol{h}}_i \cdot \check{\boldsymbol{h}}_i \qquad \text{(no summation)}\ . \tag{14.14}$$

If $\check{\boldsymbol{h}}_i$ are linearly independent, the above system of linear equations with respect to $\boldsymbol{k}$ has a unique solution. In general, the solution $\boldsymbol{k}$ may not correspond to a physical wavelength. If it happens to correspond to the wavelength used in an experiment, the intersection is, in a sense, 'accidental'. There are also 'persistent' intersections, which—for a given structure—occur for a range of wavelengths. This occurs only if the vectors $\check{\boldsymbol{h}}_i$ are linearly dependent

$$\sum_i m_i\, \check{\boldsymbol{h}}_i = 0\ . \tag{14.15}$$

One additionally has $\sum m_i\, \check{\boldsymbol{h}}_i \cdot \check{\boldsymbol{h}}_i = \sum m_i\, (2\check{\boldsymbol{h}}_i \cdot \boldsymbol{k}) = 2\left(\sum_i m_i\, \check{\boldsymbol{h}}_i\right) \cdot \boldsymbol{k}$, i.e.,

$$\sum_i m_i\, \check{\boldsymbol{h}}_i \cdot \check{\boldsymbol{h}}_i = 0\ . \tag{14.16}$$

This relationship implies that determinants of the matrices obtained from the coefficient matrix $2[\check{\boldsymbol{h}}_1\ \check{\boldsymbol{h}}_2\ \check{\boldsymbol{h}}_3]$ of the system (14.14) by replacing a column by the column of right-hand-sides of (14.14) are equal to zero. Equations (14.15) and (14.16) are necessary conditions for a persistent intersection of three lines corresponding to $\check{\boldsymbol{h}}_i$. If they are satisfied, and the lines corresponding to $\check{\boldsymbol{h}}_1$ and $\check{\boldsymbol{h}}_2$ intersect each other, then the line corresponding to $\check{\boldsymbol{h}}_3$ also passes through the intersection point, i.e., there occurs $\boldsymbol{k}(\check{\boldsymbol{h}}_1, \check{\boldsymbol{h}}_2) = \boldsymbol{k}(\check{\boldsymbol{h}}_2, \check{\boldsymbol{h}}_3) = \boldsymbol{k}(\check{\boldsymbol{h}}_3, \check{\boldsymbol{h}}_1)$. In theory, intersections of more than three lines are possible. Such a set of lines can be analyzed by partitioning it into triplets.

References

1. V.V. Lider, Precise determination of crystal lattice parameters. Physics - Uspekhi **63**, 907–928 (2020)
2. M. Born, K. Huang, *Dynamical Theory of Crystal Lattices* (Oxford University Press, London, 1954)
3. J.M. Cowley, *Diffraction Physics* (North-Holland, Amsterdam, 1981)
4. P.M. Jones, G.M. Rackham, J.W. Steeds, Higher order Laue zone effects in electron diffraction and their use in lattice parameter determination. Proc. Roy. Soc. Lond. A **354**, 197–222 (1977)
5. A. Armigliato, R. Balboni, S. Frabboni, Improving spatial resolution of convergent beam electron diffraction strain mapping in silicon microstructures. Appl. Phys. Lett. **86**, Art. No. 063508 (2005)

6. S. Frabboni, F. Gambetta, A. Armigliato, R. Balboni, S. Balboni, F. Cembali, Lattice strain and static disorder determination in Si/$Si_{1-x}Ge_x$/Si heterostructures by convergent beam electron diffraction. Phys. Rev. B **60**, 13750–13761 (1999)
7. A. Armigliato, R. Balboni, S. Frabboni, A. Benedetti, A.G. Cullis, G.P. Carnevale, P. Colpani, G. Pavia, Strain characterisation of shallow trench isolation structures on a nanometer scale by convergent beam electron diffraction. Mat. Sci. Semicon. Proc. **4**, 97–99 (2001)
8. T. Akaogi, K. Tsuda, M. Terauchi, M. Tanaka, Lattice parameter determination of a strained area of an InAs layer on a GaAs substrate using CBED. J. Electron Microsc. **53**, 11–19 (2004)
9. S.L. Toh, K.P. Loh, C.B. Boothroyd, K. Li, C.H. Ang, L. Chan, Strain analysis in silicon substrates under uniaxial and biaxial stress by convergent beam electron diffraction. J. Vac. Sci. Technol. B **23**, 940–946 (2005)
10. J. Li, R.P. Wahi, Investigation of γ/γ' lattice mismatch in the polycrystalline nickel-base superalloy IN738LC: influence of heat treatment and creep deformation. Acta Metall. Mater. **43**, 507–517 (1995)
11. R. Völkl, U. Glatzel, M. Feller-Kniepmeier, Measurement of the lattice misfit in the single crystal nickel based superalloys CMSX-4, SRR99 and SC16 by convergent beam electron diffraction. Acta Mater. **46**, 4395–4404 (1998)
12. C. Schulze, M. Feller-Kniepmeier, Transmisson electron microscopy of phase composition and lattice misfit in the Re-containing nickel-base superalloy CMSX-10. Mater. Sci. Engng A **281**, 204–212 (2000)
13. M. Yonemura, K. Sueoka, K. Kamei, Analysis of local lattice strain around oxygen precipitates in silicon crystals using CBED technique. Appl. Surf. Sci. **130–132**, 208–213 (1998)
14. H. Heinrich, A. Vananti, G. Kostorz, Strain fields at interfaces of Al–based metal matrix composites. Mater. Sci. Engng A 319–438 (2001)
15. L. Clément, R. Pantel, L.F.T. Kwakman, J.L. Rouvière, Strain measurements by convergent-beam electron diffraction: the importance of stress relaxation in lamella preparations. Appl. Phys. Lett. **85**, 651–653 (2004)
16. A. Benedetti, H. Bender, A. Lauwers, C. Torregiani, K. Maex, Effects of surface relaxation on convergent-beam electron diffraction analysis of stress in silicon. J. Microsc. **223**, 249–252 (2006)
17. K.Z. Troost, P. van der Sluis, D.J. Gravesteijn, Microscale elastic-strain determination by backscatter Kikuchi diffraction in the scanning electron microscope. Appl. Phys. Lett. **62**, 1110–1112 (1993)
18. A.J. Wilkinson, Measurement of elastic strains and small lattice rotations using electron back scatter diffraction. Ultramicroscopy **62**, 237–247 (1996)
19. A.J. Wilkinson, P.B. Hirsch, Electron diffraction based techniques in scanning electron microscopy of bulk materials. Micron **28**, 279–308 (1997)
20. A.J. Wilkinson, G. Meaden, D.J. Dingley, High-resolution elastic strain measurement from electron backscatter diffraction patterns: new levels of sensitivity. Ultramicroscopy **106**, 307–313 (2006)
21. X. Tao, A. Eades, Measurement and mapping of small changes of crystal orientation by electron backscattering diffraction. Microsc. Microanal. **11**, 341–353 (2005)
22. J.M. Zuo, J.C.H. Spence, Automated structure factor refinement from convergent-beam patterns. Ultramicroscopy **35**, 185–196 (1991)
23. C. Deininger, G. Necker, J. Mayer, Determination of structure factors, lattice strains and accelerating voltage by energy-filtered convergent beam electron diffraction. Ultramicroscopy **54**, 15–30 (1994)
24. Y. Ogata, K. Tsuda, Y. Akishige, M. Tanaka, Refinement of the crystal structural parameters of the intermediate phase of h–$BaTiO_3$ using convergent-beam electron diffraction. Acta Cryst. A **60**, 525–531 (2004)
25. A. Morawiec, An algorithm for refinement of lattice parameters using CBED patterns. Ultramicroscopy **107**, 390–395 (2007)
26. J.M. Zuo, Automated lattice parameter measurement from HOLZ lines and their use for the measurement of oxygen content in $YBa_2Cu_3O_{7-\delta}$ from nanometer-sized region. Ultramicroscopy **41**, 211–223 (1992)

27. S.J. Rozeveld, J.M. Howe, S. Schmauder, Measurement of residual strain in an Al-SiC$_w$ composite using convergent-beam electron diffraction. Acta Metall. Mater. **40**, S173–S193 (1992)
28. T. Yamazaki, T. Isaka, K. Kuramochi, I. Hashimotoa, K. Watanabe, Precise measurement of local strain fields with energy-unfiltered convergent-beam electron diffraction. Acta Cryst. A **62**, 201–207 (2006)
29. A. Morawiec, *Orientations and Rotations. Computations in Crystallographic Textures* (Springer, Berlin, 2004)
30. S. Krämer, J. Mayer, C. Witt, A. Weickenmeier, M. Rühle, Analysis of local strain in aluminium interconnects by energy filtered CBED. Ultramicroscopy **81**, 245–262 (2000)
31. H.J. Maier, R.R. Keller, H. Renner, H. Mughrabi, A. Preston, On the unique evaluation of local lattice parameters by convergent-beam electron diffraction. Philos. Mag. A **74**, 23–43 (1996)
32. A. Morawiec, Formal conditions for unambiguous residual strain determination by CBED. Philos. Mag. **85**, 1611–1623 (2005)
33. J.M. Zuo, M. Kim, R. Holmestad, A new approach to lattice parameter measurements using dynamic electron diffraction and pattern matching. J. Electron Microsc. **47**, 121–127 (1998)
34. M. Kim, J.M. Zuo, G.S. Park, High-resolution strain measurement in shallow trench isolation structures using dynamic electron diffraction. Appl. Phys. Lett. **84**, 2181–2183 (2004)
35. J.C.H. Spence, J.M. Zuo, *Electron Microdiffraction* (Plenum, New York, 1992)
36. X.Z. Li, *JECP/HOLZ* - an interactive computer program for simulation of HOLZ patterns. J. Appl. Cryst. **38**, 576–577 (2005)
37. R. Völkl, U. Glatzel, M. Feller-Kniepmeier, *Image CBED++* - an easy to use program for rapid analysis of CBED HOLZ line patterns, in *Electron Microscopy 1998: Proceedings of the 14th International Congress on Electron Microscopy, Cancun, Mexico*, ed. by H.A. Calderón Benavides, M.J. Yacamàn (Institute of Physics Publishing, Bristol, 1998), pp. 785–786
38. P. Paczkowski, M. Gigla, A. Kostka, H. Morawiec, The software tool for lattice parameters determination from nanoareas using CBED patterns. Mater. Chem. Phys. **81**, 233–236 (2003)
39. B. Li, H. Zou, J. Pan, A study of residual strain in a $K_2O.6TiO_{2W}$/Al composite by using convergent beam electron diffraction. Scripta Mater. **38**, 1419–1425 (1998)
40. A. Toda, N. Ikarashi, H. Ono, Local lattice strain measurements in semiconductor devices by using convergent-beam electron diffraction. J. Cryst. Growth **210**, 341–345 (2000)
41. A. Morawiec, A program for refinement of lattice parameters based on multiple convergent beam electron diffraction patterns. J. Appl. Cryst. **40**, 618–622 (2007)
42. D. Waasmaier, A. Kirfel, New analytical scattering-factor functions for free atoms and ions. Acta Cryst. A **51**, 416–431 (1995)
43. A. Morawiec, Determinability of complete residual strain tensor from multiple CBED patterns. Mater. Sci. Forum **524–525**, 115–120 (2006)
44. W. Kossel, V. Loeck, H. Voges, Die Richtungsverteilung der in einem Kristall entstandenen charakteristischen Röntgenstrahlung. Z. Phys. **94**, 139–144 (1935)
45. K. Lonsdale, Divergent-beam X-ray photography of crystals. Philos. Trans. R. Soc. A **240**, 219–250 (1947)
46. H.J. Ullrich, M. Schlaubitz, F. Friedel, T. Spann, J. Bauch, T. Wroblewski, S. Garbe, G. Gaul, A. Knöchel, F. Lechtenberg, E. Rossmanith, G. Kumpat, G. Ulrich, Excitation of Kossel patterns by synchrotron radiation. Nucl. Instrum. Methods A **349**, 269–273 (1994)
47. Ch. Schetelich, S. Weber, V. Geist, M. Schlaubitz, H.J. Ullrich, S. Kek, H.G. Krane, Recording of Kossel patterns using monochromatic synchrotron radiation. Nucl. Instrum. Methods B **103**, 236–242 (1995)
48. A.M. Glazer, S.P. Collins, D. Zekria, J. Liu, M. Golshan, Observation of divergent-beam X-ray diffraction from a crystal of diamond using synchrotron radiation. J. Synchrotron Rad. **11**, 187–189 (2004)
49. A.G.A. Nisbet, G. Beutier, F. Fabrizi, B. Moser, S.P. Collins, Diffuse multiple scattering. Acta Cryst. A **71**, 20–25 (2015)
50. D.L. Vieth, H. Yakowitz, Design considerations for a Kossel microdiffraction camera. Rev. Sci. Instr. **37**, 206–209 (1966)

51. D.L. Vieth, H. Yakowitz, Tensile loading device for Kossel microdiffraction and metallography. Rev. Sci. Instr. **39**, 1929–1931 (1968)
52. E. Langer, S. Däbritz, C. Schurig, W. Hauffe, Lattice constant determination from Kossel patterns observed by CCD camera. Appl. Surf. Sci. **179**, 45–48 (2001)
53. S. Berveiller, P. Dubos, K. Inal, A. Eberhardt, E. Patoor, Inter- and intra-granular strain analysis by microdiffraction Kossel. Mater. Sci. Forum **490–491**, 159–164 (2005)
54. E. Langer, R. Kurt, S. Däbritz, *KOPSKO*: a computer program for generation of Kossel and pseudo Kossel diffraction patterns. Cryst. Res. Technol. **34**, 801–816 (1999)
55. S. Weber, *KOQUA2.0*: a program for simulating divergent-beam diffraction patterns for crystals and quasicrystals. J. Appl. Cryst. **30**, 85–86 (1997)
56. P.P. Ewald, Introduction to the dynamical theory of X-ray diffraction. Acta Cryst. A **25**, 103–108 (1969)
57. P. Gielen, H. Yakowitz, D. Ganow, R.E. Ogilvie, Evaluation of Kossel microdiffraction procedures: the cubic case. J. Appl. Phys. **36**, 773–782 (1965)
58. R. Tixier, C. Wache, Kossel patterns. J. Appl. Cryst. **3**, 466–485 (1970)
59. A. Morawiec, A program for refinement of lattice parameters and strain determination using Kossel diffraction patterns. J. Appl. Cryst. **49**, 322–329 (2016)
60. A. Morawiec, A profile-based method of determining intragranular strains using Kossel diffraction patterns. Adv. Mat. Res. **996**, 52–57 (2014)
61. B.J. Isherwood, C.A. Wallace, The geometry of X-ray multiple diffraction in crystals. Acta Cryst. A **27**, 119–130 (1971)
62. N. Harris, Indexing Kossel patterns. J. Mater. Sci. **10**, 279–289 (1975)

Index

A. Morawiec, *Indexing of Crystal Diffraction Patterns*, Springer Series in Materials Science 326, https://doi.org/10.1007/978-3-031-11077-1